Mobile Technologies in Educational Organizations

Alona Forkosh Baruch
Levinsky College of Education, Israel & Tel Aviv University, Israel

Hagit Meishar Tal
Holon Institute of Technology, Israel

A volume in the Advances in Educational Technologies and Instructional Design (AETID) Book Series

IGI Global
DISSEMINATOR OF KNOWLEDGE

Published in the United States of America by
IGI Global
Information Science Reference (an imprint of IGI Global)
701 E. Chocolate Avenue
Hershey PA, USA 17033
Tel: 717-533-8845
Fax: 717-533-8661
E-mail: cust@igi-global.com
Web site: http://www.igi-global.com

Library of Congress Cataloging-in-Publication Data

Names: Baruch, Alona Forkosh, editor. | Tal, Hagit Meishar, editor.
Title: Mobile technologies in educational organizations / Alona Forkosh Baruch and Hagit Meishar Tal, editors.
Description: Hershey, Pennsylvania : Information Science Reference (an imprint of IGI Global), [2019]
Identifiers: LCCN 2018045951| ISBN 9781522581062 (Hardcover) | ISBN 9781522581079 (Ebook)
Subjects: LCSH: Mobile communication systems in education. | Computer-assisted instruction. | Organizational learning.
Classification: LCC LB1044.84 .M628 2019 | DDC 371.33/4--dc23 LC record available at https://lccn.loc.gov/2018045951

This book is published in the IGI Global book series Advances in Educational Technologies and Instructional Design (AE-TID) (ISSN: 2326-8905; eISSN: 2326-8913)

British Cataloguing in Publication Data
A Cataloguing in Publication record for this book is available from the British Library.

For electronic access to this publication, please contact: eresources@igi-global.com.

Advances in Educational Technologies and Instructional Design (AETID) Book Series

Lawrence A. Tomei
Robert Morris University, USA

ISSN:2326-8905
EISSN:2326-8913

Mission

Education has undergone, and continues to undergo, immense changes in the way it is enacted and distributed to both child and adult learners. In modern education, the traditional classroom learning experience has evolved to include technological resources and to provide online classroom opportunities to students of all ages regardless of their geographical locations. From distance education, Massive-Open-Online-Courses (MOOCs), and electronic tablets in the classroom, technology is now an integral part of learning and is also affecting the way educators communicate information to students.

The **Advances in Educational Technologies & Instructional Design (AETID) Book Series** explores new research and theories for facilitating learning and improving educational performance utilizing technological processes and resources. The series examines technologies that can be integrated into K-12 classrooms to improve skills and learning abilities in all subjects including STEM education and language learning. Additionally, it studies the emergence of fully online classrooms for young and adult learners alike, and the communication and accountability challenges that can arise. Trending topics that are covered include adaptive learning, game-based learning, virtual school environments, and social media effects. School administrators, educators, academicians, researchers, and students will find this series to be an excellent resource for the effective design and implementation of learning technologies in their classes.

Coverage

- Web 2.0 and Education
- Collaboration Tools
- Higher Education Technologies
- Game-Based Learning
- Adaptive Learning
- E-Learning
- K-12 Educational Technologies
- Hybrid Learning
- Classroom Response Systems
- Educational Telecommunications

IGI Global is currently accepting manuscripts for publication within this series. To submit a proposal for a volume in this series, please contact our Acquisition Editors at Acquisitions@igi-global.com or visit: http://www.igi-global.com/publish/.

The Advances in Educational Technologies and Instructional Design (AETID) Book Series (ISSN 2326-8905) is published by IGI Global, 701 E. Chocolate Avenue, Hershey, PA 17033-1240, USA, www.igi-global.com. This series is composed of titles available for purchase individually; each title is edited to be contextually exclusive from any other title within the series. For pricing and ordering information please visit http://www.igi-global.com/book-series/advances-educational-technologies-instructional-design/73678. Postmaster: Send all address changes to above address.

Titles in this Series

For a list of additional titles in this series, please visit: www.igi-global.com/book-series

Comparative Perspectives on Inquiry-Based Science Education
Stuart Bevins (Sheffield Hallam University, UK) Louise Lehane (St. Angela's College, Ireland) and Josephine Booth (Sheffield Hallam University, UK)
Information Science Reference • copyright 2019 • 290pp • H/C (ISBN: 9781522554394) • US $185.00 (our price)

Opening Up Education for Inclusivity Across Digital Economies and Societies
Patricia Ordóñez de Pablos (The University of Oviedo, Spain) Miltiadis D. Lytras (Effat University, Saudi Arabia) Xi Zhang (Tianjin University, China) and Kwok Tai Chui (City University of Hong Kong, China)
Information Science Reference • copyright 2019 • 352pp • H/C (ISBN: 9781522574736) • US $185.00 (our price)

Cultural and Social Implications of Artificial Intelligence in Education
Dennis Anderson (St. Francis College, USA) and Robert Niewiadomski (Hunter College, USA)
Information Science Reference • copyright 2019 • 305pp • H/C (ISBN: 9781522577935) • US $185.00 (our price)

An Invariant-Based Approach to Second Language Acquisition Emerging Research and Opportunities
Elena Orlova (Lobachevsky State University of Nizhny Novgorod, Russia)
Information Science Reference • copyright 2019 • 164pp • H/C (ISBN: 9781522582205) • US $145.00 (our price)

Performance-Based Assessment in 21st Century Teacher Education
Kim K. Winter (Western Carolina University, USA) Holly H. Pinter (Western Carolina University, USA) and Myra K. Watson (Western Carolina University, USA)
Information Science Reference • copyright 2019 • 295pp • H/C (ISBN: 9781522583530) • US $175.00 (our price)

Modern Technologies for Teaching and Learning in Socio-Humanitarian Disciplines
Larisa Alexandrovna Darinskaia (Saint Petersburg State University, Russia) and Gallina Ivanovna Molodtsova (Saint Petersburg State University, Russia)
Information Science Reference • copyright 2019 • 292pp • H/C (ISBN: 9781522578413) • US $175.00 (our price)

Critical Literacy Initiatives for Civic Engagement
Angela M. Cartwright (Midwestern State University, USA) and Emily K. Reeves (Midwestern State University, USA)
Information Science Reference • copyright 2019 • 242pp • H/C (ISBN: 9781522580829) • US $185.00 (our price)

Digital Games for Minority Student Engagement Emerging Research and Opportunities
Rupanada Misra (Rutgers University, USA) Leo Eyombo (Howard University, USA) and Floyd T. Phillips (New York Institute of Health and Behavior, USA)

IGI Global
DISSEMINATOR OF KNOWLEDGE

701 East Chocolate Avenue, Hershey, PA 17033, USA
Tel: 717-533-8845 x100 • Fax: 717-533-8661
E-Mail: cust@igi-global.com • www.igi-global.com

Table of Contents

Detailed Table of Contents

Section 4
Design of Mobile Solutions for People With Learning Difficulties

In recent years, the use of smartphones by people with intellectual disabilities (ID) has increased, leading to the development of applications designed to ease and improve their lives and independence, and help them integrate into society. Designing applications tailored to the needs of these users is complex, particularly in terms of accessibility and usability. This chapter presents a case study of the design and development process of a dedicated application aimed at helping people with mild ID in handling cash transactions. The chapter explores the design principles of mobile interfaces for people with ID. It introduces the use of the design thinking methodology, which focuses on user-centered design and helps gain an in-depth knowledge of the end users and their needs. Based on the performed user research, recommendations for the design of such applications are highlighted.

Writing plays a vital role in both daily life and academic life. Students begin learning to write in the first years in school and then they use writing skills their whole life. However, some students, such as students with dysgraphia, experience difficulties in writing. The purpose of the chapter is to develop a mobile writing application for students with dysgraphia and to reveal ideas of special education experts, educational technology experts, and classroom education experts, and a teacher about application. A pilot study was conducted with three students with dysgraphia firstly, and necessary revisions were made. After revisions, experts' views were taken. The results indicated that experts reflected positive ideas about mobile writing application for students with dysgraphia. However, the experts suggested some minor revisions for improving the mobile writing application. To sum up, final improvements were made, and a mobile writing application was developed.

A case study examined the contribution of tablet applications for English language acquisition and improvement of reading and reading comprehension skills among pupils diagnosed with learning disabilities and those not diagnosed with learning disabilities. Participants were Grade 4 pupils from a public school in central Israel in two groups. Group 1 included two boys and two girls diagnosed with learning disabilities; Group 2 included two boys and two girls not diagnosed with learning disabilities. English learning applications were used with the intent of improving the abilities of letter recognition,

Foreword

Mobile technologies not only technically enable the mobility of the individual holding a device, but also enable new learning models to grow, due to the opportunities to access the Internet anywhere and anytime. These new modes of learning facilitate ubiquitous utilization of a personal device, as opposed to restricted access via stationary devices. When mobile technologies are ready-at-hand, there are new and advanced opportunities for teachers and learners to access, share, and construct knowledge in different settings and modes, formally or informally, through serendipitous learning. Indeed, there is uniqueness in mobile learning which makes it stand out when comparing access to learning from traditional desktop computers. Empowered by fields such as learning sciences and educational technologies, teachers, and designers of learning can make use of such affordances to design innovative learning experiences that can foster rich, deep and meaningful learning.

The term "mobile technologies" is general and refers to several functions: in educational settings we focus on "mobile learning", whether at the organizational level, e.g., policy, or at the individual level, e.g., participation, acceptance and adoption of mobile learning in learning and teaching processes. Mobile learning can be referred to in several contexts and with regards to several aspects: the pedagogical aspect (e.g., location-based learning, active learning, constructivist learning); the social aspect (e.g., collaborative learning, team based learning, social networks); the learning space aspect, i.e., the connecting of physical and virtual spaces (e.g., augmented reality); and affective and cognitive aspects of our living in an era of mobile technologies which serve as an extension of human capacity. Mobile technologies indeed allow us to be constantly connected to information and people, and to develop our knowledge (whether general or personal). Therefore, they are efficient devices for educators for purposes of teaching and learning in formal and informal settings. Moreover, once utilized, they are part of our personal and professional world.

Mobile Technologies in Educational Organizations addresses these new and exciting possibilities and provides global and interdisciplinary perspectives on the strengths and challenges inherent in the implementation of mobile technologies within educational organizations – K12 and higher education. It is organized into four sections that are interdisciplinary in orientation and provide a multiplicity of perspectives: *implementation of mobile learning in educational organizations, use of mobile devices to enhance teaching and learning, global and multicultural considerations and implications of mobile learning, and design of mobile solutions for people with learning difficulties.* Each of these sections has been carefully commissioned, reviewed and organized by a group of highly regarded scholars from around the world who discuss the design of mobile technologies to support learners of all ages in diverse learning settings.

This book holds a timely contribution to some of the latest research and development areas and trends for mobile learning. The chapters focus not only on empirical studies but also stimulate theoretical discussions and address practical implications. The contributors to the book demonstrate and discuss new design and developments in the field of mobile teaching. This book provides a platform for readers to review current research and then apply these new ideas in their own educational work environment.

While there is great promise and potential in mobile learning, this field has not yet reached maturity. The challenge for research is to provide evidence of what works and how it works in mobile learning. There is still much ground to cover in terms of lowering the barriers to adoption, sustainable use in learning practices, and scaling up of the use of educational mobile technologies. This book is yet another step in this direction. *Mobile Technologies in Educational Organizations* provides powerful insights for design and implementation of educational paradigms in implementing mobile learning for teachers, learners, educational designers, professionals and decision makers in educational organizations. This book offers scholars and practitioners practical advice and theory-based guidelines on where we are now in this journey of designing and applying mobile learning. It provides up-to-date and multi-dimensional knowledge of the field which then contributes to understanding these challenges and how to cope with them.

Gila Kurtz
Holon Institute of Technology, Israel

Preface

The penetration of mobile technologies into our lives is without a doubt one of the interesting phenomena that characterize the current era, in which technology is ubiquitous, transparent, comprehensive and life-changing. A mobile environment enriched by the adequate infrastructure (e.g., broadband WiFi, free connection) facilitated utilization of mobile devices, whether smartphones, notebooks or tablets, to a degree that it indeed fits the definition of technology as an extension of human capacity. McLuhan (1966) wonderfully phrased this:

Rapidly, we approach the final phase of the extensions of man... when the creative process will be collectively and corporately extended to the whole of human society, much as we have already extended our senses and our nerves by the various media. (p. 19)

Mobile technologies, including mostly tablets and mobile devices, are being carried by us anywhere and anytime. This allows direct and prompt access to online information, but also to other users. Mobile devices revolutionize every organization in which they are integrated into and eliminate old constraints and create new possibilities. They create new power relations between users, some of which constitute empowerment of the users (Barkhuus & Polichar, 2011; Cumming, Strnadová, Knox, & Parmenter, 2014) and some of which are perceived as threat for their status and personal well-being (Bianchi & Phillips, 2005; Thomée, Härenstam, & Hagberg, 2011). Fields of usage include medicine (e.g., Dorsey et al., 2018), health system (e.g., Kao et al., 2018), marketing (e.g., Faulds, Mangold, Raju & Valsalan, 2018), the business sector (e.g., Lee & Shin, 2018), banking (e.g., Sharma & Sharma, 2019), tourism (e.g., Law, Chan & Wang, 2018), culture (e.g., Light, Bagnall, Crawford & Gosling, 2018), education (e.g., Crompton & Burke, 2018) and more. Utilization may include research and development, consumption of information - even daily news, but also professional, just-in-time, location-based information. In terms of accessibility, mobile technology holds a promise for coping with challenges in developing countries (Asongu & Boateng, 2018). It allows connectivity between members of the community and beyond. It is easily mastered by utilizing mobile, portable, easy-to-carry devices that are available constantly, in which their usability and user-interface design is easy to learn at any age and by everyone. This may result in connecting between strangers, experts in fields of interest, or colleagues, to name a few.

The information era is also characterized by change in educational organizations and in educators' roles and the need to be ready to integrate mobile devices in teaching and learning (Christensen & Knezek, 2018). While educational organizations are typically located in the physical world in a specific, fixed and generally unmovable location, learning processes are happening everywhere, in formal and especially informal settings, in a manner that is sometimes organized but many times is serendipitous, requiring pedagogical change (Crompton & Traxler, 2019). This shift in education as a result of technology utilization may pose physical, pedagogical, social and ethical challenges to educational organization, specifically to teaching and learning. This in turn raises fundamental issues regarding the purpose of schools in 21st century, especially due to the growing role of serendipitous and ubiquitous learning beyond school. (Forkosh Baruch & Erstad, 2018). Notwithstanding, mobile learning has entered all levels of the education system, from higher education to the K12 education system, as well as the teacher education system. Mobile devices pose new advantages and opportunities for learning among young adults in higher education institutes. These include quick access and instant communication, engaging learners, facilitating collaborative learning, enabling authentic learning on the move, situated learning and diverse modes of learning - to name a few. However, this new form of learning summons some challenges as well. These include, for example, coping with faculty, who are not always technology-oriented, technological challenges (e.g., the device, infrastructure) and fighting distractions distraction (Gikas & Grant, 2013).

The K12 education system has embraced mobile technology to a lesser extent, since the younger the children, the less of them actually have mobile devices. However, mobile learning continues to evolve in spite of it being "under the radar" in educational policy once educators understand its impact on the younger generation (Soloway & Norris, 2018). During the last decade there have been initiatives to implement mobile technology with children as young as elementary school. Uses of mobile technology for young ages include applications for practicing learning skills, applications for information management and applications for learning content in different subject matters (Domingo & Garganté, 2016). In secondary education most students have their own mobile phones and tablets, therefore we can expect a BYOD (Bring Your Own Device) paradigm, in which the usage of mobile learning has developed, mainly for amplification of previous learning modes, e.g., research and presentation, but also unique uses such as augmented reality (e.g., Bacca et al., 2015). Still, the structure of the curriculum is changing, e.g., aspects of classroom management and responsibility of students for their learning (Parsons & Adhikar, 2016). Therefore, the teacher education system is changing accordingly, in an attempt to prepare preservice teachers to utilize technology in general and mobile technology in particular in the training process and thereby to promote positive attitudes of future teachers regarding the utilization of mobile learning (Pinto et al., 2019).

Mobile technologies are regarded a lever for educational innovation and paradigm change, and therefore encourage mobile learning as a new educational paradigm. The added value of mobile technologies for global, multicultural learning, active learning, situated learning, location-based learning, problem-based learning, implementation of augmented or virtual reality, to name a few, is currently being researched extensively. This book is a collection of papers aimed to understand the contribution of mobile technologies in educational organizations and the change they bring to teaching and learning practices. They present a wide range of studies and examples of the use of mobile technologies for empowering learners and supporting their learning. They also refer to challenges in realizing change in organizational contexts.

The first section deals with issues of successful implementation of mobile technologies in educational organizations and presents various studies that examine how organizations cope with change that the technology entails. These articles highlight the teachers' role in the adoption and implementation of mobile technologies in educational organizations. Teachers are leaders of change, therefore, their attitudes, cooperation and capabilities are crucial to successful paradigm change. Consequently, teacher training is an important component in the implementation process.

The first chapter, "Social Innovation in Public Schools: A Case Study on the Remote Experimentation Laboratory of the Federal University of Santa Catarina," written by Isabela Silva and her colleagues presents a case of assimilation of mobile technology in a public schools in Brazil. The objective of the project was to initiate the integration of digital technologies in basic education of the Brazilian public network. The authors describe the training process carried out by RexLab and its efficiency in reducing fear and increasing motivation of teachers to use mobile technology in their schools.

The chapter "Preparing Preservice Teachers to Use Mobile Technologies" also deals with teacher training, in this case preservice teachers. In this chapter, Carolyne Obonyo presents empirical findings and examples of teaching practices across the globe on how mobile technologies are used to prepare preservice teachers for their future classrooms and adapt to the digital world. She focuses on two emerging pedagogical models, i.e., distance education and the flipped classroom model, where mobile technologies play an important role in their implementation. She argues that in order to be effective, preparations of preservice teachers to utilize mobile learning should not be based solely on learning about m-learning but should be practiced during the practicum period in partner schools.

The next chapter also deals with training related to utilizing mobile learning in teaching, this time in the context of faculty in Higher education. In chapter titled "Enhance Active Learning in Higher Education by Using Mobile Learning," Osnat Dagan and Noa Ragonis describe a faculty training project for integrating mobile learning (ML) into academic teaching as a means of accomplishing pedagogical goals. The chapter presents guidelines and principles for integrating ML into academic practices.

The second section of the book presents studies focused on unique innovative uses of mobile technologies to improve teaching and learning. Each case demonstrates a different pedagogical purpose: to improve active learning processes, facilitate knowledge construction, develop spatial awareness and contextual, location-based knowledge, or as a mean of fostering interdisciplinary student-led learning.

Teresa Coffman and Mary Beth Klinger, the authors of the chapter "Mobile Technologies for Making Meaning in Education: Using Augmented Reality to Connect Learning," discuss the potential of AR in broadening learning experience in a course, focused on situated, embedded learning and providing immersive and social experiences, and ultimately enhancing teaching and learning. AR serves as an immersive learning tool that can complement course learning goals and at the same time develop deeper skills and knowledge.

In the chapter "Strategies for Developing Mobile Location-Based Learning Activities by Teachers," Hagit Meishar-Tal examines strategies that teachers developed in the creation of mobile location-based learning activities while using a mobile game generator for creating treasure hunt like games. She found that some of the teachers that used this platform developed original and creative location-based activities and presents these activities as models and inspiration for other teachers.

Another interesting model of utilizing mobile devices for learning is described by Julie Prescott, Pippa Iliff, Daniel Edmondson, and Duncan Cross. Their chapter, "Students as Co-Creators of a Mobile App to Enhance Learning and Teaching in HE," draws on findings from a research project titled 'Student Led Technology Enhanced teaching and learning'. This project engages undergraduate students in researching their own student body to discover the needs of the student community, to design and develop two technology-based teaching and learning tools, and empowers them to train and mentor staff in using the developed mobile solutions. The student-led technology-enhanced learning and teaching project brought together staff and students from across the university in cross-disciplinary work. The authors report the impact of the project on the students in three main aspects: increasing confidence, developing employability skills and student experience.

The last study presented in this section is titled "The Role of a WhatsApp Group of a Professional Learning Community of Chemistry Teachers in the Development of Their Knowledge." In this chapter Ron Blonder and Ruth Waldman argue that using WhatsApp groups among a community of teachers enables sustained, ongoing, intensive interaction and sharing of practical knowledge that is constructivist in nature. In their analysis of a community of chemistry teachers they found that the WhatsApp group activated a unique mechanism that supports the growth of teachers' professional knowledge.

In the third section, the multicultural and intercultural aspects of mobile technology use are becoming increasingly relevant in a world of diverse cultures on one hand and globalism on the other hand. In this era, immigration and relocation are widespread and global interactions occur between users. The design and development of technological solutions, specifically mobile applications, should take into account these aspects and adapt them to users of different cultural backgrounds, e.g., for learning a foreign language. In this section we present three chapters dealing with considerations and implications related to the use of mobile technologies in global and multicultural contexts.

The chapter "Culture and Context Impact on Mobile Tech Application in Organizational Learning: Case Study of UK Higher Education Institution and Chinese State-Owned Enterprise" by Yun Chen and Yunpu Zhang describes two sharp-contrast case studies, one of a higher education institution (HEI) in the UK and the other a state-owned enterprise (SOE) in China. The two case studies were compared in terms of characteristics of learner groups, motivation mechanisms, mobile functions, and learning culture. The chapter examines the culture and context impact on mobile technology utilization in organizational learning. The authors conclude that in order to achieve the best educational experience in a learning community, culture issues should be fully considered when selecting mobile functions and platforms.

In the next chapter, "Contextualized MALL in Target and Non-Target Countries: Mobile Activity Evaluation," Anat Cohen and Orit Ezra studied differences in using Mobile Assisted Language Learning (MALL), among students in two countries – Taiwan, where the Chinese language is the spoken language, and Israel, which is predominantly a Hebrew speaking environment. The study contributes to our understanding of the benefits of studying foreign language in a cultural context and therefore in a target country.

In the chapter "The Use of Mobile Technologies by Immigrant Adolescents in Coping With the New Language and With Their Formal Studies," Gila Cohen Zilka examined the usage of mobile technologies by immigrant adolescents to support the process of their absorption in the new country, the acquisition of the new language and their learning. She found that the adolescents who participated in the study used mobile technologies extensively, e.g., translation apps, databases, various applications, and social networks. They believed that thanks to mobile technologies, they have successfully integrated into their new environment. Many of them reported that mobile technologies were a major factor in their successful absorption in the new country.

The fourth and final section deals with the contribution of mobile technologies to populations with special needs, ranging from cognitive and learning difficulties to gifted students. Mobile learning technologies served as an effective assistive means in a wide range of contexts, e.g., writing, reading, and management of learning and everyday life skills. This chapter presents a number of studies that discuss the potential of mobile technologies for populations with special needs and address design principles for mobile solutions adapted to these populations.

In her chapter "Time for a Change: Designing a Mobile Application to Assist People With Intellectual Disabilities" Rotem Israel-Foshelson presents a case study of designing an application aimed to help for people with mild intellectual deficiency manage their money and handle daily cash transactions. The chapter describes the feedback received by the users and highlights recommendations for implementation and training.

The chapter "Design and Development of a Mobile Writing Application for Students With Dysgraphia" by Sinan Hopcan and Saniye Tokel describes a mobile writing application for students with dysgraphia. The study evaluates the mobile writing application in terms of perceived ease of use, perceived usefulness, perceived enjoyment, educational content, multimedia use, and technical features. Results show that experts' views were very positive about mobile writing application for students with dysgraphia. They expressed positive opinions, saying that the application can help students with dysgraphia to write legibly and neatly within the lines.

Tami Zeifert and Veronika Zimon describe in their chapter, "Using Tablet Applications as Assistive Tools in Teaching English as a Foreign Language," a study that examined the contribution of tablet applications for the acquisition of the English language and improvement of reading and comprehension skills in two groups of 4th grade students: one including students diagnosed with learning disabilities and another including students who were not diagnosed with such disabilities. They present evidence that the use of tablet applications led to an improvement in English language skills in reading and reading comprehension among students with learning disabilities. The most outstanding improvement was in reading comprehension.

Ina Blau presents in her chapter, "Real-Time Mobile Assessment for Learning: Insights From an Experiment with Remedial, Midstream, and Excellence Middle School Students," an experiment conducted using a mobile a feedbacking system that served as assistive technology. She examined the impact of the use of this system on academic achievement, students' perceived learning, academic self-efficacy, learning motivation, and self-esteem among three groups of students: remedial, midstream and excellence middle-school students. Findings show that students in the remedial track benefit from using the system in terms of achievement, perceived learning, self-efficacy, and motivation. Students in the excellence track benefit in terms of intrinsic motivation, and students in the mainstream track in terms of self-esteem.

In their chapter, "Leveraging Mobile Technologies to Support Active Learning for All Students: Smartphones to Support Learning," Jenipher Gallop, Beverly Ray, and Cory Bennett describe a method by which all students have equitable access to learning by leveraging the potential of smartphones. They claim that the premise of incorporating mobile technology into the educational environment is to provide authentic tools that support instruction, engagement and deep learning. Technology is viewed as a cognitive and digital tool that students work with to enhance their own learning capabilities.

We hope this book will serve teachers, policymakers and other education professionals in the formal education, higher education, special and informal education systems, and will inspire practitioners to use mobile technologies in innovative and creative ways. We encourage further research on mobile technologies and mobile learning in educational organizations. The knowledge gained from the research presented in this book and the insights of practitioners from the findings may further contribute to educators in all positions in expanding the constructive use of mobile devices in learning and teaching and implementing innovative pedagogical paradigms in the utilization of mobile technologies for learning.

Alona Forkosh Baruch
Levinsky College of Education, Israel & Tel Aviv University, Israel

Hagit Meishar Tal
Holon Institute of Technology, Israel

REFERENCES

Asongu, S., & Boateng, A. (2018). Introduction to Special Issue: Mobile Technologies and Inclusive Development in Africa. *Journal of African Business*, *19*(3), 297–301. doi:10.1080/15228916.2018.1481307

Bacca, J., Baldiris, S., Fabregat, R., & Graf, S. (2015). Mobile augmented reality in vocational education and training. *Procedia Computer Science*, *75*, 49–58. doi:10.1016/j.procs.2015.12.203

Barkhuus, L., & Polichar, V. E. (2011). Empowerment through seamfulness: Smart phones in everyday life. *Personal and Ubiquitous Computing*, *15*(6), 629–639. doi:10.100700779-010-0342-4

Bianchi, A., & Phillips, J. G. (2005). Psychological predictors of problem mobile phone use. *Cyberpsychology & Behavior*, *8*(1), 39–51. doi:10.1089/cpb.2005.8.39 PMID:15738692

Christensen, R., & Knezek, G. (2018). Reprint of Readiness for integrating mobile learning in the classroom: Challenges, preferences and possibilities. *Computers in Human Behavior*, *78*, 379–388. doi:10.1016/j.chb.2017.07.046

Crompton, H., & Burke, D. (2018). The use of mobile learning in higher education: A systematic review. *Computers & Education*, *123*, 53–64. doi:10.1016/j.compedu.2018.04.007

Crompton, H., & Traxler, J. (2019). Learning with mobile devices. In M. Khosrow-Pour (Ed.), *Advanced Methodologies and Technologies in Modern Education Delivery* (pp. 793–808). IGI Global. doi:10.4018/978-1-5225-7365-4.ch062

Cumming, T. M., Strnadová, I., Knox, M., & Parmenter, T. (2014). Mobile technology in inclusive research: Tools of empowerment. *Disability & Society*, *29*(7), 999–1012. doi:10.1080/09687599.2014.886556

Domingo, M. G., & Garganté, A. B. (2016). Exploring the use of educational technology in primary education: Teachers' perception of mobile technology learning impacts and applications' use in the classroom. *Computers in Human Behavior*, *56*, 21–28. doi:10.1016/j.chb.2015.11.023

Dorsey, E. R., Glidden, A. M., Holloway, M. R., Birbeck, G. L., & Schwamm, L. H. (2018). Teleneurology and mobile technologies: The future of neurological care. *Nature Reviews. Neurology*, *14*(5), 285–297. doi:10.1038/nrneurol.2018.31 PMID:29623949

Dunaway, J., Searles, K., Sui, M., & Paul, N. (2018). News attention in a mobile era. *Journal of Computer-Mediated Communication*, *23*(2), 107–124. doi:10.1093/jcmc/zmy004

Faulds, D. J., Mangold, W. G., Raju, P. S., & Valsalan, S. (2018). The mobile shopping revolution: Redefining the consumer decision process. *Business Horizons*, *61*(2), 323–338. doi:10.1016/j.bushor.2017.11.012

Forkosh Baruch, A., & Erstad, O. (2018). Upbringing in a Digital World: Opportunities and Possibilities. *Technology. Knowledge and Learning*, *23*(3), 377–390. doi:10.100710758-018-9386-8

Kao, H. Y., Wei, C. W., Yu, M. C., Liang, T. Y., Wu, W. H., & Wu, Y. J. (2018). Integrating a mobile health applications for self-management to enhance Telecare system. *Telematics and Informatics*, *35*(4), 815–825. doi:10.1016/j.tele.2017.12.011

Law, R., Chan, I. C. C., & Wang, L. (2018). A comprehensive review of mobile technology use in hospitality and tourism. *Journal of Hospitality Marketing & Management*, *27*(6), 626–648. doi:10.1080/19368623.2018.1423251

Lee, I., & Shin, Y. J. (2018). Fintech: Ecosystem, business models, investment decisions, and challenges. *Business Horizons*, *61*(1), 35–46. doi:10.1016/j.bushor.2017.09.003

Light, B., Bagnall, G., Crawford, G., & Gosling, V. (2018). The material role of digital media in connecting with, within and beyond museums. *Convergence*, *24*(4), 407–423. doi:10.1177/1354856516678587

McLuhan, M. (1966). *Understanding Media: The Extensions of Man*. New York: McGraw-Hill.

Parsons, D., & Adhikar, J. (2016). Bring Your Own Device to Secondary School: The Perceptions of Teachers, Students and Parents. *Electronic Journal of E-learning*, *14*(1), 66–80.

Pinto, M., Caballero-Mariscal, D., Fernández-Pascual, R., Sales, D., & Guerrero-Quesada, D. (2019). The Attitudes of Teachers-in-Training Towards Information Literacy Skills and the Inclusion of Mobile Devices in the Process of Teacher Education. In S. Kurbanoğlu (Eds.), *Information Literacy in Everyday Life: Communications in Computer and Information Science* (pp. 597–606). Springer Cham. doi:10.1007/978-3-030-13472-3_56

Sharma, S. K., & Sharma, M. (2019). Examining the role of trust and quality dimensions in the actual usage of mobile banking services: An empirical investigation. *International Journal of Information Management*, *44*, 65–75. doi:10.1016/j.ijinfomgt.2018.09.013

Soloway, E., & Norris, C. (2018). Mobile Learning in K-12: Roadblocks to Adoption. In Second Handbook of Information Technology in Primary and Secondary Education (779-783). Springer.

Thomée, S., Härenstam, A., & Hagberg, M. (2011). Mobile phone use and stress, sleep disturbances, and symptoms of depression among young adults-a prospective cohort study. *BMC Public Health*, *11*(1), 66. doi:10.1186/1471-2458-11-66 PMID:21281471

Section 1

Implementation of Mobile Learning in Educational Organizations:

The Role of Training

Chapter 1
Social Innovation in Public Schools:
A Case Study on the Remote Experimentation Laboratory of the Federal University of Santa Catarina

Isabela Silva
Universidade Federal de Santa Catarina, Brazil

Karmel Nardi Silva
Universidade Federal de Santa Catarina, Brazil

Karen Lotthammer
Universidade Federal de Santa Catarina, Brazil

Simone Bilessimo
Universidade Federal de Santa Catarina, Brazil

Juarez Bento Silva
Universidade Federal de Santa Catarina, Brazil

ABSTRACT

The project "Promoting Digital Inclusion in Public Schools Through Integration of Innovative Low-Cost Technologies in the Teaching of STEM Subjects" has been carried out by the Remote Experimentation Laboratory (RExLab), Federal University of Santa Catarina, since 2008. This project has trained 363 teachers from 6 schools, since it is an initiative of technology integration in the basic education of the Brazilian public-school system. The present study focuses on the benefits of the project in relation to the teachers involved in its scope. The positive results acquired by the project over the years demonstrate that the integration of technologies in education by teachers in the public network is a practice to be encouraged.

DOI: 10.4018/978-1-5225-8106-2.ch001

INTRODUCTION

The New Information and Communication Technologies are capable of bringing benefits for many areas of knowledge, and between these educations are; through integration of technology in education, the concept of classroom is turning larger, once access to Internet ease the use to media such as online videos, images, podcasts, games, between others (Henton, 2016). Although there is an initial resistance shown by teachers in relation to the use of technology in classroom because they usually have low knowledge in relation to new technologies, the use of technology has much to offer to help them with their work, once technology can optimize many of their activities, such as correction of activities (Presby, 2017).

In this context, the project "Promoting digital inclusion in public schools through the integration of innovative technologies of low cost in the teaching of STEM subjects" was launched in 2008. This project was initiated in 2008 by the Mobile Remote Experimentation Thematic Group (GT-MRE), of the Remote Experimentation Laboratory (RExLab), belonging to the Federal University of Santa Catarina, Brazil.

The project was created with the objective of offering an initiative for the integration of digital technologies in the basic education of the Brazilian public network, in view of the lack faced by a large part of Brazilian public schools that do not have science or computer labs. To achieve this objective, the project was structured in two axes. A formative axis, aimed at teacher training in digital technologies, so that teachers learn how to use technologies as well as how to apply them in class, and, after an integrative axis, during which teachers make applications of digital technologies in the classroom with your students.

Over the years, the project has presented very positive results, being proven through the satisfaction of teachers and students involved in their actions and because of the way in which they expressed their opinions in questionnaires applied during the project.

In this way, the present work aims to present the project with focus on the benefits that this brings to the teachers who are participants of it.

METHODOLOGY

This section aims to present the methodological procedures used to carry out the present research.

Methodological Procedures

The present research was carried out using the qualitative approach, which according to Gil (2007), is perceived through an understanding of a social group to obtain the understanding of the totality of the phenomenon. According to Minayo (2001), this type of research works with a deeper version of the phenomena, considering its variables, ideal for educational environments. Prodanov and Freitas (2013) argue that qualitative research allows us to consider the dynamic relation between the real world and the subject, perceiving its inseparable link and realizing that it cannot be translated into numbers.

As for nature, the present research can be defined as applied research, since it is directed to the solution of problems in a specific way. Prodanov and Freitas (2013), consider that applied research involves local truths and interests, solving them in a practical way and exploring different factors pertaining to the subject in question.

With regard to the objectives, this study fits the exploratory research, which according to Gil (2007) is ideal to clarify a given problem better and to construct its hypotheses, that is, it aims to increase familiarity with the problematic in question. For Prodanov and Freitas (2013), this type of research facilitates the delimitation of the topic by exploring its hypotheses more broadly through flexible planning. Exploratory research often takes the form of bibliographic research or case studies.

Considering the procedures, the research is a case study, by specifically studying a group. According to Fonseca (2002), the case study arises to establish an understanding about a differentiated situation in several aspects, seeking to find its particular characteristics. This type of study presents an overview of the subject from the researcher's perspective.

Development and Application of Semipresential Course

The semipresencial course, called "Teacher Training for Integrating the Use of ICT in Basic Education", is on the InTecEdu platform (http://intecedu.ufsc.br/), developed by RExLab. This platform is a virtual learning environment based on the MOODLE system, in which several courses are offered, both teacher training and courses created by teachers already certified by this action that is in the integrative axis of teacher training, when teachers apply educational technologies learned at the time of training.

The MOODLE system is free learning software to run in a virtual learning environment. The platform is configured as a Learning Management System, which values the collaborative work and is accessed through the Internet or a local network. According to the official website of the MOODLE platform, the program is available in 75 different languages, with around 25,000 registered websites serving more than 175 countries, thus being the most widespread EMS around the world (MOODLE, 2016).

In total, the teacher training course lasts 120 hours, 90 hours of which must be carried out through activities carried out in the virtual learning environment and 30 hours of these hours are carried out through face-to-face activities such as practical classes and lectures given by the responsible extension student by the school and other members of the RExLab team, including project partners already certified by the course. In addition to completing 120 hours, for certification it is necessary that the teacher fulfills at least 75% of the proposed activities; after completing the course, the teacher receives a 120h extension course certificate. The figure below shows a screenshot of the course.

The semipresencial course "Teacher Training for Integrating the Use of ICT in Basic Education" has 6 units.

The first step of the course is presentation, functioning as a sort of introduction, in order to explain the objectives of this course, as well as details such as its modality, course duration and detailing of this.

In addition, this step counts on the application of two questionnaires, in the format of poll: TPACK Search and Teaching Profile. Through the TPACK survey, teachers report their familiarity with technologies and their level of difficulty in relation to their handling. This research has 35 questions. In relation to the questionnaire of Teacher Profile, this refers to the individual profile of the teacher, presenting questions related to their gender, age range, schooling, among others. This research has 20 questions.

After the presentation of the course, the teachers go to Unit 1, where they carry out the initial evaluation of the course, in order to compare the difference of the knowledge of the teacher before starting the course and after finishing it - in the last unit, there is a identical questionnaire, to be answered again by the teachers. In addition, there is a material for reading.

Figure 1. Screen capture of the teacher training course
Source: Authors

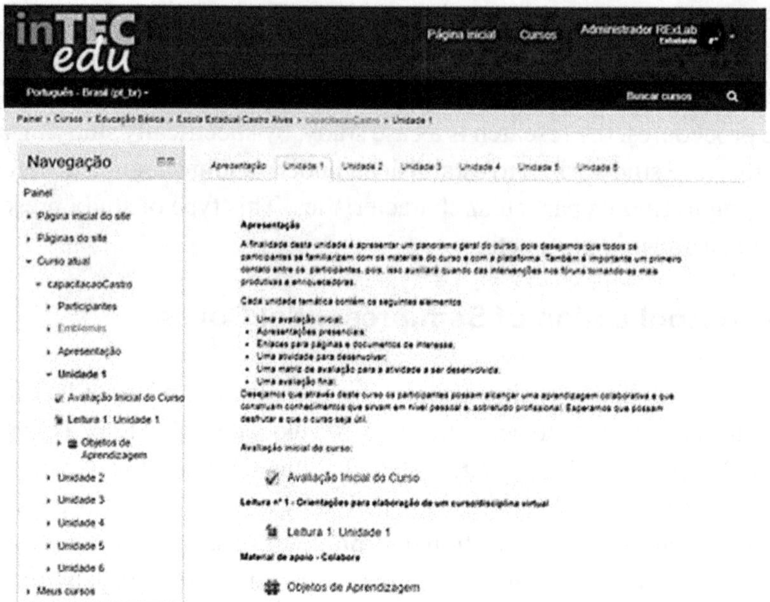

Unit 2 is called "The Educational Platform MOODLE", and it is the moment when teachers learn how to edit a course in MOODLE. From this unit, all modules are divided into three stages: the initial evaluation of the unit in question, the didactic content for reading and activities for submission and the final evaluation of the unit. As well as the initial and final evaluations of the course in general, each unit has them individually, in order to verify how much the teacher could learn from the exercises and information presented in the unit.

The next unit refers to "Planning and publication of resources of MOODLE and WEB 2.0", at which time the teacher learns to incorporate in his virtual classroom the various resources that MOODLE offers and also of WEB 2.0 as alternative material. This module has the same format as Unit 2, starting with the initial evaluation, followed by activities and material for reading and final evaluation of the unit.

With a structure similar to the two previous units, Unit 4 has as a presumption to present the "Planning and creation of educational activities in MOODLE", in order that the participants have the knowledge and skills to generate educational activities using the tools that MOODLE offers.

Following the structure of the previous units, Unit 5 is denominated "Planning and creation of communication spaces in MOODLE", aiming that the teachers have knowledge to generate spaces of learning, communication and socialization between the other participants and with their students in the future.

Finally, Unit 6, entitled "Creation and configuration of an Evaluation Laboratory", was created, so that the participants had the knowledge to generate evaluation laboratories, making possible a peer evaluation. Being the last unit, it also has the final evaluation of the course, at which point the participants' prior knowledge is compared to the post-course knowledge.

Figure 2. First timeline of RExLab
Source: Silva (2017)

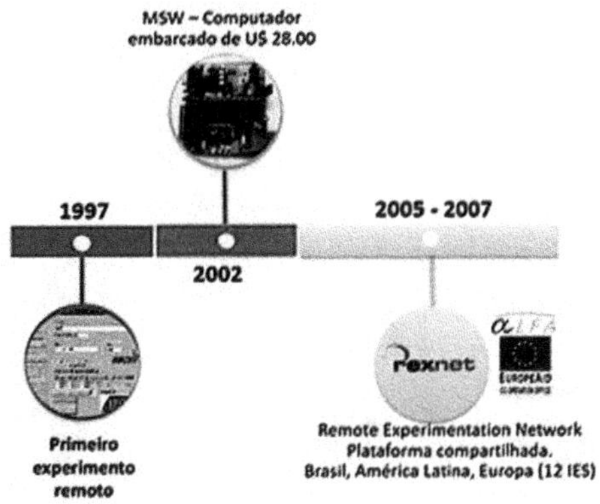

RELATED WORK

This section presents a brief history of the Remote Experimentation Laboratory (RExLab) and Continued Teacher Training for ICT Integration in Basic Education.

REXLAB: Remote Experimentation Laboratory

The Remote Experimentation Laboratory (RExLab) appeared in 1997, founded by prof. Dr. João Bosco da Mota Alves and collaborators, at the Federal University of Santa Catarina (UFSC), aiming at spreading science and technology and increasing the society's contact with these areas through the use of low-cost resources.

In order to represent the principles of the laboratory, a visual identity was developed containing the elements of science and technology. For a faithful representation of the context, a RExLab logo was developed, characterized by the authors Meyer, Pigozzi and Alves (2002) based on the structure of a tree that represents the connection between users, adding the concept of branching of the technology, transmitted by the semicircle superior greenish, which alludes to sustainability and commitment to society.

RExLab's activity consists in spreading knowledge from the area of science and technology in order to help intellectual development in members of society through transformation from the new information technologies.

One of the precepts of RExLab is to work with collaborative work, providing a constant exchange of knowledge between coordinators, collaborators and volunteers. In addition, one of the main goals is to encourage multidisciplinarity and the development of open, transparent and responsible work.

In the school context, RExLab works with teacher training with the use of courses in the virtual learning environment in order to establish the integration of technology in the classroom, without sinning in the sense of assistance to teachers, providing the necessary equipment and materials for the good work performance. According to Toda (2013), the capacities solve complex problems, establishing replicable problem solving patterns, being one of the actions of the laboratory that best represents the influence of its performance in society.

Figure 3. Second timeline of RExLab
Source: Silva (2017)

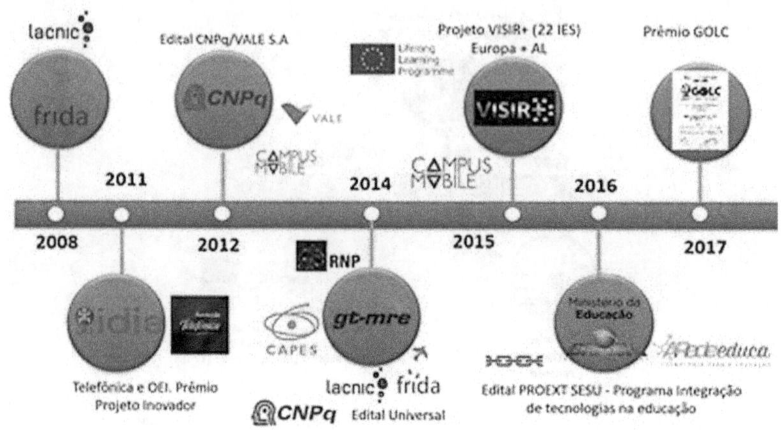

In the following figures it is possible to analyze the evolution of RExLab through the presentation in two timelines:

Figure 3 shows the timeline for the years 2008 to 2017.

From the timeline constructed by Silva (2017), it is possible to observe that since 2008, there was a good evolution in RExLab, since it was in this period, that the laboratory began to receive foments to better work. Since the introduction of the Frida prize (Regional Fund for Digital Innovation in Latin America and the Caribbean), it was possible to enter the school environment to begin a work of integrating technology that would continue in constant evolution, at least up to the present day.

In the first timeline of RExLab, it is possible to notice that in the year of its foundation, 1997, the laboratory developed its first remote experiment, an 8051 microcontroller, to complement in the teaching of microcontrollers. In 2002, the laboratory developed a low-cost embedded computer, seeking to provide access to individuals who could not afford the equipment. In the years 2005 to 2007, RExLab affirmed partnership through a platform shared between Brazil, Latin America and Europe.

In the second timeline of RExLab, it is possible to notice that in 2008, the laboratory was considered with the FRIDA (Regional Fund for Latin America and the Caribbean), which boosted the work of the laboratory from the use of the foments from the resource. In 2011, the laboratory received the innovative design award from the telephone foundation, gaining greater visibility worldwide. In the year of 2012 the laboratory was contemplated by the CNPQ VALE edict, which provided support for the development of a technology integration project in schools. In the same year, the laboratory won the Claro contest in partnership with USP, Campus Mobile, where employees had the opportunity to travel to the Silicon Valley, among other awards.

In 2014, the laboratory was approved by CNPQ Universal, which consisted of a pool of resources between the CAPES and RNP development agencies, allowing RExLab to develop the GT-MRE project, which aimed to integrate the virtual environment of learning with free content and access to remote experiments for students and teachers of basic education in the public-school system.

Table 1. Most recent awards and recognitions received by RExLab

Award or Recognition	Event or Institution	Year
GOLC Award - Online Lab Award - Remote Controlled Lab.	*Global Online Laboratory Consortium (GOLC).*	2017
Best project in the category "Exact and Earth Sciences"	27º Seminar of Scientific Initiation of UFSC	2017
Best Paper Spreading remote lab usage a system.	*Event: 2nd International Conference of the Portuguese Society for Engineering Education.*	2016
Arede Educa Award - Category Educational Platforms in the Public Sector.	Arede Educa.	2016
2nd Best Educational App - 2nd Apps Integrated Contest.	Event: V Brazilian Congress of Informatics in Education	2016
Best application in the Education category, with the block.ino project.	Event: 4th Mobile Campus	2016
Finalist of the challenge "Technology is Bridge: reducing distances in education".	Promoted by Ashoka Changemakers / Embratel / Instituto Claro with the project "Integrating technology in Basic Education	2016
Invitation to participate in the UAE Summit for Innovation in Government invitation	Unit Arab Emirates. Ministry of cabinet affairs	2016
The project "Use of remote experimentation to support teaching-learning environments in the public school system" was one of the chosen ones to represent Brazil at the "VI International Encounter EducaRed 2011" (http://encuentro.educared.org/group/ experimentation-remote).	Telefónica Foundation / Spain	2011
Selection of the project "Use of remote experimentation to support teaching-learning environments in the public-school system" as one of the four most innovative projects in Brazilian education.	The selection was carried out by the Institute for Development and Educational Innovation (IDIE) of the Organization of Ibero-American States for Education, Science and Culture (OEI) and sponsored by Fundación Telefónica.	2011
Selection of the project "Use of remote experimentation as a support for teaching-learning environments in the public-school system" at the 1st FRIDA Meeting: Research on Information and Communication Technologies and public policies in Latin America and the Caribbean, in Montevideo, gives "round table: Experiences of successful research funded by FRIDA"	Promoted by LACNIC and the Digital Innovation Fund for Latin America and the Caribbean.	2009

Source: Authors

In 2015, RExLab joined the VISIR + project, aimed at the dissemination in Latin America of the use of VISIR remote laboratories, which are intended for the use of electrical and electronic circuits.

In 2016, the laboratory was provided with support from the Ministry of Education, to promote digital inclusion in basic education, from the public-school system, using low-cost technologies for science education. In the same year, the laboratory was awarded the prize "Rede Educa", with the first place in remote mobile experimentation, and in 2017, when it was 20 years old, it received the GOLC prize of the best remote laboratory in the world.

Throughout its history, RExLab received several awards, attended to more than 6,000 people and its members published more than four hundred works. RExLab was recognized with national and international awards in the development of its projects. The table below shows the premiums received in the last two years:

CONTINUED TRAINING OF TEACHERS FOR INTEGRATING ICT IN BASIC EDUCATION

The project to be described in this work has as one of its structural axes the training of teachers to use digital technologies, thus providing instructions on how to use technological resources as an auxiliary tool in their teaching practices.

The integration of information and communication technologies (ICT) in education has become recognized as a factor that brings positive changes to classrooms and promotes the development in student learning (GETENET, 2016). Another reason for its recognition is the widespread use of mobile devices and facilitated access to information. In this way, the easy access to information by the students, causes the teacher to stop being the holder of knowledge and become a mediator in the teaching and learning process (Oliveira & Junior, 2016).

For the effective integration of ICT in education, Dalal, Archambault and Shelton (2017) cite some barriers to be overcome: lack of professional training; poor technology infrastructure; undervalued technology among managers; and, restrictive resumes.

As for vocational training, Cieb (2017) lists the competences of teachers for the use of ICT in education and its three dimensions: pedagogical; social, ethical, legal and citizen; and, professional development and responsibility. Among its elements are:

- Be able to integrate technology into students' learning experiences and their teaching strategies
- Be able to use technology to create learning experiences that meet the needs of each student
- Be able to make and promote the critical interpretation of the information available in digital media
- Be able to use technologies to maintain active, systematic and efficient communication with the actors of the educational community

However, even knowing the need to develop ICT skills in the classroom, the TIC Educação 2017 research reveals that 77% of the teachers interviewed did not take a continuing education course in the last year and 7% distance format.

In this context, there is a great need to promote courses for the continuous training of teachers so that they have instructions on how to integrate ICT into their practices and that they can also properly guide their students to their pedagogical use.

HISTORY OF THE PROGRAM "PROMOTING DIGITAL INCLUSION IN PUBLIC SCHOOLS THROUGH THE INTEGRATION OF INNOVATIVE TECHNOLOGIES OF LOW COST IN THE TEACHING OF STEM SUBJECTS"

The program is related to a number of small projects under development and constantly evolving over the years. In 2005, RExLab, together with other partner institutions, obtained the approval of the RExNet - Yippee (Remote Experimentation Network - Yielding in inter-university peer-to-peer service) project with the European Community Alpha Program. The project was attended by 12 higher education institutions from seven countries (6 European countries from 4 countries and 6 Latin American from 3 countries) and was chosen as a "case" within the aforementioned convocation.

The success and innovative nature of the RExNet project motivated the applicant to submit a proposal to the National Council for Scientific and Technological Development (CNPq), in order to obtain resources, since, although the RExNet project had its formal period closed with the Alfa II Program, the partnerships were maintained and it would be possible to capitalize in Brazil part of the experience and expertise built into the project. Thus, in 2007 a proposal related to the use of remote experimentation as a support for collaborative learning environments was submitted to CNPq. However, the final result regarding the proposal was "Rejected" and no feedback was obtained.

In 2007, the project proposal "Use of remote experimentation to support the teaching-learning environment in the public-school system" was sent to a call of the Regional Fund for Digital Innovation in Latin America and the Caribbean (FRIDA). In this case, the project was approved and obtained resources in the order of US $12,400. The success in the development of the project was formalized in an invitation to present the project in the "I FRIDA Meeting: Research in Information and Communication Technologies and Public Policies in Latin America and the Caribbean", in Montevidéo, round table "Experiences of successful research financed for FRIDA ".

In 2011, the project "Use of remote experimentation to support the teaching-learning environment in the public-school system" was selected as one of the four

more innovative projects in Brazilian education. The selection was made by the Institute for Development and Educational Innovation (IDIE) of the Organization of Ibero-American States for Education, Science and Culture (OEI) and sponsored by Fundación Telefónica. The study was carried out at the national level, and 64 projects were identified, initially considered relevant on innovation in education using ICT, where 26 were characterized as truly innovative, and of these, four projects were considered cutting edge.

In the same year, this project was one of the chosen ones to represent Brazil in the "VI International Encounter EducaRed 2011." In February 2012, it was the subject of a debate promoted at the Brazilian Campus Party on a panel discussion entitled "Digital mobility and education: the school beyond its walls", which aimed to bring the reflection on a new way of thinking about education - from the presentation of different practices of use of mobile devices such as smartphones and laptops for educational purposes.

In April 2014, the project "Use of Remote Experimentation in Mobile Devices for Basic Education in the public-school system" was approved, together with the

FRIDA, in the "Escalamientos Frida 2014" call, where it obtained approval of resources in the order of US $13,951,00 that were applied in permanent material in the

project, with a duration of one year. Currently, the Remote Experimentation Laboratory has as main project related to teacher training in schools of the public school the program "Promoting digital inclusion in schools of Basic Education of the public network from the integration of innovative technologies of low cost in teaching of Natural and Exact Sciences ". This project is carried out in partnership with the Federal University of Santa Catarina, the Brazilian Ministry of Education, the National Education and Research Network (RNP), among others. The duration of the program is from January 2016 to July 2018.

InTecEdu represents an initiative of the RExLab digital inclusion line, through the integration of technology in the context of Education and develops its projects and activities in Basic Education (Early Childhood, Elementary, Secondary), Technical Education and Higher Education (SILVA ET AL, 2018).

With its own innovative strategy to achieve its objectives, its actions are structured in two axes: a training aimed at teacher training in relation to technologies and the other one of integration of digital technologies in didactic activities. The training of teachers has its formalization through the realization of courses (semipresential and MOOCs), mini-courses, workshops and lectures that deal with themes and case studies concerning the integration of technology in education.

On the other hand, the integration of the technology takes place through the availability of online didactic contents, made available in Virtual Environment of Teaching and Learning (AVEA) customized for the project, accessed by conventional or mobile devices, that will be complemented by the interaction with remote laboratories. The architecture implemented with Remote Mobile Experimentation (MRE), uses open source and low-cost remote experiments developed in RExLab from its open hardware platform and also from the use of computer platforms such as Raspberry Pi and Arduino.

Currently, 15 schools of Basic Education, of the public-school system, in the states of Santa Catarina, Rio Grande do Sul and Minas Gerais use the AVEA of the project. In these 41 teachers, 165 classes and 5,723 students, access didactic content in the AVEA. There are 17 laboratories available in 20 instances available for use in practical activities in disciplines in the STEM areas. Regarding the integration of technology in the STEM areas, 2,211 students, 67 classes from 5 schools integrated the Remote Laboratories into classes plans in Physics and Biology (High School), Science (Elementary).

The integration of technology into the lesson plans in the STEM areas also occurred from AVEA content and tablets in classrooms, where 1,486 students from 70 classes were attended in Mathematics, Biology, Physics), Mathematics and Science (Elementary School) of 5 schools.

The Teaching Institutions currently participating in the project are:

1. Nhu Porã State Indian School (Torres / RS - Brazil)
2. Apolônio Ireno Cardoso Basic School of Education (Balneário Arroio do Silva / SC - Brazil)
3. Araranguá School of Basic Education (Araranguá / SC - Brazil)
4. Maria Garcia Pessi School of Basic Education (Araranguá / SC - Brazil)
5. Otávio Manoel Anastácio Municipal School (Araranguá / SC - Brazil)
6. Carmem Matos Borges Early Childhood Center (Araranguá / SC - Brazil)
7. Education Center for Youth and Adults (Araranguá / SC - Brazil)
8. Colégio Éthicos (Araranguá / SC - Brazil)
9. School of Basic Education Rubens de Arruda Ramos (Criciúma / SC - Brazil)
10. Castro Alves State School (Araranguá / SC - Brazil)
11. State School Professor Ignácio de Souza (Uberlândia / MG - Brazil)
12. Escola Municipal de Ensino Fundamental Jardim Atlântico (Balneário Arroio do Silva / SC - Brazil)
13. Municipal School Dilma Lúcia dos Santos (Florianópolis / SC - Brazil)
14. School of Basic Education Profª Dolvina Leite de Medeiros (Araranguá / SC - Brazil)
15. Federal Institute of Santa Catarina - IFSC (Araranguá / SC - Brazil)

RESULTS AND DISCUSSION

The "Teacher Training for Integrating the Use of ICT in Basic Education" course has been applied in six public schools located in the extreme south of Santa Catarina. In the municipality of Araranguá / SC, four schools are attended: E.E.B. Castro Alves, E.E. Profª Maria Garcia Pessi, E.M. Otávio Manoel Anastácio and E.E.B. Araranguá. In the municipality of Balneário Arroio do Silva / SC two schools are attended: E.E.B. Apolônio Ireno Cardoso and E.M.E.F. Atlantic Garden.

This course is of continuous training, being applied semesterly in the basic partner schools; until 01/10/2017, 363 teachers of the public network were trained. Taking into account that the Program whose actions grant teacher training is continuous and the number of teachers trained will certainly be expanded over time.

Regarding the way in which the training is carried out and how the teacher applies the acquired knowledge in his / her school routine, the experience of using the virtual learning environment as a complement in teaching starts from the moment the teacher asks the collaborator available for the school by the Remote Experimentation Laboratory (RExLab), instructions for using the InTecEdu platform (http://intecedu.ufsc.br/).

From this moment, the RExLab employee present at the institution presents to the teacher the course in the MOODLE of Teaching Training for Integration of the use of ICT in Basic Education, lasting 120 hours, which is developed by RExLab and offered at the school itself, demonstrating the paths necessary for the development of a satisfactory job and enabling the learning of useful work tools, such as the MOODLE platform.

Using the free MOODLE platform allows the teacher to edit courses, plan activities, perform corrections, execute student notes, and other activities. From the beginning of the application of the Teacher Training Course to Integrate the use of ICT in Basic Education, the teacher receives support and assistance to answer any doubts, in order to establish a satisfactory assimilation of the content, ensuring that it will be prepared to use the resource as an alternative to the complement of teaching. One of the important characteristics of the application of the course in question is the practical activities, which are constructed in order to familiarize the teacher with the content with which he will work.

After the foundation acquired through the training, the teacher will be able to generate content for their courses, inserting in MOODLE all the necessary items to complement the content passed in the classroom. In order for the access to be made, the RExLab collaborator available at the school will transport tablets available in the laboratory for use in class, since the institutions generally do not have the infrastructure so that the application can take place appropriately.

The students use the tablets to access the didactic sequences constructed by the teacher, after the same discourse about the content. By accessing InTecEdu through tablets, students discover videos, games and activities corresponding to the subject matter in the classroom, feeling extremely motivated by the opportunity to participate in a class with a more dynamic perspective that penetrates the technological universe to which they are inserted most of the time.

Teachers, who generally present themselves somewhat resilient at the beginning of the use, because they consider that integrating the technology could be extra work in their routine so exhausting, finally end up considering that the opposite happened, due to the fact that technology integration has assisted in the use of its time due to the useful tools of the system, among them some automatic services like: correction of questionnaires at the time and content replicated to other classes that are studying the same subjects.

CONCLUSION

The present work aims to present the benefits of the performance of the teachers participating in the project "Promoting digital inclusion in public schools of basic education through the integration of innovative technologies of low cost in the teaching of natural and exact sciences" developed by RExLab 9 years ago. These teachers work in the public-school system and were accustomed to the lack of resources and infrastructure with which public institutions of Brazilian education suffer. On the other hand, there are a multitude of technological resources of low cost and high applicability in the classroom, being alternatives with the potential to supply eventual deficits in public basic education. The big problem is that many teachers are not aware of this type of tool because they have little knowledge of digital technologies in general. In this way, one of the actions of the project developed by RExLab is to train participating teachers in relation to basic informatics and educational tools for use in the classroom.

Despite the initial reluctance of many faculty participants at the beginning of the training, because they believed that applying technology in the classroom would only give them more work to do, they soon realized that they were wrong; after all, through training they realized that in fact the use of technology optimizes their working time, since activities can be corrected automatically and materials can easily be replicated to different classes, for example.

The motivation of teachers became much greater in relation to the use of technology in the classroom, and such motivation even affected many of their colleagues, who also became interested in starting the teacher training course. This demonstrates that the training develops a kind of current event, in which, from the moment the first teacher conducts the training and presents his colleagues with the benefits he obtained from it, his colleagues feel motivated and start to feel interest in learning how to integrate technologies in the classroom.

However, school participation is also important in this process. The director or coordinator responsible for the institution should remain a partner and encourage the project in order to hold meetings with the teachers of the school to monitor and discuss the use of technologies and also with the coordinators of the project in order to inform them about the advances and problems encountered during the applications.

After all, the project must take place in a collaborative way between project coordinators, involved trainees, school directors, pedagogical staff, participating teachers and the students of these teachers. In this way, everyone will always be updated according to the advances and will share solutions to any problems, and the work will be executed in a natural and evolutionary way, which will allow the preparation of future research related to the project results.

ACKNOWLEDGMENT

The present study was funded by the Coordination for the Improvement of Higher Education Personnel (CAPES).

REFERENCES

Albugami, S. S. (2016). Developing a strategic approach to ICT implementation in Saudi secondary schools. Retrieved from https://search.proquest.com/docview/1885877026?accountid=26642

Almerich, G., Suárez-Rodríguez, J. M., Belloch, C., & Bo, R. M. (2011). Las necesidades formativas del profesorado en TIC: perfiles formativos y elementos de complejidad. Relieve, 17(2), 1-28. Retrieved from http://www.uv.es/RELIEVE/v17n2/RELIEVEv17n2_1.htm

Andrade, T. (2005). Inovação e Ciências Sociais: Em Busca de Novos Referenciais. Sistema de Información Científica: Red de Revistas Científicas de América Latina y el Caribe, 20(58), 145-156.

Caetano, L. M. D. (2015). Tecnologia e Educação: Quais os desafios? [UFSM]. *Educação, 40*(2), 295–309. doi:10.5902/19846444

Cieb. (2017). Nota technical. Retrieved from http://www.cieb.net.br/wp-content/uploads/2018/01/CIEB_NotaTecnica-8_nov2017.pdf

Coutinho, R. X., Folmer, V., & Puntel, R. L. (2014). Aproximando universidade e escola por meio do uso da produção acadêmica na sala de aula. *Ciência & Educação (Bauru), 20*(3), 765–783.

da Silva. *I.N.* (2016). *Impacto da aplicação de tecnologia no ensino de empreendedorismo para estudantes concluintes da educação básica na rede pública. 2016. Universidade Federal de Santa Catarina,* Araranguá.

da Silva, I. N., da Rocha, J. S., da Silva, J. B., & Bilessimo, S. M. S. (2018). Uso de dispositivos móveis na disciplina de guarani para estudantes de uma escola multisseriada indígena. Renote, *16*(1).

da Silva, I. N., da Silva, K. C. N., Lotthammer, K. S., & Bilessimo, S. M. S. (2017). Inclusão digital em escolas públicas através de tecnologias inovadoras de baixo custo no ensino de disciplinas stem. *Renote, 15*(2).

da Silva, O. G., & Navarro, E. C. (2012). A relação professor-aluno no processo ensino-aprendizagem. *Revista Eletrônica Interdisciplinar*, *2*(8), 95–100.

Dalal, M., Archambault, L., & Shelton, C. (2017). Professional Development for International Teachers: Examining TPACK and Technology Integration Decision Making. *Journal of Research on Technology in Education*, *49*(3-4), 117–133.

de Oliveira, L. A., & Junior, M. T. F. (2016). A inserção das tecnologias da informação e comunicação nas escolas. In *EnPED-Simpósio Internacional de Educação a Distância e Encontro de Pesquisadores em Educação a Distância.*

de Oslo, O. M. (2005). *Diretrizes para coleta e interpretação de dados sobre inovação.* OECD Publishing.

Farias Filho, M. C., & Arruda Filho, E. J. (2000). *Planejamento da pesquisa científica.* Editora Atlas, SA.

Fonseca, J. J. S. 2002(). Metodologia da pesquisa científica. Fortaleza: Apostila.

Gadotti, M. (2000). Perspectivas atuais da educação. *São Paulo em perspectiva, 14*(2), 03-11.

Getenet, S. T. (2016). Adapting technological pedagogical content knowledge framework to teach mathematics. *Education and Information Technologies*, *22*(5), 2629–2644.

Gil, A. C. (2007). *Como elaborar projetos de pesquisa.* São Paulo: Atlas.

Henton, M. C. (2016). Technology integration practices of faculty in the louisiana community and technical college system. Retrieved from https://search.proquest.com/docview/1871305453?accountid=26642

IBGE. (n.d.). Estatísticas. Retrieved from http://downloads.ibge.gov.br/downloads_estatisticas.htm

Minayo, M. C. D. S., Deslandes, S. F., & Gomes, R. (2001). *Pesquisa Social: teoria, método e criatividade* (18th ed.).

Moodle. (n.d.). Retrieved from https://moodle.org/

Neri, M.C. (2009). O Tempo de Permanência na Escola e as Motivações dos Sem-Escola. Rio de Janeiro: CPS/FVG/IBRE.

Presby, B. (2017). Barriers to reducing the digital-use divide as perceived by middle school principals. Retrieved from https://search.proquest.com/docview/1887122371?accountid=26642

Prodanov, C. C., & Freitas, E. C. (2013). *Metodologia do trabalho científico: métodos e técnicas da pesquisa e do trabalho acadêmico*. Feevale.

Restivo, M. T., Alves, G. R., Dziabenko, O., & García-Zubía, J. (2013). Acquisition of higher-order experimental skills through remote and virtual laboratories. *IT Innovative Practices in Secondary Schools: Remote Experiments*, *1*, 321–247.

Rochadel, W., do Canto, J. Z., da Silva, J. B., Delgado, M. A. G., & Rebollo, M. A. G. (2016). Ensino a distância na educação básica: A integração pedagógica de jogos digitais em ambientes virtuais. *Revista E-Tech: Tecnologias para Competitividade Industrial*, *9*(1), 32–54.

Rodrigues, A. L. (2016). A integração pedagógica das tecnologias digitais na formação ativa de professores. In *IV Congresso Internacional TIC e Educação 2016* (pp. 708-720). Instituto de Educação da Universidade de Lisboa.

Silva, *L.J.* (2017). *Internet das coisas. Curso de Engenharia Elétrica e Telemática, Unisul.* Palhoça.

Silva, I. N., da Silva, K. C. N., Lotthammer, K. S., Bilessimo, S. M. S., & da Silva, J. B. (2017). Capacitação docente semipresencial para docentes de escolas básicas da rede pública. *SUCEG-Seminário de Universidade Corporativa e Escolas de Governo*, *1*(1), 510–518.

UNESCO. (2008). *ICT competency standards for teachers: policy framework*. Paris: UNESCO.

Veiga, I. P. A. (2003). Inovações e projeto político-pedagógico: Uma relação regulatória ou emancipatória? *Cadernos CEDES*.

Viotti. E., Salerno, M. & de Negri, J.A. (2005). Inovação e Competitividade. In *Brasil: o estado de uma nação* (pp. 1–40). São Paulo: Instituto de Pesquisa Econômica Aplicada.

Chapter 2
Enhance Active Learning in Higher Education by Using Mobile Learning

Noa Ragonis
Beit Berl College, Israel

Osnat Dagan
Beit Berl College, Israel

ABSTRACT

The quality of teaching in higher education is central for preparing students for their future academic and work demands. This includes the integration of information and communication technologies, since it has the potential to enhance teaching and learning processes and is thought improve both the understanding of content knowledge and the acquisition of relevant 21st-century skills. In particular, the use of mobile devices enables active learning at anytime and anywhere. This chapter describes a project of faculty development for integrating mobile learning (ML) into academic studies to accomplish pedagogical goals, as implemented at Beit Berl College, Israel. The chapter presents fundamental guidelines and principles of integrating ML into academic practices; Nine ML workshops that were developed and implemented; A detailed description of one workshop including the learning materials used; and Evaluation of the implementation. A summary of lessons learned from the implementation experience and further recommendations are presented.

DOI: 10.4018/978-1-5225-8106-2.ch002

INTRODUCTION

The information and communication revolution affects many areas, among them the teaching and learning methods in educational frameworks. Within the framework of excellence in teaching, great attention is paid to the integration of different Information and Communication Technologies (ICT) tools into the pedagogical processes in academia. The integration of new ICT tools has become significant in order to shift the traditional "sage on the stage" framework to dynamic teaching-learning scenarios that are more relevant, increase students' attention and motivation, and better engage students in active learning processes (McAleese et al., 2013). Those processes are manifested both in schools and in academia. Using ICT tools enables running up-to-date activities in classes, and enables the development of students' up-to-date skills for their future lives (Sharples, & Spikol, 2017). Enhancement of teaching and learning processes is a central value at Beit Berl College in Israel, which is expressed in many layers. Lecturers serve as a role model for their students, particularly in a college whose main area of training is education for all age groups and of various teaching subjects. These changes and approaches, together with the significant development in the daily use of mobile technology, led to the implementation of the project described. Mobile technology enables the following pedagogies: active, constructive, authentic, personalized and reflective learning processes for students, and supports learning anywhere and at any time) Howland, Jonassen, & Marra, 2014). Hence, the college wished, and still wishes, to assimilate the use of pedagogy based on Mobile Learning (ML) among the lecturers. The need to promote and improve teaching and learning requires significant change and lecturers need support in their exposure to new approaches and practices (Hénard & Roseveare, 2012).

In this chapter we describe the process of integrating ML that took place in Beit Berl College. It is an example for an implementation model at Higher Education Institutions (HEI) that takes into account many factors, and offers ways to deal with obstacles learned through five years of activation. The college ICT academic unit is operated within the Center for Teaching Enhancement (CTE) aimed to assist faculty in the integration of ICT in pedagogical processes. The unit launched a multiyear and multilayer project targeted at integrating mobile devices to enhance active learning. The project began in 2014 with teams of pedagogical instructors and lecturers from diverse disciplines. The project goals were to foster the use of new pedagogies to better achieve the lecturers' pedagogical targets. That, while fostering active, constructive, cooperative, and authentic components to enrich the student experience in a meaningful context, and to enable the development of students' ICT skills as well. During 2016-2018, Beit Berl College was involved in the TeachEx Erasmus+ project (TeachEx, 2017) aimed at promoting teaching excellence in HIEs in Israel, led by the college CTE and the ICT unit. The project's main targets were: promoting teaching excellence; developing creative teaching methods and producing benchmark tools for teaching, including the integrating of digital tools; and creating support structures in the CTEs. The objectives are to meet the needs of diverse students, to close the generation gap, and to enable interactive and experiential learning. Some of the mobile learning (ML) workshops for faculty, which will be described in the chapter, were integrated into the development of the project's outcomes, received feedback from the other Israeli and European partners, and were improved accordingly.

The chapter first presents background on enhancing teaching in higher education and using ML in different educational levels. Following, we describe the narrative for launching the ML project, the implementation process, the developed ML workshops, and the process evaluation. We conclude with reflection on the project's main achievements, our future intentions, and recommendations for implementation in other HIEs.

BACKGROUND

Enhancing Teaching in Higher Education

Higher education is facing continuous changes in relation to international competition for student recruitment, increasing social and geographical diversity of the student body, increasing demands of value for money, a change in the requirements of graduates in the fields of employment and academia, and more (Hénard & Leprince-Ringuet, 2008). Institutions now need to ensure that the education they offer meets the expectations of students and the requirements of employers, both today and for the future. That includes principles of working with students as partners, connecting students with real-world outputs, transcending disciplinary boundaries, and connecting students with the workplace. The quality of teaching and learning determines how effectively HEIs fulfil the demands of training graduates for a changing industry and academic world that requires new skills. In view of this, HEIs become aware of their core mission and core responsibility, and considerable attention has been paid to the quality of teaching, and to the need of faculty to excel in teaching along with excellence in research (Hénard & Roseveare, 2012). Accordingly, higher education teacher training has become compulsory in many countries such as the UK, Denmark, Australia, and Vietnam (ICED, 2014; Nguyen, 2017). The goal is to be more attractive, to meet students' needs and expectations, and to develop students' skills as well. The next quote expresses the HEI's role: "The experience of learning from good teachers is appreciated not only because good teachers make students' days challenging, motivating, and rewarding, but, because quality higher education is crucial in enabling HEIs to produce the critically-thinking, creative, adaptable graduates who will shape the future" (McAleese et al., 2013, p. 4). The European University Association (EUA) went further in their purpose to examine how the actual change was made. The report, Trends 2015: Learning and Teaching in European Universities (Sursock, 2015), documents how developments in learning and teaching are perceived by institutions across the European Higher Education Area (EHEA) (451 institutions responded). The report reaffirmed the increased attention being paid to learning and teaching at universities across Europe, but also highlighted that these developments are mainly driven at the institutional or national level, and are not in accordance with the European reform agenda that wishes to enhance collaboration between institutions. In response, EUA has launched two initiatives for enhanced collaboration in teaching, and created a sustainable forum for discussion and collaboration on these issues. A new survey has been analyzed, and will be published in 2018. The current approach of student-centered learning, where knowledge is constructed in light of the constructivism approach and the development of students' skills is equally important, is expressed in different books that concern HE pedagogical approaches (e.g., Mackh, 2018; Tong, Standen, & Sotiriou, 2018). The seminal Report to the European Commission on new modes of learning and teaching in higher education)McAleese et al., 2014), indicates particularly that the higher education landscape is undergoing significant change because of technological innovations. The benefits are clear, and actions to ensure that the potential is fully realized must be taken. Integration of new technologies and associated pedagogical approaches must be strongly encouraged and be supported. This chapter comes to address the need to share good practices between HEIs in the context of teaching enhancement in academia, particularly by integrating ML technologies to enhance active learning.

Mobile Learning in Education

Mobile Learning (ML) can be defined as the processes of coming to know through exploration and interaction across multiple contexts amongst people using interactive mobile technologies devices such as smartphones and tablets (Pegrum, 2015; Sharples et al., 2007). The mobility of the learner is the power of this technology-enhanced learning. ML provides learners with the possibility to assimilate formal and a-formal learning anywhere and at any time (Crescente & Lee, 2011; Pegrum, 2015). Researchers have indicated different characteristics of ML: personal/intentional, authentic, active, constructive, and cooperative (Howland, Jonassen, & Marra, 2014; Kearney, Schuck, Burden, & Aubusson, 2012). Pedagogically, mobile devices are tools that support learning where students are able to: a) consume content through the abundance of information resources online, b) conduct investigative research, c) communicate and collaborate, d) manage tasks, and e) create multimodal learning artefacts (Ng, 2015). The ability to choose what and when to learn provides opportunities for meaningful learning that is relevant to the learner's interest and preferred format (Ng, 2015). Sharples and Spikol (2017, p. 89) pointed out that evolution of the ML field over the past decades is "from a focus on software, to embedding learning into everyday locations and the continuity of learning across locations".

During the last decade, instructional designers and educators, all over the world, recognized the potential of mobile technologies as a learning tool for students of all ages. In K-12 education, curriculum materials are developed, professional trainings for teacher are executed, and the number of schools that use ML increase every year. The use of ML in schools in Israel is limited. The Ministry of Education (MoE) rules restrict the use of ML, both in extent of time and according to grade. The parents' organizations are also against it, based on social and health reasons. Most schools also face wireless communication problems, which prevents continuous use of mobile devices. For those reasons, the number of schools that do use and promote ML is not extensive. Although, in the schools that use ML, valuable activities that enhance pupils' motivation, skills, and learning are implemented.

Mobile Learning in Higher Education

Dede and Bjerede (2011) indicated that the 21st-century skills vision could not have happened without the ML technologies, as mobile devices are not another type of computer, they are a new technology that enabled meaningful learning that is suitable to the learners' needs. The Horizon higher education reports from 2013 to 2017 specified that mobile devices have become the gateways to personalized working and learning environments that facilitate the exploration of new subjects at each user's pace (Adams Becker et al., 2017).

Students can use mobiles to practice 21st-century skills including communication, collaboration, and creating content. Mobile devices also provide more opportunities for student-instructor interaction. At a glance, there is wide acceptance of ML in higher education, and the key benefits of using ML in academia have become a subject to research. Cochrane and Bateman (2010) identified the following key benefits: a) the exploration of innovative teaching and learning practices; b) the enabling of authentic learning that facilitates anywhere, anytime student-centered learning; c) the engagement of students with the affordances of mobile Web 2.0 technologies that include connectivity, mobility, geolocation, social networking, personal podcasting, and podcasting; d) bridging the digital divide by providing access to learning contexts and affordable user content creation tools; and e) moving from a model of fixed, dedicated general computing to a mobile, wireless computing paradigm that turns any space into potential learning space.

A paper review of research on iPads use in higher education shows that, from the staff perspective, the iPad was found to offer benefits associated with electronic information dissemination, academic administration, collaboration, and a tool for motivating students and for giving them just-in-time support (Nguyen, Barton, & Nguyen, 2015; Van Wyk & Van Ryneveld, 2016). From the students' perspective, the iPad was found to enhance the learning experience since students have better access to learning material and communication. In relation to students' experience, Edmonds and Smith's (2016) research used location-based mobile games across different disciplines using varying narratives and physical locations, rich media, and augmented reality. They found that the great majority of the students enjoyed the authenticity of learning in the real world and either strongly agreed or agreed that their experience was engaging and team building. Students' attitudes towards ML, at all ages, was found to be positive, and students are motivated, encouraged, and satisfied to integrate their mobile devices into their learning (Crompton & Burke, 2018; Stanojevic & Rakic, 2018).

Much research indicated better learning processes and better achievements, and that the students benefited from this type of learning (Diemer, Fernandez, & Streepey, 2012; Edmonds & Smith, 2016; Van Wyk & Van Ryneveld, 2016; Wilkinson & Barter, 2016). But, some researchers showed that ML did not necessarily lead to better learning outcomes (e.g., Nguyen et al., 2015).

Institutional barriers for integrating ML in HEIs are also presented by Gannod & Bachman (2011). They are related to: privacy concerns, the faculty attitudes and beliefs towards technology, a lack of proper technology knowledge and skills, and a lack of shared vision and technology plan within an institution. They identify some recommendations: 1) Mobile learning should not be viewed as a replacement for existing learning techniques but rather as a supplement, 2) Mobile learning should be used to facilitate different dimensions of learning, 3) Mobile learning should be used to facilitate collaboration with instructors and other learners, 4) Mobile learning should be adopted as an institutional or broad initiative rather than a pilot by a small set of early adopters, and 5) Mobile learning adoption policy should be accompanied by protective measures to prevent abuses. In the chapter summary we look at our project in relation to these aspects.

Based on the described ML characteristics, the working definition of ML in this project focused on integrating pedagogies of active, constructive, personalized, and collaborative learning by using mobile devices. The technology is only the tool and the pedagogies that we use are most crucial (Crompton & Burke, 2018).

PROJECT DEVELOPMENT

Prolog

The ICT Academic Unit envisioned, five years ago, was to find ways to integrate ML into the college academic courses. We believed then, and we still believe, that it is the appropriate technology tool to promote our lecturers' pedagogical aims. The main reason is that ML enables active learning and that mobile instruments are available and used extensively by lecturers and students in daily life. At that time, the Ministry of Education in Israel published a call for innovation in the colleges. We sent our proposal for "Meaningful Mobile Pioneering Learning" and our program was accepted. The resulting budget has been used for training faculty, for buying the first30 iPads designated for long-term loaning to faculty and students, and for developing a supportive database. The database includes theory and practice papers on integrating ML, examples of successful use, and a collection of guides for the use of different applications. The budgeted project was a start for our journey. We experimented with different ways for training faculty, we raised additional resources from the college for faculty and students, and from 2016 we used some of the TeachEx project budget as well. The five years of implementation was accompanied by formative assessment. Changes and adjustments were made dynamically according to needs and lessons learned, some of which will be shared in the chapter.

Entry Point

Beit Berl College is a multidisciplinary college focused on training educators while spearheading new developments in education, society, and the arts. We see education as a vehicle for social mobility, equality, and justice for all sectors of Israeli society. The college faculty are diverse in seniority, in their teaching disciplines, and in their experience and confidence in using technologies. As known from the literature and from our vast experience, training faculty to integrate different technologies into their teaching process is a challenging mission, which should be treated as an ongoing progression. Any training process should give space to varying of entry points of faculty, and to adjust the training accordingly. Faculty should feel safe in the process, and must see, from the beginning, the advantages that can be gained from integrating new teaching-learning methods using one tool or another. The main addressed faculty population was pedagogical instructors who teach various didactics courses, and who thus serve as role models for their students, pre-service teachers, with regard to future implementation in schools. We wish our graduates to be relevant to the world of school students in this age. The training was not restricted to the pedagogical instructors; lecturers who teach various disciplines attended the workshops and the others attended the training venues. Our starting position is that as much as students experience good learning activities using ML, they will adopt the tool and thus develop some valuable different knowledge of 21st-century skills.

Launching the Project

The project was launched in 2014. The training program was carefully designed, first to decrease lecturers' resistance and concerns, and further to expose faculty to different patterns of activities and to relevant applications that can be used in their particular classes. The central issue we desired to deliver in faculty training was that the mobile devices can be used to achieve pedagogical aims. We wished them first to feel and experience the use of ML-based activities, and then to realize the potential gains of integrating such into their teaching and learning processes. The vast experience of the ICT academic unit was a central basis for the development of the training program, where the key point is supporting faculty along the process.

Project Design

The project design was based on the following grounds:

1. The background of the faculty in the use of ICT tools should always be respected and addressed;

2. It is important to present a range of pedagogical goals that can be promoted through integration of ML tools;
3. It is important to demonstrate ML tools on the basis of a variety of course content;
4. This is an ongoing process of assimilation, and the participation of faculty in workshops is not sufficient.
5. The campus conditions should allow convenient implementation.
6. College encouragement and internal marketing are important to the success of the project.

Grounds 1-3 were the basis for the development of the materials to be used in the training workshops and to design the training methods. Grounds 4-6 served as the basis for planning a comprehensive assimilation program, in conjunction with various factors within the campus.

PROJECT IMPLEMENTATION

Population and Aims

The target population included all the college lecturers, in order to apply the approach in the pedagogic classes as well as in the disciplinary classes.

The project's declared aims were:

1. The faculty will be able to develop learning activities that will be meaningful, active, and mobile that will change the teaching and learning processes.
2. To accompany the faculty along the implementation of meaningful, active, and mobile learning in their subject matter.
3. To develop pre-service student teachers' mobile technology skills for their future professional use.
4. To develop assessment methods that fit the ML characteristics.
5. To influence the use of ML in our partner schools, where the students practice teaching.

The First Year

In the beginning of the first year, time and resources were allocated for the professional development of the ICT unit pedagogic and technical staff. In accordance with the project's grounds and aims, training workshops were developed, a supportive database that includes theory, instructions, and examples of successful use of ML was established, and a closed Facebook community for the college faculty using ML was opened. In addition, 30 iPads were purchased to allow the implementation of the project to begin. The 30 iPads were stored at the library and each student involved in the project could borrow one for the whole semester.

After completing these steps, a call for participation was issued to faculty to join the first broad training course. Forty faculty from various college departments attended the course, which consists of 30 hours of training, during which the faculty was able to use iPads or smartphones. The training related to the following subjects, all presented and implemented using mobile devices: meaningful learning, social networks and cooperative learning, the mobile device as a window to the world of knowledge, ubiquitous learning, old games with new technologies, location-based learning, documentation and e-portfolio, creation, reflection, questionnaires, and testing augmented reality. During the training, faculty become familiar with their tablets or smartphones, learn how to use the devices for teaching and learning, and get insights regarding the advantages for using this technology for learning at anytime and anywhere. Emphasis was on the possibilities that ML enables constructive and interactive learning processes. At the end of the course, projects that were developed by the participants were presented, and valuable discussion on the evaluation of such outcomes was conducted.

The training took place during seven F2F meetings, and two on-line meetings, one synchronic and one a-synchronic. The course was accompanied by a website on Moodle and a closed Facebook site. The Moodle enabled us to maintain collaboration and to create databases that contained all the materials and outcomes that were developed during the course. Those environments were chosen as the faculty are familiar with them.

After accomplishing the course, it was recognized that it is very important to fulfil the faculty's needs and deliver subsequent and massive individual support to incorporate the application in their courses. Accordingly, the ICT unit staff held personal meetings in which they examined the pedagogical goals of the lecturer, chose the pedagogical model supported by ML appropriate for the activity, and designed guidelines for the development of the activity. In cases where faculty needed support in the classes as well, an assistant from the unit accompanied them to reduce their concerns. To encourage faculty to buy their own iPads, the college also allocated about 20% of the device price for any request.

At the end of this first year, we had nine workshops that are updated every year in accordance with feedback from the operation and also in accordance with specific workshops for units with predefined needs. A professional community of lecturers was created who were excited and curious about the upcoming implementation.

The Second Year

When the second year of the project was initiated, the ICT unit work plan included additional workshops for training other staff, and it was expected that the lecturers who participated in the first broad course would implement ML in their teaching. But, we found that even very professional lecturers who were very excited to implement these new methods in teaching, faced obstacles. Once again, it became apparent that a step-by-step personal support is necessary, and without it most lecturers would not pass the expected change. Hence, the ICT unit staff offered personal guidance, and adopted a proactive approach, and immediately turned to new lecturers who participated in the workshops to offer the subsequent personal guidance. From this point on, the unit works in this way. One of the outstanding lecturers expressed the way she sees the combination of workshops and personal guidance:

The workshops were very important to present us the possibilities and the benefit of integrating ML to our curriculum and activities but we needed the individual support to actually implement it and use it in class.

At the end of the project's second year, in order to disseminate and strengthen the use of ML by faculty, a competition for the "Outstanding use of ML" was announced. Faculty had to submit a portfolio that contains examples of ML activities they used in their classes. A panel of three senior faculty members assessed the proposals, and four outstanding faculty got iPads as prizes. The outstanding lecturers were noted in a special seminar for all the college faculty, and presented the inspiring activities they developed.

On the technological side, more iPads were purchased. And in addition, to meet the need arising from the lecturers to share students' outcomes to the whole class via a projector, some Apple TV technology was purchased as well.

The Next Three Years

Along the following years, the ML database is regularly updated with paper reviews, recommended applications and instructions on how to use them, and examples from the field that the college lecturers developed. Over time, 140 iPads were purchased.

Every year the following types of support are offered to the faculty:

1. Workshops are offered each semester for newcomers and for senior users.
2. Personal support to help lecturers find the ML pedagogical model that meets their pedagogical needs.
3. An increasing stock of iPads for loan from the college library is available for faculty and students for a semester or two.
4. Lecturers can ask the ICT academic unit staff to accompany them to the class, mainly to give technical support to the lecturer and students during the lesson.

One of the project aims was to influence the use of ML in our partner schools, where the students practice teaching. A number of attempts were made to encourage those activities with three schools. Unfortunately, this initiative did not work in a structured way between the college and the schools. Although, students reported on inspiring activities that they developed and activated with their pupils in schools.

At the end of the fifth year, the number of faculty that participated in the different activities grew to 203. Some of them loan the iPads and ask their students to do so, and others use their own mobile devices.

THE ML WORSHOPS

During the project nine workshops were developed. In this section we describe the prescribed aims and learning outcomes of the workshops, the nine workshops emphasizing their rationale, and a full description of the "Documentation to enhance learning" workshop.

Workshops' Aims and Learning Outcomes (LO)

The ML module aims are to foster the use of new pedagogies in teaching and learning processes in higher education by

1. Exposing faculty to ML theory and to different implementation frameworks,
2. Enabling opportunities for hands-on positive learning experiences using ML,
3. Raising awareness of different possibilities of using ML,
4. Increasing familiarity and developing skills through active use of ML applications.

We defined three main LOs for faculty, which express the central place we give to pedagogical aspects of using ML, and not just for using the technology per se.

1. Analyze the pedagogical contribution of using an ML activity.
2. Create assignments that include an ML activity.
3. Evaluate the use of an ML activity as being: active, constructive, collaborative, authentic, and personalized.

In each workshop, the faculty will both experience the use of ML for teaching and learning, and develop an activity. The experience aims to show the flexibility and the variety of applications to different subjects, and the development enables application in the context of each faculty member's individual instruction.

Workshops Description

The nine workshops cover the five characteristics of ML mentioned above, as presented in Table 1. The link between an activity and one of the five main ML characteristics in the table was made according to the main applied characteristic of the activity, but most of the activities relate to several aspects in practice. For example, the "Documentation to enhance reflection" activity can be done by investigation; it is about creation, and the product can be built collaboratively and shared with the entire course.

The authenticity, personal/intentional, constructivist, and active characteristics (Howland et al., 2014) are presented in all nine workshops. The five characteristics that are in Table 1 were taken from additional characteristics that were presented by Ng (2015).

Four essential learning outcomes describe what faculty will be able to do at the end of the workshop. In each workshop, it is adapted to the particular type of activity. (Each <…> can be replaced by the name of the particular workshop.)

The Learning outcomes are:

Table 1. Mapping the workshops according to the main five ML characteristics according to Ng (2015)

		Consume Content	Investigate	Collaborate	Manage Tasks	Create
1	Location-based learning	✓	✓	↑		✓
2	Documentation to enhance reflection		✓	↑		✓
3	Digital portfolio	✓			✓	✓
4	Creating a shared visual digital product	✓	✓	✓		✓
5	Quizzes and polls	✓				
6	Augmented Reality	✓	✓			✓
7	Mind Mapping	✓	✓	↑		✓
8	Gaming and Gamification		✓	↑	✓	✓
9	Seamless learning	✓	✓	✓	↑	✓

Legend: ✓- This characteristic is reflected in this workshop. ↑ - This characteristic could be in this workshop depending on the specific activity aims.

1. **Identify** pedagogical needs of <…>.
2. **Analyze** the pedagogical contribution of using <…>.
3. **Create** assignments that include <…>.
4. **Evaluate** the use of <…> as an ML activity (active, constructive, collaborative, authentic, and personalized).

For each workshop, we briefly present its rationale, give a short description of its main activity, and present an example.

Workshop I: Location-Based Learning

Rationale: Location-based learning activity refers to the physical location of a learner in or near the environment (Brown, 2010). Contextual ML can play an important role in helping people understand their context and their environment (Sharples, Milrad, Sánchez, & Vavoula, 2009).

Activity description: The activity rationale is presented in relation to the topic of accessibility. Participants will walk around campus; take pictures of places with or without accessibility for people with physical disabilities, and upload pictures to a collaborative wall. Based on the authentic collected data, the group will analyze the pictures collection, categorize them, and draw conclusions to be presented to decision makers. In the second part of the workshop attendees will develop and evaluate a location-based activity in relation to the content of one of their courses.

Examples: In Mathematics, students will walk outside the classroom, looking at different building structures, measuring them with an app and calculating their area. Physics: Students will walk outside the class, near a road, measuring the length and the speed of the movement (by apps) in order to calculate and solve movement problems.

Workshop II: Documentation to Enhance Reflection

Rationale: Reflection is an essential skill for promoting learning. While reflecting on the learning process, the learner gains a better understanding of the concepts learned, knowledge, and resources. Documenting the learning process deepens learners' understanding of the subject matter, which will make them think about their own thinking processes (metacognition).

Activity description: A problem is presented. The participants' assignment is to solve the problem in teams while documenting their problem-solving process. Each team will choose the pictures that reflect their process and produce a reflective clip, using simple mobile applications, following specific guidelines. Participants will then develop a similar activity based on their subject matter and evaluate it.

An example will be described in this workshop's detailed description (next section).

Workshop III: Digital Portfolio

Rationale: A digital portfolio provides a space for gathering students' insights from their learning experiences and outcomes. Digital portfolios can be used as an evaluation tool as well. Using mobile devices to manage a digital portfolio enables smooth anytime anywhere documentation, within and outside the classroom.

Activity description: Participants will analyze different types of digital portfolios, and discuss different approaches to evaluation. They will then develop guidelines for students in the context of one of their courses, and plan an evaluation rubric to assess it. Participants will share their outcomes and receive feedback from their peers.

For example: In computer sciences, the student will document their project development process, including obstacles and ways to overcome them, on a digital portfolio as part of the course assessment.

Workshop IV: Creating and Sharing Digital Content

Rationale: Creating and sharing diverse digital content is a key element in learning and a requirement for living and learning in the 21st century)Partnership for 21st Century Skills, 2008). Mobile devices are easy to use and accessible, hence they enable acquisition of these skills anywhere, anytime.

Activity description: Participants will build a shared concept map that represents them and their own skills as a 21st-century lecturer. Then they will develop an assignment, in the context of their own courses, that uses a shared board.

An example: In a class that raises the sociological aspect of gender. The lecturer can add a picture or a video that presents an ethical dilemma to a shared board. Students add, to the board, their opinions regarding how to solve the problem (they can use text, picture, sketch, or video) and comment to their classmates.

Workshop V: Quizzes and Polls

Rationale: Quizzes and polling tools can provide teachers and students with immediate feedback, and can also be used as an assessment tool. They can be used before, during, at the end of, or after class.

Activity description: Participants will experience the use of a "Socrative" questionnaire as learners. Different scenarios of using it will be discussed. For example, as a tool to engage students in the learning process, to explore students' opinions as a basis for class discussion, or to quickly assess students' understanding. Then participants will experience the creation of a poll in digital environments.

An example: In Civic studies, in a lesson about various government systems, students will be asked to answer about their attitudes towards and against different systems before and after the lesson. "Before" in order to identify the students' preliminary knowledge and to enable the lecturer to adjust the lesson curriculum according to that, and "after" in order to know what changes are caused as a result of the learning process.

Workshop VI: Augmented Reality – Creating and Using Layered Information

Rationale: AR enables the addition of virtual and digital layers to a real-world subject. It helps teach subjects that students cannot experience in the real world and enhances creativity and imagination (McAleese et al., 2013).

Activity description: Participants will experience an AR activity, learn the benefit of using it, and then create an AR activity for students based on their own subject matter. Sharing these activities with colleagues and evaluating it.

An example: In a lesson that relates to plants in a botanical garden, students will develop layers of information and questions on trees and flora that exist in the garden. Another group of students will learn using their classmates' developed material.

Workshop VII: Concept Mapping

Rationale: Concept mapping is a graphic tool for knowledge representations in systematic methods that could be hierarchic. A concept map contains concepts and connectors. Concept maps assist learning and knowledge organizing by identifying the concepts and organizing them in a meaningful way, from the most general concept to the most specific one, and can be enriched by giving names to connections between concepts. A concept map reflects the subjective understanding of its creator. Therefore, there could be different outcomes for the same context (Novak & Gowin, 1984). Students could build concept maps implementing ubiquitous learning using mobile devices.

Activity description: Participants will understand the benefits of using concept maps, experience creating a concept map for a specific subject, and create a concept map activity for their students based on the subject matter they teach.

An example: Sciences: Chemistry, creating a mind map of a chemical process that includes the various involved concepts and the relations and effects between them.

Workshop VIII: Gaming and Gamification

Rationale: Gaming and gamification enable active learning with fun and competition. Games enable the students to practice their knowledge and enjoy it. Games allow players to restart and play again, and the freedom to fail. It allows students to experiment without fear, and increases students' engagement (Lee & Hammer, 2011). Gamification is the use of game-play mechanics for learning. It has some common elements with the behaviorist learning theory, such as superiority of positive reinforcements, small step-by-step tasks, immediate feedback, and progressive challenges (Biró, 2014).

Activity description: Participants will experience gaming and gamification activities, they will explore apps for creating games, they will experience existing games and open apps that enable integrating gamification, and then create an activity for students based on their own subject matter that integrates a game and gamification.

An example: English: integrating games for learning tenses, and adding gamification for fun and competition such as getting points for using the tense in the right way and getting badges at the end of the game.

Workshop IX: Seamless Learning

Rationale: Seamless learning means that the learning environment enables continuous learning, formal and non-formal, personal and social, with various technologies inside or outside the classroom and across time and location (Chan et al., 2006; Sharples & Spikol, 2017).

Activity description: Participants will experience a seamless learning activity, studying what is seamless learning, and then create a learning activity for students based on their own subject matter.

An example: English as second language: learning idioms. Students are asked to choose some idioms (in groups), take pictures (with tablets or smartphones) that describe the idioms (individually and within and outside the classroom), process them (using laptops), and share the pictures with the class and collaborate to produce a joint project.

The Documentation to Enhance Reflection Workshop: A Detailed Example of *Workshop II*

Workshop II focuses on the use of ML while documenting the learning process and reflecting on it for better understanding and for constructing knowledge and skills.

The rationale of this module is that reflection is an essential skill for promoting learning. While reflecting on the learning process, the learner gains a better understanding of the concepts learned, on the skills they used, and on their lack of knowledge, resources, or skills. The mobile devices enable documentation of the learning process by taking pictures throughout a meaningful learning activity. Based on the collection of documentation it enables reflection processes while choosing the pictures that best represent the process, and then creates an authentic outcome – a video clip with subtitles that describes the process. Being active in choosing the documents and editing the video clip deepens the learners' understanding, and promotes their ability to reflect on their own thinking processes (metacognition).

The main goal of this workshop is that faculty will be able to use the developed activities with their students, based on the use of mobile devices. During these activities, faculty will first experience the use of documentation as part of the reflection process. After experiencing the activity as an active, constructive, and collaborative process, the faculty will develop a similar activity appropriate for implementation in one of their own courses. The participants will present their ideas and receive feedback on how to improve and refine their planned activity. Some suggestions for application in different disciplines are presented in the continuation of the current section.

The learning outcomes of this particular workshop are in accordance with the general LO presented previously.

Faculty will be able to:

1. **Identify** pedagogical needs of documentation for reflection.
2. **Analyze** the pedagogical contribution of using documentation for reflection.
3. **Create** assignments that lead and enable students to document process in real time by taking pictures and videos, and reflect on the processes using those documents.
4. **Evaluate** the use of documentation for reflection as an ML activity (active, constructive, collaborative, authentic, and personalized).

Table 2 details the workshop timeline. The workshop is planned for four academic hours. The planning includes referral to slides in the presentation that serve the workshop. The detailed plan and the presentation are at the following link: https://sites.google.com/gordon.college/mlteachex/documentation-reflection

Examples From Implementation in Various Disciplines

This workshop was developed in the project since it could be useful in many disciplines. This activity can be used in any learning situations where the emphasis is on a process, where the learning objectives relate to concepts in the subject matter, but also to thinking and performing as a process. The example used in the workshop was taught successfully in a "Technology Thinking" course at Beit Berl College,

Table 2. Detailed timeline for the "Documentation to enhance reflection" workshop

Duration and Materials	Activity
10 min Slides 1-3	Introduction: ML characteristics and advantages, presentation, and discussion.
10 min Slides 4-5	Activity explanation: We chose an activity appropriate for implementation in each group that always creates motivation to succeed. This activity is an example of using "documentation and reflection" while studying design thinking in technology education or engineering. It can be applied in various disciplines. The mission for the participants is: work in teams to build the tallest tower possible, using 50 light disposable cups. Each team is guided to appoint one team member as documenter.
30 min Slide 6	The Activity: Participants work in teams of 4-5 to build the tallest tower. The documenter documents the process by taking pictures and videos of the designing process, using the tablet camera.
5-10 min Slides 7-8	Guidance: Presenting some video editing tools, such: • WeVideo • Animoto Video Maker • Perfect Video • Spark video • iMovie
40 min Slides 7-8	Produce a reflective video clip: Using one of the editing tools, the teams produce a reflective video clip, based on the pool of documents they have collected. In this stage, the team participants browse through their collection of images and videos, decide which best reflect the teamwork process, create the movie, and add subtitles that explain the process.
20-30 min. Slide 9	Presentation: Each team presents the product (the tower) and the video clip to all workshop participants. After determining together which tower is indeed the tallest, the participants discuss the design and undergo a reflective thinking process at the personal and team levels.
20 min	Development: Divide the participants into new teams of faculty from the same or similar disciplines. The teams will develop an activity in the same structure that suits their students.
20 min	Evaluation: The teams will present their planned activities, share ideas, give feedback, and make recommendations to their colleagues.
15 min Slides 10-13	Discussion and reflection: The discussion will focus on two main aspects: 1. The advantages or disadvantages of this type of activity for enhancing disciplinary knowledge and skills. 2. The advantages and disadvantages of the documentation and reflection process: a. Why it is important? b. How does documentation enhance reflection? c. Does the process of producing a video clip help one reflect on the learning processes? d. As a lecturer, what could I do better and what have I learned from my personal experience in this workshop? In the discussion, it is important to address how this type of assignment enables each student's team to acquire the knowledge and skills differently – according to personal skills, needs, preferences, etc. Each participant has the opportunity to construct his own knowledge. It is fruitful to point to the fact that even though each team had the same assignment with the same tool-kit, they used it in a different way/path/order, and their products (the tower itself and the video clip) are different.
15 min Slides 14-15	Summary: Give references to a TED movie about design and make a tower and creativity (see for example: Wujec, 2010).

and achieved its learning objectives. In what follows, we present some examples from a variety of disciplines that can implement the activity in a similar manner.

- **Chemistry:** Document and reflect on conducting an experiment. The collected documents could be pictures of materials, measurements, chemical reactions in progress, final states of materials, etc. The video clip enables better understanding of the process of designing an experiment and following its execution. The video clip is a concrete outcome that can serve further learning.
- **Pedagogy:** Teacher training. Document and reflect on practice teaching. The collected documents could be pictures of the board organization, listening to pupils' questions, pupils that are engaged in the lesson and others that are not, etc. The student can reflect on their teaching behavior, and edit a video that characterizes their own views about the authentic scenarios. Similarly, it could be used in various professions such as with medical or social work students.
- **Technology Education/Engineering:** Troubleshooting process. Engineering students, for example, must be able to identify flaws in order to be able to correct them. Documenting and reflecting on real world situations enhances the students' knowledge of the subject matter and enables them to reflect on their behavior and thereby develop the necessary professional skills. It is about creating a troubleshooting mental model.
- **Business Administration:** Long-term documentation and reflection can be implemented with a manager entering a new role and examining his/her acclimatization.
- **Art:** Planning and designing a sculpture. Documenting and reflecting on such processes in the field of art enables personal growth in relation to the different stages of planning and designing. Constant observation of the creative process that goes forward and backward.

In each of the above suggestions, the documentation and the creation of a relevant authentic video clip can enhance reflection and thereby enhance learning. The learning process is active, constructive, collaborative, authentic, and personalized, thereby exploiting all the advantages of using ML.

PROJECT EVALUATION

Throughout the implementation of the project, a consistent informal formative evaluation was conducted. Regular meetings of the development and implementation team took place, and feedback from the participating lecturers and from the workshop instructors were collected. Accordingly, adjustments were made and influenced the workshops and the database materials. Further, it leads to the development of new initiatives mentioned, such as college participation in the purchase of iPads for faculty, competition for the most significant application of ML in the teaching and learning process, and a seminar in which the outstanding applications are presented to all college lecturers. At the end of the fourth year a feedback questionnaire was developed and disseminated to follow the actual implementation in classes.

The Feedback Questionnaire

The main target of the feedback questionnaire was to follow up the actual implementation in college classes. After five years of implementation through the upcoming years, it was important to learn about:

faculty preferences for venues of training, the most used pedagogical models, reaction of students from the lecturer's point of view, and their intentions and needs for future implementation in teaching.

The questionnaire was distributed online, and 28 lecturers gave their feedback. Among them 43% were pedagogical instructors, 39% were disciplinary lecturers, and 18% teach both pedagogy and disciplinary courses. This dispersal of respondents allows drawing conclusions, since the two main college lecturer populations were represented similarly. The research applied a mixed method approach; analysis of questions 1, 2, 4, and 5 was by quantitative descriptive statistics, and the analysis of questions 3 and 6 used using qualitative methods for categories discovery and to further quantitate them into percentages (Creswell, 2018).

Question #1: The question related to the type of training the faculty used: 47% used personal guidance by a pedagogical consultant of the ICT academic unit, 32% only attended workshops, and 21% used a combination of workshops and personal guidance. In fact, 68% of the lecturers used personal guidance. This is in line with the implementation process, as experienced by the ICT unit staff, and influenced the training process. The main insights along the project progression were that it was not enough to participate in the workshops, but it is important to accompany the lecturers with a personal instruction process that is relevant to each, and relates to their teaching subject and to their personal preferences. This was applied from the second year of the project.

Question #2: In this question there was an attempt to determine the extent of the implementation of ML in classes. Lecturers' responses were very encouraging: 29% stated that they integrate ML to a small extent, 61% stated that they combine ML activities significantly according to their pedagogical needs, and 10% stated they incorporate ML into all their lessons, as an integral part of learning.

Question #3: In order to learn about the suitability of the pedagogical models applied in the variety of workshops, lecturers were asked to specify which pedagogical activities they use, and their answers were categorized. The data is presented in Figure 1.

It can be seen that some of the activities follow the workshops described above, particularly workshops #1,2,4,5. Some lecturers used the mobile devices for other activities such searching and sharing

Figure 1. Frequency of the use of different pedagogical models of ML
Legend: W#1 means Workshop#1

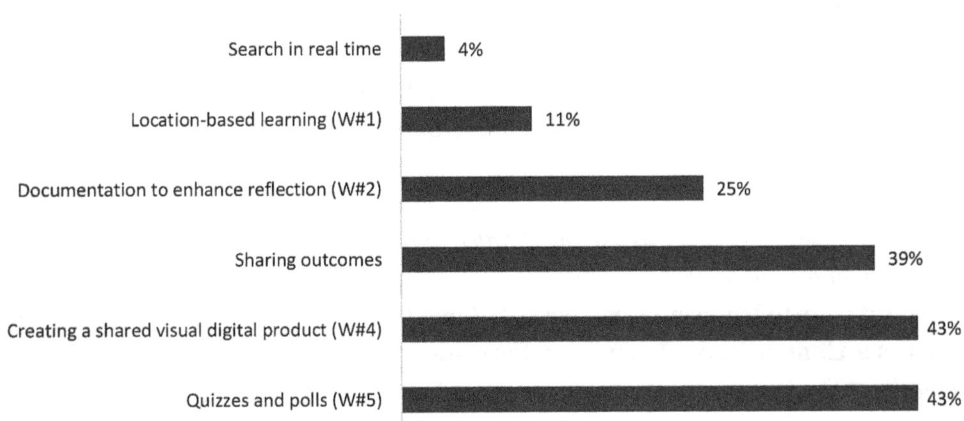

students' outcomes. Workshops #6,7,8,9 are not used by the lecturers that answered the questionnaire. Those workshops offered more advanced methods of teaching and learning, and hence lecturers need more confidence to apply them. In relation to workshop #3: digital portfolio, we believe that it is most appropriate for the pedagogical courses, where students undergo a professional development process and develop many materials. The missing application among respondents requires further consideration about improving the assimilation of this pedagogical model.

Question #4: ML is widely used with the purpose to diversify learning processes, and this is also an important goal. However, our main aim was to enhance active and meaningful learning processes. Accordingly, we asked to what extent the integration of ML contributes to the courses. We asked whether the contribution is to a large extent, to a moderate extent, or to a small extent. The positions of the lecturers were divided almost equally, 32% believe that the contribution is to a large extent, 40% specify that the contribution is moderate, and 28% see the contribution as limited. These data reflect the process of change, different lecturers adopt the new pedagogy differently. This issue will be discussed in the summary section.

Question #5: The last closed question related to the degree of students' satisfaction as perceived by the lecturers. The positions of the lecturers were as follows: 32% of the lecturers trust that students believe that the contribution of integrating ML activities is to a large extent, 50% think that students see the contribution as moderate, and 18% think that students see the contribution only as limited. Similar to the lecturers' answers to question #5, we see the process of change. It can be assumed that the degree of faith and confidence of the lecturers influences the quality of the activities implemented and how they are perceived by the students.

Question #6: In the last question, lecturers were asked to describe the change in their teaching philosophy and their perception of competence in relation to the integration of ML into academic teaching. The answers were categorized and are presented in Figure 2.

The top five categories express the lecturers' positions in relation to their own attitudes. Some did not realize the potential, as shown in the three red-colored categories. For example: *"There was no change", "I should see that this does contribute greatly to the empowerment of students' knowledge, and not only to the variety of means of learning"*. Some feel that they are still lacking in self-confidence in relation to the use of mobile technology, and some express fear of the work load required for developing ML activities. Unlike those, as expressed in the two green-colored categories, lecturers relate to the extensive support that enables the implementation, and beyond that, mentioned they are willing to learn new tools, that it excited them, and made them more relevant.

The other positions categories, colored blue, address different valuable aspects that the integration of ML can improve. In what follows we present some quotes from lecturers' answers to express their original positions:

- **Modeling for Teaching:** *"I try to use the iPads in many classes, with the intention and awareness of being a model for students."*
- **Allows Adaptation to Different Learners:** *"Adapted to the existing differences between learners."*
- **Implements a Change in the Lecturer's Position:** *"Allows to disconnect from the frontal teaching method", "The lecturer moves to the position of advisor, guide and mentor".*

Figure 2. The frequency of reference categories in relation to lecturers' changes in perception of teaching

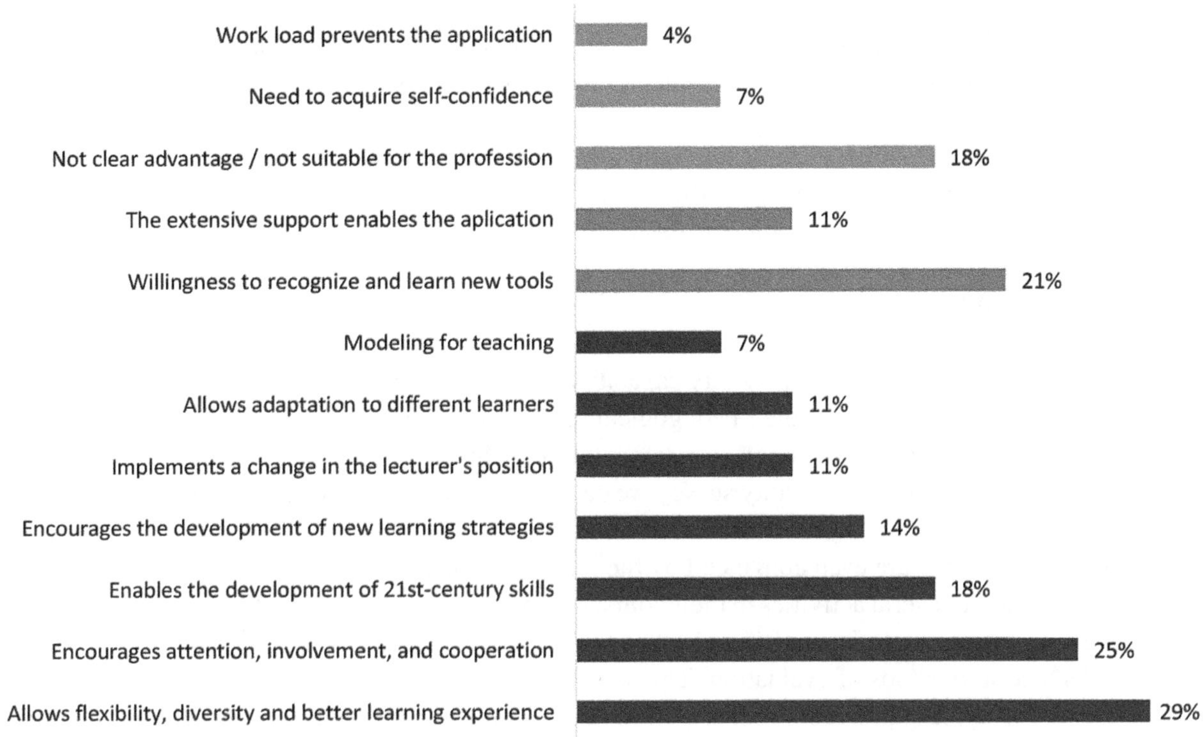

- **Encourages the Development of New Learning Strategies:** *"Encouraged a new way of thinking and planning for me and also for the students", "Type of learning which is appropriate for the digital world".*
- **Enables the Development of 21st-Century Skills:** *"The integration of ML is very important and central in a world where the media is mobile based."*
- **Encourages Attention, Involvement, and Cooperation:** *"More attention and more cooperation", "The teaching becomes active and activated", "Getting out of the routine encourages students to cooperate".*
- **Allows Flexibility, Diversity and a Better Learning Experience:** *"Experience and didactic lessons become more diverse and experiential", "More creative, authentic and meaningful", "Allows for diversity and flexibility".*

The varied categories expressed by lecturers evidence their own obstacles, and their needs to first be convinced of the advantages of integrating ML. The lecturers that move over this stage explore all the well-known advantages of integrating ML activities in any class. About 50% of the respondents related at least two benefits in integrating ML.

SUMMARY AND RECOMMENDATIONS

Implementation of new pedagogies in general, and in higher education in particular, are complex, and require great efforts to achieve a change. The chapter presents the vision and implementation of a training program for lecturers aimed at fostering active, constructive, and cooperative components, using ML, to enrich the students' experience in a meaningful and authentic context, and to enable them to develop their ICT skills. To finalize the chapter, we discuss the main achievements of the project implementation, the main training approach, and recommendation for HEIs who wish to widely adopt ML.

Main Achievements

The project's first aim was to develop faculty capacity to develop meaningful and active learning activities, using mobile devices, to meet their pedagogical targets. In the fifth year of the project it is evident in the ICT unit weekly work plan that the implementation of ML into the academic teaching in BBC is extensive and valuable. From the faculty survey we can see the process of the adaption of the new pedagogy, and its range. Some lecturers are not convinced that the investment is worthwhile, while most see its advantages, and some are even convinced by the magic of constructive and active learning. Most of the lecturers integrate several activities in their courses, and some use it consistently. Those who adapted it meaningfully made a significant change in the structure of their courses, in the lessons flow, in the course tasks, and in methods of evaluation. This is evident in faculty answers to questions 2, 3, and 4. We are pleased with the significant progress, but wish to enlarge the number of courses where students can experience different learning and acquire new skills.

The faculty reports on their changes in perception of teaching show that they understand the power of teaching with ML as identified in the literature: Modeling for teaching allows adaptation to different learners, implements a change in the lecturer's position, encourages the development of new learning strategies, enables the development of 21st-century skills, encourages attention, involvement, and cooperation, and allows flexibility, diversity, and a better learning experience (Dede & Bjerede, 2011; Howland et al., 2014; Kearney et al., 2012; Ng, 2015; Sharples & Spikol, 2017). This is evident in faculty answers to question 6.

Approximately 208 lecturers (out of 580 part-time faculty) attended at least one training session. All 140 iPads are in use by more than 53 lecturers and their students, and many more were interested. Other lecturers use their own devices, and rely on the tools that the students have.

Examination of the impact on students in relation to the project's third aim has not yet ended. Positive feedback from the students can be indicated in the lecturers' attitudes, question 5. Most lecturers (72%) thought that the ML activities contributed to students to a large or moderate extent.

In relation to aspects of evaluation and assessment, project aim number four: Three different tools were developed by the TeachEx ML team where we were deeply involved, aimed at evaluating the process from different aspects: (1) Feedback questionnaire for faculty that attend an ML workshop,[1] (2) Post-workshop reflective questionnaire for the workshop trainer,[2] (3) Reflective questionnaire for faculty that integrated ML activities in their courses,[3] and (4) Feedback questionnaire for students that learn using ML activities.[4] Those tools are open to use by the community. In relation to assessment, we led a community of colleagues to develop guidelines for faculty about how to assess students' learning outcomes to ML activities.

The fifth project aim was regarding the use of ML in our partner schools, where the students practice teaching. As presented above, students reported on inspiring activities they implemented in schools, but there was no formal process between the college and the schools. It appears that it was hard for the schools to maintain deep cooperation since teachers were engaged in the project assimilation and were not available for further guidance and tasks. We believe that in the following years there will be a place for that.

The Personal Guidance Approach

Our experience shows that for the "first step" the lecturer should be exposed to the field by participating in an exploratory workshop that stimulates interest, or by exposure to implementations by other lecturers and their achievements. Based on that, lecturers are willing to deepen their knowledge and skills, and to invest the necessary resources for planning, developing, implementing, and evaluating an ML activity. We wish to emphasize the personal guidance approach that is central to the process, and is a key component of its success, as presented in the project second aim. The one-on-one guidance dealt with pedagogical aspects, but not only those. The support consisted of holistic assistance, which sometimes included help in lending the iPads to students from the library, or accompanying the lecturer to the class so they will feel confident in case of any technological obstacle. The one-on-one meetings are conducted with a pedagogical consultant who specializes in integrating ML activities into the academic framework. The meeting includes planning of learning activities according to the lecturer's needs and ideas, which are grounded in the course content and in the pedagogical objectives and challenges. This assistance includes choosing an appropriate application to be used and, if needed, developing the activity with the chosen application, hand-in-hand with the lecturer. This type of assistance led most of the lecturers to become committed to the process and to make innovative changes in their courses. This approach has proved itself along the years.

Recommendations for Implementation in HEIs

Our recommendations for successful implementation of ML in Higher Education are presented in three categories:

1. **Institutional:**
    a. **Cooperation of the Institution's Management:** The management support is crucial for the progression and success. The advantages should be highlighted and evidence based.
    b. **Dissemination:** It is recommended to lead the process with department heads. When the head of the department understands the advantages, and is willing to promote this theme in the department, the process is more acceptable and tailored to the population and content.
    c. **Motivation in the Training Team:** The training team should be curious and even excited to initiate the project, and be dedicated to the process.
    d. **Influence the Partner Schools:** In order to increase the involvement and success in the assimilation of mobile learning in schools, it is worthwhile to establish structured processes between the college and the management of the partner schools.

2. **Technological:**
   a. **Environment That Enables Implementation:** The quality of the network that will allow access by many devices at the same time is important, the existence of tools that enable sharing of outcomes from mobile devices, the possibility of loaning equipment, and technical support.
   b. **Devices:** The opportunity to loan devices, or any assistance to purchase them, is needed. Lecturers do not like to purchase a device before they are sure that it helps them, so the possibility to borrow one enables its use until they become confident.
3. **Training:**
   a. **Workshops Content:** It is important to develop workshops that require different pedagogical skills and different technological skills, and to think at the stage of development on their marketing according to the institution. The nine presented workshops can be used with or without changes.
   b. **Workshops Activities:** Workshops are important in order to demonstrate the benefit of using ML. They give the faculty a broader picture, and let them experience and raise first thoughts in a supportive framework.
   c. **New Initiatives and Challenges:** Add new and up-to-date applications and pedagogical approaches to the workshops menu – keep lecturers interested and curious.
   d. **Personal Support:** The personnel accompanying and supporting each lecturer is crucial for the successes. Lecturers are facing changes and it is important to be there with them, to support, encourage, and to give useful feedback.
   e. **Database:** An updated database of curriculum materials, literature review, and examples of successful activities in various subjects is important to keep faculty updated.

Some of our recommendations are in relation to the institutional barriers for integrating ML in HEIs that were presented by Gannod and Bachman (2011) and are introduced in the literature survey.

To the End

It takes time to become "native" in using ML in teaching processes. The change needs to undergo a pedagogical transformation since the lecturer is no longer the one that provides the knowledge, but rather provides environments in which knowledge can be built. Faculty should feel safe in the process, and must see, from the beginning, the advantages that can be gained from integrating new teaching-learning methods using one tool or another. The process develops important skills in relation to the Technological Pedagogical Content Knowledge – TPACK – presented by Koehler and Mishra (2009). This change is a big challenge since lecturers have to overcome their traditions and their fear of technology. But, it could be done, and it is done.

Epilog

The pictures presented below are taken from faculty experiences in the different workshops presented. Our main teaching guidelines are evident in the workshops where the lecturers themselves are active in the process of learning. To finalize the chapter, we present the

Figure 3.
Publication of the images is with the approval of all who appear in the photos.

ACKNOWLEDGMENT

We wish to acknowledge the broad support we received from the college president and administration, to our colleagues who have given us confidence that we will lead them to new and innovative ways of teaching, and to our TeachEx team partners for the valuable feedback in the development process and lessons learned from the piloting part of the workshops.

REFERENCES

Adams Becker, S., Cummins, M., Davis, A., Freeman, A., Hall Giesinger, C., & Ananthanarayanan, V. (2017). *NMC Horizon Report: 2017 Higher Education Edition*. Austin, TX: The New Media Consortium.

Bíró, G. I. (2014). Didactics 2.0: A pedagogical analysis of gamification theory from a comparative perspective with a special view to the components of learning. *Procedia: Social and Behavioral Sciences, 141*, 148–151. doi:10.1016/j.sbspro.2014.05.027

Brown, E. (2010). Introduction to location-based mobile learning. In E. Brown, M. Sharples, G. Clough, B. Tangney, J. Wishart, M. Wijers, & M. Specht (Eds.), *Education in the wild: Contextual and location-based mobile learning in action*. Nottingham, UK: Learning Sciences Research Institute, University of Nottingham.

Chan, T. W., Roschelle, J., Hsi, S., Kinshuk, Sharples, M., Brown, ... Hoppe, U. (2006). One-to-one technology-enhanced learning: An opportunity for global research collaboration. *Research and Practice in Technology Enhanced Learning*, *1*(1), 3–29. doi:10.1142/S1793206806000032

Cochrane, T., & Bateman, R. (2010). Smartphones give you wings: Pedagogical affordances of mobile Web 2.0. *Australasian Journal of Educational Technology*, *26*(1), 1–14. doi:10.14742/ajet.1098

Crescente, M. L., & Lee, D. (2011). Critical issues of m-learning: Design models, adoption processes, and future trends. *Journal of the Chinese Institute of Industrial Engineers*, *28*(2), 111–123. doi:10.1080/10170669.2010.548856

Creswell, J. W. (2018). *Research design: Qualitative, quantitative, and mixed methods approaches* (5th ed.). Thousand Oaks, CA: Sage Publications, Inc.

Crompton, H., & Burke, D. (2018). The use of mobile learning in higher education: A systematic review. *Computers & Education*, *123*, 53–64. doi:10.1016/j.compedu.2018.04.007

Dede, C., & Bjerede, M. (2011). *Mobile learning for the 21st century: Insights from the 2010 Wireless EdTech Conference*. San Diego, CA: Qualcomm.

Diemer, T. T., Fernandez, E., & Streepey, J. W. (2012). Student perceptions of classroom engagement and learning using iPads. *Journal of Teaching and Learning with Technology*, *1*(2), 13–25.

Edmonds, R., & Smith, S. (2016). Location-based mobile learning games: Motivation for and engagement with the learning process. In *Proceedings of the 15th World Conference on Mobile and Contextual Learning; mLearn 2016, Mobile Learning Futures–Sustaining Quality Research and Practice in Mobile Learning* (pp. 72-76).

Gannod, G. C., & Bachman, K. M. (2011). Integrating M-Learning in a broad context: issues and recommendations. In I.A. Sanchez, & P. Isaias (Eds.), *Proceedings of the IADIS International Conference Mobile Learning*, Avita, Spain (pp. 291-294).

Hénard, F., & Leprince-Ringuet, S. (2008). *The path to quality teaching in higher education*. OECD. Retrieved from https://www1.oecd.org/edu/imhe/44150246.pdf

Hénard, F., & Roseveare, D. (2012). *Fostering quality teaching in higher education: Policies and practices*. OECD. Retrieved from https://www.oecd.org/edu/imhe/QT%20policies%20and%20practices.pdf

Howland, J. L., Jonassen, D. H., & Marra, R. M. (2014). *Meaningful learning with technology*. Melbourne, Australia: Pearson Education Australia.

ICED. (2014). The Preparation of University Teachers Internationally. Retrieved from http://icedonline.net/wp-content/uploads/2011/08/The-Preparation-of-UniversityTeachers-Internationally.pdf

Kearney, M., Schuck, S., Burden, K., & Aubusson, P. (2012). Viewing mobile learning from a pedagogical perspective. *Research in Learning Technology*, *20*(1). doi:10.3402/rlt.v20i0.14406

Koehler, M., & Mishra, P. (2009). What is technological pedagogical content knowledge (TPACK)? *Contemporary Issues in Technology & Teacher Education*, *9*(1), 60–70.

Lee, J., & Hammer, J. (2011). Gamification in education: What, how, why bother? *Academic Exchange Quarterly*, *15*(2), 146.

Mackh, B. M. (2018). *Higher Education by Design - Best Practices for Curricular Planning and Instruction*. Abingdon: Taylor and Francis Press. doi:10.4324/9781351133715

McAleese, M., Bladh, A., Berger, V., Bode, C., Muehlfeit, J., Petrin, T., . . . Tsoukalis, L. (2013). Improving the Quality of Teaching and Learning in Europe's HEIs. *European Commission for Education, Culture, Multilingualism, Youth & Sport*. Retrieved from http://ec.europa.eu/dgs/education_culture/repository/education/library/reports/modernisation_en.pdf

McAleese, M. (2014). *New modes of learning and teaching in higher education*. Retrieved from http://ec.europa.eu/dgs/education_culture/repository/education/library/reports/modernisation-universities_en.pdf

Ng, W. (2015). *New digital technology in education, conceptualizing professional learning for educators*.

Nguyen, L., Barton, S. M., & Nguyen, L. T. (2015). iPads in higher education—Hype and hope. *British Journal of Educational Technology*, *46*(1), 190–203. doi:10.1111/bjet.12137

Nguyen, T. S. (2017). *Conceptual changes in Higher Education Teaching and Learning: Insights from a compulsory teacher training program for higher education teachers in Vietnam* [PhD Thesis]. Retrieved from https://ses.library.usyd.edu.au/bitstream/2123/17657/1/Nguyen_tvs-thesis.pdf

Novak, J. D., & Gowin, D. B. (1984). *Learning how to learn*. Cambridge, UK: Cambridge University Press. doi:10.1017/CBO9781139173469

Partnership for 21st Century Skills. (2008). *21st century skills, education & competitiveness*. Retrieved from http://www.p21.org/storage/documents/21st_century_skills_education_and_competitiveness_guide.pdf

Pegrum, M. (2015). Teaching and Digital Technologies Mobile learning: what is it and what are its possibilities? In M. Henderson & G. Romeo (Eds.), *Teaching and Digital Technologies* (pp. 142–154). Cambridge, UK: Cambridge University Press.

Sharples, M., Lonsdale, P., Meek, J., Rudman, P., & Vavoula, G. N. (2007). An evaluation of MyArtSpace: A mobile learning service for school museum trips. *Paper presented at the mLearn Conference*, Melbourne, Australia.

Sharples, M., Milrad, I., Sánchez, A., & Vavoula, G. (2009). Mobile learning: Small devices, big issues. In N. Balacheff, S. Ludvigsen, & T. de Jong et al. (Eds.), *Technology Enhanced Learning: Principles and Products* (pp. 233–249). Springer. doi:10.1007/978-1-4020-9827-7_14

Sharples, M., & Spikol, D. (2017). Mobile Learning. In E. Duval, M. Sharples, & R. Sutherland (Eds.), *Technology Enhanced Learning* (pp. 89–96). New York: Springer International Publishing. doi:10.1007/978-3-319-02600-8_8

Stanojevic, L., & Rakic, B. (2018). Mobile technology in Higher Education – A students' perspective on learning with mobile computing devices. In Z. Nikitovic, D. Markovic, & C. D'Arconte (Eds.), *Implementation of Information technology: IT, Marketing, education and business working together for business success*. London: Silver and Smith Publishers. Retrieved from http://www.eee-conference.com/_img/arhiva/2018/knjiga_ii_-_it_marketing_prelom.pdf#page=8

Sursock, A. (2015). Trends 2015: Learning and Teaching in European Universities. Brussels, Belgium: The European University Association (EUA). Retrieved from http://eua.be/Libraries/higher-education/trends-2015.pdf?sfvrsn=0

TeachEx. (2017). TeachEx – Teaching Excellence in Israel – brochure. Retrieved from https://www.teachex.eu/wp-content/uploads/Teachex-Brochure-Final-1.pdf

Tong, V. C. H., Standen, A., & Sotiriou, M. (Eds.). (2018). *Shaping Higher Education with Students: Ways to Connect Research and Teaching*. London: UCL Press. Retrieved from http://discovery.ucl.ac.uk/10044671/1/Shaping-Higher-Education-with-Students.pdf doi:10.2307/j.ctt21c4tcm

Van Wyk, M., & Van Ryneveld, L. (2016). Introducing mobile videos for academic support. *Mobile Learning* Futures–sustaining quality research and practice in mobile learning. In *Proceedings of the15th World Conference on Mobile and Contextual Learning, mLearn 2016* (pp. 273–282). Sydney, Australia: The University of Technology, Sydney. Retrieved from https://www.researchgate.net/profile/Sue_Gregory2/publication/309485997_The_Use_of_Wearable_Technologies_in_Australian_Universities_Examples_from_Environmental_Science_Cognitive_and_Brain_Sciences_and_Teacher_Training/links/58784fbc08ae6eb871d19331/The-Use-of-Wearable-Technologies-in-Australian-Universities-Examples-from-Environmental-Science-Cognitive-and-Brain-Sciences-and-Teacher-Training.pdf#page=273

Wilkinson, K., & Barter, P. (2016). Do mobile learning devices enhance learning in higher education anatomy classrooms? *Journal of Pedagogic Development*.

Wujec, T. (2010). *Build a tower, build a team. TED ideas worth spreading*. Retrieved from https://www.ted.com/talks/tom_wujec_build_a_tower?utm_source=email&source=email&utm_medium=social&utm_campaign=ios-share

ADDITIONAL READING

Alrasheedi, M., & Capretz, L. F. (2018). Determination of critical success factors affecting mobile learning: A meta-analysis approach. arXiv:1801.04288

Cheon, J., Lee, S., Crooks, S. M., & Song, J. (2012). An investigation of mobile learning readiness in higher education based on the theory of planned behavior. *Computers & Education*, *59*(3), 1054–1064. doi:10.1016/j.compedu.2012.04.015

Dunleavy, M., & Dede, C. (2014). Augmented reality teaching and learning. In *Handbook of research on educational communications and technology* (pp. 735–745). New York: Springer. doi:10.1007/978-1-4614-3185-5_59

Dunleavy, M., Dede, C., & Mitchell, R. (2009). Affordances and limitations of immersive participatory augmented reality simulations for teaching and learning. *Journal of Science Education and Technology*, *18*(1), 7–22. doi:10.100710956-008-9119-1

Herrington, A., & Herrington, J. (2007). *Authentic mobile learning in higher education*. Hershey, PA: IGI Global.

Sharples, M., Taylor, J., & Vavoula, G. (2005). Towards a theory of mobile learning. *Paper presented at the mLearn 2005 4th World conference on mLearning*, Cape Town.

Spektor-Levy, O., Plutov, I., Israeli, N., & Perry, N. (2018). Integrating Technology in Preschool Science and Inquiry. In *Information and Technology Literacy: Concepts, Methodologies, Tools, and Applications* (pp. 1470–1493). Hershey, PA: IGI Global. doi:10.4018/978-1-5225-3417-4.ch076

Valk, J. H., Rashid, A. T., & Elder, L. (2010). Using mobile phones to improve educational outcomes: An analysis of evidence from Asia. *The International Review of Research in Open and Distributed Learning*, *11*(1), 117–140. doi:10.19173/irrodl.v11i1.794

Wu, W. H., Wu, Y. C. J., Chen, C. Y., Kao, H. Y., Lin, C. H., & Huang, S. H. (2012). Review of trends from mobile learning studies: A meta-analysis. *Computers & Education*, *59*(2), 817–827. doi:10.1016/j.compedu.2012.03.016

ENDNOTES

1 https://docs.google.com/forms/d/e/1FAIpQLSdCqhKRLmrTPXB8LGFgwMZBktINMLc8s4kncunISDAbpqlkPA/viewform

2 https://docs.google.com/forms/d/1_gDZEvNijCt175PCemWZsnodq79Es5FNQnonCHyXV9c/edit

3 https://docs.google.com/forms/d/e/1FAIpQLSdYKQGuKDfyBr-tgzXDN2V3c7-5wXmPzUM-rxtOkIK4XnsmkA/viewform

4 https://docs.google.com/forms/d/e/1FAIpQLSd1_WCphU2ty1xNxzcE-rWGAPmpMW5Tm-F14AKCUEhSLicXnxg/viewform

Chapter 3
Preparing Preservice Teachers to Use Mobile Technologies

Carolyne Nekesa Obonyo
University of Canterbury, New Zealand

ABSTRACT

The use of mobile technologies to enhance 21st century learning is increasing in K-12 schools and teacher education institutions. Thus, there is a need to effectively prepare preservice teachers to use mobile technologies in their future classrooms. This chapter explores the effective use of mobile technologies in teacher preparation in ways that are transferred to K-12 teaching and learning. It goes on to look at two major organizations: the university and partner school involved in the preparation of preservice teachers. Additionally, the purposes of incorporating information and communication technologies in teacher preparation as identified by Davis are explored to understand how mobile technologies align with these purposes. Common challenges of using mobile technologies in teacher preparation are also presented.

DOI: 10.4018/978-1-5225-8106-2.ch003

INTRODUCTION

The use of mobile technologies is changing educational landscapes. Accordingly, K-12 schools and teacher education institutions are using mobile technologies to enhance their practices and meet the needs of 21st century learners. This is due to the increasingly ubiquitous and pervasive nature afforded by mobile technologies that has resulted in teacher education and K-12 classrooms where bring your own device (BYOD) plays a significant role (e.g. Crompton, Burke & Gregory, 2017; Burns-Sardone, 2014). In view of this, preparing preservice teachers (PTs) for K-12 classrooms where BYOD flourishes is an important practice and it also provides additional opportunities to increase the flexibility of teacher preparation. In this chapter PTs refer to student teachers being prepared to gain qualified teaching status in primary or secondary schools upon graduating from their program.

This chapter presents examples of empirical findings and teaching practices across the globe to provide a background on how mobile technologies are used to further prepare PTs for their future classrooms and the digital world as well. First, it identifies PTs' multiple learning contexts which have been enhanced by mobile technologies. This is followed by a summary of existing literature about preparing PTs to use mobile technologies in educationally effective ways. In particular, two major organizations involved in the preparation of PTs are presented. These are initial teacher education (ITE) programs where PTs take their university-based courses and K-12 schools (partner schools) where PTs gain field experiences. Given the fact that ITE takes place in universities and colleges as well as in partner schools, m-learning would appear to be a natural fit with numerous pedagogical affordances assuming there is infrastructure that support the use of mobile technologies.

The next section presents case studies identified in the literature review focusing on how mobile technologies have fostered pedagogical innovations in teacher preparation. In addition, the chapter discusses different ways mobile technologies have enhanced delivery and accessibility of electronic learning materials especially for distance learning PTs who are separated by time and place. This chapter ends by identifying common challenges of using mobile technologies, and future research directions in teacher preparation. Presenting a comprehensive and contemporary body of knowledge regarding this topic, will increase our understanding about how teacher preparation programs are responding to the call of effectively preparing PTs to use mobile technologies to support K-12 students' learning (U.S. Department of Education, Office of Educational Technology, 2017).

BACKGROUND

Mobile Learning Defined

There is no single definition of mobile learning (m-learning) in the literature. Keskin and Metcalf (2011), for example, define m-learning in terms of ubiquity; access; immediacy of communication; individuality; convenience; mobility in physical and social spaces including virtual spaces. Pegrum, Howitt and Striepe (2013), however, view m-learning in terms of portability of devices due to their smaller sizes and light weight. A different approach is from Crompton (2013) who argues that "definitions of m-learning in the past decade appear to have been constructed around four central constructs—learning pedagogies, technological devices, context, and social interactions" (p.48). Another view focuses on key features such as (1) the mobility of learners, (2) interactive learning that occurs in multiple contexts, and (3) construction of knowledge supported by mobile technologies (e.g. West & Visloo, 2013). In essence, the flexibility and convenience afforded by mobile devices have enhanced learning because mobile learners are not constrained within a predetermined location and time. Khaddage, Muller and Flintoff (2016) also stress the mobility of learners but indicate that m-learning definition is shifting from emphasizing the devices being used to context-awareness. This is due to the affordances of mobile devices to facilitate in and out of classroom learning activities (Pegrum et al., 2013) and they are also extensively being used in other organizations outside of school learning.

This chapter considers m-learning based on Cavanaugh, Maor and McCarthy (2018) definition (from K-12 handbook), who indicate that "m-learning includes school learning experiences and environments that are accessible to students in and out of school with devices and services that go with students when and where they learn, including in blended and online programs" (p. 575). Ally also has a similar perspective and defines it as "the delivery of electronic learning materials, with built-in learning strategies, on mobile computing devices to allow access from anywhere and at any time" (as cited in Ally, Grimus & Ebner, 2014, p. 45). My perspective of m-learning is similar to those described above, but my emphasis is on learning experiences that are supported by mobile technologies throughout multiple contexts due to mobility of learners (PTs). PTs interact through different media with a sequence of learning activities either independently or collaboratively at any time. Moreover, PTs own a range of mobile devices that mediate their learning across multiple contexts (Maher, 2018). This also includes PTs enrolled in distance education programs that deliver electronic learning materials via mobile technologies. The focus is to provide PTs with 'anytime anywhere' access to learning resources.

Learning Contexts With M-Learning

The variety of contexts in which m-learning occurs, helps to clarify the complexity of interactions that have become possible in teacher preparation. Context has been identified as one of the important factors of m-learning, and Chung, Hwang and Lai (2019) highlighted its four components— independent, formalized, physical and socializing contexts. Pegrum et al. (2013) pointed out that m-learning can support seamless learning, which entails situating learning across multiple contexts, hence "learning is not limited by location or time" (Chung et al., 2019, p. 4). Therefore, its increasing relevance in teacher preparation is not only due to the rapid advancement of m-learning in K-12 classrooms as indicated by Burden and Kearney (2018), but more so the mobility of PTs who learn from different contexts anytime, provided they have continuous access to the Internet. It is useful to consider the contexts in which PTs find themselves while preparing to become a teacher, because mobile technologies have led to the blurring of boundaries between these contexts (Pegrum et al., 2013) adding a third space that travels with the PTs in their mobile devices.

Thus, at times, PTs can experience an overlap of continuous learning both formal and informal while interacting with people, across time, and multi-layered contexts. These contexts (physical, social and virtual spaces) include:

- The university campus within the classroom or other learning spaces such as the library, or the laboratory. In addition, a field trip organized by the TE is a university space.
- Partner school campus during their practicum experience, could be within the classroom, at the library, in a school playground or local school community, including schools in remote locations (Mac Mahon, Grádaigh, & Ghuidhir, 2016).

- PTs' own personal learning spaces either at home, a public library, or in cafes.

Figure 1 illustrates ways in which learning contexts overlap. These contexts are interconnected and PTs learn seamlessly through the support afforded by mobile devices. For example, PTs can choose to learn independently from their own personal spaces during their placements in partner schools or at the university. The center represents when PTs find themselves in all the three contexts (physically and/or virtually) where they can collaborate with their peers, TEs or experts. This figure helps to envisage the challenges for TEs to design PT education so that mobile technologies can better be used to support technology-enhanced learning across these multiple contexts. For example, Burden and Kearney (2018) noted that TEs should design new learning ecologies using appropriate m-learning activities that "genuinely exploit the unique pedagogical affordances of mobile technologies rather than replicating existing patterns" (p. 88).

The organizational context for any vocational preparation is complex even within one type of organisation but, as shown in Figure 1, in the preparation of K-12 school teachers it is particularly complex. This is because of the need for congruence between evolving practices of both the higher education organisations and K-12 schools with which it partners, both of which are influenced by regional and national changes in policy. Goodlad (1994) used the term *"simultaneous renewal"* to emphasize that both organizational contexts must evolve together. This need for partnership is recognized worldwide. However, this will always be challenging and the continuing disparity is recognized in teacher preparation where PTs are prepared for both current and future K-12 schooling (Cavanaugh et al., 2018).

In 2010, Davis provided a concise overview of technology in ITE in which she identified that information and communication technologies (ICTs) had been incorporated since the 1980s for three main purposes:

Figure 1. Initial teacher education learning contexts

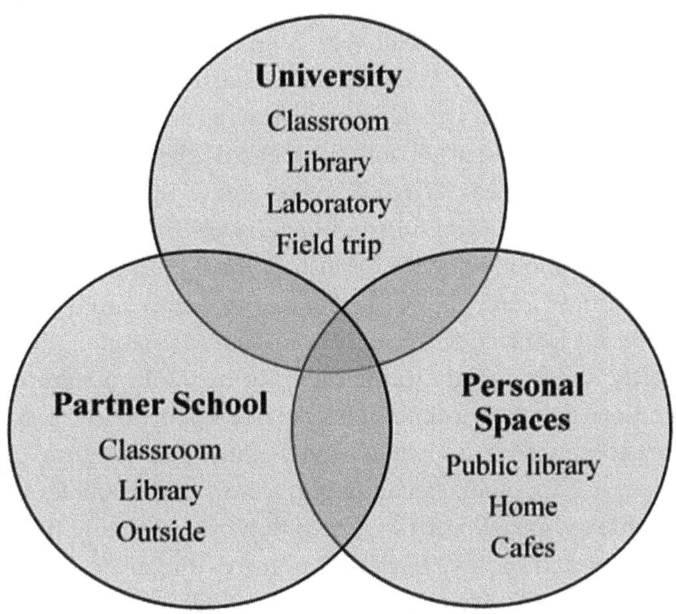

(1) Preparing teachers to use ICTs in educationally effective ways; (2) preparing K-12 teachers to teach ICT-related content; and (3) applying ICTs to serve teacher education. In essence technology is a cross-curricular-theme, a content area, and a tool that can be applied to learning and teaching, which includes the use of telecommunications to improve access to education (Davis, 2010, p. 217).

Since 2010, mobile technologies have begun to permeate these programs as well as K-12 schools. Therefore, in this chapter it is necessary to indicate the ways in which mobile technologies align with these purposes, as well as the additional challenges faced. Essentially mobile technologies are viewed as tools that can be applied to learning and teaching to support PTs' learning in both universities and partner schools so purposes (1) and (3) are most relevant. In addition, purpose (2) is largely irrelevant because mobile technologies do not add anything to a content area in K-12 school curriculum.

Preparing PTs to Use Mobile Technologies in Educationally Effective Ways

Worldwide, teacher education providers are required to effectively prepare PTs to use technology to support teaching and learning in their future classrooms. For example, in the US, the National Educational Technology Plan (NETP) stressed effective integration of technology (mobile technologies not an exception) into classroom pedagogies, so that PTs "leave their teacher preparation programs with a solid understanding of how to use technology to support learning" (U.S. Department of Education, Office of Educational Technology, 2017, p. 35). In Ireland, an action plan from the National Digital Strategy for Schools 2015-2020 (Department of Education and Skills, 2015), indicated that teacher education providers should embed ICT "in the planning, design and delivery of all teacher education courses and programmes" (p. 33). The UNESCO framework on m-learning emphasizes that "without guidance and instruction teachers [TES] will often use technology to 'do old things in new ways' rather than transform and improve approaches to teaching and learning" (West & Visloo, 2013, p. 31). Countries such as Japan, United Arab Emirates, and Malaysia (including New Zealand, see Ministry of Education, 2014) have adopted m-learning as a strategy in their K-12 school systems in order to become high-performing (Cavanaugh et al., 2018). Given the presence of mobile technologies in K-12 classrooms, it is essential to determine how PTs can be supported to effectively use mobile technologies as pedagogical tools.

There is a need for pedagogical shift if PTs have to be adequately prepared for their future classrooms, and the challenging 21st century life. Over two decades, a growing body of evidence indicated that PTs feel uncomfortable and not fully prepared to teach with the technologies that are present in their classrooms (e.g. Sang, Valcke, van Braak & Tondeur, 2010). Thomas and O'Bannon (2013) surveyed 92 PTs' and found that they had difficulty in identifying instructional benefits of mobile phones since 96% of PTs did not experience modelling of mobile phones for instructional purposes. This finding aligns with the findings of a BYOD study (Burns-Sardone, 2014) that reported 99% of PTs (n=58) did not know how to integrate smart phone use for learning purposes. Thomas and O'Bannon (2013) stressed the need for TEs to effectively model the use of mobile technologies to help PTs integrate them into pedagogical practice. Effective integration of mobile technologies provide opportunities for pedagogical innovation (Tondeur et al., 2017) that is realized in high-performing education systems (Cavanaugh et al., 2018).

Modelling of appropriate uses of mobile technologies across all the courses is essential. This strategy has been perceived to be able to impact on PTs' preparation by motivating them to incorporate mobile technologies into their teaching and learning because it transfers directly to K-12 classrooms (Ottenbreit-Leftwich, Glazewski, Brush, Aslan & Zachmeier, 2017). In addition, TEs can design their courses so as to provide PTs with structured tasks through which they are supported to engage in meaningful learning activities involving mobile technologies (Naylor & Gibbs, 2018). In this way the TEs develop one or more courses that are authentic and more easily transferred to K-12 contexts. Such activities in turn stimulate PTs to support one another to develop skills with relevant devices and software. Organizations may be geared towards responding to rapid changes with ongoing redesign of pedagogical approaches in their own sector (university or K-12 school). However, TEs continue to be challenged to model integration of technology. NETP asserted that in order for PTs to use technology in meaningful ways, TEs should model integration of technology in all courses instead of a single-educational technology course (U.S. Department of Education, Office of Educational Technology, 2017). Since preparation of PTs takes place both at the universities and in partner schools, it is worth exploring both contexts to provide a comprehensive view of teacher preparation and mobile technologies.

Integrating M-Learning Into University-Based Courses

Literature presents that preparation of PTs at the university should support them acquire knowledge and skills to meaningfully use mobile technologies in their learning and teaching (Ottenbreit-Leftwich et al., 2017). This can be realized where innovative pedagogies in ITE are also supported by organizational developments in both higher education and K-12 school organizations (Avidov-Ungar & Forkosh-Baruch, 2018). Effective integration of mobile technologies can provide PTs with opportunities to develop understanding of content and pedagogy (Husbye & Elsener, 2013); improve their learning achievements (Mac Mahon et al., 2016); connect and collaborate as they construct knowledge; experience authentic learning based on real-world situations (Naylor & Gibbs, 2018), which enhances their professional learning. It is evident that ITE programs which teach how to incorporate mobile technologies into teaching and learning can help in addressing lack of innovative use of technology by PTs (Ottenbreit-Leftwich et al., 2017).

Researchers have been exploring how to prepare PTs to develop their pedagogical repertoire in order to use mobile technologies while they learn to teach. Akçayır, Akçayır, Pektaş and Ocak (2016) conducted a quasi-experimental study in Turkey to demonstrate how 38 PTs (experimental group) used their smart phones with augmented reality (AR). They highlighted that during an electrolysis of water experiment, PTs controlled their learning by using their phones to watch molecules move (via the animation link they had been provided). Findings indicated that the experimental group improved on their laboratory skills and attitude towards physics laboratory. In particular they reported that the TE mentioned that AR applications motivated the PTs, increased their interest in the course, and allowed PTs to easily comprehend complex physics experiment. In the US, in a study conducted by Burns-Sardone (2014), 58 PTs used AudioBoo (a mobile app installed in their smart phones) to verbally record their book reviews of children's stories, then converted the recordings to QR codes. PTs used QR Reader to scan the QR codes which enabled them to listen and learn about their peers' book reviews. The findings indicated that this activity enhanced the learning experiences of PTs and that they were willing to use their smart phones for learning purposes. Burns-Sardone (2014) make the point that mobile applications (apps) can be used to engage PTs and support their learning. More examples of how TEs are modelling the use of mobile technologies are presented in the next section that examines case studies in teacher preparation.

M-Learning During School-Based Practicum

Gaining field experiences in partner school settings is a vital component of any teacher preparation program (Cheng, 2013). Every PT has at least one opportunity of being mentored by experienced teachers to develop their identity as a teacher and also apply and practise what they are learning in their university-based courses. Some ITE programs have been designed to include school-based practicum throughout, rather than placing it at the end of the program. This enables PTs to take their university-based courses concurrently with their school-based practicum to enhance their learning. It is argued that to effectively prepare PTs, a school-based practicum should exhibit the following features:

- Clarity of goals, including the use of standards guiding the performances and practices to be developed
- Modeling of good practices by more-expert teachers in which teachers make their thinking visible
- Frequent opportunities for practice with continuous formative feedback and coaching
- Multiple opportunities to relate classroom work to university course work
- Graduated responsibility for all aspects of classroom teaching
- Structured opportunities to reflect on practice with an eye toward improving it. (Darling-Hammond & Baratz-Snowden, 2007, p. 124)

The following two examples illustrate how mobile technologies can be used to provide opportunities for PTs to reflect on their teaching practices. These examples include providing formative assessment, and using social networking sites such as Facebook.

Studies have examined how mobile technologies can enhance formative feedback, which is one of the main mentoring roles. There are concerns about the feedback PTs receive from their mentors. Dann and Allen (2015) asserted that during field experience PTs receive "inadequate, ill-directed and inappropriately timed" (p.6) feedback that negatively impacts on PTs' achievement, certification and employment. On the other hand, they argued that "mentor teachers complain of attending to feedback and assessment procedures that are timely and labour intensive" (p.8). Given this concern, Dann and Allen (2015) focused on how smartphones with a video-capture system could be used to improve formative assessment of PTs. Mentor teachers used videos of PTs' teaching lessons to provide feedback. A key finding was that this approach enhanced feedback from mentor teachers. In addition, PTs reviewed the feedback by watching their video recordings as often as they wanted (outside the school setting and teaching time), which improved their learning experiences and reflective practices.

In a more recent example with similar findings, Mac Mahon et al. (2016) reported on their study with 38 PTs from the National University of Ireland. PTs used iPads to video record and reflect on their own teaching then shared their files with the mentor teachers. Mentor teachers used their iPads to provide PTs with immediate written feedback. A key finding revealed that over 80% of PTs indicated that analysing their practices using videos was effective. PTs also used the iPads as pedagogical tools to motivate and engage their students. In Canada, Vaughan and Lawrence (2013) found that mobile technologies can enhance assessment of PTs during their practicum. In this study, TEs (83%) indicated that mobile devices can be used to video tape PTs to give richer feedback. These studies reveal that mentor teachers and PTs can view video recordings during their appropriate time and place, and immediately engage in reflective conversation about the feedback. Providing PTs with effective feedback and feed-forward enables them to monitor, evaluate and reflect on their performance, an integral part of self-regulated learning during school-based practicum (see Zimmerman, 2002).

As described by Maher (2018), PTs can be prepared to use Facebook to support social practices, and communicate with their peers and TEs. Further, studies have investigated the use of Facebook as a platform to facilitate mobile collaborative learning during school-based practicum. The aim was to support PTs to share their experiences and critically reflect on their teaching practices. For example, Liu (2016) designed four tasks for 153 PTs in different partner schools in Taiwan. PTs were inspired to use their mobile devices to interact with their peers on Facebook Group. The first task involved PTs providing information about their schools. The second task entailed two tasks being assigned every month about educational practices (school administration, classroom management, lesson preparation, and teaching and reflection), and PTs were required to post their opinion about each task item. This was followed by a task where all PTs were encouraged to practice critical thinking and reflection before they could respond at least once to any of their peers' postings. Lastly, PTs were encouraged to post at any time any problem they encountered during their teaching practicum. Findings suggested that numerous postings and responding to Facebook messages could positively facilitate PTs' collaborative learning regarding their practices in partner schools, and establish a sense of community.

This was consistent with the study of Deng and Tavares (2013) in which PTs took their own initiative to use their Facebook Group despite having been encouraged to use Moodle (mainly used for university-based courses). PTs felt that Facebook interface enabled them to interact freely with their peers at any time and exchange views about various challenges they encountered in their own classrooms. In terms of learning purposes, PTs learnt more from their peers' responses on Facebook than on Moodle, although they regarded Moodle with formal learning under their lecturers' control. Therefore, this may suggest that Moodle was not an ideal platform for their intended use within school-based context during their practicum, since they associated it with homework, assignments and assessment. Social networking sites such as Facebook, and Google+ can be used to maximize PTs' educational potential through mobile collaborative learning (Liu, 2016).

The importance of preparing PTs for K-12 classrooms during their school-based practicum, has been reinforced so that TES and PTs can link theory and practice. Since PTs join different partner schools, when given the opportunity, they can engage in interactions with one another about their individual experiences. This will enable PTs to learn more about different school practices through continuous conversations (enhanced by mobile technologies) and reflection in real time. These studies found that PTs used mobile technologies to exchange teaching ideas, share resources, and learn collectively beyond their structured class time, and location. In addition, researchers argued that mentor teachers who integrate mobile technologies into their teaching practices, collaborate with PTs to discuss their practices, and provide them with an opportunity for hands-on learning enable PTs to learn more about technology use in the classroom (Admiraal et al., 2017). Darling-Hammond and Baratz-Snowden (2007) asserted that "where the vision of teaching in the placement aligns with the practices being taught in courses - have been found to make a difference in teachers' [PTs] practices, confidence, effectiveness, and long-term commitment to teaching" (p. 124).

LEARNING WITH MOBILE TECHNOLOGIES: CASE STUDIES IN TEACHER PREPARATION

Research shows that ubiquitous mobile technologies support new and innovative ways of learning and teaching (West & Visloo, 2013). This section aims to highlight how mobile technologies have fostered pedagogical innovations in teacher preparation. The author considers empirical findings where TEs modelled pedagogical integration of mobile technologies for PTs in ways that are transferred to K-12 teaching and learning. Four case studies are presented to illustrate some of these strategies.

A literature search was conducted of refereed journal articles published between years 2010-2018. The aim was to guarantee a level of research quality, and access the most up-to-date information from research on emerging mobile technologies in teacher education. This search was conducted using both electronic and manual searches of articles in the following library databases: ERIC, JSTOR and Academic Search Premier including Google Scholar. The following search terms were used: "Mobile learning," and "initial teacher education." Under "mobile learning" search phrases such as "m-learning," "mobile devices," "digital technologies," and "mobile technologies," were included. Under "initial teacher education" search phrases such as "teacher educators," "preservice teachers," "mentor teachers", and "teacher preparation" were included. To ensure reliability, priority was given to articles published in high-ranked teacher education and educational technology journals based on guidelines suggested by Webster and Watson (2002). Manual searches were also conducted in mobile-specific journals.

The four case studies were selected from a search of the published literature because, out of the 44 articles identified for this chapter, only four case studies (Naylor & Gibbs, 2018; Admiraal et al., 2017; Naylor & Gibbs, 2015; Pegrum et al., 2013) met the author's inclusion criteria for discussion in this section. A primary focus for inclusion was empirical studies in ITE that covered two or more discipline areas and/or ITE programs, and were located in varied geographical locations or countries. In addition, studies that were not about how TEs and PTs used mobile technologies for learning and teaching were excluded. The aim is to help us understand the dynamics of preparing PTs to use mobile technologies from a pedagogical perspective.

The first study involved a collaboration of English and Science PTs to develop e-books based on their field trip to the seashore (Naylor & Gibbs, 2015). The TE introduced the task by demonstrating how to make an e-book using video clips, photos and sounds of the seashore. PTs were provided with numerous learning activities across a variety of settings that supported their learning. This pilot study indicated that PTs used their iPads mainly to gather resources for their e-books by capturing images, sound and recording videos at the seashore. The built-in camera, audio recordings and video capture features in the iPads (highly portable) enabled PTs to take photos and record sounds of the seashore. Within the classroom, the PTs of English created poems for their e-books, received support from an external expert to edit their poems and shared their e-books with school students. In contrast, the Science PTs were directed to analyse the distribution of the flora and fauna they had chosen and represent it in diagrams. They also calculated population estimates of limpets and sea anemones, which they displayed in their e-books in the form of bar graphs or pie charts.

In both contexts, the learning activities were open-ended to give PTs a sense of autonomy and control to decide what to research and include in their e-books. PTs were also required to discuss and present their findings. Furthermore, the main activity situated learning within a highly realistic setting that reflected real-life context. The TE provided a constructive environment for location-based learning that was more beneficial for Science PTs. The use of iPads supported this location-based learning outside the classroom setting, which is a key feature of mobile technologies. In addition, PTs who felt challenged with the task had an opportunity for peer-mentorship through negotiations of meaning and collaboration. The study documented that the TE modelled how to make an e-book in ways that are transferable to K-12 teaching and learning. For example, 82% of PTs "considered that the use of mobile technologies provided learning strategies that might be used in their future teaching" (Naylor & Gibbs, 2015, p. 445). PTs need be prepared to construct e-books since it is a key learning activity in K-12 schools where students use e-books to practise language learning anywhere and anytime (Cavanaugh et al., 2018). It is noteworthy that a survey of 107 primary and secondary school teachers from Australia (64%) and Europe (20%) revealed that e-book creation apps amongst other apps were found to be frequently used by school teachers for students to create their own digital content (see Kearney, Burden & Rai, 2015).

The second study, also conducted by (Naylor & Gibbs, 2018) involved the same cohort of 37 PTs in UK and 16 college students (studying both subjects) in Norway. The PTs implemented ideas from the pilot study where they used iPads (which had Book Creator apps) as learning tools to co-construct e-books with the college students. The use of Book Creator apps on their iPads made it easier for them to construct their e-books. This international collaboration study aimed at enabling PTs to develop their pedagogy as future teachers by engaging the college students in meaningful learning experiences. Within the classroom they accessed the Wi-Fi to research and identify maritime topics, and instantly translate scientific words. They also visited the submarium to explore information relevant to their topics and used iPads to collect data. Findings indicated that PTs "used their subject knowledge and pedagogical knowledge, together with the technology, to ensure that the college students were involved in learning experiences" (Naylor & Gibbs, 2018, p. 73). PTs mentioned "that the use of mobile technologies provided learning strategies that they could use in their future teaching" (p. 71). Ally et al. (2014) argued that "when mobile devices are used for instruction in pre-service training, teachers [PTs] have the opportunity to observe concrete examples of pedagogical strategies for mobile learning and to consider what that learning is like from a student's perspective" (p. 47). In addition, Naylor and Gibbs (2018) concluded that the technological pedagogical content knowledge of PTs improved. These findings suggest that situating the tasks in real world context support meaningful and authentic learning of PTs. This is one of the pedagogical affordances of mobile technology used beyond those of a desktop computer.

The third study (Pegrum et al., 2013) was conducted at the University of Western Australia. In this study, eight PTs enrolled in the Early Childhood and Primary programs were provided with iPads for their use. The study aimed to explore how PTs used iPads as learning tools, and if so, how they experienced a new sense of learning spaces. Three TEs supported PTs to use their "iPads to make audio and video recordings, develop mind maps, create lesson plans and deliver class presentations" (p. 467). The results indicated that PTs used their iPads to develop an understanding of content and pedagogy, in addition to staying connected, and organised. For example, one PT used her iPad to audio record her teaching, then used the recording to reflect on her questioning skills. Another PT mentioned that she used her iPad to search for online resources to extend her understanding of mathematical concepts. PTs also used iPads to access their peers' multimedia virtual records in different contexts. Based on the results, Pegrum et al. (2013) recommended four best practices when using iPads: Modelling the use of iPads in pedagogically-appropriate manner, encouraging higher levels of students' reflection, supporting seamless learning across formal and informal contexts, and using iPads for appropriate purposes and contexts.

The fourth study examined preparation of 52 PTs in two technology-infused courses within one ITE program, in the Netherlands (Admiraal et al., 2017). Apart from exploring how PTs used technology, this study also focused on how technology intersects with pedagogical and content knowledge, which aligns with the TPACK framework (see Mishra & Koehler, 2006). During a social studies course, PTs were expected to design flipped lessons and use them in their classes with the school students. The main aim of this activity was to connect theory and practice. The anytime anywhere nature of m-learning enabled PTs to create instructional videos for this assignment, during their workshop at the university and continued at home. PTs also used audio or graphics to integrate into their teaching materials. Findings indicated that all PTs experimented their flipped lessons with the school students and positively evaluated this assignment. In another course that prepared PTs for secondary school pedagogy, TEs used video clips to model technology-infusion approach. The study showed that modelling the use of technology in the classroom motivated PTs to integrate technology into their teaching.

However, a balance in PTs' preparation between university-based courses and school-based practicum is crucial. PTs also need opportunities to develop knowledge about meaningful technology integration into their teaching during school-based practicum. This experience aims to empower them to implement their theoretical learning into classroom teaching. In Admiraal's et al. (2017) study, PTs implemented technology infusion in K-12 schools. PTs worked on their assignments to design flipped lessons, and were required to implement them during their teaching practice. The participants of this study also included 392 secondary school students, who were expected to provide their feedback about PTs' use of technology amongst other teaching activities. An evaluation of this class activity revealed a disparity between what PTs were taught at the university and how they applied it in their classrooms. Preparing PTs with practices that are transferable to K-12 settings is essential since their preparation in universities should improve their performance in K-12 schools.

Reflecting on the four case studies above: Firstly, the author notes that these studies have presented clear examples of how PTs can be prepared to use mobile technologies in ways that apply to their future classrooms. These studies also illustrate how TEs modelled the use of mobile technologies (especially iPads) using meaningful tasks to engage PTs in authentic learning experiences that enhanced their learning. Researchers have indicated that modelling the use of mobile technologies in the classroom is associated with PTs' positive perceptions towards mobile technologies (Ottenbreit-Leftwich et al., 2017). Secondly, it is notable that in all the four case studies, learning activities were situated in the transformation stage of the SAMR model (see Puentedura, 2012). In addition, two pedagogical features of the iPAC framework authenticity and collaboration featured predominantly (see for example Burden & Kearney, 2018). Thirdly, three case studies (Naylor & Gibbs, 2018; Naylor & Gibbs, 2015; Pegrum et al., 2013) explored how PTs used iPads to engage in collaborative tasks with their peers across multiple learning contexts. Pegrum et al. (2013) noted that the use of "iPads supported students' [PTs] seamless learning across formal and informal contexts" (p. 475).

Finally, the fourth study (Admiraal et al., 2017) explored how TEs provided hands-on learning to PTs during their university-based courses and school-based practicum where they were evaluated by their students. The majority of studies conducted during PTs' school-based practicum have not been focusing on school students. For example, Lawson, Çakmak, Gündüz, and Busher (2015) conducted a systematic review of 114 research studies on teaching practice and revealed that only two studies included school students' perspectives. Admiraal et al. (2017) noted that the feedback from the school students and PTs' experiences during their teaching practice enabled them to change their teaching, which led to school students being actively engaged in class activities. In this literature review, this is the only study which investigated preparation of PTs (both at the university and in partner schools) to use mobile technologies to support K-12 students' learning. Even so, it did not expound on how mentor teachers modelled the use of technology. Maher (2018) has also noted that:

Many of the studies that have been conducted to date tend to focus on mobile learning exclusively in the university context or the school context. There is a need to examine how mobile learning is supported in universities and how such learning is then applied to schools by pre-service teachers during their professional experience placements (p. 170).

The author's ongoing research study (Obonyo, Davis & Fickel, 2018) at the time of writing explored how TEs incorporate mobile technologies into their courses. Preliminary findings indicated that TEs are adapting their pedagogies using mobile technologies in multiple contexts for meaningful learning of PTs which are likely to transfer to K-12 contexts. Given the need for a diversity of preservice programs, it will be valuable to have more studies describing how TEs integrate mobile technologies across a range of disciplines and in ways that are transferable to the diverse range of K-12 schooling worldwide. More so, it will help understand how PTs are being prepared from multiple viewpoints, both on and off the university campus.

APPLYING MOBILE TECHNOLOGIES TO SERVE PRESERVICE TEACHER EDUCATION

Mobile technologies play a significant role in how preservice programs have responded to internal and external changes (Gustafson, 2015). These technologies have been used to support redesign of the courses and instruction leading to opportunities such as convenience, flexibility, and accessibility (Cavanaugh et al., 2018). This section focuses on two emerging pedagogical models, distance education and flipped classroom model where mobile technologies play an important role in their implementation. These strategies are commonly being used in ITE to provide flexible and accessible teaching and learning environments of both PTs and the TEs. Until the ubiquity of mobile technologies these strategies were challenging to implement.

Distance Education

Responding to PTs' learning needs and preferences including being cost-effective, have caused a number of teacher education programs to employ distance education technology. Distance education covers many alternative distance learning methods such as e-learning, open learning, flipped learning, distributed learning, and blended learning (Gustafson, 2015). In these learning methods courses are delivered using different technology such as video conferencing, Adobe Connect, Zoom, Skype and the use of Learning Management Systems (LMS) for PTs to access learning resources online at their own time and pace. A number of researchers have contributed to how ITE programs have employed distance education.

The study conducted by Gustafson (2015) aimed to experiment with teacher education as distance education with a cohort of seven digitally skilled PTs. The study documented how PTs navigated through their learning context using technology for course participation, communication and collaboration. PTs recorded their teaching lessons using a digital video camera and uploaded them to a video sharing site for their TEs to view and provide feedback. They also used Google Drive and Facebook Group to participate and collaborate during their asynchronous coursework. On the other hand, TEs used iPads and Facetime to facilitate remote observations of PTs. The findings revealed that being tech savvy enabled PTs to use Web 2.0 tools, could easily multi-task and accomplish tasks quickly. Furthermore, in this study smaller group activities using Skype, Facebook, and Google Hangouts were seen to be more beneficial. However, PTs mentioned that they did not participate as much as they would have during their face-to-face class. This can be associated with lack of instructors to prompt them for participation or get assistance when they had difficulties. Gustafson (2015) concluded that technology enhanced learning requires scaffolding and support so that PTs can understand and be knowledgeable about pedagogical applications to effectively use technology.

Similarly, in their study with 14 PTs and six TEs, Vaughan and Lawrence (2013) investigated how mobile technologies supported professional practice competencies in a blended ITE program that was launched to connect theory with practice through on-going field placements. All the participants owned laptops while 92% of PTs owned smartphones. Both TEs and PEs reported that mobile technologies were helpful in supporting professional responsibilities, and facilitating authentic learning. PTs used mobile devices for lesson planning, video record their teaching, and provided audio assessment feedback to their peers. PTs recommended the following strategies for effective use of mobile devices in a blended preservice program: linking mobile devices to learning outcomes and assignments, using appropriate mobile device for the learning activities, and improved Internet connectivity (Vaughan & Lawrence, 2013).

Flipped Classroom Model

Flipped classroom model (FCM) is another emerging pedagogical model which falls under the broad definition of blended learning. FCM is an innovative teaching style that some TEs are implementing to engage PTs in learning activities inside and outside the classroom. In this model, in-class time is used to engage PTs in interactive learning activities, problem solving, and attempting quizzes. Lectures are delivered before class time where PTs take the lead and responsibility for their learning by watching video lectures at their own pace, during their convenient time, and location (Zainuddin & Attaran, 2016). In addition, traditional lectures are replaced with Podcasts delivered on mobile devices since Podcasts are a versatile way of communication.

Fraga and Harmon (2014) examined 51 PTs' perspectives about FCM and its impact on their achievement by comparing the FCM with the conventional model. The majority of PTs used their laptops and tablets to view narrated PowerPoint presentations while at home. The results indicated that PTs preferred FCM due to opportunities for a flexible time to study, freedom to work independently, ability to control their own learning, and the support they received during class time. In addition, this study identified that topic consideration, PTs' learning environments and learning styles are likely to influence the effectiveness of FCM. However, this study did not find significant differences between PTs in the FCM and the conventional model in terms of their academic performance.

Other supporting evidence associated with the benefits of FCM came from Kurt's (2017) study in Turkey. In a quasi-experimental study about implementation and effectiveness of the FCM, an experimental group of 32 PTs in the flipped model demonstrated an improvement in their self-efficacy, had higher academic achievements, expressed positive perceptions of the FCM, and said would apply it to their future classrooms. This study indicated that PTs watched lecture podcasts, completed an online quiz, and read assigned resources outside the classroom. During class time they engaged in discussions, analyzed case scenarios, and collaborated with their peers to act out the scenarios. Kurt (2017) noted that before the study, PTs had taken a classroom management course and found the lectures difficult and boring to follow which had an adverse effect on their self-confidence.

Research suggests that FCM results in active learning, increased learner engagement and interactions (Zainuddin & Attaran, 2016). It mainly relies on using technology especially mobile technologies in a flexible way to create a student-centred learning environment so that PTs can experience individualized learning (Fraga & Harmon, 2014) and also develop awareness of and skills with mobile technologies. However, recently Admiraal et al. (2017) revealed that secondary school students suggested that in order for FCM to be effective, video lessons should be announced in different ways, the importance of watching the lecture videos should be emphasized, and the important content of a lesson should be focused more in the lecture videos. Distance education and FCM present possibilities for using mobile technologies that go beyond traditional models of teaching and learning.

CHALLENGES OF USING MOBILE TECHNOLOGIES IN TEACHER EDUCATION

There appear to be numerous benefits of integrating mobile technologies into teacher education. However, factors surrounding its implementation can adversely affect the quality of PTs' preparation, and their zeal to practise what they believe regarding technology integration. Burden and Hopkins (2016) classified these challenges as first order barriers (external to the teacher). They determined issues such as lack of time and resources for training, lack of technical support, infrastructure problems, and hostile school culture. Second order barriers (internal to the teacher) included confidence levels, skills, pedagogical beliefs and attitudes towards technology. Burden and Hopkins (2016) noted that first order barriers were most frequently reported by PTs. Further, Pegrum et al. (2013) identified that PTs face challenges such as limitations with devices, time, and attitude. These barriers negatively impact on PTs' ability to integrate mobile technologies into their practise although they strongly believe mobile technologies can help them to improve their technology skills and pedagogical thinking (Burden & Hopkins, 2016).

To incorporate mobile technologies successfully into pedagogical practices will require training TEs and mentor teachers. Tondeur et al. (2017) observed that "teacher educators do not feel adequately prepared to effectively integrate technology into their classrooms" (p. 174). This has been associated with the lack of quality training on specific technologies (Burns-Sardone, 2014), and how to use mobile devices for teaching and learning. The main challenge TEs experience is the need to continually adapt and change their pedagogical practices (Kearney et al., 2015) to support PTs and be able to work in their future classrooms. Likewise, mentor teachers need to be trained to effectively support PTs since they have a greater impact on PTs' perceptions about teaching (Cheng, 2013). UNESCO guidelines for m-learning outlines the need for training if the unique educational benefits of m-learning have to be realized (West & Visloo, 2013).

PTs have raised several issues during their professional practice including mobile technologies not being common amongst some school mentors, infrastructure problems, school policies and cultures, and time limitations (Burden & Hopkins, 2016). They also find inadequate technology skills one of the barriers towards effective use of technology in their future classrooms (Ottenbreit-Leftwich et al., 2017). O'Bannon and Thomas (2015) conducted a study with 245 PTs to determine their views about barriers to the use of mobile phones in K-12 classrooms. Results indicated that the majority of PTs identified texting, playing games, tweeting, and searching the Internet as the most crucial disruptions. Additional concerns included cheating, cyber bullying, and accessing inappropriate information on the Internet.

FUTURE RESEARCH DIRECTIONS

Research studies have focused on the relationship between PTs and their mentor teachers (for example see Izadinia, 2016), and PTs' concerns during their teaching practice in partner schools (Lawson et al., 2015). However, further research is needed on the use of mobile technologies when PTs are on practicum, and how mentor teachers model the use of mobile technologies. In addition, exploring K-12 school students' views about PTs' practices. Admiraal et al. (2017) asserted that lack of attention in this area especially in secondary education, causes PTs to learn to teach with technology integration when they have graduated and entered their profession. They suggest that "more attention to technology in teacher preparation programs might make this learning process of teachers in school practice more efficient and effective … increasing attention to technology in teacher preparation programs is not enough; it is the 'how' what matters" (pp. 105-106). In addition, research and/or literature specifically addressing TEs' mobile learning pedagogical approaches is limited. In an on-going research study, I am examining PTs' preparation to teach with mobile technologies during their practicum in K-12 school settings.

Based on the literature search, minimal empirical research on non-intended serendipitous learning calls for further studies in this area. Crompton et al. (2017) reviewed 113 studies about the use of m-learning in PK-12 education and found that none of them took place in non-formal contexts. She noted that "the lack of studies in non-formal environments appear to reflect the difficulty m-learning researchers face in accessing private spaces of learners" (p. 58). This review revealed that 50% of studies took place in a formal setting, 23% were conducted in informal setting, while 27% in both. Although Crompton et al. (2017) focused on PK-12 education, potentially this suggests that a similar trend is likely to be a replicate in teacher education. This view is supported by Maher (2018) who recommended that future m-learning research should explore how PTs learn in formal and informal settings. Therefore, there is a need to explore m-learning in informal settings, specifically in PTs' personal spaces, to understand how they use or do not use mobile technologies as learning tools.

TEs need guidelines towards effective pedagogical approaches while incorporating mobile technologies into their teaching. This is particularly noted in the research by Ng'ambi (2013) where he proposed five-phase pedagogical guidelines that TEs may use: The first phase entailed outlining an educational goal to drive the instruction. The second phase indicated involving students (individually or collaboratively) to create artefacts using technology. The third phase is students' presentation of their learning, followed by reflections that are visible to the entire class to enhance discussion and feedback. Lastly students write a reflective essay that is submitted for assessment.

CONCLUSION

The use of mobile technologies to enhance 21st century learning is increasing in K-12 schools and teacher education institutions. Teacher preparation programs are expected to effectively prepare PTs to use mobile technologies for their future classrooms and the digital world. Although PTs experience continuous learning across multiple contexts, there appears to be a tendency to think about m-learning in isolated aspects of PTs lives and not across physical, social and virtual spaces. Studies have examined opportunities and challenges of incorporating mobile technologies in teacher preparation. Four case studies presented, have explored different ways mobile technologies have been used to effectively prepare

PTs for their future classrooms. However, majority of these studies have been conducted when PTs are taking their university-based courses.

Because preparation of PTs takes place at the university and in partner school, I argue that exploring how these organizations use mobile technologies in ways that are transferred to K-12 teaching and learning is essential. Little is known of how PTs use mobile technologies during their school-based practicum and how mentor teachers model the use of mobile technologies. TEs have taken up this point suggesting that collaboration between partner schools and ITE programs is vital for effective preparation of PTs (Fickel & Mackey, 2013). Practices in both organizations need be coherent so that PTs can be able to apply what they have learned. Providing PTs with concrete examples of how mobile technologies can be used in the classroom is a strategy that should be incorporated both at the university and partner school.

M-learning is definitely a significant part of ITE programs, however from research it appears to be focused on learning about m-learning and is not experiential part of learning to be a teacher. The main focus should be to prepare PTs for their future classrooms to teach with mobile technologies. Therefore, two purposes of incorporating ICTs in teacher preparation as identified by Davis (2010) have been presented to understand how mobile technologies align with these purposes. These include: Preparing PTs to use mobile technologies in educationally effective ways, and applying mobile technologies to serve PT education. Furthermore, challenges of using mobile technologies in teacher education have been examined. The National Educational Technology Plan asserted that:

Schools should be able to rely on teacher preparation programs to ensure that new teachers come to them prepared to use technology in meaningful ways. No new teacher exiting a preparation program should require remediation by his or her hiring school or district. Instead, every new teacher should be prepared to model how to select and use the most appropriate apps and tools [mobile technologies] to support learning. (U.S. Department of Education, Office of Educational Technology, 2017, pp. 35-36).

ACKNOWLEDGMENT

This chapter would not be possible without the contributions of my senior supervisor Distinguished Professor Niki Davis, Associate Professor Cheryl Brown, and the anonymous reviewers for their comments on earlier draft of this chapter.

REFERENCES

Admiraal, W., van Vugt, F., Kranenburg, F., Koster, B., Smit, B., Weijers, S., & Lockhorst, D. (2017). Preparing pre-service teachers to integrate technology into K-12 instruction: Evaluation of a technology-infused approach. *Technology, Pedagogy and Education*, *26*(1), 105–120. doi:10.1080/1475939X.2016.1163283

Akçayır, G., Akçayır, M., Pektaş, H. M., & Ocak, M. A. (2016). Augmented reality in science laboratories: The effects of augmented reality on university students' laboratory skills and attitudes toward science laboratories. *Computers in Human Behavior*, *57*, 334–342. doi:10.1016/j.chb.2015.12.054

Ally, M., Grimus, M., & Ebner, M. (2014). Preparing teachers for a mobile world, to improve access to education. *Prospects*, *44*(1), 43–59. doi:10.100711125-014-9293-2

Avidov-Ungar, O., & Forkosh-Baruch, A. (2018). Professional identity of teacher educators in the digital era in light of demands of pedagogical innovation. *Teaching and Teacher Education*, *73*, 183–191. doi:10.1016/j.tate.2018.03.017

Burden, K., & Hopkins, P. (2016). Barriers and challenges facing pre-service teachers use of mobile technologies for teaching and learning. *International Journal of Mobile and Blended Learning*, *8*(2), 1–20. doi:10.4018/IJMBL.2016040101

Burden, K., & Kearney, M. (2018). Designing an educator toolkit for the mobile learning age. *International Journal of Mobile and Blended Learning*, *10*(2), 88–99. doi:10.4018/IJMBL.2018040108

Burns-Sardone, N. (2014). Making the case for BYOD instruction in teacher education. *Issues in Informing Science and Information Technology, 11*, 191-201. Retrieved from http://iisit.org/Vol11/IISITv11p191-201Sardone0505.pdf

Cavanaugh, C., Maor, D., & McCarthy, A. (2018). Mobile learning. In K. Kennedy & R. E. Ferdig (Eds.), *Handbook of Research on K-12 Online and Blending Learning* (2nd ed.; pp. 575–591). Pittsburgh, PA: ETC Press. doi:10.1184/R1/6686813.v1

Cheng, E. C. (2013). Enhancing the quality of pre-service teachers' learning in teaching practicum. *Education in Science*, *1*, 6–16.

Chung, C., Hwang, G., & Lai, C. (2019). A review of experimental mobile learning research in 2010–2016 based on the activity theory framework. *Computers & Education*, *129*, 1–13. doi:10.1016/j.compedu.2018.10.010

Crompton, H. (2013). Mobile learning: New approach, new theory. In Z. L. Berge & L. Muilenburg (Eds.), *Handbook of mobile learning* (pp. 47–57). New York: Routledge.

Crompton, H., Burke, D., & Gregory, K. H. (2017). The use of mobile learning in PK-12 education: A systematic review. *Computers & Education*, *110*, 51–63. doi:10.1016/j.compedu.2017.03.013

Dann, C., & Allen, B. (2015). Mobile video collection in preservice teacher practicum placements. *Journal of Technology and Teacher Education*, *23*(1), 5–27. Retrieved from https://www.learntechlib.org/primary/p/129814/

Darling-Hammond, L., & Baratz-Snowden, J. (2007). A good teacher in every classroom: Preparing the highly qualified teachers our children deserve. *Educational Horizons*, *85*(2), 111–132. Retrieved from https://www-jstor-org.ezproxy.canterbury.ac.nz/stable/42926597

Davis, N. E. (2010). Technology in preservice teacher education. In P. Peterson, E. Baker, & B. McGaw (Eds.), *International Encyclopaedia of Education* (3rd ed.; Vol. 8, pp. 217–221). Oxford: Elsevier. doi:10.1016/B978-0-08-044894-7.00748-X

Deng, L., & Tavares, N. J. (2013). From moodle to facebook: Exploring students' motivation and experiences in online communities. *Computers & Education*, *68*, 167–176. doi:10.1016/j.compedu.2013.04.028

Department of Education and Skills. (2015). Digital strategy for schools 2015-2020: Enhancing teaching, learning and assessment. Author.

Fickel, L., & Mackey, J. (2013). Rapporteurs' report: What should initial teacher education programs for 2022 look like and why? *Waikato Journal of Education*, *18*(1), 117–119. doi:10.15663/wje.v18i1.145

Fraga, L. M., & Harmon, J. (2014). The flipped classroom model of learning in higher education: An investigation of preservice teachers' perspectives and achievement. *Journal of Digital Learning in Teacher Education*, *31*(1), 18–27. doi:10.1080/21532974.2014.967420

Goodlad, J. (1994). *Educational renewal: better teachers, better schools*. San Francisco, CA: Jossey-Bass.

Gustafson, T. M. (2015). *Teacher education as distance education: Pre-service experience and the uses of technology*. Michigan State University.

Husbye, N. E., & Elsener, A. A. (2013). To move forward, we must be mobile: Practical uses of mobile technology in literacy education courses. *Journal of Digital Learning in Teacher Education*, *30*(2), 46–51. doi:10.1080/21532974.2013.10784726

Izadinia, M. (2016). Student teachers' and mentor teachers' perceptions and expectations of a mentoring relationship: Do they match or clash? *Professional Development in Education*, *42*(3), 387–402. doi:10.1080/19415257.2014.994136

Kearney, M., Burden, K., & Rai, T. (2015). Investigating teachers' adoption of signature pedagogies. *Computers & Education*, *80*, 48–57. doi:10.1016/j.compedu.2014.08.009

Keskin, N. O., & Metcalf, D. (2011). The current perspectives, theories and practices of mobile learning. *Turkish Online Journal of Educational Technology*, *10*(2), 202–208.

Khaddage, F., Müller, W., & Flintoff, K. (2016). Advancing mobile learning in formal and informal settings via mobile app technology: Where to from here, and how? *Journal of Educational Technology & Society*, *19*(3), 16–26.

Kurt, G. (2017). Implementing the flipped classroom in teacher education: Evidence from turkey. *Journal of Educational Technology & Society*, *20*(1), 211–221.

Lawson, T., Çakmak, M., Gündüz, M., & Busher, H. (2015). Research on teaching practicum - a systematic review. *European Journal of Teacher Education*, *38*(3), 392–40. doi:10.1080/02619768.2014.994060

Liu, S. (2016). The perceptions of participation in a mobile collaborative learning among pre-service teachers. *Journal of Education and Learning*, *5*(1), 87. doi:10.5539/jel.v5n1p87

Mac Mahon, B., Grádaigh, S. Ó., & Ghuidhir, S. N. (2016). iTE: Student teachers using iPad on a second level initial teacher education programme. *International Journal of Mobile and Blended Learning*, *8*(2), 21–34. doi:10.4018/IJMBL.2016040102

Maher, D. (2018). Supporting pre-service teachers' understanding and use of mobile devices. In J. Keengwe (Ed.), *Handbook of Research on Mobile Technology, Constructivism, and Meaningful Learning* (pp. 160–177). Hershey, PA: IGI Global. doi:10.4018/978-1-5225-3949-0.ch009

Ministry of Education. (2014). *Future-focused Learning in connected communities.* Wellington, New Zealand: Author.

Mishra, P., & Koehler, M. J. (2006). Technological pedagogical content knowledge: A framework for teacher knowledge. *Teachers College Record, 108*(6), 1017–1054. doi:10.1111/j.1467-9620.2006.00684.x

Naylor, A., & Gibbs, J. (2015). Using iPads as a learning tool in cross-curricular collaborative initial teacher education. *Journal of Education for Teaching, 41*(4), 442–446. doi:10.1080/02607476.2015.1081718

Naylor, A., & Gibbs, J. (2018). Deep learning: Enriching teacher training through mobile technology and international collaboration. *International Journal of Mobile and Blended Learning, 10*(1), 62–77. doi:10.4018/IJMBL.2018010105

Ng'ambi, D. (2013). Effective and ineffective uses of emerging technologies: Towards a transformative pedagogical model. *British Journal of Educational Technology, 44*(4), 652–661. doi:10.1111/bjet.12053

O'Bannon, B. W., & Thomas, K. M. (2015). Mobile phones in the classroom: Preservice teachers answer the call. *Computers & Education, 85*, 110–122. doi:10.1016/j.compedu.2015.02.010

Obonyo, C., Davis, N., & Fickel, L. (2018). Mobile learning practices in initial teacher education: illustrations from three teacher educators. In E. Langran & J. Borup (Eds.), *Proceedings of Society for Information Technology & Teacher Education International Conference* (pp. 2424-2429). Washington, DC: Academic Press.

Ottenbreit-Leftwich, A. T., Glazewski, K. D., Brush, T. A., Aslan, S., & Zachmeier, A. (2017). Addressing technology integration concerns: Asynchronous video mentoring between pre-service teachers and exemplary technology-using in-service teachers. *Australasian Journal of Educational Technology, 29*(4). doi:10.14742/ajet.3246

Pegrum, M., Howitt, C., & Striepe, M. (2013). Learning to take the tablet: How pre-service teachers use iPads to facilitate their learning. *Australasian Journal of Educational Technology, 29*(4), 464–479. doi:10.14742/ajet.187

Puentedura, R. (2012). *The SAMR Model: Background and exemplars.* Retrieved from http://www.hippasus.com/rrpweblog/archives/2012/08/23/SAMR_BackgroundExemplars.pdf

Sang, G., Valcke, M., Braak, J., & Tondeur, J. (2010). Student teachers' thinking processes and ICT integration: Predictors of prospective teaching behaviors with educational technology. *Computers & Education, 54*(1), 103–112. doi:10.1016/j.compedu.2009.07.010

Thomas, K., & O'Bannon, B. (2013). Cell phones in the classroom: Preservice teachers' perceptions. *Journal of Digital Learning in Teacher Education, 30*(1), 11–20. doi:10.1080/21532974.2013.10784721

Tondeur, J., Pareja Roblin, N., van Braak, J., Voogt, J., & Prestridge, S. (2017). Preparing beginning teachers for technology integration in education: Ready for take-off? *Technology, Pedagogy and Education, 26*(2), 157–177. doi:10.1080/1475939X.2016.1193556

U.S. Department of Education, Office of Educational Technology. (2017). Reimagining the Role of Technology in Education: 2017 National Education Technology Plan Update. Author.

Vaughan, N., & Lawrence, K. (2013). Investigating the role of mobile devices in a blended pre-service teacher education program. *Canadian Journal of Higher Education*, *43*(3), 56–77.

Webster, J., & Watson, R. T. (2002). Analyzing the past to prepare for the future: Writing a literature review. *Management Information Systems Quarterly*, *26*(2), xiii–xxiii.

West, W., & Vosloo, S. (2013). *UNESCO policy guidelines for mobile learning*. Paris: UNESCO. Retrieved from http://unesdoc.unesco.org/images/0021/002196/219641e.pdf

Zainuddin, Z., & Attaran, M. (2016). Malaysian students' perceptions of flipped classroom: A case study. *Innovations in Education and Teaching International*, *53*(6), 660–670. doi:10.1080/14703297.2015.1102079

Zimmerman, B. J. (2002). Becoming a self-regulated learner: An overview. *Theory into Practice*, *41*(2), 64–70. doi:10.120715430421tip4102_2

ADDITIONAL READING

Baran, E. (2014). A review of research on mobile learning in teacher education. *Journal of Educational Technology & Society*, *17*(4), 17–32. Retrieved from http://www.jstor.org.ezproxy.canterbury.ac.nz/stable/jeductechsoci.17.4.17

Caldwell, H. (2018). Mobile technologies as a catalyst for pedagogic innovation within teacher education. *International Journal of Mobile and Blended Learning*, *10*(2), 50–65. doi:10.4018/IJMBL.2018040105

Chee, K. N., Yahaya, N., Ibrahim, N. H., & Hasan, M. N. (2017). Review of mobile learning trends 2010-2015: A meta-analysis. *Journal of Educational Technology & Society*, *20*(2), 113–126. doi:10.1016/j.compedu.2012.03.016

Cochrane, T. D. (2014). Critical success factors for transforming pedagogy with mobile web 2.0. *British Journal of Educational Technology*, *45*(1), 65–82. doi:10.1111/j.1467-8535.2012.01384.x

Hunter, J., Daly, N., (2013). iPads in teacher education: Integrating a range of language learning knowledges through a technological medium. *Computers in New Zealand Schools: Learning, teaching, technology, 25*(1–3), 96-116.

Ibáñez, M., & Delgado-Kloos, C. (2018). Augmented reality for STEM learning: A systematic review. *Computers & Education*, *123*, 109–123. doi:10.1016/j.compedu.2018.05.002

Lindsay, L. (2016). Transformation of teacher practice using mobile technology with one-to-one classes: M-learning pedagogical approaches. *British Journal of Educational Technology*, *47*(5), 883–892. doi:10.1111/bjet.12265

Luo, T., Murray, A., & Crompton, H. (2017). Designing authentic learning activities to train pre-service teachers about teaching online. *International Review of Research in Open and Distributed Learning*, *18*(7). doi:10.19173/irrodl.v18i7.3037

Minicozzi, L. L. (2018). iPads and pre-service teaching: Exploring the use of iPads in k-2 classrooms. *The International Journal of Information and Learning Technology*, *35*(3), 160–180. doi:10.1108/IJILT-05-2017-0032

Şad, S., & Göktaş, Ö. (2014). Preservice teachers' perceptions about using mobile phones and laptops in education as mobile learning tools. *British Journal of Educational Technology*, *45*(4), 606–618. doi:10.1111/bjet.12064

Schuck, S. (2016). Enhancing teacher education in primary mathematics with mobile technologies. *Australian Journal of Teacher Education*, *41*(3), 126–139. doi:10.14221/ajte.2016v41n3.8

Wetzel, K., Buss, R., Foulger, T. S., & Lindsey, L.-A. (2014). Infusing educational technology in teaching methods courses: Successes and dilemmas. *Journal of Digital Learning in Teacher Education*, *30*(3), 89–103. doi:10.1080/21532974.2014.891877

KEY TERMS AND DEFINITIONS

Bring Your Own Device (BYOD): A term used to encourage learners to bring their personal mobile devices to the classroom for learning purposes.

Mentor Teacher: An experienced teacher who supports PTs by coaching, providing feedback, and allowing them to reflect on their practices during their field experience.

Mobile Learning: The process of acquiring knowledge and skill by being taught, studying, and practicing inside and outside the classroom using mobile technologies.

Partner School: K-12 school where PTs gain their field experience by being mentored by experienced teachers based on the agreement with the university.

Preservice Teacher: A student teacher studying in either a public or private ITE institute to gain qualified teaching status in primary school or secondary school upon graduating from their program.

School-Based Practicum: A period when PTs get an opportunity to practice teaching.

Teacher Educator: An instructor at the institute of higher education who teaches PTs.

Section 2
Use of Mobile Devices to Enhance Teaching and Learning

Chapter 4
Mobile Technologies for Making Meaning in Education:
Using Augmented Reality to Connect Learning

Teresa L. Coffman
University of Mary Washington, USA

Mary Beth Klinger
College of Southern Maryland, USA

ABSTRACT

This chapter examines the use of mobile technologies to integrate technology into the classroom environment so that students can experience real-world learning. The use of augmented reality is emphasized to enhance the learning process and provide engaging and authentic student-centered experiences. Using augmented reality, personalized learning is possible that encourages a constructivist approach and where the learning process is the main focus. Students are able to construct knowledge through augmented experiences that support the curricular content and then share this newfound knowledge with others. The use of augmented reality as a cognitive tool can connect learning and allow students to work collaboratively with deeper and higher-level meaning as a result.

DOI: 10.4018/978-1-5225-8106-2.ch004

INTRODUCTION

Educational research has focused on the cognitive and personal context from which educators apply augmented reality (AR) to learning experiences. It is considered a constructivist learning tool and is used to engage students with digital technologies (Dunleavy, 2014; Saidin, Halim, & Yahaya, 2015). AR is able to immerse students in meaningful exchange and provide contextual options for students to interact with the content and their own learning (Azuma, et al, 2001; Dede, 1995; Efstathiou, Kyza, & Georgiou, 2018; Neville, 2010).

Improvements in mobile technologies and other digital tools such as smart phones, tablets, and smart glasses create an opportunity for educators to integrate digital technologies into the teaching and learning process (Chang, Lee, Liang, & Wu, 2013; Dede, 2011; Squire & Klopfer, 2007). Students are able to use virtual reality (VR), augmented reality (AR), and/or mixed (hybrid) reality (MR) to collaborate on authentic, real-world problems and provide solutions, thereby enhancing their own learning in a creative and innovative approach (Dunleavy, Dede, & Mitchell, 2009; Geer & Sweeney, 2012).

While VR and MR are increasing in educational popularity, the focus of this chapter is on augmented reality to connect learning in the classroom with new material and experiences (Carbonell, Saorin, & De la Torre Cantero, 2018). AR is explored throughout this chapter using a social cognitive instructional design and a constructionist model to support student-centered learning experiences. Educators can use the interactive capabilities presented by AR to engage learners, encourage reflection, and support active inquiry as they design instruction to maximize learning (Dunleavy et al., 2009; Mathison & Gabriel, 2012).

The objective of this chapter is to examine the use of augmented reality as a cognitive tool to create authentic learning experiences for students to explore and apply their learning in situated experiences (Brown, Collins, & Duguid, 1989; Dunleavy, 2014). It focuses on the design of instruction to yield a student-centered approach in the creation of augmented reality products. Emphasis is placed on helping students make connections and construct content around their own knowledge and interests in a cooperative learning environment.

BACKGROUND

Augmented reality is the process of superimposing layers of information such as text, video, and/or primary source data on top of a physical object to support or extend the object (Craig, 2013; Curcio, Dipace, & Norlund, 2016). AR is an interactive process. Whether as a user or a creator, learners are interacting with digital information as it relates to a real-world object. Through this interaction, context between what a student is learning and the object itself can create an immersive approach for students to explore the world in an authentic way (Harley, Poitras, Jarrell, Duffy, & Lajoie, 2016). At the same time, new knowledge can be integrated within the learning experience. This creates an opportunity for students to develop new information and practice essential skills in applied ways (Agarwal, Finley, Rose, & Roediger, 2017).

Augmented reality has the potential to create a student-driven design whereby students can solve content specific problems in collaboration with classmates through applied practical experiences that utilize technology to extend the learning experience (Bitter & Corral, 2014). This extension has the potential to transform learning, especially when students take an active role in designing and constructing the virtual augmented product (Bailey, 2019).

There are multiple advantages to using AR in the classroom to support learning. Augmented reality can help students learn, process, and remember information more readily. Because of its real-world component, it is also more engaging and fun.

Multiple benefits exist when integrating augmented reality into the educative process, such as:

- Higher student interest through interactive learning;
- Improved collaboration capabilities and teamwork skills;
- More effective and efficient learning using visualization and full immersion in the subject; and
- Practical, real-world learning that is safe and inexpensive (Aleksandrova, 2018).

MAIN FOCUS OF THE CHAPTER

The premise of incorporating technology into the educational environment is to provide authentic tools that support instruction and compliment students' ability to learn and engage deeply with the content. Technology is viewed as a cognitive and digital tool that students work with to enhance their own learning capabilities (Jonassen, 1996).

The use of technology in an AR context suggests that technology is much more than just an apprentice or add-on, but rather functions as an intelligent partner. It serves as a tool that the learner is working with to achieve specific goals. Technology is used to:

- Build upon prior learning,
- Interact with the material, and
- Develop meaningful understanding around course goals in authentic ways (Bailey, 2019).

When instruction includes the use of digital technologies as a cognitive tool, the shift involves the learner taking ownership of the technology tool and his or her learning (Herrington, & Parker, 2013). Not only does this enhance technological capabilities, but the learning tasks allow students to construct knowledge in tangible and meaningful ways (Nesbit, Niu & Liu, 2019; Papert & Harel, 1991). Ultimately, the use of AR has the potential of heightening the learner's educative experience (Billinghurst & Dunser, 2012; Cuendet, Bonnard, Do-Lenh, & Dillenbourg, 2013; İbili & Şahin, 2013).

AR in Education

Educative experiences integrating AR technology support learning outcomes through the process of incorporating natural interactions with a physical object by extending or augmenting it with virtual content (Chiang, Yang & Hwang, 2014; Hung, Chen, & Huang, 2017). This extension encourages learners to connect their learning and additionally requires more complex student thinking about what was learned during the lesson than traditional stand-and-deliver instruction (Dunlosky, Rawson, Marsh, Nathan, & Willingham, 2013; Lam & Muldner, 2017).

Oftentimes in traditional instruction, the learned concepts may seem independent of a student's own learning and connections to the world around them. This learning might not be viewed as easily transferred to other contexts (Sayed, Neven, Zayed & Sharawy, 2011; Yuen, Gallayanee, & Johnson, 2011). However, by providing an opportunity for learners to experience what they learned with constructive ways for them to engage with their learning around specific content through the creation and design of digital representations that can be augmented, students are able to access and interpret information in meaningful ways (Klopfer & Sheldon, 2010; Yuen et al., 2011).

As a result, more opportunities are created to develop deeper meaning and connections to their own learning (Bers, Flannery, Kazakoff, & Sullivan, 2014; Nesbit et al., 2019; Ritchhart, Church, & Morrison, 2011). Students create stronger connections between theory and application, as well as advance non-linear capabilities with learning-by-doing (Baeten, Struyven, & Dochy, 2013; DeLucia, Francese, Passero, & Tortoza, 2012; Papert & Harel, 1991; Wang, 2012).

An Illustration

When a device such as a smart phone or smart glasses is pointed at an object with a built-in trigger or overlay, an augmentation is displayed over the object. This augmentation can be multimedia or an interactive element with the purpose of extending or enhancing the object (Efstathiou et al., 2018; Osberg, 1992). This immersive technology can be incorporated into a learning experience to extend content and thereby enhance learning (Hornecker & Dünser, 2009; Le, Pedro, Pham & Park, 2016).

This merging of digital content with a physical object creates a sense of transparency with the object and the information being explored. It also has the potential to create opportunities for learners to interact with content in meaningful ways around the context of learning (Dede, 1995; Moyer, Bolyard & Spikell, 2002; Neville, 2010). This active engagement with the object through the augmentation allows learners to actively control their own learning and to participate actively in the learning process (Ozdamli & Uzunboylu, 2015; Serafin, Adjorlu, Nilsson, Thomsen, & Nordahl, 2017). Students do this through the exploration and engagement with abstract concepts and situations created by the augmentations (Curcio et al., 2016; Fernandez, 2017; Jerry & Aaron, 2010; Shelton & Hedley, 2004).

Throughout this process, appropriate instruction structures the immersion with the object and augmentations to help learners make necessary connections to the classroom teaching (Hung et al., 2017; Jonassen, 1996). Students are introduced to a variety of digital technologies in order to create virtual content that can be an overlay for an object. Technologies must be integral to the learning process, add to the teaching process and overall teaching strategies, and align to the curricular needs and goals of the discipline in order to be successful (Coffman, 2017; Overbay, Patterson, Vasu & Grable, 2010).

Constructivism and the Use of AR as a Cognitive Tool

Constructivist practice allows students to make meaning around learning objectives and real-world issues or problems by providing relevant experiences, information, and tasks (Coffman, 2017; Jonassen, Wilson, Wang, & Grabinger, 1993). It identifies structured and meaningful conflict between a learner's past knowledge and current understanding to challenge beliefs using a social context that elicits shared understanding (Phillips, 2000; Resnick, 1989; Toledo, Collazos, Cantero, & Redondo, 2018). AR activities facilitate this purposeful learning to stimulate continued knowledge creation and transfer lessons learned to new situations (Luckin & Stanton, 2011; Metcalf, Kamarainen, Grotzer, & Dede, 2013; Nesbit et al., 2019).

A constructivist design to instruction means that students are actively engaged with the learning experience (Overbay et al., 2010). This includes the content, the process of learning, and the digital AR technology. Throughout the activity, students understand that the information they are creating helps to define or extend the physical object, providing both context and meaningful learning (Bujak et al., 2013; Hoffmann, Meisen, & Jeschke, 2014; Ozdamli, & Uzunboylu, 2015).

Students are encouraged to question and apply both abstract and concrete skills to identify information needed for their instructional task (Ritchhart et al., 2011; Saitta, Gittings, & Geiger, 2011). They then plan and design virtual augmentations that use multilayered artifacts to enhance or explain a real-world object in new ways (Coffman, 2017; Tarng, & Ou, 2012).

Transparency in the learning process is an important consideration when utilizing technology as a cognitive tool (Jonassen et al., 1993). Augmented reality allows for this transparency due to its complementing of content and its ability to extend or support learning (Saidin et al., 2015; Serafin et al., 2017). This extension of learning is enhanced if students choose an authentic task, such as creating a multidimensional virtual experience (Fischer, Greiff, & Funke, 2012). Students are in control of their learning as they interact with the content and imagine the task in new ways to develop supplemental content that can be augmented and experienced by others (Cai, Wang, Chiang, 2014; Sung, Hwang, & Chang, 2016).

When students have opportunities to create their own AR experiences, it empowers them (Geer & Sweeney, 2012; Mathison & Gabriel, 2012; Saidin et al., 2015). As students identify a problem that needs to be solved with the newly created augmentation, they develop a project of study and identify a solution that enhances or extends the real-world object and content learned. Students create virtual content that they augment onto a chosen object to enhance the physical space that it exists within. They consider how this new information will extend the meaningfulness of the object and participate in a hands-on experience. Learners are immersed into the context of their learning and physical surroundings (Carbonell & Bermejo, 2017; Coffman & Klinger, 2014; Papert & Harel, 1991; Yuen, et al., 2011).

The integration of technology into the learning process has the potential to deepen the learning experience. Instruction can incorporate digital tools that support complex cognitive processing tasks around learning goals (Coffman, 2017). Through the process of creating virtual objects that are anchored to objects in the real-world, AR provides a new way for students to interact with information. This ubiquitous experience challenges student to be critical thinkers and complex problem solvers (Ferrington & Loge, 1992; Nincarean, Alia, Halim, & Rahman, 2013; Swan & Marshall, 2010). Thinking is extended beyond a simple visual enhancement of information toward one that supports complex thinking and an understanding of how the information compliments and relates to the real-world (Dunleavy et al., 2009; Juan, Furió, Alem, Ashworth & Cano. 2011).

An Illustration

When students are taught to solve a mathematical problem using a specific equation, traditional instruction focuses on applying factual knowledge to fill in the equation to solve the intended problem. To create an abstract understanding of solving the mathematical equation, students can solve the problem by using technology to research, organize information, and then create a virtual AR experience that connects the mathematical equation to an authentic experience in the real-world (Bujak, et al., 2013; Jacka, 2018; Ke & Hsu, 2015; Swan & Marshall, 2010).

Cooperative Groups and AR

Utilizing cooperative groups allows for an engaged process of team-based created AR. Students work as a cooperative entity and challenge each other to think and perform to a higher standard than each would individually (Holden, 2014; Jonassen et al., 1993; Totten, Sills, Digby, & Russ, 1991; Vygotsky, 1978).

Placing students in a constructionist collaborative environment promotes self-discovery and a problem-solving process that encourages an active approach to learning (Brunner, 1997; Papert, 1990; Papert & Harel, 1991; Vygotsky, 1978). In cooperative groups, roles are identified, questions are asked, research and interpretations are debated, information is gathered, and new knowledge is configured and organized to be presented to others in a meaningful way. This encourages engagement in the learning process (Fischer et al., 2012; Lee, 2012; Ozdamli & Uzunboylu, 2015; Regenbrecht, Baratoff, & Wilke, 2005).

In a cooperative team, students research a topic and design AR content that adds to the body of knowledge. The process of extending learning and developing a product using relevant technology in a group environment supports an authentic learning process which ultimately helps students develop team building skills, as well as critical and creative thinking, extending their learning in a real-world context (Leinonen, Keune, Veermans, & Toikkanen, 2016; Sung et al., 2016).

An Illustration

Instead of having students individually write a review about a novel they just read, students can be introduced to a cooperative group experience using AR to provide deeper discovery and engagement about the themes, characters, and locations within the novel. Through triggers, students embed text and digital information, to include primary source data and multimedia to augment multisensory elements from the novel. They enhance the text as well as extend their own critical thinking and learning about the novel itself (Billinghurst, Kato, & Poupyrev, 2001).

Instructional Requirements With AR

As part of designing an AR project, a pedagogical framework is identified that incorporates exploration, discovery, meaningful student interactions, and both individual and team learning as students create interactive content (Annetta, Burton, Cheng, Chmiel, & Frazier, 2012; Craig, 2013). Instruction must clearly define expectations and key check points should be developed throughout the experience (Bacca, Baldiris, Fabregat, & Graf, 2014; Billinghurst & Dunser, 2012).

Differences in student skill sets should also be considered. As students explore what information best complements their chosen real-world object and then what AR trigger to embedded on the physical object, it is incumbent upon the instructor to determine what knowledge students need and what skill development strategies should be incorporated throughout the learning process to provide optimal opportunities for reflection (Bailey, 2019; Bers et al., 2014).

Students should be encouraged to personalize learning and create something unique. The learning environment is immersive and opportunities should be provided for feedback and practice (Billinghurst & Dunser, 2012; Cuendet et al., 2013). When instruction supports learners in the creation process, AR is viewed as a construction tool (Brennan, Valverde, Prempeh, Roque, & Chung, 2011). The *process* of learning is the main focus, which in turn promotes collaboration and connection to others (Darling-Hammond, Austin, Orcutt & Rosso, 2001; Dewey, 1938; Hung et al., 2017).

Throughout the AR project design process, the instructor introduces new technology tools and facilitates ways in which students can use multiple means of representation and interactivity to support their learning as they develop digital content to enhance a physical object (Bower, Howe, McCredie, Robinson, & Grover, 2014; Bujak et al., 2013). Students utilize discovery, creation, and evaluation skills as they advance their understandings about the new information. Ultimately, this allows learners to collectively extend their knowledge gained from the instruction (Curcio et al., 2016).

Reducing Cognitive Load and Scaffolding the Learning Process

In the process of creating learning experiences in the product design process, there is also a need to reduce the burden students might experience. The complex thinking that is required by students in designing and creating augmented reality presents the possibility of overwhelming cognitive capabilities and overloading a learner (Darling-Hammond et al., 2001). This could occur when trying to connect learning outcomes to real-world objects or when trying to engage in deeper, more complex thinking (Azuma et al., 2001; Regenbrecht et al., 2005).

The instructional strategies incorporated with using AR must aid in reducing cognitive load. Thus, it is important to scaffold the learning process. Properly executed scaffolding both prepares as well as encourages students to engage in deeper level thinking around AR learning tasks, concepts, and diverse resources (Dunleavy et al., 2009; Kamarainen et al., 2016; Kesim & Ozarslan, 2012; Luckin, 2006).

Incorporating scaffolding throughout the AR experience:

- Increases learning outcomes (Jerry & Aaron, 2010; Lee, 2012; Tarng & Ou, 2012),
- Enhances motivation and a focus on learning goals (Akçayır & Akçayır, 2017; Billinghurst & Dunser, 2012; Carbonell et al., 2018; Kettle & Häubl, 2010), and
- Assists in reducing cognitive load with content, digital technologies, and the learning process (Ambrose, Bridges, DiPietro, Lovett, & Norman, 2010; Cristancho, Moussa, & Dubrowski, 2011; Toh et al., 2015).

Mobile Technologies and Pre-Packaged AR

Mobile technologies such as cell phones, tablet computers, and laptops have become more affordable and accessible. In addition, free drag-and-drop app access allows for the design and creation of content more easily (Ke & Hsu, 2015). The result is that both create greater accessibility to AR technologies in teaching and student learning. As a result, the learning curve of these technologies is decreasing due to the drag-and-drop capabilities inherent in each (Metcalf et al., 2013; Ozdamli, & Uzunboylu, 2015).

One pivotal reason for the increased use and integration of student-centered AR into teaching is the ease of use of AR software. To design and share AR content, free-to-use software with a drag-and-drop interface is becoming more readily available. Students can learn this software with some instruction or in a discovery approach through access to online tutorials and by exploring examples of existing AR content and apps (Mitchell, 2011; Ozdamli, & Uzunboylu, 2015).

This software can be used on mobile devices such as smart phones or tablet computers. The content can then be easily shared with others by printing, placing on a physical object, or posting on the Internet (Chang, Morreale & Medicherla, 2010; Johnson et al., 2014).

Pre-packaged AR content is one mode of integration that can be customized around an instructor's pedagogy, curricular requirements, and student needs. Pre-packaged augmented reality has been used in teaching subjects such as:

- Science (Chiang et al., 2014; Hsiao, Chen & Huang, 2011; Kerawalla, Luckin, Seljeflot, & Woolard, 2006; Tarng & Ou, 2012; Toh et al., 2015)
- Mathematics ((Jerry & Aaron, 2010; Mitchell, 2011; Moyer et al., 2002)
- Games (Annetta et al., 2012; Juan et al., 2011; Kirkley, & Kirkley, 2004; Klopfer & Sheldon, 2010)
- Literature (Billinghurst et al., 2001)

Currently, the most popular application for augmented reality in education is the use of AR apps. Available in almost any academic subject, AR apps can be introduced directly into the classroom to help explain a concept, provide a visual representation of material, and/or help students test their knowledge in practice (Aleksandrova, 2018; Shelton & Hedley, 2004; Young, Ellobody & Hu, 2012).

Pre-packaged products demonstrate the capabilities of AR technology. Technologies available in support of pre-packaged AR include:

- A video camera to create short video clips
- Image editing software to enhance or edit images to create a message
- Storage space to collect, organize, and save virtual information
- The Internet to research and share dynamic resources
- Software that students can use to develop an AR product by creating their own triggers and overlays, such as HP Reveal, Layer, or Buildar (Chang et al., 2010; Johnson et al., 2014)

The pre-made AR experience is enhanced by instruction to help students better achieve the intended learning goals. The AR overlay provides an opportunity for the learner to engage with the content and the connections identified by the instructor which in turn extends the overall learning experience in a more meaningful and memorable manner (Akçayır, Akçayır, Pektaş, & Ocak, 2016; Baeten et al., 2013; Luckin & Stanton, 2011).

An Illustration

In each of the following pre-made AR examples, information is interfaced creating a sense of seamlessness between the virtual environment and the real object (Azuma et al., 2001, Bacca et al., 2014).

An AR marker on a textbook page is pre-made AR content. A marker or trigger contains a data set of information. This information can contain media to include multimedia. To use, a student opens an AR app and points the camera on the mobile device at the marker on the textbook page. This marker triggers a message to open a universal resource locator (URL) and displays the content in a floating window that shows over the textbook page. For the student, the impression is that the content is fixed over the object in a three-dimensional space and is emerging from the object off the static object, such as a page in a textbook or from a computer monitor (Shelton & Hedley, 2004; Young et al., 2012). This has the potential to expose the learner to additional relevant resources. It also demonstrates the capabilities of technology and provides opportunities for the user to dig deeper into the content.

In a second example, once the Google Translate app is downloaded onto a smartphone or tablet, a mobile device camera can be used to translate words. Google Translate connects the letters within the word to decipher the translation into whatever language is specified. As Google Translate displays the translation, it adds a virtual layer with the translated word seemingly superimposed over the word (Siu, 2018).

SOLUTIONS AND RECOMMENDATIONS

Augmented reality that is student created is a meaningful way for learners to organize and present new information around relevant real-world objects. This process of learning provides opportunities for learners to accommodate new knowledge around their current mental models and enhances as well as challenges their thinking about their learning, the content, and the world around them (Angelo & Cross, 1993; Fink, 2013; Johnson, Johnson, & Smith, 1991; Piaget, 1977; Vygotsky, 1978).

When learners create and design their own AR artifacts and experiences in cooperation with their peers, learning becomes more personalized, meaningful, and tangible. This ability to augment information supports learning by creating engagement with the content, the environment, and others (Billinghurst & Dunser, 2012; Dede, 2011; Wang, 2012).

Throughout this process, students are learning by doing. In organized and well-crafted lessons, instructors can integrate AR technology in experiential learning activities and promote diversity in thinking differently about the learning objectives (Huang, Chen, & Chou, 2016; Wenger, 1999). Students use augmented reality to design an immersive educative experience that is student-centered with both a virtual and real-world experience interwoven throughout to learn new concepts.

Throughout this educative AR experience, instruction should:

- Be scaffolded throughout the learning process
- Focus learners on important questions
- Create opportunities that allow learning to become meaningful and experiential
- Allow for the integration of AR technology to be relevant for the learner (Ausubel, Novak & Hanesian, 1978; Ke & Hsu, 2015)

Organizing instruction around authentic problems, timely research, and working in cooperative groups for learners to create their own augmentations allows for more relevant and meaningful learning (Neville, 2010; Toledo et al., 2018). This triangulation provides an opportunity for students to design AR content around new representations of knowledge and extend this body of knowledge to a broader audience (Carbonell et al., 2018; Chiang et al., 2014).

Students think critically about their new understandings, physical objects, and a virtual augmentation that they create to extend the object and its surrounding physical space (Mitchell, 2011; Serafin et al., 2017; Yuen et al., 2011). This innovative thinking blends reality with a computer-generated layer of information hovering virtually over the object with the purpose of enhancing its meaning and accessibility (Bower et al., 2014). Students then strengthen higher level thinking skills around curricular goals and the real-world (Leinonen, 2016).

FUTURE RESEARCH DIRECTIONS

Within the current educative environment, technology is viewed as an immersive element within instruction. It is not an add-on nor is it a tool-based experience (Coffman, 2017). Rather, it can afford more complex and meaningful opportunities for learners to analyze, access information, interpret and organize their knowledge (Shelton & Hedley, 2004). It allows students to design, build, experience, and share with others (Antonioli, Blake, & Sparks, 2014; Bonwell & Eison, 1991; Derry, 1990; Herrington & Parker, 2013; Kommers, Jonassen & Mayes, 1992).

Infusing technology throughout the learning process rather than simply incorporating its use in discrete units of study requires a shift in instruction. Thus, educators must be cognizant of their own instructional strengths as well as recognize gaps in knowledge and expertise with available technology (Metcalf et al., 2013; Ozdamli & Uzunboylu, 2015). Future research should focus on best practices for educators in using technology as a cognitive tool and instructional support for faculty with augmented reality.

The biggest issue which impacts AR implementation is lack of teacher training (Aleksandrova, 2018). It is incumbent upon the instructor to understand AR potentialities in the classroom and to learn how best to integrate this technology into teaching, learning, and the overall curriculum in a manner that best assists instruction and student learning of the subject matter (Akçayır & Akçayır, 2017; Bailey, 2019; Dunleavy, 2014; Holden, 2014).

To have students work with augmented reality, instructors should also have a sense of comfort with AR and this can be achieved by exploring the technology - both pre-made AR experiences and the process of developing AR products. This personal experience and self-efficacy with augmented reality will aid in providing transferable knowledge to students who can then take an active role in their own AR construction (Dunleavy et al., 2009; Klopfer & Sheldon, 2010; Toh, Miller, & Simpson, 2015).

It is not required that the instructor be a high-level expert with the technology or its capabilities. Rather, the expectation is that there is a literacy with the technologies, an awareness of critical thinking capabilities, and a self-efficacy with the technology, content knowledge, and one's own pedagogical skills (Butler & Winne, 1995; Toh et al., 2015). In this manner, the teacher can create learning proficiencies and then scaffold those experiences creating opportunities for students to use technology to support their learning, as well as develop and extend their own knowledge (Harris & Hofer, 2011; Sang, Valcke, van Braak, & Tondeur, 2010).

In addition, an instructor must always consider where in the curriculum AR best fits into his or her instruction. Specific learning goals and curricular needs may need to be adapted so that students can be entrepreneurial in creating AR and teachers can feel confident in their ability to deliver competent instruction (Caine & Caine, 2011; Holden, 2014; Mitchell, 2011).

Future research can help support this shift in teaching and learning, as well as provide opportunities for educators to receive quality training in technology integration for students to create AR experiences. Educational institutions can strive to serve as learning organizations that support innovative teaching and encourage educators to extend their own conceptual knowledge and improve the instructional process with their students as it relates to technology use and augmented reality (Clifford & Friesen, 2001; Serafin et al., 2017).

CONCLUSION

As discussed in this chapter, augmented reality can function as a multisensory experience that allows both exploration of complex topics and the process of extending one's own understanding (Dunleavy et al., 2009; Jerry & Aaron, 2010; Klopfer & Sheldon, 2010; Tarng & Ou, 2012). Incorporating AR into the curriculum can help students think critically and use knowledge in new and creative ways.

Augmented reality as a cognitive tool can broaden course content beyond abstract knowledge to situated, embedded learning experiences that are both immersive and social, and ultimately enhance teaching and learning (Brown et al., 1989; Hung et al., 2017; Lave & Wenger, 1998). AR is being utilized more readily as an immersive learning tool that can complement course learning goals and at the same time develop deeper skills and knowledge (Kamarainen, Metcalf, Grotzer, Brimhall & Dede, 2016; Metcalf et al., 2013).

As educators move from producing and delivering digital learning experiences and/or using pre-made digital learning experiences to augment instruction to more student centric approaches whereby learners create their own digital content around learning goals, digital technologies will continue to evolve to provide an opportunity for deeper learning around immersive experiences (Le et al., 2016). Augmented reality as a cognitive tool incorporates complex cognitive processing and offers an exceptional opportunity for learners to facilitate deeper learning around curricular goals as they contemplate and attempt to solve real-world problems in authentic ways (Jonassen, 1996; Kesim & Ozarslan, 2012).

As an interactive digital technology, AR has the potential for educators to create a more immersive, relevant, and student-driven instructional design around context specific knowledge (Annetta et al., 2012; Dunleavy et al., 2009). The integration of augmented reality should not distract from the process of learning. Rather, AR provides a practicality for both theoretical and applied topics and helps to provide context around the essential stages of the learning process. At every step, digital technology should be aligned to the instructor's own pedagogy and course learning goals.

This chapter explored immersive and authentic uses of AR technology as a cognitive tool used by learners to create situation-based experiences. In coordination with the accessibility of mobile technologies and applications that allow learners to create virtual augmentations, incorporating augmented reality into instruction can provide rich and meaningful experiences both for learning content as well as developing complex cognitive skills.

Learning experiences can be designed that challenge students to think differently about the world around them and then transfer this new knowledge to new situations and contexts (Chen & Lin, 2016; Huang et al., 2016; Klinger & Coffman, 2011; Toledo et al., 2018). In this manner, AR connects learning and supports the goal of mobile technologies providing meaning within education and to the learning organization.

REFERENCES

Agarwal, P. K., Finley, J. R., Rose, N. S., & Roediger, H. L. III. (2017). Benefits from retrieval practice are greater for students with lower working memory capacity. *Memory (Hove, England)*, *25*(6), 764–771. doi:10.1080/09658211.2016.1220579 PMID:27531308

Akçayır, M., & Akçayır, G. (2017). Advantages and challenges associated with augmented reality for education: A systematic review of the literature. *Educational Research Review*, *20*, 1–11. doi:10.1016/j.edurev.2016.11.002

Akçayır, M., Akçayır, G., Pektaş, H. M., & Ocak, M. A. (2016). Augmented reality in science laboratories: The effects of augmented reality on university students' laboratory skills and attitudes toward science laboratories. *Computers in Human Behavior*, *57*, 334–342. doi:10.1016/j.chb.2015.12.054

Alexsandrova, M. (2018). Augmented reality in education: The hottest EdTech trend 2018 and how to apply it to your business. *Eastern Peak*. Retrieved from https://easternpeak.com/blog/augmented-reality-in-education-the-hottest-edtech-trend-2018-and-how-to-apply-it-to-your-business/

Ambrose, S. A., Bridges, M. W., DiPietro, M., Lovett, M. C., & Norman, M. K. (2010). *How learning works: Seven research-based principles for smart teaching*. Hoboken, NJ: Wiley.

Angelo, T. A., & Cross, K. P. (1993). *Classroom assessment techniques* (2nd ed.). San Francisco: Jossey-Bass.

Annetta, L., Burton, E. P., Cheng, R., Chmiel, M., & Frazier, W. (2012). Augmented reality games: Using technology on a budget. *Science Scope*, *36*(3), 54–60.

Antonioli, M., Blake, C., & Sparks, K. (2014). Augmented reality applications in education. *The Journal of Technology Studies*, 96-107.

Ausubel, D. P., Novak, J. D., & Hanesian, H. (1978). *Educational psychology: A cognitive view* (2nd ed.). New York: Holt, Rinehart and Winston.

Azuma, R., Baillot, Y., Behringer, R., Feiner, S., Julier, S., & MacIntyre, B. (2001). Recent advances in augmented reality. *IEEE Computer Graphics and Applications*, *21*(6), 34–47. doi:10.1109/38.963459

Bacca, J., Baldiris, S., Fabregat, R., & Graf, S. (2014). Augmented reality trends in education: A systematic review of research and applications. *Journal of Educational Technology & Society*, *17*(4), 133.

Baeten, M., Struyven, K., & Dochy, F. (2013). Student-centered teaching methods: Can they optimize students' approaches to learning in professional higher education? *Studies in Educational Evaluation*, *39*(1), 14–22. doi:10.1016/j.stueduc.2012.11.001

Bailey, L. W. (2019). Virtual and augmented reality: Exploring the potential for instructional application. In L. W. Bailey (Ed.), *Educational technology and the new world of persistent learning* (pp. 12–23). Hershey, PA: IGI Global. doi:10.4018/978-1-5225-6361-7.ch002

Bers, M. U., Flannery, L., Kazakoff, E. R., & Sullivan, A. (2014). Computational thinking and tinkering: Exploration of an early childhood robotics curriculum. *Computers & Education*, *72*, 145–157. doi:10.1016/j.compedu.2013.10.020

Billinghurst, M., & Dunser, A. (2012). Augmented reality in the classroom. *Computer*, *45*(7), 56–63. doi:10.1109/MC.2012.111

Billinghurst, M., Kato, H., & Poupyrev, I. (2001). The magic book: Moving seamlessly between reality and virtuality. *IEEE Computer Graphics and Applications*, *21*, 6–8.

Bitter, G., & Corral, A. (2014). The pedagogical potential of augmented reality apps. *International Journal of Engineering Science Invention*, *3*(10), 13–17.

Bonwell, C. C., & Eison, J. A. (1991). *Active learning: Creating excitement in the classroom*. Washington, DC: The George Washington University, School of Education and Human Development.

Bower, M., Howe, C., McCredie, N., Robinson, A., & Grover, D. (2014). Augmented reality in education – cases, places and potentials. *Educational Media International*, *51*(1), 1–15. doi:10.1080/09523987.2014.889400

Brennan, K., Valverde, A., Prempeh, J., Roque, R., & Chung, M. (2011). More than code: The significance of social interactions in young people's development as interactive media creators. In T. Bastiaens, & M. Ebner (Eds.), *Proceedings of World Conference on Educational Multimedia, Hypermedia and Telecommunications 2011* (pp. 2147-2156). Chesapeake, VA: AACE.

Brown, J. S., Collins, A., & Duguid, S. (1989). Situated cognition and the culture of learning. *Educational Researcher*, *18*(1), 32–42. doi:10.3102/0013189X018001032

Bruner, J. S. (1977). *The process of education*. Cambridge, MA: Harvard University Press.

Bujak, K. R., Radu, I., Catrambone, R., MacIntyre, B., Zheng, R., & Golubski, G. (2013). A psychological perspective on augmented reality in the mathematics classroom. *Computers & Education*, *68*, 536–544. doi:10.1016/j.compedu.2013.02.017

Butler, D., & Winne, P. (1995). Feedback and self-regulated learning: A theoretical synthesis. *Review of Educational Research*, *65*(3), 245–281. doi:10.3102/00346543065003245

Cai, S., Wang, X., & Chiang, F. K. (2014). A case study of augmented reality simulation system application in a chemistry course. *Computers in Human Behavior*, *37*, 31–40. doi:10.1016/j.chb.2014.04.018

Caine, R. N., & Caine, G. (2011). *Natural learning for a connected world: Education, technology, and the human brain*. New York: Teachers College Press.

Carbonell, C. C., & Bermejo, L. A. A. (2017). Augmented reality as a digital teaching environment to develop spatial thinking. *Cartography and Geographic Information Science*, *44*(3), 259–270. doi:10.1080/15230406.2016.1145556

Carbonell, C. C., Saorin, J. L., & De la Torre Cantero, J. (2018). Teaching with AR as a tool for relief visualization: Usability and motivation study. *International Research in Geographical and Environmental Education*, *27*(1), 69–84. doi:10.1080/10382046.2017.1285135

Chang, G., Morreale, P., & Medicherla, P. (2010). Applications of augmented reality systems in education. In *Society for Information Technology & Teacher Education International Conference,* San Diego, CA (pp. 1380–1385).

Chang, H., Lee, S. W., Liang, J. C., & Wu, H. K. (2013). Current status, opportunities and challenges of augmented reality in education. *Computers & Education*, *62*, 41–49. doi:10.1016/j.compedu.2012.10.024

Chen, C. C., & Lin, P. H. (2016). Development and evaluation of a context-aware ubiquitous learning environment for astronomy education. *Interactive Learning Environments*, *24*(3), 644–661. doi:10.1080/10494820.2014.915417

Chen, C. H., Ho, C. H., & Lin, J. B. (2015). The development of an augmented reality game-based learning environment. *Procedia: Social and Behavioral Sciences*, *174*, 216–220. doi:10.1016/j.sbspro.2015.01.649

Chiang, T. H., Yang, S. J., & Hwang, G. J. (2014). An augmented reality-based mobile learning system to improve students' learning achievements and motivations in natural science inquiry activities. *Journal of Educational Technology & Society*, *17*(4), 352–365.

Clifford, P., & Friesen, S. (2001). *The stewardship of the intellect: Classroom life, educational innovation and technology. Issues in the Integration of Technology into Teaching, Learning, and School Culture.* Calgary, AB: Detselig Enterprises.

Coffman, T. (2017). *Designing instruction to promote higher level thinking* (3rd ed.). Lanham, MD: Rowman & Littlefield Education.

Coffman, T., & Klinger, M. B. (2014). Collaboration and communication in the online classroom through a brain-based approach. *International Journal of Information Communication Technologies and Human Development*, *6*(3), 42–52. doi:10.4018/ijicthd.2014070104

Craig, A. B. (2013). *Understanding augmented reality: Concepts and applications*. London, UK: Elsevier.

Cristancho, S. M., Moussa, F., & Dubrowski, A. (2011). A framework-based approach to designing simulation-augmented surgical education and training programs. *American Journal of Surgery*, *202*(3), 344–351. doi:10.1016/j.amjsurg.2011.02.011 PMID:21871988

Cuendet, S., Bonnard, Q., Do-Lenh, S. D., & Dillenbourg, P. (2013). Designing augmented reality for the classroom. *Computers & Education*, *68*, 557–569. doi:10.1016/j.compedu.2013.02.015

Curcio, I. D. D., Dipace, A., & Norlund, A. (2016). Virtual realities and education. *Research on Education and Media*, *8*(2), 60–68. doi:10.1515/rem-2016-0019

Darling-Hammond, L., Austin, K., Orcutt, S., & Rosso, J. (2001). How people learn: Introduction to learning theories. *The learning classroom: Theory into practice. A Telecourse for Teacher Education and Professional Development.* Retrieved from https://web.stanford.edu/class/ed269/hplintrochapter.pdf

Dede, C. (1995). The evolution of constructivist learning environments: Immersion in distributed, virtual worlds. *Educational Technology*, *35*(5), 46–52.

Dede, C. (2011). Emerging technologies, ubiquitous learning, and educational transformation. In C. D. Kloos, D. Gillet, R. M. Crespo García, F. Wild, & M. Wolpers (Eds.), *Towards Ubiquitous Learning* (pp. 1–8). Springer. doi:10.1007/978-3-642-23985-4_1

DeLucia, A., Francese, R., Passero, I., & Tortoza, G. (2012). A collaborative augmented campus based on location–aware mobile technology. *International Journal of Distance Education Technologies*, *10*(1), 55–71. doi:10.4018/jdet.2012010104

Derry, S. J. (1990). Flexible cognitive tools for problem solving instruction. *Paper presented at the annual meeting of the American Educational Research Association*, Boston, MA.

Dewey, J. (1938). *The experience of education*. New York: Simon & Schuster.

Dunleavy, M. (2014). Design principles for augmented reality learning. *TechTrends*, *58*(1), 28–34. doi:10.100711528-013-0717-2

Dunleavy, M., Dede, C., & Mitchell, R. (2009). Affordances and limitations of immersive participatory augmented reality simulations for teaching and learning. *Journal of Science Education and Technology*, *18*(1), 7–22. doi:10.100710956-008-9119-1

Dunlosky, J., Rawson, K. A., Marsh, E. J., Nathan, M. J., & Willingham, D. T. (2013). Improving students' learning with effective learning techniques: Promising directions from cognitive and educational psychology. *Psychological Science in the Public Interest*, *14*(1), 4–58. doi:10.1177/1529100612453266 PMID:26173288

Efstathiou, I., Kyza, E., & Georgiou, Y. (2018). An inquiry-based augmented reality mobile learning approach to fostering primary school students' historical reasoning in nonformal settings. *Interactive Learning Environments*, *26*(1), 22–41. doi:10.1080/10494820.2016.1276076

Fernandez, M. (2017). Augmented virtual reality: How to improve education systems. *Higher Learning Research Communications*, *7*(1), 1–15. doi:10.18870/hlrc.v7i1.373

Ferrington, G., & Loge, K. (1992). Virtual reality: A new learning environment. *The Computing Teacher*, *9*(7), 16–19.

Fink, D. L. (2013). *Creating significant learning experiences: An integrated approach to designing college courses*. San Francisco: Jossey-Bass.

Fischer, A., Greiff, S., & Funke, J. (2012). The process of solving complex problems. *The Journal of Problem Solving*, *4*(1), 19–42. doi:10.7771/1932-6246.1118

Geer, R., & Sweeney, T. A. (2012). Students voice about learning with technology. *Journal of Social Sciences*, *8*(2), 294–303. doi:10.3844/jssp.2012.294.303

Harley, J. M., Poitras, E. G., Jarrell, A., Duffy, M. C., & Lajoie, S. P. (2016). Comparing virtual and location-based augmented reality mobile learning: Emotions and learning outcomes. *Educational Technology Research and Development*, *64*(3), 359–388. doi:10.100711423-015-9420-7

Harris, J. B., & Hofer, M. J. (2011). Technological pedagogical content knowledge (TPACK) in action: A descriptive study of secondary teachers' curriculum-based, technology-related instructional planning. *Journal of Research on Technology in Education*, *43*(3), 211–229. doi:10.1080/15391523.2011.10782570

Herrington, J., & Parker, J. (2013). Emerging technologies for authentic learning. *British Journal of Educational Technology*, *44*(4), 607–615. doi:10.1111/bjet.12048

Hoffmann, M., Meisen, T., & Jeschke, S. (2014). Shifting virtual reality education to the next level: Experiencing remote laboratories through mixed reality. In S. Frerich, T. Meisen, & A. Richert et al. (Eds.), *Engineering Education 4.0* (pp. 235–249). New York: Springer.

Holden, C. (2014). Homegrown augmented reality. *TechTrends*, *58*(1), 42–48. doi:10.100711528-013-0719-0

Hornecker, E., & Dünser, A. (2009). Of pages and paddles: Children's expectations and mistaken interactions with physical–digital tools. *Interacting with Computers*, *21*(1–2), 95–107. doi:10.1016/j.intcom.2008.10.007

Hsiao, K. F., Chen, N. S., & Huang, S. Y. (2011). Learning while exercising for science education in augmented reality among adolescents. *Interactive Learning Environments*, *20*(4), 331–349. doi:10.1080/10494820.2010.486682

Huang, T. C., Chen, C. C., & Chou, Y. W. (2016). Animating eco-education: To see, feel, and discover in an augmented reality-based experiential learning environment. *Computer Education*, *96*, 72–82. doi:10.1016/j.compedu.2016.02.008

Hung, Y. H., Chen, C. H., & Huang, S. W. (2017). Applying augmented reality to enhance learning: A study of different teaching materials. *Journal of Computer Assisted Learning*, *33*(3), 252–266. doi:10.1111/jcal.12173

İbili, E., & Şahin, S. (2013). Software design and development of an interactive 3D geometry book using augmented reality: ARGE3D Afyon Kocatepe University. *Journal of Science and Engineering*, *13*(1), 1–8.

Jacka, L. (2018). *Using virtual worlds in educational settings: Making learning real.* London: Taylor and Francis. doi:10.4324/9780203728901

Jerry, T., & Aaron, C. (2010). The impact of augmented reality software with inquiry-based learning on students' learning of kinematics graph. In *2nd International Conference on Education Technology and Computer (ICETC), 2010* (pp. V2-1–V2-5). Shanghai, China: IEEE. 10.1109/ICETC.2010.5529447

Johnson, D. W., Johnson, R. T., & Smith, K. (1991). *Active learning: Cooperation in the college classroom.* Edina, MN: Interaction.

Johnson, L., Adams Becker, S., Estrada, V., Freeman, A., Kampylis, P., Vuorikari, R., & Punie, Y. (2014). *NMC Horizon Report Europe: 2014 Schools Edition.* Austin, TX: The New Media Consortium. Retrieved from https://www.learntechlib.org/p/182011/

Jonassen, D. H. (1996). *Computers in the classroom: Mindtools for critical thinking.* Columbus, OH: Merrill/Prentice-Hall.

Jonassen, D. H., Wilson, B. G., Wang, S., & Grabinger, R. S. (1993). Constructivist uses of expert systems to support learning. *Journal of Computer-Based Instruction*, *20*(3), 86–94.

Juan, M. C., Furió, D., Alem, L., Ashworth, P., & Cano, J. (2011). ARGreenet and Basic-Greenet: Two mobile games for learning how to recycle. In *Proceedings of the 19th International Conference on Computer Graphics, Visualization and Computer Vision 2011* (pp. 25-32).

Kamarainen, A., Metcalf, S., Grotzer, T. A., Brimhall, C., & Dede, C. (2016). Atom tracker: Designing a mobile augmented reality experience. *International Journal of Designs for Learning*, *7*(2), 111–130. doi:10.14434/ijdl.v7i2.19430

Ke, F., & Hsu, Y. C. (2015). Mobile augmented-reality artifact creation as a component of mobile computer-supported collaborative learning. *The Internet and Higher Education, 26*, 33–41. doi:10.1016/j.iheduc.2015.04.003

Kerawalla, L., Luckin, R., Seljeflot, S., & Woolard, A. (2006). Making it real: Exploring the potential of augmented reality for teaching primary school science. *Virtual Reality (Waltham Cross), 10*(3-4), 163–174. doi:10.100710055-006-0036-4

Kesim, M., & Ozarslan, Y. (2012). Augmented reality in education: Current technologies and the potential for education. *Procedia: Social and Behavioral Sciences, 47*, 297–302. doi:10.1016/j.sbspro.2012.06.654

Kettle, K. L., & Häubl, G. (2010). Motivation by anticipation: Expecting rapid feedback enhances performance. *Psychological Science, 21*(4), 545–547. doi:10.1177/0956797610363541 PMID:20424099

Kirkley, B. S. E., & Kirkley, J. R. (2004). Creating next generation blended learning environments using mixed reality, video games and simulations. *TechTrends, 49*(3), 42–53. doi:10.1007/BF02763646

Klinger, M. B., & Coffman, T. (2011). Emphasizing diversity through 3D multi-user virtual worlds. In G. Kuruback & T. V. Yuzer (Eds.), *Handbook of research on transformative online education and liberation: Models for social equality*. Hershey, PA: IGI Global. doi:10.4018/978-1-60960-046-4.ch005

Klopfer, E., & Sheldon, J. (2010). Augmenting your own reality: Student authoring of science-based augmented reality games. *New Directions for Youth Development, 128*(128), 85–94. doi:10.1002/yd.378 PMID:21240956

Kommers, P. A. M., Jonassen, D. H., & Mayes, J. T. (1992). *Cognitive tools for learning*. Heidelberg, Germany: Springer. doi:10.1007/978-3-642-77222-1

Lam, R., & Muldner, K. (2017). Manipulating cognitive engagement in preparation-to-collaborate tasks and the effects on learning. *Learning and Instruction*.

Lave, J., & Wenger, E. (1998). *Situated learning: Legitimate peripheral participation*. Cambridge, UK: Cambridge University Press.

Le, Q. T., Pedro, A., Pham, H. C., & Park, C. S. (2016). A virtual world based construction defect game for interactive and experiential learning. *International Journal of Engineering Education, 32*(1B), 457–467.

Lee, K. (2012). The future of learning and training in augmented reality. *InSight: A Journal of Scholarly Teaching, 7*, 31-40.

Leinonen, T., Keune, A., Veermans, M., & Toikkanen, T. (2016). Mobile apps for reflection in learning: A design research in K-12 education. *British Journal of Educational Technology, 47*(1), 184–202. doi:10.1111/bjet.12224

Luckin, R. (2006). The learner centric ecology of resources: A framework for using technology to scaffold learning. *Computers & Education, 50*(2), 449–462. doi:10.1016/j.compedu.2007.09.018

Luckin, R., & Stanton, D. F. (2011). Limitless or pointless? An evaluation of augmented reality technology in the school and home. *International Journal of Technology Enhanced Learning, 3*(5), 510–524. doi:10.1504/IJTEL.2011.042102

Mathison, C., & Gabriel, K. (2012). Designing augmented reality experiences in authentic learning environments. In P. Resta (Ed.), *Proceedings of Society for Information Technology & Teacher Education International Conference 2012* (pp. 3470-3473). Chesapeake, VA: Association for the Advancement of Computing in Education (AACE).

Metcalf, S. J., Kamarainen, A. M., Grotzer, T. A., & Dede, C. J. (2013). Teacher perceptions of the practicality and effectiveness of immersive ecological simulations. *International Journal of Virtual and Personal Learning Environments*, *4*(3), 66–77. doi:10.4018/jvple.2013070105

Mitchell, R. (2011). Alien contact: Exploring teacher implementation of an augmented reality curricular unit. *Journal of Computers in Mathematics and Science Teaching*, *30*(3), 271–302.

Moyer, P. S., Bolyard, J. J., & Spikell, M. A. (2002). What are virtual manipulatives? *Teaching Children Mathematics*, *8*(6), 372–377.

Nesbit, J., Niu, H., & Liu, Q. (2019). Cognitive tools for scaffolding argumentation. In O. O. Adesope & A. G. Rud (Eds.), *Contemporary technologies in education: Maximizing student engagement, motivation and learning*. London, U.K.: Palgrave Macmillan. doi:10.1007/978-3-319-89680-9_6

Neville, M. (2010). Meaning making using new media: Learning by design case studies. *E-Learning and Digital Media*, *7*(3), 237–247. doi:10.2304/elea.2010.7.3.237

Nincarean, D., Alia, M. B., Halim, N. D. A., & Rahman, M. H. A. (2013). Mobile augmented reality: The potential for education. *Procedia: Social and Behavioral Sciences*, *103*(0), 657–664. doi:10.1016/j.sbspro.2013.10.385

Osberg, K. (1992). Virtual reality and education: A look at both sides of the sword. *HIT Lab*.

Overbay, A., Patterson, A. S., Vasu, E. S., & Grable, L. L. (2010). Constructivism and technology use: Findings from the impacting leadership project. *Educational Media International*, *47*(2), 103–120. doi:10.1080/09523987.2010.492675

Ozdamli, F., & Uzunboylu, H. (2015). M-learning adequacy and perceptions of students and teachers in secondary schools. *British Journal of Educational Technology*, *46*(1), 159–172. doi:10.1111/bjet.12136

Papert, S. (1990). Introduction by Seymour Papert. In I. Harel (Ed.), *Constructionist learning*. Boston: MIT Laboratory.

Papert, S., & Harel, I. (1991). *Constructionism*. Westport, CT: Ablex Publishing.

Phillips, D. (Ed.). (2000). *Constructivism in education*. Chicago: University of Chicago Press.

Piaget, J. (1977). *To understand is to invent: The future of education*. New York: Penguin.

Regenbrecht, H., Baratoff, G., & Wilke, W. (2005). Augmented reality projects in the automotive and aerospace industries. *IEEE Computer Graphics and Applications*, *25*(6), 48–56. doi:10.1109/MCG.2005.124 PMID:16315477

Resnick, L. B. (1989). Introduction. In L. B. Resnick (Ed.), *Knowing, learning, and instruction: Essays in honor of Robert Glaser* (pp. 1–24). Hillsdale, NJ: Erlbaum.

Ritchhart, R., Church, M., & Morrison, K. (2011). *Making thinking visible; How to promote engagement, understanding and independence for all learners*. San Francisco, CA: Wiley.

Saidin, N. F., Halim, N. D. A., & Yahaya, N. (2015). A review of research on augmented reality in education: Advantages and applications. *International Education Studies*, *8*(13), 1–8. doi:10.5539/ies.v8n13p1

Saitta, E. K. H., Gittings, M. J., & Geiger, C. (2011). Learning dimensional analysis through collaboratively working with manipulatives. *Journal of Chemical Education*, *88*(7), 910–915. doi:10.1021/ed100019f

Sang, G., Valcke, M., van Braak, J., & Tondeur, J. (2010). Student teachers' thinking processes and ICT integration: Predictors of prospective teaching behaviors with educational technology. *Computers & Education*, *54*(1), 103–112. doi:10.1016/j.compedu.2009.07.010

Sayed, E., Neven, A. M., Zayed, H. H., & Sharawy, M. I. (2011). ARSC: Augmented reality student card -An augmented reality solution for the education field. *Computers & Education*, *56*(4), 1045–1061. doi:10.1016/j.compedu.2010.10.019

Serafin, A. S., Adjorlu, N., Nilsson, A. N., Thomsen, L., & Nordahl, R. (2017). Considerations on the use of virtual and augmented reality technologies in music education. In *2017 IEEE Virtual Reality Workshop on K-12 Embodied Learning through Virtual Augmented Reality (KELVAR)* (pp. 1–4). 10.1109/KELVAR.2017.7961562

Shelton, B. E., & Hedley, N. R. (2004). Exploring a cognitive basis for learning spatial relationships with augmented reality. *Technology, Instruction. Cognition and Learning*, *1*(4), 323–357.

Siu, D. N. (2018). Google translate is using AI to power a robust offline mode. *Mashable*. Retrieved from https://mashable.com/2018/06/12/google-translate-offline-mode-ai/#abl8MTBhPsqg

Squire, K., & Klopfer, E. (2007). Augmented reality simulations on handheld computers. *Journal of the Learning Sciences*, *16*(3), 371–413. doi:10.1080/10508400701413435

Sung, H. Y., Hwang, G. J., & Chang, Y. C. (2016). Development of a mobile learning system based on a collaborative problem-posing strategy. *Interactive Learning Environments*, *24*(3), 456–471. doi:10.1080/10494820.2013.867889

Swan, P., & Marshall, L. (2010). Revisiting mathematics manipulative materials. *Australian Primary Mathematics Classroom*, *15*(2), 13–19.

Tarng, W., & Ou, K. L. (2012). A study of campus butter fly ecology learning system based on augmented reality and mobile learning. In *IEEE Seventh International Conference on Wireless, Mobile and Ubiquitous Technology in Education (WMUTE),* Takamatsu, Japan (pp. 62–66). IEEE. 10.1109/WMUTE.2012.17

Toh, C., Miller, S., & Simpson, T. (2015). The impact of virtual product dissection environments on student design learning and self-efficacy. *Journal of Engineering Design*, *26*(1–3), 48–73. doi:10.1080/09544828.2015.1018149

Toledo, J. A., Collazos, C. A., Cantero, M. O., & Redondo, M. A. (2018). Collaborative strategy with augmented reality for the development of algorithmic thinking. In V. Agredo-Delgado & P. H. Ruiz (Eds.), *Human-Computer Interaction: 4th Iberoamerican Workshop* (pp.70-82). New York: Springer.

Totten, S., Sills, T., Digby, A., & Russ, P. (1991). *Cooperative learning: A guide to research.* New York: Garland.

Vygotsky, L. S. (1978). *Mind in society*. Cambridge, MA: Harvard University Press.

Wang, X. (2012). Augmented reality: A new way of augmented learning. *eLearn, 10.*

Wenger, E. (1999). *Learning in doing: Social, cognitive and computational perspectives.* New York: Cambridge University Press.

Young, B., Ellobody, E., & Hu, T. W. (2012). 3D visualization of structures using finite-element analysis in teaching. *Journal of Professional Issues in Engineering Education and Practice, 138*(2), 131–138. doi:10.1061/(ASCE)EI.1943-5541.0000093

Yuen, S. C., Gallayanee, Y., & Johnson, E. (2011). Augmented Reality: An overview and five directions for AR in education. *Journal of Educational Technology Development and Exchange, 4*(1), 119–140. doi:10.18785/jetde.0401.10

ADDITIONAL READING

Caudell, T. P., & Mizell, D. W. (1992). Augmented reality: An application of heads-up display technology to manual manufacturing processes. In *Proceedings of the Twenty-fifth Hawaii International Conference on System Sciences,* Kauai, HI (pp. 659–669). IEEE Computer Society. 10.1109/HICSS.1992.183317

Chen, C., & Wang, C. H. (2015). Employing augmented reality embedded instruction to disperse the imparities of individual differences in earth science learning. *Journal of Science Education and Technology, 24*(6), 835–847. doi:10.100710956-015-9567-3

Derry, S. J., & Lajoie, S. P. (2013). A middle camp for (Un) intelligent instructional computing: An introduction. In *Computers as Cognitive Tools.* New York: Routledge.

Dörrenbächer, L., & Perels, F. (2016). Self-regulated learning profiles in college students: Their relationship to achievement, personality, and the effectiveness of an intervention to foster self-regulated learning. *Learning and Individual Differences, 51*, 229–241. doi:10.1016/j.lindif.2016.09.015

Huizenga, J., Admiraal, W., Akkerman, S., & Dam, G. T. (2009). Mobile game-based learning in secondary education: Engagement, motivation and learning in a mobile city game. *Journal of Computer Assisted Learning, 25*(4), 332–344. doi:10.1111/j.1365-2729.2009.00316.x

Johnson, L., Smith, R., Levine, A., & Haywood, K. (2010). *The 2010 Horizon report: Australia - New Zealand.* Austin, TX: T.N.M. Consortium.

Karamanoli, P., & Tsinakos, A. (2015). Use of augmented reality in terms of creativity in school learning. *Paper presented at the Make2Learn 2015 Workshop ICEC'15*, Trondheim, Norway.

Lin, C. Y., & Chang, Y. M. (2015). Interactive augmented reality using Scratch 2.0 to improve physical activities for children with developmental disabilities. *Research in Developmental Disabilities, 37*, 1–8. doi:10.1016/j.ridd.2014.10.016 PMID:25460214

Lin, T. J., Duh, H. B. L., Li, N., Wang, H. Y., & Tsai, C. C. (2013). An investigation of learners' collaborative knowledge construction performances and behavior patterns in an augmented reality simulation system. *Computers & Education, 68*, 314–321. doi:10.1016/j.compedu.2013.05.011

Zahn, C., Pea, R., Hesse, F. W., & Rosen, J. (2010). Comparing simple and advanced video tools as supports for complex collaborative design processes. *Journal of the Learning Sciences, 19*(3), 403–440. doi:10.1080/10508401003708399

KEY TERMS AND DEFINITIONS

Active Learning: An immersive experience whereby students engage in the process of learning by actively participating with the content, their peers, and the digital technology.

Complex Cognitive Processing: The conscious and mental capabilities that a person uses to think, reason, learn, understand, and remember.

Constructionism: Learning is an active process where learners use cognitive tools to create knowledge structures through the act of making artifacts.

Constructivism: Meaningful learning that allows students to construct knowledge and from these experiences connect to new understandings of the world around them; provides context to their learning.

Knowledge Construction: An active process that involves the learner personally constructing knowledge from prior knowledge and gaining new understanding from experience either individually or socially.

Multilayered Content: Virtual information that is overlaid on a physical object extending that object to provide an immersive experience.

Personalize Learning: Providing appropriate instructional tools, activities, pacing, and content that are meaningful and relevant to learner needs and interests.

Social Constructivism: Knowledge construction that occurs by interacting and engaging with others.

Trigger: Marker or a sensor that provides a visual cue to computer generated information over a real-world object.

Chapter 5
Strategies for Developing Mobile Location-Based Learning Activities by Teachers

Hagit Meishar Tal
Holon Institute of Technology, Israel

ABSTRACT

This chapter analyses the educational activities created by teachers on an online game generator, Treasure-HIT. The analysis focused on two main components that exist in every game: (1) the location clue, which leads the players to the different stations of the game, and (2) the station task, which the players have to perform when they reach the right station. This qualitative research was based on content analysis of the activities accumulated on the system's server. A total of 112 games were analyzed, including 1272 clues, 1005 stations and 1862 station tasks. Five types of location-dependent tasks were identified: identifying information found on site, camera documentation, receiving an answer from someone on site, physical activities dependent on the location, measurements using a mobile phone, and leaving a personal mark.

DOI: 10.4018/978-1-5225-8106-2.ch005

INTRODUCTION

In recent years, smartphones have become popular learning tools (Prensky, 2005). The mobility of wireless devices produces new learning opportunities. The main one is "location-based learning", namely, learning depends on the place and context in which the learner is situated. In our current study we utilized Treasure-HIT, a free platform for teachers to create treasure hunts that use smartphones in order to locate stations and manage learning activities in each station (Kohen-Vacs, Ronen, & Cohen, 2012). Users can develop location- based learning activities using Treasure-HIT. The aim of this study was to examine what strategies the teachers developed in the creation of mobile location-based learning activities on this platform.

Location-Based Mobile Learning

The concept of mobile learning has become a term widely used by educators who use mobile phones in teaching and learning. This concept has several definitions and each emphasizes different options for learning with these devices. First and foremost, mobile learning makes use of mobile devices such as phones, tablets, smart watches, etc. (Kukulska-Hulme & Traxler, 2005). According to this definition, mobile learning can also occur as part of the regular classroom where cell phones are used to increase interaction and encourage active learning in the classroom (Traxler, 2007; Zadok & Meishar-Tal, 2015).

Another definition of mobile learning is that the learning takes place when the learner is in on the move (Sharples, Taylor, & Vavoula 2010; Lee et al. 2005). Mobile devices are light, easy to carry, require no wiring and allow an Internet connection anywhere, anytime, thus enabling user mobility in space. This allows expansion of learning activities beyond the boundaries of the traditional classroom (Giemza, Verheyen, & Hoppe, 2012). Learners can collect and analyze information using a mobile device. Learning can occur in diverse learning spaces such as museums (Sung et al., 2010), urban areas (Morrison et al., 2009), archaeological sites (Costabile et al., 2008), and outdoors (Medzini, Meishar-Tal, & Sneh 2014). This is authentic learning (Rieger & Gay, 1997) that generates motivation and pleasure (Jones et al, 2006; Meishar-Tal & Ronen, 2016).

Getting out of the classroom is an opportunity to implement principles of experiential and contextual learning (Kolb, 2014; Sharples, Taylor, & Vavoula 2010). Mobile technology to be exploited optimally to the requirements of out-of -class learning should focus on direct interaction of learners with the environment. The physical place of the activity should be closely tied to the activity itself (Medzini, Meishar-Tal, & Sneh, 2013).

The mobile device could be used to enhance environmental awareness (FitzGerald, 2012). The use of built-in mobile applications such as camera, video or audio recorders enable documentation and collection of situated information via the smartphone. Using other applications that use smartphone sensors, such as a noise meter, a compass etc., can be used for measuring physical phenomena on the spot (Meishar-Tal & Gross, 2014).

One way to integrate mobile technologies into out-of- class learning is by developing location-based games (Avouris and Yiannoutsou, 2012). These are typically multi-player games played out on city streets and in built up urban environments. Some of these mobile games transcend place and time and can be played in many diverse places and extended to long periods of time, while others are designed to be event-based, to be played in specific places at specific times, like during visits in museums and other non-traditional game venues.

Location-based mobile learning games bring together four of the most popular and current eLearning trends: mobile learning, digital storytelling, gamification, and location-based learning (Johnson, Adams, Becker, Estrada, & Freeman, 2015). Avouris and Yiannoutsou (2012) identify three categories of mobile location based games: "ludic tradition" games, which their main purpose is enjoyment, including treasure hunts, action games and role playing games, "pedagogic tradition" games, which there main purpose is education and learning, including participatory simulators, situated language learning and educational action games, and "hybrid tradition", including museum games and mobile fiction. All games can be characterized by three types of interactions they generate: interaction between users (collaboration/competition), interaction with physical objects, and interaction with the game engine. Although Avouris and Yiannoutsou (2012) classify treasure hunt games as "ludic tradition" games, there are several examples in the literature for using treasure hunt games in educational contexts and for pedagogical purposes (Lai et al., 2013; Wu et al., 2010, Edmonds & Smith, 2017). The general framework of the treasure hunt game is that the players have to go through a set of stations by solving clues indicating their location and performing activities like riddle solving or physical activities that involve interaction with the environment and with the game engine. The game has a competitive element that contributes to motivation and enjoyment of learning (Ihamäki, 2014). Nevertheless, Meishar-Tal and Ronen (2016) and Nadarajah et al. (2017) found that in games organized as competition between groups the player who holds the mobile device, whether it is a smartphone or an iPad, enjoys the game more than the other players.

Studies show that such games also improve the participants' spatial knowledge and navigation skills (Winter et al., 2011). In a research conducted by Nadarajah et al. (2017) two groups of participants were compared, one navigating by an iPad based "riddle solving system" and the other by a 2D map with GPS. The findings show that in the groups that used the iPad riddle solving system participants paid more attention to the environment when solving riddles than when using a map.

Research focused on the "QRace" generator, a treasure-hunt based on QR codes, showed that only about 20% of the games created on this platform were location-based. In most cases the tasks developed in the games were not based on information located where the game was taking place, neither did they require information related to it. They used the game platform only as tool for creating playful competitive games. (Israel et al., 2014), In this manner, it could be regarded as "ludic" use and not "pedagogic" use according to Avouris and Yiannoutsou (2012) terminology. These findings indicate that the pedagogical challenge in creating location-based learning activities is complex and requires teacher training, guidance and pedagogical support to improve their overall design.

The Treasure-Hit Platform

Treasure-HIT was developed as an environment that enables teachers to create treasure hunt games that can be conducted both indoors and outdoors. The rules of a Treasure-HIT are very similar to those of the traditional, non-technological game. Each game is composed of stations where the players are presented with one or more tasks. The tasks require exploration of the environment in order to achieve place awareness and contextual learning (Kohen-Vacs, Ronen & Cohen, 2012). Completion of all tasks at a station is required in order to proceed to the next station. By playing with Treasure-HIT, the clues which lead the players through the different stations on their way to the treasure are presented on their mobile device, and the verification of location is also managed by the application, either by QR codes or by GPS. Nevertheless, similar to the traditional version of the game, Treasure-HIT requires physical presence of the players at the different locations in search for the treasure (Cohen, 2015).

The goal of this game, beyond enjoyment, is to create location-based learning opportunities. During the game students examine objects in the physical environment and delve into details and aspects related thereto. The game can be conducted as a competition between groups or individuals, or as a personal challenge.

The Treasure-HIT platform can be used by teachers in many disciplines, such as history, art, geography, biology, archeology or architecture. A game created by Treasure-HIT is considered successful if the game is challenging and sparks curiosity, and is also relevant to the place where it is played.

RESEARCH QUESTIONS AND METHODOLOGY

The study was designed to examine the characteristics of activities developed by teachers on the Treasure Hit platform in 2015 and to identify strategies they developed in this environment in order to implement the principle of location-based learning. The main research question was: To what extent do the Treasure-HIT games developed by teachers use the principles of location-based learning?

The research focused on two main components that exist in every game: (1) The location clue - which leads the players to the different stations of the game. (2) The station task- which the players have to perform when they reach the right station. Accordingly, two specific questions were defined:

1. What location-based learning strategies did teachers develop for providing location clues?
2. What location-based learning strategies did teachers develop to set station tasks?

This qualitative research was based on content analysis of the activities accumulated on the system's server. In the first phase of the research, all the activities created by teachers during 2015 were filtered from the system's database. Then, in-depth content analysis and coding of activities in each game relating to the location clues and the station tasks were conducted. The location clues were examined to identify location-based strategies used in the games (Figure 1). The station tasks were analyzed to identify whether the tasks were location dependent, and which location-based strategies had been used. The findings were quantified to identify the frequency of each strategy identified.A total of 112 games were analyzed, including 1272 clues, 1005 stations and 1862 station tasks.

FINDINGS

This chapter is divided into two parts: findings regarding location clues and findings regarding location-based tasks.

Location-Dependent Clues

The total number of clues stored in the system was 1272 in 1005 stations. More than half of the clues were location-dependent (63.36%). The remaining clues were lacking any connection to a specific place (For example: "Move to the next station").

Table 1. Strategies for location-based clues

Clue Type	Example
Riddle	[Benjamin well] Only after Baron Rothschild saw us suffering carrying water, did he donate money for the construction of the well. Who am I? Come to the site and answer the questions.
Indirect address	[Library] go to Osnat's work place
Image	See Figure 1 Please reach the place in the photo
Direct address	Please come to the center of town near the bank
Direction Guidance	You must get to "Aloni Isaac" clearing (Amikam intersection, turn left, turn right onto a dirt path leading to Yitzhak Aloni grove. Follow the arrows and flags to the clearing where the next task awaits you.
Map Reading	See Figure 2
Using multimedia	Use the following video clip to identify the next station

Figure 1. Using an image as a clue

Strategies for Providing a Location-Based Clue

Seven types of clues were identified: a riddle (36%), an indirect address (24%), an image (22.8%), a direct address (11.2%), direction guidance (4%), reading a map (1%), and use of multimedia (0.3%). In 43% of the cases there was a combination of more than one strategy.

Location-Based Tasks

The total number of tasks stored in the system in 2015 was 1862 in 1005 stations. 56% of the tasks were not location-based, i.e., tasks that were not related to the place where there were performed. These tasks focused on subject matter such as mathematics or other fields of knowledge (Table 2).

Among the tasks, 18% were location related, in other words, the questions required knowledge that related to the station but the answer could be given without physical presence on site (Table 3).

Only 26% of the tasks were location-dependent, in the sense that in order to complete the task successfully the players had to be in the specific place where the station was located (Table 4).

Five types of location-dependent tasks were identified (Figure 3): identifying information found on site (71%), using camera documentation (14%), receiving an answer from someone on site (11%), physical activities dependent on the location (2%), measurements using a mobile phone (2%), and leaving a personal mark (0.5%)

Table 4 presents some examples for each type of task identified.

Figure 2. Map reading

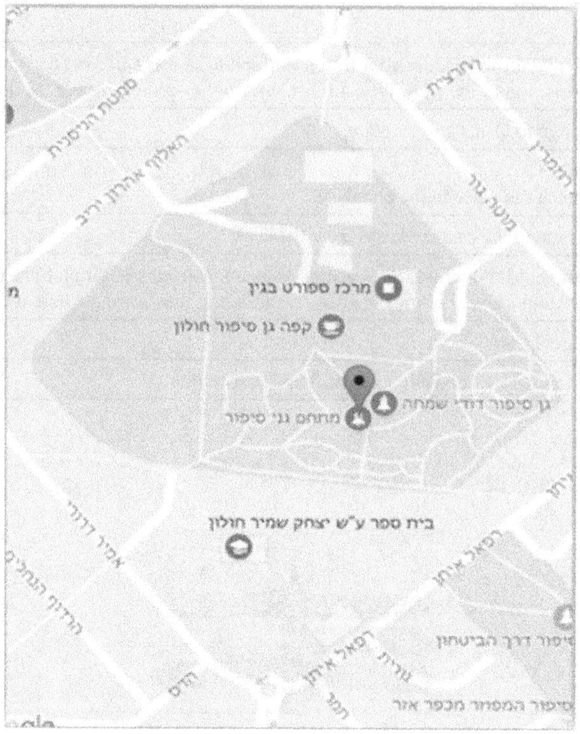

Table 2. Examples of non-location-based tasks

Non-Location-Based Tasks	
Focus on learning content	[#6 station] Where do you usually see flooring? More than 1 answer is allowed ☐ everywhere ☐ at home ☐ at school ☐ in the kitchen ☐ in the street
	[new teachers' lounge] Cellular respiration is carried out only in single-cell organisms. ☐ Yes ☐ No

The vast variety of examples presented here show that developing a location-based task can be creative and fun. These examples can serve as an inspiration for teachers to develop their own activities, adapted to their students' needs.

DISCUSSION

The aim of this study was to examine strategies for developing location-based learning activities on the Treasure-HIT platform. The research revealed that teachers who used the Treasure-HIT platform

Table 3. Examples of location-related learning activities

Location Related Tasks	Examples
Based on previous knowledge or online information related to the location	[The Menorah station] What is the name of the candle that lights the Menorah? ☐ The lighting candles ☐ The main candle ☐ The shamash ☐ The king of lights
	[The algae puddle] What is the best temperature for the development of algae? ☐ 36 degrees ☐ 50 degrees ☐ 26 degrees ☐ 24 degrees
	[Caesarea Amphitheatre] Which singer appeared in the Caesarea Amphitheatre? Dudu Aharon

Figure 3. Five types of location-based tasks

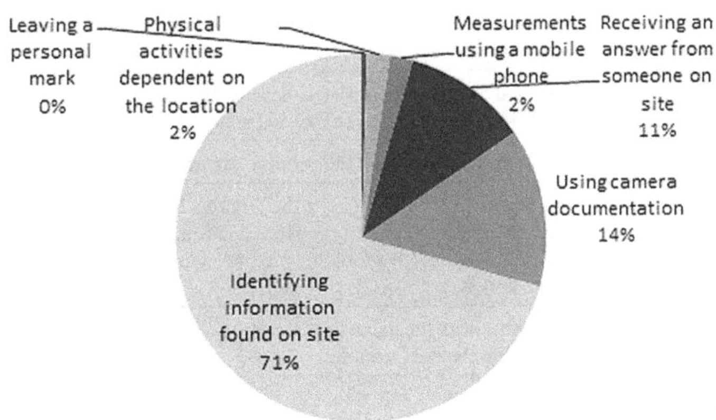

for developing mobile location-based games used original and creative strategies in the development process. Nevertheless, the findings show that most of the teachers used the Treasure-HIT platform to create non-location-based games.

Following the Avouris and Yiannoutsou (2012) terminology the findings show that most educational activities developed in this environment lack the interaction with physical objects that require location-based awareness and thereby should be considered as non-location-based. This finding supports previous findings for similar systems (Israel et al., 2015. Teachers often use the system to create a playful and fun competitive learning experience but do not exploit the full potential of the tool to create an interaction between the learners and their environment, in order to raise awareness of place and explore spaces around us with the means that smartphones provide. This finding could imply that major number of teachers lack the required knowledge for developing location-based activities. In the era of mobile and contextual learning (Sharples, Taylor, & Vavoula 2010; Lee et al. 2005), the educational system should train teachers to design and develop high- quality location-based activities.

However, the games created by teachers that could not be considered as location-based activities were also created for worthy learning purposes. Therefore, it could be argued that treasure hunts, like mobile games, could not be classified as "ludic" as stated by Avouris and Yiannoutsou (2012). The

Table 4. Examples of location-based tasks

Location-Based Tasks	Examples
Identifying information found on site	[The fountain] How many cycles are on the fountain? ☐ 1 ☐ 2 ☐ 3 ☐ 4
	[Meteorological station] What tools of weather measurement does this station have? ☐ rain gauge, wind sock ☐ rain gauge, humidity meter, temperature gauge ☐ rain gauge, Globisens, vane
	To strengthen their spirit, Israeli government invited the children of Shaar HaGolan to light the torch of independence. Look at the walls and answer the question: When did this event take place?
Receiving an answer from someone in place	Insert the code that the station manager gave you: 777777
Using camera documentation	End point! Take a Selfie photo near the bougainvillea plant and show it to the teacher to get the code You must take a picture one of the children petting a cow. To continue write the word: Photographed
Physical activities dependent on location	[Basketball Station] Shoot five times into the basketball hoop. If you succeed you will receive the code that will lead you to the next station.
Measurements using a mobile phone	Measure the azimuth from the cellular antenna to the playground sign
Leaving a personal mark	In this station you have to create the Dimona Municipality logo using natural materials you find on site. Take a photo of the piece and send it to the phone number 054-5555555. Write down the following "We took the picture" and sent it to receive the code that will lead you to the next station.
	Write with stones the name of your group and send the photograph to the teacher to get a code to proceed to the next station.

teachers used the generator in a flexible pedagogical manner: as a generator for fun and enjoyable learning games, without consideration of location-based awareness, as location related games, though not location dependent and in the highest level, as location-based learning games. All three options can be regarded as pedagogical uses.

Looking carefully into the location-based games created in the system, the current research identified innovative and original strategies developed by some teachers to create interesting location-based activities. These strategies can provide a model that may be emulated and learned by other teachers.

In terms of supplying clues, that direct the players to the next station, the use of direct and indirect referrals characterized most tasks. The more rarely used strategies are more original and these include use of map-reading and use of multimedia. Map-reading as part of the location-based activity is valuable primarily because this is an era in which navigation applications are replacing map-reading and impairing independent spatial orientation (Twietmeyer, 2014). Integrating map-reading in the clues is good utilization of an opportunity to practice map-reading as a mean of orientation. Another example of wise use of mobile device capabilities in clues is the utilization of multimedia, such as short videos that the teachers created and used as a mean of referrals to stations. Such use contributes to a rich and varied learning experience.

We also identified new and original strategies of location-based activities in the tasks developed by users. In addition to asking questions related to the place where the task is performed, teachers developed some interesting ways to facilitate interaction between the learner and the environment. This may include taking advantage of the built-in camera in mobile phones or by communication with people on the spot. The rarer and more original activities were the most interesting ones. Teachers used tools that mobile phones are equipped with to enhance the activity experience on the site. Means of measurement such as the GPS, the compass, and various applications for measuring distance, temperature, etc., had been utilized constructively. Integrating these tools into the location-based activity enriched the learning experience, developed additional e-skills and contributed to raising the learner's awareness of the environment.

The Treasure-HIT environment provides a location-based learning opportunity, but like any other technological tool, the opportunity should be used wisely. Analyzing the Treasure-HIT clues and activities shows that the teachers developed different types of games for learning, ranging from non-location-based activities to interesting and original strategies for location-based activities. Presentation of innovative and unique examples of the use of the environment for location-based learning, as shown in this paper, may contribute to the expansion of knowledge necessary for teachers to develop future games in Treasure HIT or similar location-based game generators.

REFERENCES

Avouris, N. M., & Yiannoutsou, N. (2012). A Review of Mobile Location-based Games for Learning across Physical and Virtual Spaces. *Journal of Universal Computer Science*, *18*, 2120–2142.

Costabile, M. F., De Angeli, A., Lanzilotti, R., Ardito, C., Buono, P., & Pederson, T. (2008, April). Explore! Possibilities and challenges of mobile learning. In *Proceedings of the SIGCHI Conference on Human Factors in Computing Systems*, Florence, Italy (pp. 145-154). ACM.

Edmonds, R., & Smith, S. (2017). From playing to designing: Enhancing educational experiences with location-based mobile learning games. *Australasian Journal of Educational Technology*, *33*(6).

FitzGerald, E. (2012), Creating user-generated content for location-based learning: An authoring framework. *Journal of Computer Assisted Learning*, *28*(3), 195–207.

Giemza, A., Verheyen, P., & Hoppe, H. U. (2012, March). Challenges in scaling mobile learning applications: The example of quizzer. In *2012 IEEE Seventh International Conference on Wireless, Mobile and Ubiquitous Technology in Education* (pp. 287-291). 10.1109/WMUTE.2012.69

Ihamäki, P. (2014). The potential of treasure hunt games to generate positive emotions in learners: Experiencing local geography and history using GPS devices. *Int. J. Technology Enhanced Learning*, *6*(1), 5–20. doi:10.1504/IJTEL.2014.060025

Israel, R., Ronen, M., Kalderon, S., & Shoham, H. (2014) Are we ready for this QRace? In Y. Eshet, I. Blau, A. Caspi et al. (Eds.), *Proceeding of the 10th Chais Conference for Research in Innovation and Learning Technologies,* The Open University of Israel, Raanana (pp. 56-63).

Johnson, L., Adams Becker, S., Estrada, V., & Freeman, A. (2015). *NMC Horizon Report: 2015 Higher Education Edition*. Austin, TX: The New Media Consortium.

Jones, A., Issroff, K., Scanlon, E., Clough, G., & McAndrew, P. (2006). Using mobile devices for learning in informal settings: Is it motivating? In *IADIS International Conference on Mobile Learning*, Dublin (pp. 251–255). IADIS Press.

Kohen-Vacs, D., Ronen, M., & Cohen, S. (2012). Mobile treasure hunt games for outdoor learning. *Bulletin of the IEEE Technical Committee on Learning Technology*, *14*(4), 24–26.

Kolb, D. A. (2014). *Experiential learning: Experience as the source of learning and development*. FT Press.

Lee, I., Yamada, T., Shimizu, Y., Shinohara, M., & Hada, Y. (2005). In search of the mobile learning paradigm as we are going nomadic. In P. Kommers & G. Richards (Eds.), *Proceedings of EdMedia: World Conference on Educational Media and Technology 2005* (pp. 2888-2893). Association for the Advancement of Computing in Education (AACE). Retrieved from http://www.editlib.org/p/20521

Liu, T. C., Wang, H. Y., Liang, J. K., Chan, T. W., Ko, H. W., & Yang, J. C. (2003). Wireless and mobile technologies to enhance teaching and learning. *Journal of Computer Assisted Learning*, *19*(3), 371–382. doi:10.1046/j.0266-4909.2003.00038.x

Meishar-Tal., Sneh, Y., & Medzini, A. (2014). Designing mobile learning activities for outdoor learning. In *First International Conference on the use of iPads in Higher Education 2014*.

Meishar-Tal, H., & Gross, M. (2014). Teaching sustainability via smartphone-enhanced experiential learning in a botanical garden. *International Journal of Interactive Mobile Technologies*, *8*(1), 10–15. doi:10.3991/ijim.v8i1.3441

Meishar-Tal, H., & Ronen, M. (2016). Who Enjoys More? Factors Related to Enjoyment in an Educational Mobile Game. In *European Conference on Games Based Learning* (p. 442). Academic Conferences International Limited.

Nadarajah, S. G., Overgaard, B. N., Pedersen, P. W., Schnatterbeck, C. G., & Rehm, M. (2017). Gamifying Navigation in Location-Based Applications. EAI Endorsed Trans. *Creative Technologies*, *4*, e2.

Prensky, M. (2005). What can you learn from a cell phone? Almost anything. *Innovate (North Miami Beach, Fla.)*, *1*(5), 1–8.

Rieger, R., & Gay, G. (1997). Using mobile computing to enhance field study. In R. Hall, N. Miyake, & N. Enyedy (Eds.), *Computer Supported Collaborative Learning Conference (CSCL '97)* (pp. 215-223). Charlottesville, VA: AACE. 10.3115/1599773.1599800

Sharples, M., Taylor, J., & Vavoula, G. (2010). A theory of learning for the mobile age. In B. Bachmair (Ed.), *Medienbildung in neuen Kulturräumen* (pp. 87–99). Wiesbaden: VS Verlag für Sozialwissenschaften. doi:10.1007/978-3-531-92133-4_6

Traxler, J. (2007). Defining, discussing and evaluating mobile learning: The moving finger writes and having writ..... *The International Review of Research in Open and Distributed Learning*, *8*(2). doi:10.19173/irrodl.v8i2.346

Twietmeyer, M. (2014). *GPS: Is it affecting our sense of place? A case study* (Doctoral dissertation). California State University, Northridge, CA.

Winter, S., Richter, K. F., Baldwin, T., Cavedon, L., Stirling, L., Duckham, M., ... Rajabifard, A. (2011). Location-based mobile games for spatial knowledge acquisition. *Cognitive Engineering for Mobile GIS*, 1-8.

Wu, S., Chang, A., Chang, M., Yen, Y. R., & Heh, J. S. (2010, April). Learning historical and cultural contents via mobile treasure hunting in the five-harbor district of Tainan, Taiwan. In *2010 6th IEEE International Conference on Wireless, Mobile and Ubiquitous Technologies in Education (WMUTE)* (pp. 213-215). IEEE.

Zadok, Y., & Tal, H. M. (2015). Engaging students in class through mobile technologies–Implications for the learning process and student satisfaction. *Research Highlights in Technology and Teacher Education, 2015*, 105.

Chapter 6
Students as Co-Creators of a Mobile App to Enhance Learning and Teaching in HE

Julie Prescott
https://orcid.org/0000-0001-8612-2495
University of Bolton, UK

Pippa Iliff
University of Bolton, UK

Daniel J. T. Edmondson
University of Bolton, UK

Duncan Cross
https://orcid.org/0000-0001-8558-2262
University of Bolton, UK

ABSTRACT

This chapter considers how technology can enhance teaching and learning through technology within the higher education setting. The chapter considers findings and draws conclusions from a recent project conducted by the authors involved in this technologically innovative project. In particular this chapter aims to consider how mobile apps can be beneficial for learning and education purposes, consider the pedagogic value of technology use within higher education (HE), look at how students can be co-creators in their own learning, and provide an overview of the mobile app and the research findings that it was developed from. The chapter will also take into consideration the challenges associated with developing and implementing a technological intervention in HE.

DOI: 10.4018/978-1-5225-8106-2.ch006

INTRODUCTION

This chapter will draw on findings from a recent HEFCE catalyst funded research project 'Student Led Technology Enhanced teaching and learning'. This project successfully aimed to engage undergraduate students in researching their own student body to discover the needs of the student community to design and develop two technology-based teaching and learning tools, as well as empower them to train and mentor staff in the use of the interventions. This was achieved through an action research approach, which was more of a participatory action research approach as it was collaborative in nature and sought to improve the quality of the learner's experience (Cohen, Mannion and Morrison, 2018). The research designed two interventions. A university-wide mobile app and a discipline-specific virtual reality therapeutic game. In light of the book's aims and focus, this chapter will focus on the mobile app intervention. The chapter will provide a review of the current literature in the area of technology use in teaching and learning in Higher Education with a focus on mobile apps and students as creators in their own learning, the chapter will also discuss results from the research. In particular, this chapter will examine student's currents use of technology for learning and attitudes towards new technology with a view of using the information to create an application/game that can be embedded into a higher education establishment. These student views come from the context of a small university in the North West of England. The project results will aim to take into account all aspects that can enhance student engagement and experience.

Following the literature, this chapter will consider a technological intervention designed and developed by, and for students from prior research findings, a university-wide mobile app, uBolt. The chapter will also consider the challenges encountered in the development and implementation of the uBolt app as well as the benefits and challenges associated with students as co-creators.

MOBILE APPS FOR LEARNING AND EDUCATION PURPOSES

Mobile devices are becoming increasingly popular with research indicating that Smartphone ownership now exceeds laptop ownership (Zydney & Warner, 2016). New updated versions of these devices with their innovative features make them more convenient and affordable for users (Zydney & Warner, 2016). Apps are constantly designed to make our lives easier with unprecedented opportunities for communicating, interacting and sharing of content (Cook & Pachler, 2012).

West (2013) suggests that the younger generation is accustomed to personalised content and synchronous communication. The Millennial generation born between 1977 and 1995 are said to have adopted mobile devices to the centre of their everyday lives, with 97% owning a mobile device (Nielsen Reports, 2016). Similarly, generation Z (Gen Z), people born after 1995, are suggested to be the most digitally oriented generation to date and have adopted more immersive digital behaviours at the pace of technological advancement (Boruff & Storie, 2014).

The introduction of mobile devices has had a positive impact in various fields, one, in particular, being health (M-Health) (Li et al., 2018) and the use of apps for therapeutic purposes (Rathbone & Prescott, 2017; Rathbone, Clarry & Prescott, 2017). The education sector is also witnessing a rise in the use of technology and mobile apps. The New Media Consortiums 2013 Horizon Report predicted that mobile applications and tablet computing would have a time-to-adoption of less than one year in higher education (Johnson et al., 2018). The increasing ownership of smart devices by students and their integral part of their educational experience is forcing educators to adapt their teaching style and utilize these technologies as tools for teaching and learning, due to their potential in enhancing student engagement (Yu-Chang & Yu-Hui, 2013). Due to this rise is ownership a significant number of Universities now use mobile technologies or create mobile-optimized versions of their websites that can be downloaded from mobile application stores. These mobile technologies play a vital role in University life (Mansour, 2016).

The availability of mobile apps for educational use, the reduction of cost of using mobile networks to go online, and the increase of mobile bandwidth capacity are becoming the enablers for the era of the mobile learning revolution (Kim et al. 2013). Portable devices like Smartphone's, tablets, and e-book readers are powerful tools for students' learning. Thus, the concept of "mobile learning" (m-learning) has been established to link up "mobile" and "learning", joining two formerly separate concepts (El-Hussein & Cronje, 2010).

M-learning (mobile learning) comprises three elements: mobility of technology, mobility of learner and mobility of the learning processes (El-Hussein & Cronje, 2010). All of these elements have been shown to contribute to the effective delivery of learning experiences (El-Hussein & Cronje, 2010). Due to the growing popularity of m-learning, research into the positive characteristics in terms of learning is being carried out (Aharony, 2014). Mobile devices facilitate more effective communication between staff and students and access to course material is easily accessible, and communication is not restricted to office hours or a specific location (Agawal & Karahanna, 2000; Shen, Wang & Pan, 2008). This has been evidenced in research from Richardson & Lenarcic (2008) that was conducted within a university to examine the use of mobile technology to enhance students' ability to access both general university information and course material. The results evidenced that the use of mobile technology improved students' experience, by "increasing the effectiveness of 'student to student', 'student to staff', 'staff to student' and 'university to student' communication".

Furthermore, the continued surge in ownership of mobile devices has given rise to a new range of teaching resources that facilitate the flow of information and advance the developments of applications for the diverse student population. The emergence of mobile technologies has increased the development of apps and software that supports m-learning. Over 180 million apps have been downloaded from the Apple store alone, with the educational sector being the third most popular category after games and business (Apple Store Downloads, 2018).

Research also suggests that perceived control and subjective norms drove the adoption of mobile technology amongst students and that the influence of peers embracing m-learning increased students ability and confidence (Yeap, Ramayah, and Soto-Acosta, 2016). Research on student's use of apps has found that perceived usefulness and the need for an element of fun or positive emotions attached with the use of the app are key elements that drive students' intentions to use mobile apps in their studies (Carter & Yeo, 2017; Ding & Chai, 2015). Past research has also evidenced that students who use mobile technologies alongside their studies are more likely to complete their undergraduate course and continue on to further study (Russell-Bennett, Rundle-Thiele, & Kuhn, 2010). Prior research (Nevin, 2009) also shows that the use of mobile apps in learning and teaching can help to engage students in lectures, and apps have significant impacts on the improvement of learning by students.

Traditional classroom settings are being transformed, becoming more interactive, whilst promoting learning outside of the traditional classroom environment and allowing learning to occur holistically (Huang, Lin & Cheng, 2010). Educators are also able to connect with students on a more personal level as these are devices that students use in their daily lives (Ward, Finley, Keil & Clay, 2013). Research on the use of Smartphone technologies and mobile applications have been examined in their ability for educators to set weekly quizzes and it was found that this increased student participation and encouraged continuous learning (Mercedes & Radel, 2016). Revealing that perceived convenience, flexibility, engagement and interactivity are all factors that encourage and promote mobile learning within students (Chen & Denoyelles, 2013).

Similarly to mobile applications for learning, research by Connelly et al. (2012) found that playing computer games was linked to positive; knowledge acquisition, perceptual and cognitive skills, behavioural change, affective and motivational outcomes, and physiological outcomes. Mobile gaming apps have appealing designs which are attractive to learners and therefore heighten the replay value which translates into better learning outcomes. The interactivity of mobile apps increases the learner participation and motivation, resulting in more positive learning outcomes (Facer et al, 2004). Trends in the literature on mobile apps for learning have reported that mobile learning is motivating for students and has a positive effect on learning outcomes (Hsu & Ching, 2013; Hwang & Wu, 2014). Studies have evidenced positive improvement in students understanding plus application, of theories and concepts (Hsu & Ching, 2013).

However, it has been argued that there is not sufficient evidence regarding mobile games improving learning outcomes. This is due to the lack of studies which have looked into the design or specific features of mobile applications that contribute to their suggested effectiveness (Schmitz et al. 2012). Studies have also been criticized for not thoroughly examining the specific theoretical foundations underlying mobile learning research (Cheung & Hew, 2009). Although students' attitudes towards mobile devices for learning have been found to be generally positive, this does not imply sustainability (Kenny et al. 2012). An eight-week study showed that engagement was high and maintained at the beginning, concluding that sustainability is difficult if students are disengaged (Li & Wang, 2012).

The rapid growth of m-learning apps means that the industry is highly competitive and M-learners are presented with a wide variety of choice and learning options, many of which are free. One outcome of this development of m-learning apps is that those that require payment to download or involve in-app purchases are much less likely to succeed than free ones (Statista, 2015).

Therefore, an understanding of the key elements which affect an m-learners' decisions to use/purchase paid m-learning apps is vital form-learning app developers. Successful m-learning apps need to have an understanding of the strengths and weaknesses of mobile technology, whilst at the same time understanding the needs of the m-learner from their perspective.

The Pedagogic Value of Technology in HE

There is the implication that competence in living technologies will mean competence in learning technologies. However, students differ in their ability, access and the application of technology (Corrin, Lockyer and Bennett, 2010). A number of researchers challenge the idea of a homogeneous technological generation especially in regards to how students use technology as part of their academic studies (Kennedy et al., 2008; Corrin, Lockyer and Bennett, 2010). According to quantitative research by Corrin, Lockyer and Bennett (2010) of Australian undergraduate students across a number of facilities; students tend to use technology for their academic life in terms of; accessing information online, using email and using a computer to create presentations. The study found the frequency of all the technology investigated in the study was higher in everyday life than in academic study. The authors conclude that students do not, therefore, have the ability, or desire, to transfer their aptitude of technology in everyday life to an academic setting. Indeed, Prescott et al. (2013a) posit that Facebook has educational potential and that today's learners should be encouraged and supported to use Web 2.0 technologies for learning in formal education. However, there is growing evidence that the value of Facebook in the educational context does not relate to formal learning at all, but to the 'social networking' elements of the site (Prescott et al., 2013b) as well as having implications for online professionals (Prescott et al. 2013c).

There is a paucity of research in the area of student use of technology for learning, and further research is needed to investigate how and why students use technology in the way(s) they do. There is also a difficulty in defining 'educational use'. Corrin, Lockyer and Bennett (2010) suggest that further in-depth research, aside from survey-based research methods, are needed to see how students use technology to suit their lives and how technology can effectively support teaching and learning. Hosein, Ramanau and Jones (2010) suggest that technology needs to be seamlessly integrated into pedagogical environments and viewed by Learners as a hybrid between recreational and educational tools that can be used successfully as a learning technology. Interestingly, the study also found that older students were able to close the digital divide with the Net Generation students as younger students are not able to transfer their skills automatically to studying and learning (Hosein, Ramanau and Jones, 2010; Corrin, Lockyer and Bennett, 2010; Kirkwood and Price, 2005). Prescott (2014, 2016) investigated university staff use of and attitudes towards Facebook as an educational tool. In a study of how teaching staff at one UK university use Facebook and their attitudes towards Facebook and online professionalism, Prescott (2014, 2016) further considered how attitudes towards the use of Facebook as an academic tool differed between teaching staff with a teacher-centred style and teaching staff with a learning-centred style. Differences in attitudes were found, which indicates those with a teacher-centred style do not view online and offline identities blurring as much.

A significant number of Universities are now using mobile technologies and have created mobile-optimised versions of their websites, which are playing a vital role in the academic life of students (Mansour, 2016). Advances in the technological design of smartphones have led to the potential for innovative use in education. In addition to this, the popularity of devices such as smartphones and iPhone makes mobile application on an easy to use option (Yu-Chang & Yu-Hui, 2013).

Educational computer games have become a valuable teaching tool due to their effectiveness in increasing student motivation. Research has shown that games are most effective for learning when they are not only educational but fun to play (Wilson et al., 2017). Students who are viewed to be intrinsically motivated often engage in academic behaviours that lead to more successful learning outcomes (Sampayo-Vargas et al., 2013). Research has indicated that computer games are found to be significantly more effective in promoting motivation. Alongside this, students who played multiplayer games indicated the following motivations for playing an educational game in an HE context; competition, cooperation and recognition (Hainey et al., 2016).

Students as Co-Creators in Learning

Student-staff partnerships in higher education re-frame the ways that students and staff work together as active collaborators in teaching and learning' Mercer-Mapstone et al. (2018, p12)

The landscape of Higher Education is evolving, especially in the UK, reflecting both external and internal influences, which includes increased competition for students, new technology, campuses, and political shifts; with a neoliberal emphasis on education resulting in a perception of students as customers. As such, Institutions are being driven to re-evaluate their 'service delivery' and create value (Chalcraft et al., 2015). Within this concept of service delivery, heavy emphasis has been placed on 'student engagement' and increasingly 'students as partners' with Healey, Flint, & Harrington (2014) identifying staff and student partnership as one of the most important issues facing higher education in the twenty-first century.

There has been much debate about the terminology used and how that is interpreted with the broad and ambiguous concepts providing a sense of power for appropriate context and adaptation, and yet on the other hand bringing a lack of clarity with regards to rationales, motivations, and a shared understanding of the concepts (Flint, 2016). Though Matthews (2016) provides the differentiation of student engagement as outcome driven based on what students undertake during their time at university with student's as partners being process orientated and focussed on furthering common educational goals of staff and students, there are still different interpretations. Some have related it to the 'student voice' concerning their experiences, satisfaction with services, and feedback on course content or a collective representation. Educational staff have focused more on trying to make the student experience more engaging based on the notion that positive student engagement leads to enhanced resilience, persistence, learning an academic achievement (Celuch et al., 2018). Some advocates of student engagement have begun to promote the uniting of voices and students being actively involved in their own educational experience and described as 'agents of change', co-creators or 'students as partners'.

Essentially students as partners should be a positive, meaningful relationship between students and staff. Mercer-Mapstone et al. (2017a) identified a range of interpretations of students as partners in their systematic literature review of student's partners in higher education. It can be defined through the concept of a reciprocal relationship and process that engages all participants equally though not automatically in the same manner (Cook-Sather, Bovill, & Felten 2014), and a relationship which provides benefits to those engaged in the process of working and learning together (Healey, Flint, & Harrington, 2014). The importance of values and partnership principles was a strong theme within the systematic review and highlighted the emerging literature on the positive outcomes of engaging in partnerships (Cook-Sather et al., 2014; Bovill et al., 2010; Werder, Thibou, and Kaufer, 2012; Werder & Otis, 2010; Cook-Sather & Abbot, 2016; Cook-Sather & Agu, 2013)

It should be recognised that there are challenges associated with student partnership working in higher education. Healey et al. (2014) recognised that the shifting of 'traditional' hierarchies in HE could be transformational, yet there are challenges associated with the reconfiguration of boundaries and structures. Bovill, Cook-Sather, Felten, Millard, & Moore-Cherry (2016) identified three themes (customs and culture of higher education, institutional context, inclusive approaches to partnership) that which all of which the authors encountered when undertaking the project. Mercer-Mapstone, Marquis, and McConnell (2018) discuss partnership identity and the transition of moving from 'us and them' to 'we', which mirrors many of the experiences that authors were familiar with when undertaking the students as partners project. Mercer-Mapstone et al. (2017b) discuss the issues around power and the tensions between labels of 'staff' and 'student' which impacted early on our project, and while they were eased some tensions remained due to manner in which the project was initially proposed and the students being recruited to work on the project, so despite it being a partnership there were reporting structures which reinforced the hierarchical structures. Even though there were tensions throughout the project the team used Cook-Sather et al.'s (2014) partnership principles (respect, reciprocity and shared responsibility in learning and teaching) to displace the traditional structures and roles to have a balanced and fluid relationship between staff and students, which was also an important factor in Mercer-Mapstone et al.'s (2017a) review.

ABOUT THE STUDY

The staff that formed the research team were identified through their mutual research interests in teaching and learning, and in particular students as partners (Healey et al., 2014) through engaging them in research for learning and teaching. They both had an interest in enhancing their own teaching through two particular themes: (1) the use of technology to enhance the student learning experience (2) to ensure students had the strongest voice in expressing their learning support needs.

The team successfully gained funding from HEFCE 2016 Catalyst Fund for innovations in teaching and learning for a student-led project. This Student Led Technology Enhanced teaching and learning aimed to engage undergraduate students in researching their own student body, to discover the needs of the student community. A particular focus of the research was to design and develop two technology-based teaching and learning tools, as well as empower students to train and mentor staff in the use of the two interventions that emerged. A participatory action research approach was selected as it was collaborative in nature and sought to improve the quality of the learner's experience (Cohen, Mannion and Morrison, 2018, p444). At the start of the project there was some initial concern that the students would not be able to take the lead on the project with a staff member from the research team commenting *'I was initially concerned that the students were not going to be able to lead as much as we hoped, they grew into the role and engaged the wider student population in a way we couldn't.'* it was clear as the research progressed and the students gained the confidence in their research skills and project management abilities that the staff team could let them take more of the lead.

OVERVIEW OF FINDINGS

As part of the research, the research team sought the views of undergraduate students in order to gain an understanding of student attitudes towards technology for teaching and learning. This would allow the research team the ability to gain some insight into how HE students at the university currently use technology for their education. A short questionnaire was developed and disseminated across the university by student researchers. The aim of the questionnaire was to help the research team decide what interventions the students may want and may eventually use.

The students led the design and the face validity of questionnaires (supported by the staff team). They carried out university-wide data collection to gain insights into the student populations' requirements of technology for learning and teaching. Over 200 students engaged in this initial phase of research, however, the team recognised a gap in the research data with regard to potential incoming students and the relatively speedy advancement of social media and other digital tools. To bridge this gap, the student researchers suggested engaging local college students in the research. The lecturers facilitated this arrangement with the local college, and the student researchers undertook a series of focus groups with 20 students from the college that were looking to transition to higher education. The student researchers were coached on focus group techniques before undertaking the focus groups, and they then transcribed and analysed the data, again with support.

Questionnaire results indicate that students at the University of Bolton have a positive attitude towards using technology to support learning, that they use a variety of technology resources to support their studies and that they consider technology to be helpful to their studies. In particular students have a positive attitude towards using educational apps. Students have apps on their smartphones and use them. However, they do not currently use educational apps in the classroom or for teaching and learning purposes. Results also revealed that the majority of students at the University are positive about using an educational computer game and agree that they would find learning more enjoyable if computer games were used in teaching and learning. Students stated that they have a positive attitude to the idea of a university-based educational game and that they feel confident with playing computer games. Overall, the results indicate that students at the University have a positive attitude to the idea of using technology to enhance their educational experience and that they think that education should utilise new media technologies. In general, students had a positive attitude to the use of technology in education and considered it a part of everyday life and had a positive attitude to the educationalal value of computer games specifically. It was therefore deemed important to embed a games element to learning within the mobile app.

The questionnaire gave the student the option to offer some qualitative input if they wanted to expand on their answers and offer an insight into what technology they would use and why. It was revealed that students wanted a technology intervention that would have a fun element, with some form of reward. They said that they would want the intervention to have academic value and the opportunity to learn something that wasn't necessarily going to be included in lectures.

Focus Group Feedback

As previously mentioned, three focus groups took place to gain input from students considering coming to university. Twenty 6th form students took part in the focus groups which lasted approximately one hour each and were facilitated by two of the student researchers. Each of the focus groups were digitally recorded and transcribed verbatim. The analysis was conducted independently by two of the student researchers on Nvivo software (version 11) to allow for triangulation. A content analysis approach was employed allowing the research team to gauge what it is that students might want when transitioning into university life and how technology could be used to engage students in their own teaching and learning. It was clear from the focus groups with the 6th form students that they wanted technology embedded within their learning. When discussing the possibility of a mobile app what appealed to the students was the app helping them to find their way around the university and meeting new people. The college students were concerned about the transition into university life and felt an app could help them to socialise as well as provide wayfinding functions. In support of the quantitative findings from the university students, the college students also wanted technology in their education to be fun and engaging.

UBolt

On completion of the initial research phase, the whole team worked together to provide a project brief that would develop a university-wide mobile application ('app') to support learning and teaching. This gave students the opportunity to develop a range of employability skills including soft skills such as team working and communication, but also managing others, summarising and reporting data, and presenting briefs for development. The brief was based on the outcomes of the research data and included elements such as; study skills, way-finding, socialisation, and competitiveness.

Four undergraduate creative technology students were employed on a short-term consultancy basis over 3 months to develop a university-wide app "UBolt" which was successfully launched at the start of the 2017-18 academic year. The student app development team and student researchers engaged with facilities, library and support staff, and academics to ensure that location and wayfinding functions were

Figure 1. The uBolt mobile app: homepage

accurate and that the generic study skills content was accurate and supportive. Adopting the principles of 'Gamification', gamified study skills content was accessed via clues in the app that led users to QR codes across campus to further support wayfinding. The Student Union supported the app by providing funding for prizes such as coffee vouchers for those using the QR codes and games. Over a quarter of the University's first-year undergraduates downloaded the app as well as students from across other years.

EVALUATION OF THE APP

After three months the research team evaluated the app by looking at what students had engaged with and gaining feedback from students across the university on their experience of using the app.

As can be seen in table 1 students partially engaged with their timetable and the interactive map of the university campus. LEAP online is the University's award-winning innovative online learning support facility that provides students with academic information (such as referencing), videos and activities that enable students to earn digital badges. LEAP online was also another popular function of the app.

Figure 2. The uBolt mobile app: computer availability

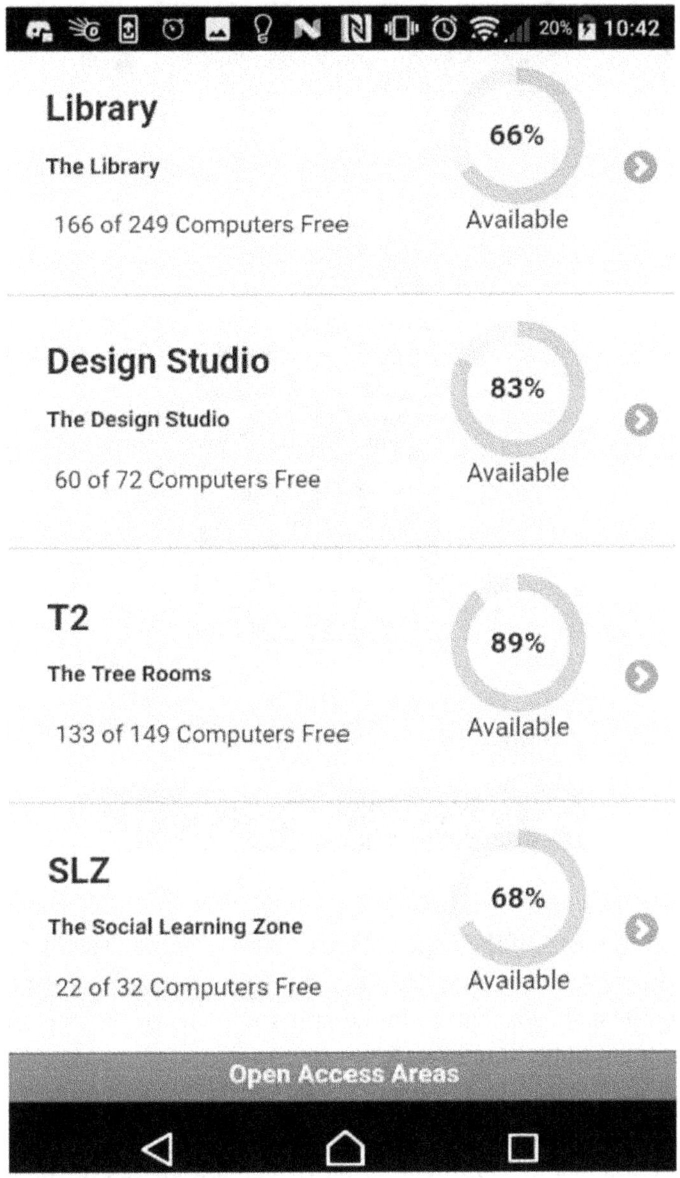

Interestingly, although the initial research had suggested students wanted to interact with each other and socialize through technology this was not a function that students engaged with (3.8%). The app had a number of study skills games designed and built into it, again based on the initial research findings, which proved not as engaging to students in the final app.

Staff team members took this insight forward as a member of the university-wide curriculum design working group. An evaluation of the app, through informal feedback and an evaluation questionnaire, was undertaken by the students with over 51 users of the app engaging in this process. The app was received positively by students with some areas of improvement suggested such as more user-generated input. Positive comments included that the app was easy to use and logical, useful to the students, entertain-

Figure 3. The uBolt mobile app: QR codes

ing to use and interactive. In terms of the negatives, the main negative was that it did not allow for user input. Other suggested improvements included, that the app it needed to be more visually appealing, that there was not enough visual information in the app, the goals of the app could be clearer and that the feedback from the app could be improved.

The app was demonstrated to senior members of university staff including the project lead for 'CampusM', a nationally used app, which the University decided to adopt after the HEFCE bid was won. The CampusM project lead stated that the app was *'far more technologically advanced than they expected'*. With another member of the CampusM team asking *'can we use some of the content, as it would save us about 18 months of work'* and implying the university will save a significant amount of money.

STUDENTS AS PARTNERS: CHALLENGES AND OPPORTUNITIES

The University of Bolton, in its strategic plan, is a Teaching Intensive Research Informed institution that aims to be a local and regional asset. The University's strategic priorities include intensive teaching and learning, and enhancing the student experience alongside increasing the reputation and changing perceptions of the organisation (Strategic Plan 2015-2020).

Figure 4. The uBolt mobile app: Quiz

The university implementation of enhancing the student experience has at times been a top-down process with schools and departments implementing policies and strategies in localised silos to enhance the student experience at the departmental level, rather than taking a more integrated approach. However, the student-led technology-enhanced learning and teaching project brought together staff and students from across the university in cross-disciplinary work. The staff members of the team are based in the Education, and Psychology departments, with the student team members being from Education, Psychology, Sports Rehabilitation and Creative Technologies. Colleagues from across the university were approached to engage with the project and acted as critical friends and consultants from the School of Computing and Creative Technologies, Student Experience, Student Union, IT, and Marketing and Recruitment. The project has had a marked impact on the students involved in the project in a number of ways, in terms of the impact on the team this can be evidenced through three main themes; increasing confidence, developing employability skills and student experience.

Students presented the project at a number of internal and external conferences, which helped to increase their confidence in not only the project but also in themselves as they gained external validation that what we were doing was of interest to other people. They were the first cross-disciplinary group of students at the university to present at a conference together. The staff worked collaboratively in preparing the students to present ensuring that they were confident about the conference with one student stating:

Figure 5. The uBolt mobile app: Leaderboard

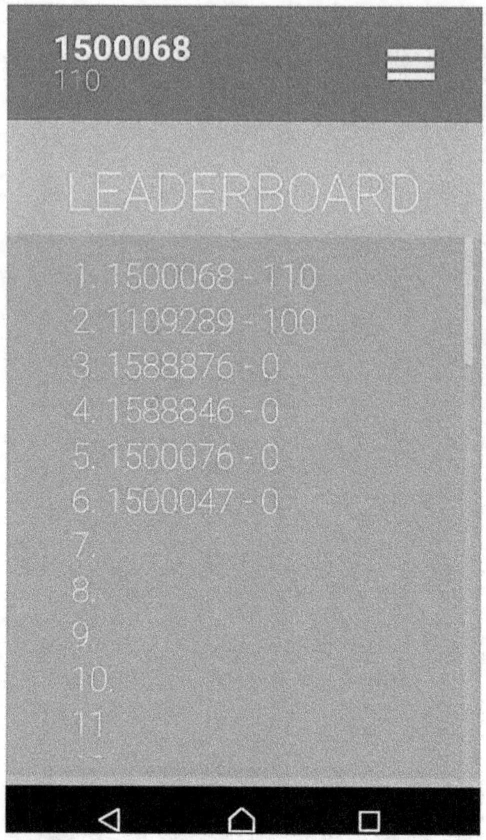

Table 1. Percentage of students that used the different functions of the app

App Functions	(%)
Access to your timetable	73.6
Interactive university map	50.9
Leap online	49.1
Prizes	34
Games based on digital literacy	26.4
Search feature for room location and staff location	26.4
Contacting staff email and Tel contact	26.4
Links to leap online, Moodle, Union and university events	24.5
Live cross campus CPU availability	22.6
Live leader board	20.8
Games based on the university infrastructure	18.9
Games based on generic study skills	17
Social interaction and ability to meet new people	3.8

'Presenting at a conference in front of academics who are not your teachers was a really scary experience but a valued experience as not only where they from different subject areas but from different institutions.' (Student comment)

The student also found that it made a positive difference to their performance in their studies.

Employability and research skills were gained and enhanced not only from the conferences the student researchers presented at but also through being part of a collaborative funded project, working as part of a team, having the opportunity to lead and manage others as well as learning from each other. One student stated

'Through this collaborative process, I enhanced my ability to not only listen to others but also understand this from their perspective. From learning to understand each other, it became apparent the different skills that each of us had and how we complimented each other for the benefit of the research but the mutual understanding and passion in terms of wanting to produce something that we were proud of'. (Student comment)

Another reflected that:

'I have gained an enormous amount from this experience, including research skills that will help me in my third year for my dissertation and also the confidence to build professional networks for when I graduate.' (Student comment)

The student researchers also highlighted that the project supported them in making the links from purely theoretical research in controlled environments, to real life work and the adaptability and resilience needed when undertaking research which the following quote exemplifies 'the *ability to adapt research and work quickly is a skill I now value a lot, I hadn't appreciated the real world context before. Which will serve me well when I get a job' (Student comment).*

The project also impacted on the students' perceptions of their futures, helping to focus them on what they enjoyed and what they really wanted to do in their futures. Expressed clearly in the following reflected quote by one of the students

'The project for me has been an experience that has had a major impact not only by enhancing my knowledge and skills but has become an amazing journey of personal development. From the start of this project to date I have changed as a person and have a clear vision for the future. I have enjoyed the project so much that I now want to carry on with research in the same field and apply for a PhD studentship. I would like to see more opportunities for students to get involved with research as it provides invaluable experience'. (Student comment)

Continuous student-led conversations ensured that all members of the team were managing and matching their own expectations as well as the collective team expectations and objectives. Mentoring/coaching through the initial research process started as a supervisor-led process, to ensure that the student researchers were comfortable with applying their skills to a live project. Moreover, the collaborative discourse ensured students were using their skills in the correct manner supporting their transition towards independent researchers.

The collaborative work gave the students a greater insight into working in research and academia and the scope of the roles that academics undertake. All of the students involved in the project now plan to go on to further study with two now interested in undertaking doctoral research. One of those students stated

'This project has helped me feel more confident in my ability with regards to my future aspirations, studying for a PhD wasn't on my radar. It has provided a great insight into the type of research that I will be required to do in my postgraduate studies and given me a better understanding than I previously had. I feel more prepared, less intimidated and less like an imposter.' (Student comment)

All of the students felt that the collaborative project had enhanced their student experience beyond what they were expecting from their university experience, increasing not only their academic skills and performance but through creating opportunities to '*work with likeminded people from other courses*' but also socially with one student saying '*we'd never have been friends before working on the project together, but I feel like I've made a friend for life.*' The students also valued and recommended the opportunity to work closely with staff, with staff members of the team valuing the student input and perspectives for teaching and learning. The staff members of the team felt the collaborative project *'enhanced and supported the student experience through transitions and applicability of knowledge to work through an authentic employment experience'. From the back of this project,* staff have presented a proposal to the senior management team to enable short-term undergraduate and staff research collaborations to engage more students in research at the teaching-intensive University and to enhance the student experience across the institution.

Challenges Associated With Developing and Implementing a Technological Intervention

Despite the success of the overall project and the technology-based intervention developed to engage students in their own learning and teaching there was a range of challenges the team encountered that included team dynamics, stakeholder engagement, technical issues.

Team dynamics were perhaps the most manageable of the challenges encountered. Personalities could be understood, and situation managed to ensure that the team were working efficiently, though other barriers included disciplinary differences in terms of base research knowledge and understanding, and work ethic and work patterns. There were expectations that the students would all have a similar grounding in research methodology, this proved to be incorrect with their being very wide differences between the four disciplinary areas. This was resolvable in terms of providing training and support to mitigate for the differences, but the staff were surprised at the gaps in knowledge and were unsure whether this was solely a disciplinary difference. Work ethic and work pattern were an issue in one particular scenario which caused many issues within the team dynamic, especially when reasons were made regarding availability for work, e.g. studying or lecturing but then discussed on social media regarding parties or other social events.

Stakeholder engagement was at times challenging as departmental priorities and agendas were often engaged to provide barriers to progress. Learning to navigate the political waters was a challenge and not one that we completely succeeded. Some stakeholders appreciated the work undertaken and how it could scaffold other projects to provide more advanced technological interventions that were beneficial to the organisation, though others felt that our successes were detrimental to their own workload or employment.

The main challenges involved in implementing a mobile app within an institution such as the university revolved around barriers to access, in terms of access to institutional accounts. Whilst the intervention had appropriate support at the highest levels there were issues that we were unaware of that could impede on other workstreams in a negative way especially regarding security checking of operation systems, and reputational risk if the IT department felt that our work was subpar.

CHAPTER SUMMARY

Literature encourages academic staff to embrace and, where possible, embed technology within the curriculum. Mobile apps are an increasing feature in many areas of our lives including for health, including mental health, support and information (i.e. Rathbone & Prescott, 2017) and increasingly for educational purposes. A number of universities have adopted Campus M for their institutions and the university involved in this study is no exception.

It is evident from this research that engaging students to design and develop a technological intervention aimed at students is a good method of making sure the technological intervention is appropriate to the demographic. This HEFCE funded project developed a mobile app that provided students with games, wayfinding functions and the ability to socialize as well as the more practical and beneficial functions of the app such as interactive maps and timetables. Students as partners allows the student voice to be heard without the barriers and constraints that may pertain if the research and the design of such technological interventions was staff led.

REFERENCES

Agarwal, R., & Karahanna, E. (2000). Time flies when you're having fun: Cognitive absorption and beliefs about information technology usage. *Management Information Systems Quarterly*, *24*(4), 665–694. doi:10.2307/3250951

Aharony, N. (2014). Library and Information Science students' perceptions of m-learning. *Journal of Librarianship and Information Science*, *46*(1), 48–61. doi:10.1177/0961000613518819

Apple Store Downloads Statistics. (n.d.). *Statista*. Retrieved from www.statista.com/statistics/263794/number-of-downloads-from-the-apple- app-store/

Boruff, J. T., & Storie, D. (2014). Mobile devices in medicine: A survey of how medical students, residents, and faculty use smartphones and other mobile devices to find information. *Journal of the Medical Library Association: JMLA*, *102*(1), 22–30. doi:10.3163/1536-5050.102.1.006 PMID:24415916

Bovill, C., Aitken, G., Hutchison, J., Morrison, F., Roseweir, K., Scott, A., & Sotannde, S. (2010). Experiences of learning through collaborative evaluation from a postgraduate certificate in professional education. *The International Journal for Academic Development*, *15*(2), 143–154. doi:10.1080/13601441003738343

Bovill, C., Cook-Sather, A., & Felten, P. (2011). Students as co-creators of teaching approaches, course design, and curricula: Implications for academic developers. *The International Journal for Academic Development*, *16*(2), 133–145. doi:10.1080/1360144X.2011.568690

Bovill, C., Cook-Sather, A., Felten, P., Millard, L., & Moore-Cherry, N. (2016). Addressing potential challenges in co-creating learning and teaching: Overcoming resistance, navigating institutional norms and ensuring inclusivity in student-staff partnerships. *Higher Education*, *71*(2), 195–208. doi:10.100710734-015-9896-4

Carter, S., & Yeo, A. C. M. (2017). From hedonism and utilitarianism to anticipated actual behavior: Do UK and Malaysian postgraduate students behave differently to mobile apps? *Economics, Management, and Financial Markets*, *12*(2), 11–34. doi:10.22381/EMFM12220171

Celuch, K., Bačić, D., Chen, M. W., Maier-Lytle, J., & Smothers, J. (2018). The potential of student co-creation in extracurricular experiences. *Marketing Education Review*, 1–14. doi:10.1080/10528008.2017.1419432

Chalcraft, D., Hilton, T., & Hughes, T. (2015). Customer, collaborator or co-creator? what is the role of the student in a changing higher education servicescape? *Journal of Marketing for Higher Education*, *25*(1), 1–4. doi:10.1080/08841241.2015.1044790

Chen, B., & Denoyelles, A. (2013). Exploring students' mobile learning practices in higher education. *EDUCAUSE Review*, 7.

Cheung, W. S., & Hew, K. F. (2009). A review of research methodologies used in studies on mobile handheld devices in K-12 and higher education settings. *Australasian Journal of Educational Technology*, *25*(2). doi:10.14742/ajet.1148

Cohen, L., Mannion, L., & Morrison, K. (2018). *Research Methods in Education*. London: Routledge.

Connolly, T. M., Boyle, E. A., MacArthur, E., Hainey, T., & Boyle, J. M. (2012). A systematic literature review of empirical evidence on computer games and serious games. *Computers & Education*, *59*(2), 661–686. doi:10.1016/j.compedu.2012.03.004

Cook, J., & Pachler, N. (2012). Online people tagging: Social (mobile) network (ing) services and work-based learning. *British Journal of Educational Technology*, *43*(5), 711–725. doi:10.1111/j.1467-8535.2012.01346.x

Cook-Sather, A., & Abbot, S. (2016). Translating partnerships: How faculty-student collaboration in explorations of teaching and learning can transform perceptions, terms, and selves. *Teaching & Learning Inquiry*, *4*(2), 1–14. doi:10.20343/teachlearninqu.4.2.5

Cook-Sather, A., & Agu, P. (2013). Students of color and faculty members working together toward culturally sustaining pedagogy. *To Improve the Academy, 32*, 271-285.

Cook-Sather, A., Bovill, C., & Felten, P. (2014). *Engaging students as partners in learning and teaching: A guide for faculty*. San Francisco, CA: Josey-Bass.

Corrin, L., Lockyer, L., & Bennett, S. (2010). Technological diversity: An investigation of students' technology use in everyday life & academic study2. *Learning, Media and Technology*, *35*(4), 387–401. doi:10.1080/17439884.2010.531024

Ding, Y., & Chai, K. H. (2015). Emotions and continued usage of mobile applications. *Industrial Management & Data Systems*, *115*(5), 833–852. doi:10.1108/IMDS-11-2014-0338

El-Hussein, M., Osman, M., & Cronje, J. C. (2010). Defining mobile learning in the higher education landscape. *Journal of Educational Technology & Society*, *13*(3).

Facer, K., Joiner, R., Stanton, D., Reid, J., Hull, R., & Kirk, D. (2004). Savannah: Mobile gaming and learning? *Journal of Computer Assisted Learning*, *20*(6), 399–409. doi:10.1111/j.1365-2729.2004.00105.x

Fleischman, D., Raciti, M., & Lawley, M. (2015). Degrees of co-creation: An exploratory study of perceptions of international students' role in community engagement experiences. *Journal of Marketing for Higher Education*, *25*(1), 85–103. doi:10.1080/08841241.2014.986254

Flint, A. (2016). Moving from the fringe to the mainstream: opportunities for embedding student engagement through partnership. *Student Engagement in Higher Education Journal, 1*(1), 1-6. Retrieved from https://journals.gre.ac.uk/index.php/raise/article/view/382/340

Hainey, T., Connolly, T, M., Boyle, E, A., Wilson, A., & Razak, A. (2016). A systematic literature review of games-based learning empirical evidence in primary education. *Computers & Education.* doi:10.1016/j.compedu.2016.09.001

Healey, M., Flint, A., & Harrington, K. (2014). *Engagement through partnership: Students as partners in learning and teaching in higher education.* York, UK: Higher Education Academy. Retrieved from https://www.heacademy.ac.uk/engagement-throughpartnership-students-partners-learning-and-teaching-higher-education

Hosein, A., Ramanau, R., & Jones, C. (2010). Learning & living technologies: A longitudinal study of first-year students' frequency & competence in the use of ICT. *Learning, Media and Technology*, *35*(4), 403–418. doi:10.1080/17439884.2010.529913

Hsu, Y. C., & Ching, Y. H. (2013). Mobile app design for teaching and learning: Educators' experiences in an online graduate course. *The International Review of Research in Open and Distributed Learning*, *14*(4). doi:10.19173/irrodl.v14i4.1542

Huang, Y. M., Lin, Y. T., & Cheng, S. C. (2010). Effectiveness of a mobile plant learning system in a science curriculum in Taiwanese elementary education. *Computers & Education*, *54*(1), 47–58. doi:10.1016/j.compedu.2009.07.006

Hwang, G. J., & Wu, P. H. (2014). Applications, impacts and trends of mobile technology- enhanced learning: A review of 2008–2012 publications in selected SSCI journals. *International Journal of Mobile Learning and Organisation*, *8*(2), 83–95. doi:10.1504/IJMLO.2014.062346

Johnson, L., Adams Becker, S., Cummins, M., Estrada, V., Freeman, A., & Ludgate, H. (2018). *Technology outlook for community, technical, and junior colleges 2013-2018: An NMC horizon project sector analysis*. Austin, TX: The New Media Consortium.

Kennedy, T., Wellman, B., & Clement, K. (2008). Gendering the digital divide. *IT & Society*, *1*(5), 72–96.

Kenny, R. F., Park, C. L., Neste-Kenny, V., Jocelyne, M. C., & Burton, P. A. (2012). *Mobile Self-Efficacy in Canadian Nursing Education Programs-Replication*. Academic Press.

Kim, J., Ilon, L., & Altmann, J. (2013). Adapting smartphones as learning technology in a Korean university. *Journal of Integrated Design & Process Science*, *17*(1), 5–16.

Kirkwood, A., & Price, L. (2005). Learners & learning in the twenty-first century: What do we know about students'' attitudes towards & experiences of information & communication technologies that will help us design courses? *Studies in Higher Education*, *30*(3), 257–274. doi:10.1080/03075070500095689

Li, K. C., Lee, L. Y. K., Wong, S. L., Yau, I. S. Y., & Wong, B. T. M. (2018). Effects of mobile apps for nursing students: learning motivation, social interaction and study performance. *Open Learning: The Journal of Open, Distance and e-Learning, 33*(2), 99-114.

Li, K. C., & Wang, F. L. (2012). Student engagement: Meanings, approaches and ideas for educators interested in ICT. In *International Conference on ICT in Teaching and Learning* (pp. 1-10). Springer. 10.1007/978-3-642-31398-1_1

Luna-Nevarez, C., & McGovern, E. (2018). On the Use of Mobile Apps in Education: The Impact of Digital Magazines on Student Learning. *Journal of Educational Technology Systems*, 0047239518778514.

Mansour, E. (2016). Use of smartphone apps among library and information science students at South Valley University, Egypt. *The Electronic Library*, *34*(3), 371–404. doi:10.1108/EL-03-2015-0044

Matthews, K. E. (2016). Students as partners as the future of student engagement. *Student Engagement in Higher Education Journal*, *1*(1) 1-5. Retrieved from https://journals.gre.ac.uk/index.php/raise

Mercedes, B., & Radel, B. (2016). Smartphones: From distraction to attraction. *Journal of Educational Technology Systems*, *45*(1), 93–102. doi:10.1177/0047239516639720

Mercer-Mapstone, L., Dvorakova, L. S., Matthews, K. E., Abbot, S., Cheng, B., Felten, P., ... Swaim, K. (2017a). A Systematic Literature Review of Students as Partners in Higher Education. *International Journal for Students as Partners*, *1*(1), 143–154. doi:10.15173/ijsap.v1i1.3119

Mercer-Mapstone, L., Dvorakova, S. L., Groenendijk, L., & Matthews, K. E. (2017b). Idealism, conflict, leadership, and labels: Reflections on co-facilitation as partnership practice. *Teaching and Learning Together in Higher Education, 1*(21). Retrieved from http://repository.brynmawr.edu/tlthe/vol1/iss21/8/

Mercer-Mapstone, L., Marquis, E., & McConnell, C. (2018). The 'Partnership Identity' in Higher Education: Moving From 'Us' and 'Them' to 'We' in Student-Staff Partnership. *Student Engagement in Higher Education Journal*, *2*(1), 12–29.

Nevin, R. (2009). Supporting 21st century learning through Google Apps. *Teacher Librarian*, *37*(2), 35.

Nielsen Reports. (2016). *Millennials are top smartphone users*. Retrieved from http://www.nielsen.com/us/en/insights/news/2016/millennials-are-top-smartphone-users.html

Prescott, J. (2014). Faculty/educators use of and attitudes towards Facebook. *Active Learning in Higher Education*. doi:10.1177/1469787414527392

Prescott, J. (2016). Teaching Style and Attitudes towards Facebook as an Educational Tool. In Learning and Teaching in Higher Education. Sage.

Prescott, J., Stodart, M., Wilson, S., & Becket, G. (2013 a). *The experience of using Facebook as an educational tool*. Health and Social Care Education; doi:10.11120/hsce.2013.00033

Prescott, J., Wilson, S., & Becket, G. (2013b). Facebook use in the learning environment: Do students want this? *Learning, Media and Technology, 38*(3), 345–350. doi:10.1080/17439884.2013.788027

Prescott, J., Wilson, S., & Becket G. (2013c). Students' attitudes towards Facebook and online professionalism: subject discipline, age and gender differences *Health and Social Care Education, 2*(2), 1-8. doi:10.11120/hsce.2013.00032

Rathbone, A., Clarry, L., & Prescott, J. (2017). Assessing the Efficacy of mHealth Applications Using the Basic Principles of CBT: A Systematic Review. *JMIR Mental Health.* doi:10.2196/jmir.8598

Rathbone, A. L., & Prescott, J. (2017). The use of mobile applications and SMS messaging as physical and mental health interventions: A systematic review. *Journal of Medical Internet Research, 19*(5). doi:10.2196/jmir.7740

Richardson, J., & Lenarcic, J. (2008). Text Messaging as a Catalyst for Mobile Student Administration: The" Trigger" Experience. *International Journal of Emerging Technologies & Society, 6*(2).

Russell-Bennett, R., Rundle-Thiele, S. R., & Kuhn, K. A. (2010). Engaging marketing students: Student operated businesses in a simulated world. *Journal of Marketing Education, 32*(3), 253–263. doi:10.1177/0273475310377758

Sampayo-Vargas, S., Cope, C. J., He, Z., & Byrne, G. J. (2013). The effectiveness of adaptive difficulty adjustments on students' motivation and learning in an educational computer game. *Computers & Education, 69*, 452–462. doi:10.1016/j.compedu.2013.07.004

Schmitz, B., Klemke, R., & Specht, M. (2012). Effects of mobile gaming patterns on learning outcomes: A literature review. *International Journal of Technology Enhanced Learning, 4*(5-6), 345–358. doi:10.1504/IJTEL.2012.051817

Seale, J. (2009). Doing student voice work in higher education: An exploration of the value of participatory methods. *British Educational Research Journal, 36*(6), 995–1015. doi:10.1080/01411920903342038

Shen, R., Wang, M., & Pan, X. (2008). Increasing interactivity in blended classrooms through a cutting-edge mobile learning system. *British Journal of Educational Technology, 39*(6), 1073–1086. doi:10.1111/j.1467-8535.2007.00778.x

Statistics and facts about mobile app usage. (2015). Retrieved from https://www.statista.com/topics/1002/mobile-app-usage/

Ward, N. D., Finley, R. J., Keil, R. G., & Clay, T. G. (2013). Benefits and limitations of iPads in the high school science classroom and a trophic cascade lesson plan. *Journal of Geoscience Education, 61*(4), 378–384.

Werder, C., & Otis, M. (Eds.). (2010). *Engaging student voices in the study of teaching and learning.* Sterling, VA: Stylus.

Werder, C., Thibou, S., & Kaufer, B. (2012). Students as co-inquirers: A requisite theory in educational development. *Journal of Faculty Development, 26*(3), 34–38.

West, D. M. (2013). Mobile learning: Transforming education, engaging students, and improving outcomes. *Brookings Policy Report*, 1-7.

Wilson, A. S., Broadbent, C., McGrath, B., & Prescott, J. (2017). Factors associated with player satisfaction and educational value of serious games. In Serious Games and Edutainment Applications (vol. 2). Springer. doi:10.1007/978-3-319-51645-5_23

Wright, S. (2015). *Deep Learning Isn't about Technology*. Retrieved hhtp://plpnetwork.com/2012/09/24/deeper-learning-technolgy

Yeap, J. A., Ramayah, T., & Soto-Acosta, P. (2016). Factors propelling the adoption of m-learning among students in higher education. *Electronic Markets*, *26*(4), 323–338. doi:10.100712525-015-0214-x

Yu-Chang, H., & Yu-Hui, C. (2013). Mobile app design for teaching and learning: Educators' experiences in an online graduate course. *International Review of Research in Open and Distance Learning*, *14*(4).

Zydney, J. M., & Warner, Z. (2016). Mobile apps for science learning: Review of research. *Computers & Education*, *94*, 1–17. doi:10.1016/j.compedu.2015.11.001

Chapter 7
The Role of a WhatsApp Group of a Professional Learning Community of Chemistry Teachers in the Development of Their Knowledge

Ron Blonder
Weizmann Institute of Science, Israel

Ruth Waldman
Weizmann Institute of Science, Israel

ABSTRACT

The authors analyze chemistry teachers' discourse in a WhatsApp group. This online communication platform is used for continually studying the communication behavior of leading chemistry teachers who are members of a professional learning community (PLC). They describe the network of chemistry teachers' PLC in Israel, which provides the context for the study. WhatsApp enables sustained ongoing, intensive interaction, and sharing of knowledge that is practical, directly related to the members' needs, and is participant driven and constructivist in nature. A theoretical perspective of teachers' knowledge and professional development (PD) was developed in 2015 by Gess-Newsome, which was applied to examine the mechanism underlying teachers' knowledge development.

DOI: 10.4018/978-1-5225-8106-2.ch007

INTRODUCTION

The McKinsey report (Barber & Mourshed, 2007), which investigated successful education systems, describes two basic requirements needed in order develop a high-quality education system: (1) to attract high-quality teachers to the system; and (2) teachers need to continuously develop their professionalism. Teachers are the key for achieving good education, and therefore, effective ways in which they can develop their knowledge and skills are essential to improve education systems.

This chapter discusses the knowledge development of leading chemistry teachers who participate in a Professional Learning Community (PLC) of chemistry teachers; it focuses on the contribution of a WhatsApp group to this development. In Israel, chemistry PLCs were initiated in 2014 and operate at the Weizmann Institute of Science with support from the Ministry of Education and the Trump Foundation. Chemistry teachers meet in the PLC regularly to explore their teaching practice and their students' learning achievements. The PLC facilitators try to create a feeling of trust (Tschannen-Moran, 2009) among the teachers in the PLC. In practice, however, not all the communities work—many fail. The trust and the created sense of community (McMillan & Chavis, 1986) lead to creating a safe environment in which teachers can share their questions regarding teaching, their difficulties, and also their successes (Booth, 2012). The main emphasis in the PLC is on developing the teachers' pedagogical content knowledge (PCK) in an environment, which for the PLC members, supports a sense of belonging. These activities focus around the students (Rogers, 1983). Using students' data is a key component of effective professional development design (Guskey, 2003; Little, 2012) Namely, teaching is discussed less than the influence of a selected pedagogy on students' learning. These goals are achieved by including the following components in the chemistry teachers' PLC meetings:

- An opening exercise that deepens the social and personal acquaintance of the chemistry teachers in the PLC.
- Investigation of students' understanding and misconceptions in chosen chemistry topics using diagnostic tools that were designed for the chemistry teachers' PLC.
- Discussion of pedagogical ways to deal with the misconceptions that teachers find in their own classes (based on evidence revealed by analyzing the diagnostic tools in class).
- **"My corner":** A section in which one of the PLC teachers presents a short activity that the other teachers can easily adapt to their own class. This section can be a lab experiment, a demonstration, or a technological tool that can be integrated into the chemistry lesson.
- **Chemistry Laboratory:** New experiments and their integration into chemistry class.
- **"Teachers' Questions":** A section in which the teachers in the PLC raise a problem that they face in their own chemistry class, for which they need help from the PLC.

The chemistry teachers' PLC operates in a net model (shown in Figure 1). A leading PLC, which includes leading chemistry teachers, meets every other week at the Weizmann Institute of Science. In the other week, the leading chemistry teachers lead in pairs a regional chemistry teachers' PLC. Eight regional chemistry teachers' PLCs have operated close to the teachers' home (CtH), which have spread throughout Israel in the last 4 years. About 200 chemistry teachers, namely, 30% of the chemistry teachers in Israel, are members of these PLCs.

Figure 1. The chemistry teachers' PLC net model of operation. The leading chemistry teachers create the central PLC, and lead in pairs the regional close to home (CtH) chemistry teachers' PLCs

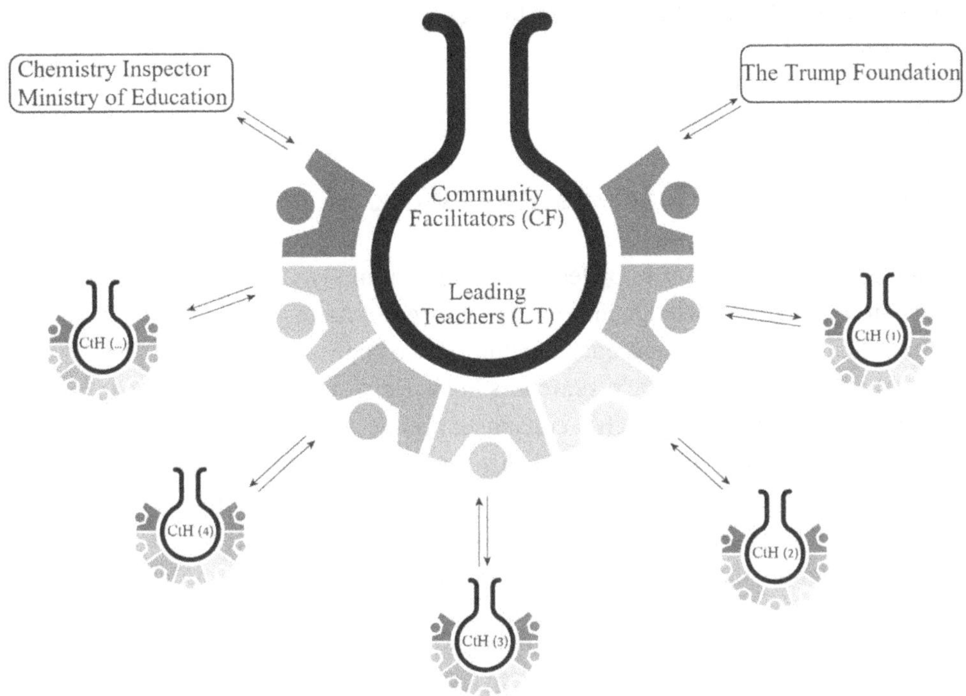

Other important elements that should be considered from the components of the PLC meeting are the direction of influence and the direction in which knowledge flows in the PLC. Part of the activities are designed by the facilitators (e.g., the diagnostic questionnaires, some of the lab experiments). In this part the direction of knowledge is top-down: from the facilitators to the teachers who are PLC members. However, there is another part in which a bottom-up direction dictates the activities in the PLC meetings. This part includes sections such as "My corner", "teacher's questions", and the results of the students' diagnostic questionnaires, which lead to pedagogical discussions during the meetings.

The activity of the PLCs is supported by a philanthropic foundation (the Trump Foundation), which supports the development of science and mathematics education in Israel, and by the Ministry of Education. This double-source of support enables us to receive a budget to operate the PLC net, and also to be part of the Israeli educational system, which is essential for sustaining the PLC in the future.

In this chapter, we analyze the knowledge development of the leading chemistry teachers' PLC. More specifically, we examine the knowledge development of the individual teachers in the group. In addition to the face-to-face meetings of the PLC teachers, a PLC WhatsApp group was initiated by the PLC facilitators. The WhatsApp group includes all the PLC teachers and the facilitators that lead the activity. The WhatsApp group helps the PLC members remain in contact and enhances the effectiveness of teachers' professional learning as well as continues the interactions among the PLC members between meetings, effectively creating an online community (Waldman & Blonder, submitted).

The goal of this work is to examine how the WhatsApp group contributes to the process of developing teachers' knowledge in the PLC. It is important to emphasize that the model of professional development (PD) includes participating in the chemistry teachers' PLC with the aid of the WhatsApp group. In this study we especially assigned the contribution of the WhatsApp group to the process of developing teachers' knowledge. For that purpose, a current model of teachers' knowledge is used as the theoretical perspective of the study, which will be presented next.

BACKGROUND

Theoretical Framework of Teachers' Knowledge

The concept of pedagogical content knowledge (PCK), originally proposed by Shulman (1986), distinguishes between teachers' knowledge and the knowledge of an expert in the field. Since then, many models describing PCK, especially in science teaching, have been developed (e.g., Magnusson, Krajcik, and Borko (1999)). A summit of leading researchers in the field of teachers' knowledge and professional development (PD) was organized to reach a consensual model that will serve the research community (Gess-Newsome, 2015). In this study, we used this model to examine the development of chemistry teachers' knowledge in their WhatsApp group (the individuals in the group). According to this model (Gess-Newsome, 2015), and compared to previous models (Kind, 2009), teachers' knowledge and skills are perceived in a more holistic manner and include components that add to the teachers' professional knowledge bases. Two main differences distinguish this model from the previous PCK (Kind, 2009). In the ensuing years since PCK was defined by Shulman and adopted by the community of science educators, the absence of the aspect of teachers' beliefs and attitudes related to knowledge development has hindered the usefulness of the PCK concept. In the new suggested model (Gess-Newsome, 2015), teachers' beliefs were introduced to act as amplifiers and filters in the process of teachers' knowledge construction. The second difference is the perception of PCK as personal pedagogical content knowledge and skills (as opposed to the knowledge that is known by the science education community about how to teach a specific scientific content). The new model is presented in the black printed parts in Figure 2.

The suggested model includes several components and will now be briefly explained (Gess-Newsome, 2015). The teachers' professional knowledge bases (TPKB) are as follows: assessment knowledge, pedagogical knowledge, content knowledge, knowledge of the students, and curricular knowledge. There is a wide consensus regarding those components that are essential for good teaching. These TPKBs are generic; knowledge derived from the TPKB informs users and is updated by the topic-specific professional knowledge (TSPK). The TSPK is the knowledge shared by the science education community regarding the teaching and learning of a specific topic. This category of knowledge contributes in three aspects: "1) It makes explicit that content for teaching occurs at the topic level (i.e., force and motion) and not at the disciplinary level (i.e, physics or science); 2) this knowledge blends subject matter, pedagogy, and context; and 3) it is recognized as a public knowledge, or knowledge held by the profession, allowing it to assume a normative role…" (Guess-Newsome, 2015, p. 30). These two knowledge components, TPKB and the TSPK (presented in the top two frames of Figure 2), are context free, whereas the remaining components in the model refer to classroom context.

Traditionally, "the focus of models for teacher pedagogical knowledge and skills has been on the knowledge that is held by [the] profession and accessed by teachers. But in order to personalize knowledge, it must pass through the lens of the teachers. As [a] free agent, a teacher has the opportunity to embrace, reject, or modify new knowledge skills and practices" (Guess-Newsome, 2015, p. 34). The lenses used by teachers are (1) their own beliefs and (2) their interpretation of students' reactions to their teaching

Figure 2. Model of teachers' professional knowledge and skills including PCK, as proposed by Guess-Newsome (2015). The black parts are taken from the original model and the blue parts show how the teachers' PLC WhatsApp group contributed to the development of the teachers' professional knowledge.

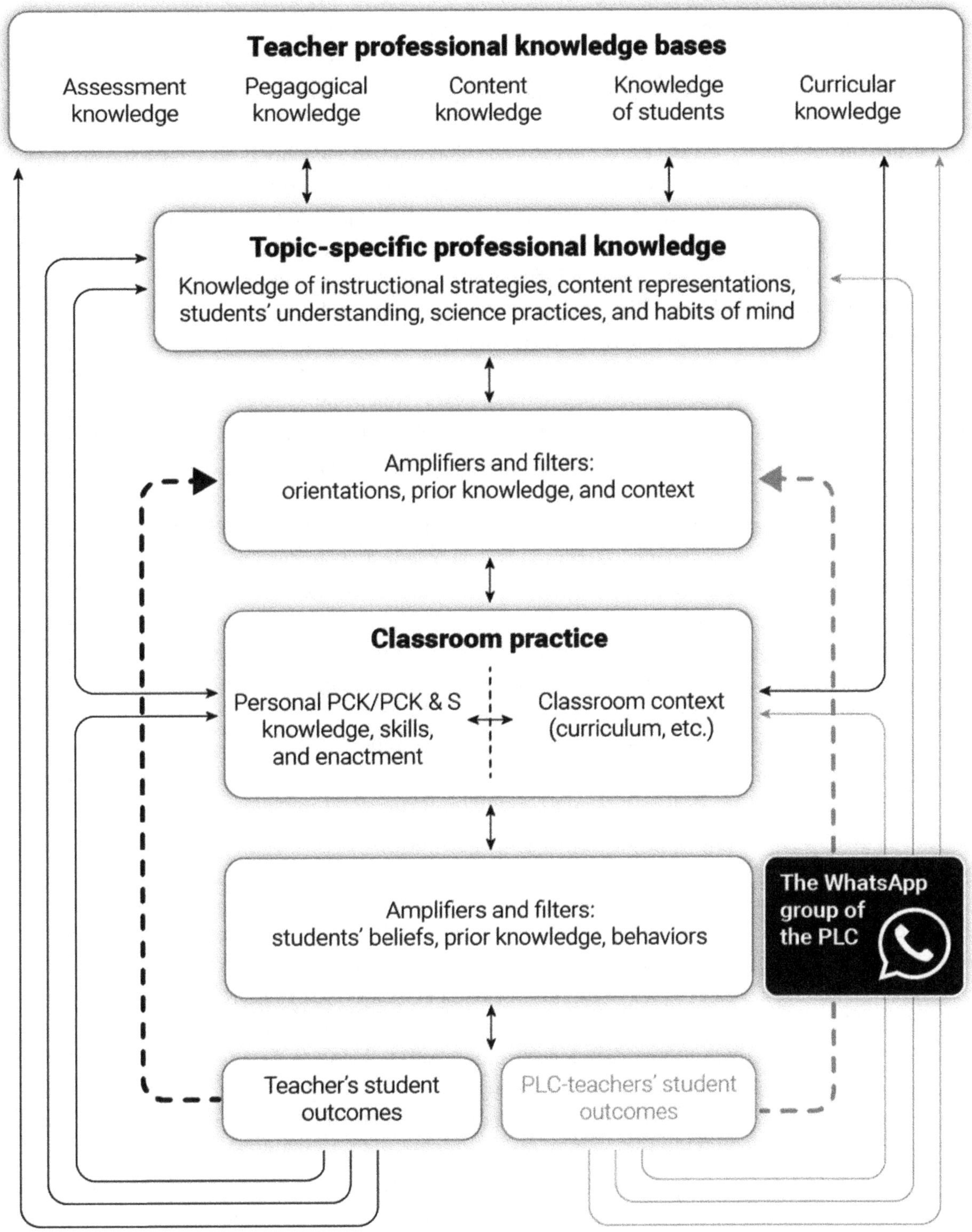

(presented in Figure 1) as the "amplifiers and filters" that can support or inhibit the development of teachers' knowledge bases. There are many reasons why teachers do NOT implement new knowledge and skills in their teaching. The good-old teaching that has been using for years is working well and a new method/approach/textbook might not work. There is a need to invest a huge amount of time to adapt new materials to the classroom's context, and that time might be wasted if the students will not positively react and will not learn as they do using the regular methods (Ron Blonder & Mamlok-Naaman, 2016).

The amplifiers and filters are very influential in the process of teachers' knowledge development and promote or inhibit teachers from implementing new topic-specific professional knowledge that teachers have learned through their own teaching. For example, if a chemistry teacher thinks that using technology in the lesson distracts students' attention, she will not try a new animation that she learned in the PD. A strong factor that can influence and modify these amplifiers and filters is students' reactions to the implementation. Continuing with the above example, if the teacher eventually decides to try animation in her chemistry class and saw its positive influence on her students' understanding, it might modify her previous beliefs about the negative influence of integrating technology into chemistry lessons as well as her topic-specific professional knowledge and her knowledge bases. This influence is caused by mastery experience (Bandura, 1986). However, this mechanism is restricted to those new experiences that the teachers eventually decided to bring to their class. According to social-cognitive theory, there are other sources that can influence teachers' beliefs (Bandura, 1986; Britner & Pajares, 2006) besides mastery experience. Using vicarious experience, namely, watching a teacher who is similar to you while she tests the simulation in her own class, can influence one's own beliefs. However, there are not many occasions, if at all, in which a teacher can watch her colleague testing new pedagogical approaches and can observe its influence on students' learning. This is true in face-to-face meetings, but by using the potential of online platforms, we might find that these limited occasions can turn into a source of opportunities. Online communities play an important role in the sense that teachers voluntarily participate in learning, reflect on their own practices, share them with other teachers, and provide mutual emotional support (Macià & García, 2016). In the current study we aim to identify the mechanism underlying the teachers' knowledge development that occurs in these PLCs.

Although the teacher is surrounded by many people (her students), she is isolated, since she is teaching alone in class. Even when it comes to disciplinary teams in school, which can provide support to teachers, chemistry teachers in Israel cannot benefit from this resource since most schools have only one or two chemistry teachers. This aspect of chemistry teachers' professional isolation is studied using quantitative tools and will be described elsewhere. In order to support teachers' professional growth, the teacher needs to have access to other professionals with whom she can discuss her doubts, share her difficulties, reflect upon her experiences in class, and more. The literature suggests several methodologies to overcome teachers' isolation. Using video-recorded lessons, which have become a resource for pedagogical guided-discussion, is one solution (Arcavi & Karsenty; Karsenty & Arcavi, 2017; Karsenty & Sherin, 2017). The video-recorded lessons provide a wide platform for teachers' reflection. Another approach is the evidence-based PD. When teachers are asked to bring to the PD meeting data they have collected from their students, which provide evidence for the learning process and achievements, they actually "open the walls of their classroom" to their colleagues in the PD (Taitelbaum, Mamlok-Naaman, Carmeli, & Hofstein, 2008). This approach is usually integrated into the teachers' PLC (Little, 2002). The chemistry teachers' PLC described here works according to this approach, which is based on the socio-cultural aspects of learning (Cobb, Yackel, & Wood, 1992; Sfard, 1998). The chemistry teachers' PLC provides the community that decreases teachers' isolation and it is where teachers can learn and develop their professionalism. The mobile online platform that was used in the study provides a means of continuing teachers' interactions beyond the PLC meetings (Ndlovu & Hanekom, 2014).

The WhatsApp Online Communication Platform in Education Settings

As mentioned before, teachers' professional learning communities (PLCs) are groups of teachers who regularly meet face to face in order to examine their knowledge and practices and thus, improve their teaching. Since face-to-face meetings typically take place only once every few weeks, the WhatsApp mobile online platform was integrated to maintain the communication among the PLC members between the face-to-face meetings; the WhatsApp group provides an informal setting for the PLC members to communicate. The use of the mobile application does not replace formal PLC meetings; rather, it offers a way to extend the support of learning outside the formal meetings, to the conversations and interactions of everyday life (Sharples, Taylor, & Vavoula, 2010).

Online communities or networks, empowered by social media technologies, have the potential to respond to teachers' individual and collective needs; thus, they facilitate teachers in creating learning processes that can help them adapt to changes, with an atmosphere of openness and colleagueship (Lieberman & Mace, 2010), and which create a sense of community (Waldman & Blonder, submitted). It has been found that social networks promote knowledge sharing and establish dialogues, which facilitate obtaining suitable resources and finding solutions (Forkosh-Baruch & Hershkovitz, 2012; Rap & Blonder, 2016; Wenger, 2011).

During the last few years, the WhatsApp application has become one of the most heavily used applications for online immediate communication (Seufert, Hoýfeld, Schwind, Burger, & Tran-Gia, 2016). One of its main characteristics is its ability to conduct and enhance group communication (Bouhnik & Deshen, 2014). A recent study among Israelis found that WhatsApp was used by 67% of the respondents and 52% reported that they send messages in WhatsApp groups on a daily basis (Avidar, Ariel, Malka, & Levy, 2013).

WhatsApp groups are characterized by unique features that make them suitable for use by the PLC teachers (Bouhnik & Deshen, 2014; Church & de Oliveira, 2013). These characteristics can be sorted into technical, communication, and educational aspects. *The technical aspects* of WhatsApp groups include their simplicity of operation, and that they are free of charge, free of advertisements, available on cellular phones, and are immediate. In addition, they maintain the privacy of the group participants and they are very popular in Israel (Avidar, Ariel, Malka, & Levy, 2013). The aspects related to *group communication* include the following: the WhatsApp feature to create a group, the option to communicate within the group's boundaries, the ability to conduct an ongoing conversation with many friends simultaneously, and to share content such as text messages, attached images, audio files, video files, and links to web addresses. The third aspect is related to the *educational potential* of WhatsApp: It enables learning any time and anywhere, the possibility to correct mistakes immediately, along with the presence of the teacher. Having accessibility to online learning materials, it is a platform for sharing materials, and it provides a secure environment for the group discourse. However, WhatsApp has several challenges that should be also addressed when considering its use in education settings. The group might cause the flooding of irrelevant or nonsensical messages, messages that are sent at late hours in a way that might burden and annoy the users, especially if they have more than one group or groups that have more than 15 members.

Several studies and review papers have examined the potential of harnessing the advantages of the WhatsApp mobile online platform for teachers' professional development. Teachers share on WhatsApp groups field knowledge, pedagogical content knowledge, in-school teaching practices, and provide emotional support to their colleagues (Cansoy, 2017). A high portion of the discussions can be classified into Shulman's categories of teachers' knowledge (Ndlovu & Hanekom, 2014). However, according to the literature review by Baran (2014), "more research is needed to understand how teachers and teacher educators' professional development can be supported with mobile learning."

RESEARCH GOALS AND QUESTIONS

In a previous study (Waldman & Blonder, submitted) in which the discourse in the PLC WhatsApp group was analyzed, we showed that about a third of the goals of opening a WhatsApp discourse dealt with asking for help, clarification, and consultation regarding chemistry, chemistry teaching, and management (Waldman & Blonder, submitted). In all cases, the members received satisfactory answers from the group. However, we did not analyze the contribution of the WhatsApp group to the development of professional knowledge of chemistry teachers participating in the PLC. The socio-cognitive stance for learning, in which learning is described as participation in discourse, was used in the present study (Lave & Wenger, 1991; Sfard, 2008). We will therefore analyze the teachers' discourse in the WhatsApp group by referring to the following research questions:

1. What teachers' professional knowledge is developed in the PLC's WhatsApp group?
2. What development mechanism of teachers' professional knowledge is unique to the WhatsApp group?

By finding the answer to the first question, we will be able to compare the contents of the discussions in the present PLC's WhatsApp group to other teachers' discussions via mobile online platforms. By studying the second question, we will add to the existing literature a theoretical understanding of the development mechanism.

METHODOLOGY

The *research population* includes a PLC of 14 chemistry teachers and 5 community facilitators (CFs). The PLC meets every other week at the Weizmann Institute of Science. In addition to the actual meetings, the community has a very active WhatsApp group in which members continue to discuss different aspects related to the professional chemistry community. The activity in this WhatsApp group was analyzed throughout two academic years (2015-2017).

Research Methods

The study qualitatively analyzes the WhatsApp discourse and utilizes an inductive approach in which data were analyzed using a qualitative research methodology (Strauss & Corbin, 1998). The posts, discussions, and shared information in the WhatsApp group were divided into Discourse Episodes

(DEs), which were defined as follows: A discourse event that was opened in the WhatsApp group and that deals with a certain topic. Each DE includes all the posts in the discourse event until a new DE is initiated. During the two years of research, 6,100 posts were posted. These posts were assembled into 300 DEs. For each DE, the topic of the discourse was identified using a bottom-up approach: First, each DE was coded according to its topic. These initial codes were then classified into general categories (Miles, Huberman, & Saldana, 2013). For example, the DE topic of "Chemistry teaching" included the initial codes of content knowledge, pedagogical content knowledge, curriculum knowledge, student diagnostics, technological pedagogical content knowledge, questions regarding matriculation exams, and in-school teaching practices. Then, 25% of the DEs were given to a second researcher, who used the coding scheme of the general categories. In cases of disagreement, the two researchers discussed the gaps. These discussions led to modifications of the categories scheme, and the DEs were re-classified using the new coding scheme.

Here we will also present two DEs that will enable further discussion of the mechanism underlying professional knowledge development enabled by the WhatsApp group.

RESULTS

Topics of the Discourse Episodes in the PLC WhatsApp Group

The DEs comprise episode topics that include professional knowledge discussions, and other episodes that are related to other aspects in the teachers' PLC, as shown in Table 1. First, we refer to professional knowledge discussion episodes: We found that 37 and 27% (in the first and the second years, respectively) of the DE topics deal with chemistry teaching. The chemistry teaching episode topic is composed of the following sub-categories: content knowledge, pedagogical content knowledge, curriculum knowledge, student diagnostics, technological pedagogical content knowledge, questions regarding matriculation exams, and in-school teaching practices. The second sub-category of professional knowledge discussion is "Leading the CtH regional PLCs" (Table 1), which received 25 and 21% (in the first and the second years, respectively). In this subcategory, the PLC teachers share via the WhatsApp group unique activities they conducted that are related to their role as PLC leaders. In this chapter, we will not focus on those aspects related to their role in leading the regional PLC. In addition, the DEs that were categorized "PLC meetings" (Table 1) include issues that were discussed in the face-to-face PLC meetings and are also rich in chemistry teaching topics. The last subcategory of professional knowledge discussion deals with laboratory experiments (8 and 9% (in the first and the second years, respectively). We found discourses dealing with the experiment itself, discussions of the phenomenon being explored, as well as ways to design experiments for enhancing students' motivation to learn and for demonstrating the theoretical principles being studied. Table 1 shows the professional knowledge discussions divided into four different topics.

In addition to episodes that dealt with professional knowledge, two other topics were found and categorized as "other": LMS, and social. The DEs that were categorized under the Learning management systems (LMS) topic include technical and organizational parts of the PLC (e.g., when is our next meeting? Can someone pick me up from the train station?). The social DEs deal with social aspects of PLC members, such as birthday greetings.

Table 1.The DE topics in the WhatsApp group during the academic years 2015-6 and 2016-7. Each DE can include more than one topic and can be counted more than once.

Discourse Episode Topic	2015-2016	2016-2017
	Number of DE (%)	Number of DE (%)
Professional Knowledge Discussions		
Chemistry teaching	55 (37%)	75 (27%)
Leading the CtH regional PLCs	38 (25%)	58 (21%)
PLC meetings	29 (19%)	46 (17%)
Chemistry laboratory	15 (8%)	27 (9%)
Other		
LMS	35 (23%)	87 (32%)
Social	26 (17%)	41 (15%)
Total DE	149	273

Examining the DE topics revealed that many of the DEs contain chemistry and chemistry education content. According to (Rap & Blonder, 2016; Sfard, 2006), the teachers' discussion that is related to their professional knowledge constitutes learning. However, how does the discussion actually promote the PLC teachers' professional knowledge? In order to address this question, a detailed analysis of two DEs will be presented. These DEs were chosen based on their representative nature, and will serve here as two vignettes that will allow us to discuss the way these discussions influence teachers' professional knowledge.

Knowledge Development of PLC Teachers

First Vignettes: A New Laboratory Experiment

Several DEs that dealt with a specific laboratory experiment called "Colored arm-wrestling" are presented (the experiment is detailed in Appendix 1). These DEs occurred during one month (January 30-February 27, 2018). The first DE concerning this experiment was conducted on January 30, 2017, during the PLC meeting. The experiment was developed by the PLC facilitators. During the face-to-face PLC meetings, several photos of teachers conducting the experiment were posted in the WhatsApp group (one of these photos is shown in Figure 3, A).

The experiment was conducted during the PLC meeting, which all the PLC teachers attended. They discussed how the experiment is connected to different parts in the chemistry curriculum and posed suggestions for possible inquiry questions that could be carried out as a possible continuation of the experiment. Note that the teachers are not obligated to conduct the experiment in class.

Two days after the PLC meeting, some of the teachers further explored the experiment in their school labs (before using it in their teaching). On February 2, Ben, one of the PLC teachers, examined the experiment and shared it with the PLC by posting 8 photos (two of them are presented in Figure 3, B & C in the WhatsApp group with the text: "Different variations – trying to adapt to the glassware I have in my school lab".

Figure 3. PLC teachers conducting the "Colored arm-wrestling" experiment during the face-to-face meeting. Different stages of knowledge building are reflected in the photos. A) Conducting the experiment during the PLC meeting. B-C) Ben checking the experiment in the school lab and determining the conditions needed to conduct it in school. D) Abed (a PLC teacher who is an expert in the lab experiment) conducted the experiment in different educational settings. E-F) Conducting the experiment with students or with other PLC teachers they lead. G) Hanna presenting her successful results.

Discourse Episode DE 1: *Ben tried to adapt the experiment to the glassware available in his school lab, and the responses from three members of the PLC (February 2, 2017).*

1. Different variations – trying to adapt to the glassware I have in my school lab (Ben, after posting photos B-D, Figure 3).
2. Did you try with an Erlenmeyer? (Karin)
3. I have liter Erlenmeyers, but not for 6 groups…(Ben)
4. What about the pickle jar? (Mira, Facilitator)
5. I don't have them and I don't know who can eat so many pickles in a short time [excited face]. (Ben).
6. You need to have a high flask volume. (Mira, Facilitator)

7. I can see that CO_2 is created...and the air doesn't leave the flask. So what I don't understand is if I'll decrease the quantity of the materials according the volume of the flask, should it work? The glove should be sucked into the flask? Right? (Ben)
8. It is supposed to... you are right. I think that it's better to let the air go out first and only then cover it with the glove. (Mira, Facilitator)
9. I tried. This really works better. (Ben)
10. It also works with a 1 litter Erlenmeyer and a balloon instead of a glove. The glove works with the pickle jar. I prefer that because of the effect of a glove being sucked in. (Sandy, Facilitator)
11. Try it with the smaller flask and decrease the amount of chemicals. (Sandy, Facilitator)
12. I think that you can buy an empty jar in the grocery story; it doesn't cost much. (Sandy, Facilitator)
13. I think that this is what I'll do. (Ben)

Discourse Episode DE 2: *One of the teachers posted a successful trial of the experiment, which led to a discourse involving 8 PLC members. (February 27, 2017).*

1. Successful at last... Yeyyy (Hanna)
2. <Photo G, Figure 3> (Hanna)
3. What I was just trying didn't work again!!! What's the secret? (Karin)
4. LOL (Karin)
5. I bought myself a pickle jar yesterday. (Shir)
6. <A photo of a pickle jar> (Shir)
7. Hanna, share with us step by step. What did you do? (Dalit, Facilitator)
8. I let the air out, I took 6 instead of 7 g of sodium hydroxide. I don't have powder, so I used grains that I crushed. (Hanna)
9. Thank you Hanna, I will try it again now (Tali)
10. [Thumbs up] (Hanna)
11. <again the same photo as in line 2, photo G Figure 3> (Hanna)
12. Your persistence was worthwhile! Great (Rose, Facilitator)
13. Great! (Avi)
14. How much air did you let out? And for how long? (Mira, Facilitator)
15. Just a little bit for one second. (Hanna)
16. Thanks! (Mira, Facilitator)

DE1 was initiated by Ben, other teachers in the PLC, and the PLC facilitators responded, as presented in DE 1 (posts 1-5). They asked about other laboratory glassware that Ben tried to use. However, his school lab does not have enough of these items. The PLC facilitator (Sandy) recommended that he use the same jar of pickled vegetables that was used at the PLC meeting, but this solution was also not practical for Ben.

Then, participants discussed the conditions that lead to a successful experiment (DE 1, posts 6-11). During the discourse, they asked questions and tried to explain the chemistry reactions that occurred. Throughout the discussion, they realized that the volume of the glassware used in the experiment influenced the surprising part of the experiment, in which the glove is impressively sucked-in. The discussion dealt with additional aspects of the experiment such as the role of the acid-base indicator used, and replacing the glove with a balloon, which might be sound technical, but actually influenced the experiment since it also influenced the volume of the reaction system. The DE contains several posts that provide the participants with a sense of appreciation and social support. This was reflected in the positive feedback that Hanna received after she posted the photo of her successful experimental system as well as PLC members' questions regarding the experimental details.

Several days later, on February 7, one of the PLC members (Abed) conducted the experiment in a different educational setting. Abed, who has much experience in the chemistry lab, posted a photo showing him conducting the experiment with a chemistry teachers' PLC that he leads in northern Israel. In addition, he received very supportive feedback from a PLC member in the WhatsApp group. This DE is not presented in the paper.

In the next DE (DE 2), which is presented here, on February 27, another teacher (Hanna) posted a photo of herself successfully conducting an experiment in her school lab, with the text: "*Successful at last…Yeyyy*". This post resulted in opening a DE that includes enthusiastic posts from PLC members and also requested detailed explanations from Hanna, so that everyone will be able to conduct the experiment successfully in their school. She replied to all the questions and provided the information based on her experience in successfully conducting the experiment.

Later on, during the year, additional teachers posted different photos related to the experiment: buying pickle jars, trying other conditions, and also included photos after they had conducted the experiment at school (Figure 3, E-F). Each of these posts in the WhatsApp group received comments from the PLC members.

What professional knowledge was involved in the discussion? Several DEs in the PLC WhatsApp group dealt with different aspects of an experiment that was presented during the PLC face-to-face meeting. At first, they recorded themselves during the meeting. Then, they discussed the conditions needed for conducting the experiment with the available school laboratory equipment. Finally, they posted photos showing themselves conducting the experiment with their students or in their other educational settings. We found that specific professional knowledge regarding the experiment was generated in the discourse. This includes a detailed understanding of the specific chemical reaction and the conditions that influence the amount of gas products. This is an acid-base reaction that involves the use of an acid-base indicator, phenolphthalein, which changes color during the experiment, and the generation of $CO_{2(g)}$, which is generated and then reacts, thus influencing the pressure in the reaction. We also noted how different teachers personalize this general knowledge to their own school laboratory context, and relate that aspect to the teachers' personal PCK. Feedback interviews that were conducted at the end of the academic year provided additional support indicating that the WhatsApp discourse during this experiment helped the PLC teachers conduct the experiment in their own chemistry class.

Second vignette: Using candies for chemistry teaching. In the second vignette, we describe teachers' professional knowledge development that dealt with developing experiential teaching. Using a variety of teaching methods is a recommended approach (R. Blonder & Sakhnini, 2012). In science teaching, it is used to increase students' interest in science, to motivate them to learn new challenging topics in science, and to support their understanding. There are many ways used for experiential teaching, e.g., integrating dance or body movement into the learning process (Zohar, Bagno, Eylon, & Abrahamson, 2017), and the creation of visual art products as part of the learning process (Sharaabi-Naor, Kesner, & Shwartz, 2014). In the example presented, we show how candies were used in teaching chemistry.

During the last meeting of the CtH PLC in the first research year, one of the PLC teachers (Erez) initiated DE 2, and posted 4 photos from an activity that he had conducted in which the CtH PLC teachers built models of molecules using different sweets, candies, and snacks as the building blocks of the model (photo A, Figure 4). Erez also explained the photos: "In the activity we asked the teachers to build the structural formula for different molecules with candies". One of the facilitators commented: "Today we're getting out of the box ☺".

It is important to note that the idea of using candies in chemistry teaching, which was presented by Erez, was an original activity initiated by him and was not shown in any other professional development framework for chemistry teachers in Israel. After Erez posted his activity with the teachers, we discussed the idea of using candies in chemistry lessons with teachers (not in the WhatsApp group). The PLC chemistry teachers did not express enthusiasm about the idea of bringing candies for group work during the lesson. They raised many concerns. For example, students would not take the chemistry class seriously, students might play and eat the candies instead of learning chemistry, and the group work with the candies could create a mess in class. In the next CtH PLC meeting, which was at the beginning of

Figure 4. Using candies in chemistry teaching. Different stages of knowledge building are reflected in the photos: A). The first time, the idea was suggested by Erez, one of the PLC teachers. B) Another PLC teacher conducts a similar activity with the PLC she leads. C-D) Activities with students.

the next academic year, one of the PLC teachers posted photos showing her conducting a similar activity using candies to build models of molecules with the text: "Building sweet molecules and exposing our knowledge about materials" (photo B, Figure 4). This teacher adopted Erez's activity, which he posted in the WhatsApp group. She used candies to conduct experiments within a chemistry teacher's PLC that she leads in central Israel. In her photos, she also includes the activity sheet that was used, which she later shared with the PLC members by Email. After one week, one of the PLC facilitators posted a photo showing an activity she conducted with her students to address their difficulties in understanding what happens to bi-atom molecules in the evaporation process. She used two different candies and asked her students to use the candies to build two different materials ($He_{(g)}$, and $H_{2(g)}$) that would resemble the gas phase under the same pressure and temperature conditions (photo C, Figure 4). This teacher used experiential teaching methods to deal with a misconception she had discovered in her class, using one of the diagnostic questionnaires that were presented at the PLC meetings. At the beginning of the next activity year, another PLC teacher conducted a variation of the candies activity with the CtH PLC she leads. She posted 4 photos of her PLC teachers while they used candies to build the molecules (e.g., photo D, Figure 4). In this activity, she connected the use of candies to a different part of the chemistry curriculum in Israel, which is related to functional groups. In the topic, the students are requested to remember the structure of different functional groups (e.g., carbocyclic acid, ketones, and alcohol) and to be able to recognize them. Many students have difficulties in recalling the different functional groups; however, the activity with the candies, since it is experiential, supports better and longer learning. As was mentioned in the previous example, every photo that was posted received many supportive responses as well as discussions regarding the exact way the teaching method should be used in class. This example shows how an idea proposed by one of the PLC teachers was discussed in the PLC WhatsApp group and from there, reached the chemistry classes and the chemistry students of other PLC teachers.

DISCUSSION

In the chemistry PLC, a group of chemistry teachers systematically reflect upon their knowledge and practice in order to improve their teaching. The teachers' PLC has already been shown to be a supportive environment in which teachers can build their professional knowledge and practices (Cochran-Smith & Lytle, 1999; Darling-Hammond & Sykes, 1999). We will now discuss how the WhatsApp group (or other similar online immediate message platforms that share the same features, as mentioned in the introduction section) contributed to the development of the PLC teachers' professional knowledge.

According to the analysis of the topics that were discussed in the PLC WhatsApp group (Table 1), much of the discourse in the WhatsApp group included chemistry teaching issues and chemistry contents. Most of the discourse dealt with professional aspects of chemistry teaching. However, other topics of discourse were also found and include DEs that were more social and DEs that dealt with organizational aspects of the PLC (LMS). In addition, the discourse in the WhatsApp group combined the presence of the PLC facilitators with PLC teachers' posts, as demonstrated in the vignettes discussed. Previous studies of online communities indicated that the presence of discourse regarding the learned content as well as social, and teaching aspects is essential in order to achieve learning with an online community (Garrison, Anderson, & Archer, 1999; Rap & Blonder, 2016). In the present study, three components were identified: the discourse includes a discussion about chemistry teaching (here – the content learned), and social and teaching presence is shown here by the presence of the PLC facilitators. Therefore, this online PLC fulfills the conditions that are needed in order to achieve learning for its participants.

In this study, we re-examined the model of teachers' professional knowledge and skills (Guess-Newsome, 2015), and suggested an additional path by which teachers develop their knowledge and skills, which is denoted by the blue path in Figure 2. According to the original model (presented in black, Figure 2), the teacher uses her amplifiers and filters before she decides whether to introduce a new pedagogy or a new activity to her class, for example, the laboratory experiment that was discussed in the Results section. Although the experiment was presented to the PLC teachers by the PKC facilitators using a top-down approach, the PLC teachers thought about the experiment, and some of them tried it in their school lab with their lab technician, and then decided whether they would introduce the experiment to their students. However, in the PLC supported by the WhatsApp group the teachers who investigated the experiment in their schools and discussed the conditions in the WhatsApp group, the "blue path" was opened for other teachers. In doing so, they deepened their understanding of the experiment's chemistry explanations, and experienced its technical challenges before introducing it to their students. When one of the PLC teachers actually tried the new lab experiment in his class or with the teachers in the CtH PLC he leads, and posted his experience to the group, he provided the other PLC members of the WhatsApp group with a personal example of successfully conducting the experiment (R. Blonder et al., 2013). The WhatsApp group members undergo vicarious experience (Bandura, 1986) by seeing their colleagues' results when trying out the new lab experiment. They also received the reactions of students in the teacher's class to the experience, which can also influence their attitudes and decision to perform the new experiment in their own class. This additional path is denoted in blue in Figure 1. Teachers who hesitate to conduct a new experiment with their students, because they might believe that it technically will not work, or they are afraid that their students might not be able to conduct the experiment or understand it, can receive authentic information about the effectiveness of the experiment in a real class context. They can read the discussion in the PLC WhatsApp group and receive artifacts from PLC members, which arrive directly from their class. This path (indicated in blue, in Figure 2) provides an additional route by which teachers' filters and amplifiers can be influenced and it can also influence teachers' PCK&S regarding the new experiment without teachers trying the new experiment in their own class.

Let us examine the second vignette, which is provided in the Results section, in light of the suggested path (the blue path, Figure 2). One of the teachers posted an example of an experiment he conducted with his students as part of experiential teaching pedagogy. He used candies for experientially teaching the topic of the structural formula of molecules. This activity is an example of a pedagogical approach in which a variety of teaching methods are used by a teacher in responding to each student's needs. Candies were used after the traditional teaching of structural formula. This enables many students to study chemistry in ways that are more aligned with their own interests and learning styles (Young, Klemz, & Murphy, 2003). Implementing a wide spectrum of instructional techniques and ways of measuring students' achievement and progress requires matching an appropriate assessment tool for each technique (R. Blonder, 2018). In practice, however, it is often difficult to employ a wide repertoire of instructional methods (Hofstein, Mamlok, & Rosenberg, 2006). In addition, teachers might feel that adopting a different pedagogy is beyond their capabilities and therefore, they would not introduce it to their class, and would never receive their students' responses to the pedagogy. In the example provided in the second vignette, the different pedagogy is the group work with candies. By using the new path provided by the PLC WhatsApp group (the blue path, Figure 2), the teachers can experience the trials of other PLC members in their classes and can see the reactions of the other teachers' students to the activity. Again, the vicarious experience influenced teachers' attitudes, and consequently, their professional knowledge development. Exposure to an idea that works for other teachers, offered by their PLC peers, not only broadens the teachers' repertoire, but also makes this idea accessible. In this example, there is no directed hierarchy of knowledge flow. The idea originated from one of the PLC teachers and was accepted, adapted, and implemented by other PLC members in a bottom-up process. This process contributed to the building of social capital residents in the PLC, which leads to behavioral change—a change that results in greater knowledge sharing, which in turn, positively influences teachers' performance (Lesser & Storck, 2001).

Knowledge creation in PLCs occurs when people participate in the problem solution process and share the knowledge needed to solve them (Wenger, 1999). The community enables members to jointly deal effectively in finding solutions in innovative ways (Manville & Foote, 1996). A study that was conducted using a Delphi methodology found that 42% of the knowledge that experts need in order to conduct their work comes from other people's advice, opinions, and answers (Uriarte, 2008). This finding supports the educational use of teachers' PLC, especially if we consider the isolated situation of disciplinary teachers (e.g., chemistry teachers) in schools. However, using the WhatsApp group to present evidence from teachers' classes regarding a specific activity (e.g., trying a new lab experiment, using candies for experiential teaching) was found to contribute significantly to the knowledge-building process of teachers' personal PCK&S via the suggested mechanism. We found that the PLC teachers raised questions that they faced while teaching (including how to vary their teaching) and this initiated a bottom-up process of knowledge development. This path also included practical and technical issues, as well as innovative teaching ideas. We also found that different professional knowledge requires the top-down direction of knowledge flow. New lab experiments, a new pedagogical approach (such as using diagnostic tools, or a personalized teaching approach) were initiated by the PLC facilitators using a top-down approach. Using this mechanism of providing teachers with the opportunity to present their teaching in front of their colleagues was found to be a very effective approach to enhance the PLC teachers' self-efficacy beliefs, and was used in previous studies during face-to-face meetings, and thus, influenced their filters and professional knowledge (R. Blonder et al., 2013). In the current study, the online platform of the PLC WhatsApp group provided a method by which this process can continue between the meetings in a very accessible and useful way.

Research shows that social factors are important in the development and maintenance of communities. They facilitate the maturing of the community, knowledge sharing, longer engagements, and the community's outcome expectations (Macià & García, 2016). In addition, Grossman, Wineburg, and Woolworth (2001) claimed that holding real-life meetings in PLCs is not enough to ensure the development of a sense of community. It has also been shown that participation in face-to- face meetings does not always involve cooperation between community members, especially in those cases that require greater commitment from participants, e.g. a collaborative creation of artefacts (Coutinho & Lisbôa, 2013). In addition, the literature regarding online PLCs distinguishes between two types of participants in the online communication platforms: the active users that post photos and text and react to others (such as Ben in DE1 and DE2), and the passive participants, who usually read the contents that are posted but do not contribute their own ideas and perspectives. Most of the studies reported that the percentage of passive participants is high (Ardichvili, Maurer, Li, Wentling, & Stuedemann, 2006; Preece, Nonnecke, & Andrews, 2004). Regarding the chemistry teachers' PLC studied here, the norms of the group created a WhatsApp group that is very actively used by the vast majority of the PLC members. These norms of trust are essential for the sharing of contents and for the active participation of the PLC teachers in the discourse. Teachers initiated discourses (as shown in DE2, which dealt with the candies) and felt confident enough to post their doubts and their unsuccessful trials (as was shown in DE regarding the experimental system of the new lab experiment). This is an essential aspect for developing teachers' professional knowledge. The two vignettes presented in the current chapter represent many other DEs in which the development of teachers' professional knowledge was demonstrated. Other examples include discussions about other topics and other dimensions in chemistry teaching. We wish to stress that the attention given in the PLC meetings on developing trust among the PLC members created a sense of community that supported the development of teachers' knowledge.

CONCLUSION AND IMPLICATIONS

In this study we linked the model for teachers' professional knowledge development (Gess-Newsome, 2015) to an online platform used for group communication. We showed that the WhatsApp group, when used by the teachers' PLC, provided an additional means of developing teachers' professional knowledge by adding the shared vicarious experiences of other teachers in the PLC for the benefit of the entire PLC community. PLC teachers could read the discourse in the online mobile WhatsApp group and discover how their peers experienced a new pedagogy in their class. This process influences the PLC teachers' attitudes towards this pedagogy and could affect their personal PCK&S. It is therefore recommended to add the technologically supported path to the theoretical model of teachers' professional knowledge and to further investigate its contribution and its working mechanism.

We would like to suggest that better understanding the mechanism by which the group communicates via WhatsApp could help managers of online groups enhance this process. For example, they can invite the group members to share their classroom experiences with the group. This can be done when one of the PLC members shares with the facilitators that she or he tried out a new activity in chemistry class. In this way, we can transform our research insights into practical contributions to develop teachers' knowledge through online platforms.

ACKNOWLEDGMENT

This research was supported by the Trump Foundation [grant numbers 154]; the National Chemistry Teachers' Center and was supported by the Ministry of Education [Tender: 09.07.13]. We would like to thank the leading team of the chemistry PLC: Dr. Dvora Katchevich, Dr. Malka Yayon, Dr. Rachel Mamlok-Naaman, Mrs. Sara Ekons, and Mrs Enas Easa.

REFERENCES

Arcavi, A., & Karsenty, R. (n.d.). *Enhancing Mathematics Teachers' Knowledge and Reflection Through Peer-Discussions of Videotaped Lessons: A Pioneer Program in Israel*. Academic Press.

Ardichvili, A., Maurer, M., Li, W., Wentling, T., & Stuedemann, R. (2006). Cultural influences on knowledge sharing through online communities of practice. *Journal of Knowledge Management, 10*(1), 94–107. doi:10.1108/13673270610650139

Avidar, R., Ariel, Y., Malka, V., & Levy, E. C. (2013). Smartphones and young publics: A new challenge for public relations practice and relationship building. *Public Relations Review, 39*(5), 603–605. doi:10.1016/j.pubrev.2013.09.010

Bandura, A. (1986). The explanatory and predictive scope of self-efficacy theory. *Journal of Social and Clinical Psychology, 4*(3), 359–373. doi:10.1521/jscp.1986.4.3.359

Baran, E. (2014). A Review of research on mobile learning in teacher education. *Journal of Educational Technology & Society, 17*(4), 17–32.

Barber, M., & Mourshed, M. (2007). *How the world's best-performing school systems come out on top.* McKinsey & Company.

Blonder, R. (2018). Student-curated exhibitions: Alternative assessment in chemistry education in Israel. In C. Cox & W. E. Schatzberg (Eds.), *International perspectives on chemistry education research and practice* (Vol. 1293, pp. 39–55). American Chemical Society.

Blonder, R., Jonatan, M., Bar-Dov, Z., Benny, N., Rap, S., & Sakhnini, S. (2013). Can You Tube it? Providing chemistry teachers with technological tools and enhancing their efficacy beliefs. *Chemistry Education Research and Practice, 14*(3), 269–285. doi:10.1039/C3RP00001J

Blonder, R., & Mamlok-Naaman, R. (2016). Learning about teaching the extracurricular topic of nanotechnology as a vehicle for achieving a sustainable change in science education. *International Journal of Science and Mathematics Education, 14*(3), 345–372. doi:10.100710763-014-9579-0

Blonder, R., & Sakhnini, S. (2012). Teaching two basic nanotechnology concepts in secondary school by using a variety of teaching methods. *Chemistry Education Research and Practice, 13*(4), 500–516. doi:10.1039/C2RP20026K

Booth, S. E. (2012). Cultivating knowledge sharing and trust in online communities for educators. *Journal of Educational Computing Research, 47*(1), 1–31. doi:10.2190/EC.47.1.a

Bouhnik, D., & Deshen, M. (2014). WhatsApp goes to school: Mobile instant messaging between teachers and students. *Journal of Information Technology Education: Research, 13*, 217–231. doi:10.28945/2051

Britner, S. L., & Pajares, F. (2006). Sources of science self-efficacy beliefs of middle school students. *Journal of Research in Science Teaching, 43*(5), 485–499. doi:10.1002/tea.20131

Cansoy, R. (2017). Teachers' professional development: The case of WhatsApp. *Journal of Education and Learning, 6*(4), 285–293. doi:10.5539/jel.v6n4p285

Church, K., & de Oliveira, R. (2013). *What's up with whatsapp?: comparing mobile instant messaging behaviors with traditional SMS.* Paper presented at the 15th international conference on Human-computer interaction with mobile devices and services, Munich, Germany. 10.1145/2493190.2493225

Cobb, P., Yackel, E., & Wood, T. (1992). Interaction and learning in mathematics classroom situations. *Educational Studies in Mathematics, 23*(1), 99–122. doi:10.1007/BF00302315

Cochran-Smith, M., & Lytle, S. L. (1999). Relationships of knowledge and practice: Teacher learning in communities. *Review of Research in Education, 24*, 249–305.

Coutinho, C. P., & Lisbôa, E. S. (2013). Social networks as spaces for informal teacher professional development: Challenges and opportunities. *International Journal of Web Based Communities, 9*(2), 199–211. doi:10.1504/IJWBC.2013.053244

Darling-Hammond, L., & Sykes, G. (1999). *Teaching as the learning profession: Handbook of policy and practice*. San Francisco: Jossey-Bass.

Forkosh-Baruch, A., & Hershkovitz, A. (2012). A case study of Israeli higher-education institutes sharing scholarly information with the community via social networks. *The Internet and Higher Education, 15*(1), 58–68. doi:10.1016/j.iheduc.2011.08.003

Garrison, D. R., Anderson, T., & Archer, W. (1999). Critical inquiry in a text-based environment: Computer conferencing in higher education. *The Internet and Higher Education, 2*(2–3), 87–105. doi:10.1016/S1096-7516(00)00016-6

Gess-Newsome, J. (2015). A model of teacher professional knowledge and skill including PCK: results of the thinking from the PCK summit. In A. Berry, P. J. Friedrichsen, & J. J. Loughran (Eds.), *Re-examining pedagogical content knowledge in science education* (pp. 28–42). New York, NY: Routledge.

Grossman, P., Wineburg, S., & Woolworth, S. (2001). Toward a theory of teacher community. *Teachers College Record, 103*(6), 942–1012. doi:10.1111/0161-4681.00140

Guskey, T. R. (2003). What makes professional development effective? *Phi Delta Kappan, 84*(10), 748–750. doi:10.1177/003172170308401007

Hofstein, A., Mamlok, R., & Rosenberg, O. (2006). Varying instructional methods and students assessment methods in high school chemistry. In M. McMahon, P. Simmons, R. Sommers, D. DeBaets, & F. Crawley (Eds.), *Assessment in Science* (pp. 139–148). Arlington, VA: NSTA press.

Hofstein, A., Shore, R., & Kipnis, M. (2004). Providing high school chemistry students with opportunities to develop learning skills in an inquiry-type laboratory: A case study. *International Journal of Science Education, 26*, 47–62. doi:10.1080/0950069032000070342

Karsenty, R., & Arcavi, A. (2017). Mathematics, lenses and videotapes: A framework and a language for developing reflective practices of teaching. *Journal of Mathematics Teacher Education, 20*(5), 433–455. doi:10.100710857-017-9379-x

Karsenty, R., & Sherin, M. G. (2017). Video as a catalyst for mathematics teachers' professional growth. *Journal of Mathematics Teacher Education, 20*(5), 409–413. doi:10.100710857-017-9387-x

Lave, J., & Wenger, E. (1991). *Situated learning: Legitimate peripheral participation*. Cambridge, UK: Cambridge University Press. doi:10.1017/CBO9780511815355

Lesser, E. L., & Storck, J. (2001). Communities of practice and organizational performance. *IBM Systems Journal, 40*(4), 831–841. doi:10.1147j.404.0831

Lieberman, A., & Mace, D. (2010). Making practice public: Teacher learning in the 21st century. *Journal of Teacher Education, 61*(1-2), 77–88. doi:10.1177/0022487109347319

Little, J. W. (2002). Locating learning in teachers' communities of practice: Opening up problems of analysis in records of everyday work. *Teaching and Teacher Education, 18*(8), 917–946. doi:10.1016/S0742-051X(02)00052-5

Little, J. W. (2012). Professional community and professional development in the learning-centered school. *Teacher learning that matters: International perspectives*, 22–46.

Macià, M., & García, I. (2016). Informal online communities and networks as a source of teacher professional development: A review. *Teaching and Teacher Education, 55*, 291–307. doi:10.1016/j.tate.2016.01.021

Magnusson, S., Krajcik, J., & Borko, H. (1999). Nature, sources, and development of pedagogical content knowledge for science teaching. In J. Gess-Newsome & N. G. Lederman (Eds.), *Pedagogical content knowledge and science education* (pp. 95–132). Dordrecht: Kluwer Academic Publishers.

Manville, B., & Foote, N. (1996). Harvest your workers' knowledge-It's a brave new world out there. Call it post-modern reengineering, and to make your organization perform you'll have to build systems that support knowledge--not. *Datamation-Highlands Ranch, 42*(13), 78–83.

McMillan, D. W., & Chavis, D. M. (1986). Sense of community: A definition and theory. *Journal of Community Psychology, 14*(1), 6–23. doi:10.1002/1520-6629(198601)14:1<6::AID-JCOP2290140103>3.0.CO;2-I

Miles, M. B., Huberman, A. M., & Saldana, J. (2013). Qualitative data analysis. *Sage (Atlanta, Ga.).*

Ndlovu, M., & Hanekom, P. W. (2014). Overcoming the limited interactivity in telematic sessions for in-service secondary mathematics and science teachers. *EDULEARN14 Proceedings*.

Preece, J., Nonnecke, B., & Andrews, D. (2004). The top five reasons for lurking: Improving community experiences for everyone. *Computers in Human Behavior, 20*(2), 201–223. doi:10.1016/j.chb.2003.10.015

Rap, S., & Blonder, R. (2016). Let's Face(book) it: Analyzing interactions in social network groups for chemistry learning. *Journal of Science Education and Technology, 25*(1), 62–76. doi:10.100710956-015-9577-1

Rogers, C. R. (1983). *Freedom to learn for the 80's*. Merrill Publishing.

Seufert, M., Hoýfeld, T., Schwind, A., Burger, V., & Tran-Gia, P. (2016). *Group-based communication in WhatsApp*. Paper presented at the IFIP Networking Conference (IFIP Networking) and Workshops. 10.1109/IFIPNetworking.2016.7497256

Sfard, A. (1998). On two metaphors for learning and the dangers of choosing just one. *Educational Researcher, 27*(2), 4–13. doi:10.3102/0013189X027002004

Sfard, A. (2006). Participation discourse on mathematics learning. In J. Maass & W. Schlöglmann (Eds.), *New mathematics education research and practice* (pp. 153–170). Rotterdam, The Netherlands: Sense.

Sfard, A. (2008). *Thinking as communicating*. Cambridge, UK: Cambridge University Press. doi:10.1017/CBO9780511499944

Sharaabi-Naor, Y., Kesner, M., & Shwartz, Y. (2014). enhancing students' motivation to learn chemistry. *Sisyphus-Journal of Education*, *2*(2), 100–123.

Sharples, M., Taylor, J. A., & Vavoula, G. (2010). A theory of learning for the mobile age. In B. Bachmair (Ed.), *Medienbildung in neuen Kulturräumen: Die deutschprachige und britische Diskussion* (pp. 87–99). Wiesbaden: VS Verlag für Sozialwissenschaften. doi:10.1007/978-3-531-92133-4_6

Shulman, L. S. (1986). Those who understand: Knowledge growth in teaching. *Educational Researcher*, *15*(2), 4–14. doi:10.3102/0013189X015002004

Strauss, A., & Corbin, J. (1998). *Basics of qualitative research: Techniques and procedures for developing grounded theory* (2nd ed.). Thousand Oaks, CA: Sage Publications.

Taitelbaum, D., Mamlok-Naaman, R., Carmeli, M., & Hofstein, A. (2008). *Evidence for teachers' change while participating in a continuous professional development programme and implementing the inquiry approach in the chemistry laboratory*. Academic Press.

Tschannen-Moran, M. (2009). Fostering teacher professionalism in schools: The role of leadership orientation and trust. *Educational Administration Quarterly*, *45*(2), 217–247. doi:10.1177/0013161X08330501

Uriarte, F. A. (2008). Knowledge Management: Konsep, Arsitektur dan Impelementasi Introduction to Knowledge Management: A brief introduction to the basic elements of knowledge management for non-practitioners interested in understanding the subject. In Sense of community in a professional learning community of chemistry teachers: A study of an online platform for group communication. Springer.

Wenger, E. (1999). *Communities of practice: Learning, meaning, and identity*. Cambridge University Press.

Wenger, E. (2011). *Communities of practice: A brief introduction*. Academic Press.

Young, M. R., Klemz, B. R., & Murphy, J. W. (2003). Enhancing learning outcomes: The effects of instructional technology, learning styles, instructional methods, and student behavior. *Journal of Marketing Education*, *25*(2), 130–142. doi:10.1177/0273475303254004

Zohar, R., Bagno, E., Eylon, B.-S., & Abrahamson, D. (2017). Motor skills, creativity, and cognition in learning physics concepts. *Journal of Functional Neurology, Rehabilitation, and Ergonomics*, *7*(3), 67–76.

APPENDIX

Lab-Manual of the Experiment: "Colored Arm-Wrestling"

The experiment was adapted by Mrs. Sara Akons from (Hofstein, Shore, & Kipnis, 2004).

Note: Protective glasses and gloves must be worn!

General Instructions

- Read all the instructions well before beginning the experiment.
- Check that you have all the necessary equipment and materials at your disposal in order to conduct the experiment.

Equipment and Materials

- 6 g $NaOH_{(s)}$
- 10 g of $NaHCO_{3(s)}$
- 10 g of citric acid
- 100 ml of red cabbage solution
- A 100 ml cylinder
- A 3-liter pickle jar
- A lab glove
- 1 clip
- A rubber band
- 3 plastic spoons

Stage A: The Course of the Experiment

Pay attention to:

- Follow the instructions for carrying out Stage A.
- Record as many observations as possible.
- Have all members of the group participate in carrying out the various tasks.
- Use correct and precise scientific language throughout the process.
- Take an empty lab-glove and add 6 g NaOH powder to one of the glove fingers. Close the finger with clips.
- Carefully add to the same glove (not to the finger) 10 g of $NaHCO_3$ powder and 10 g of citric acid.
- Measure with the cylinder a 100 ml red cabbage solution.
- Add the red cabbage solution to the pickle jar.
- Carefully close the jar with a glove. Make sure that it is closed well and use a rubber band to tighten the glove to the jar.
- The powder that was added to the glove in stage 3 will be poured into the jar.
- Observe the system until no change can be seen.

- Open the clips that closed the finger with the powder from stage 2, and pour the powder into the jar (keep the jar closed).
- Mix intensively until all the powder in the jar is dissolved.

Stage B: The Steps of the Inquiry

1. Formulate 5 varied, relevant questions that arose following the observations.
    a. Choose one of these questions that you want to investigate.
    b. Formulate this question clearly as an inquiry question and if possible, as a link between two variables.
    c. Formulate clearly a hypothesis for the question you chose to investigate.
    d. Give reasons for your hypothesis, based on correct, relevant scientific knowledge.
2. Plan an experiment that will check the validity of your hypothesis.
    a. Detail all the stages of the experiment, including the control stage.
    b. List the equipment and materials needed on an equipment request form.
    c. Consult with the teacher and, if necessary, make changes.
    d. Submit the list of equipment and materials to the laboratory technician.
3. Get the teacher's approval for your proposed experiment.
    a. Carry out the experiment that you proposed after receiving your teacher's approval.

Section 3

Global and Multicultural Considerations and Implications of Mobile Learning

Chapter 8
Culture and Context Impact on Mobile Tech Application in Organizational Learning:
Case Study of UK Higher Education Institution and Chinese State-Owned Enterprise

Yun Chen
University of Salford, UK

Yunpu Zhang
China Nuclear Engineering Corporation Limited, China

ABSTRACT

While the significance of mobile technologies to support deep learning has been recognized, the impact factors of applying these technologies into various organizational learning environments still remain poorly understood. This chapter has explored the community of inquiry (CoI) framework as the lens to understand culture and context impact on mobile technology application to facilitate learners' engagement and educational experiences. Two sharp-contrast case studies are selected to provide the practical insight of the research. Results infer that usage of mobile technology has a positive impact on organizational learning, which leads to greater learning engagement as long as culture and context are considered as key issues to select appropriate mobile technologies. These factors decide the mobile native functions, delivery platforms, and provided learning contents, which are key elements in the proposed CoI framework in the context of m-learning.

DOI: 10.4018/978-1-5225-8106-2.ch008

INTRODUCTION

Mobile technology is becoming increasingly pervasive and has spread at a rapid pace over the past few years. The concept of using mobile technologies in learning has been known to facilitate learning processes, collaboration and sharing among learners. Mobile learning (M-learning) is therefore defined as the use of portable mobile devices with the Internet in the learning process (Anohah et al, 2017; Crompton & Burke, 2018). Since this form of learning is accessible through a mobile device by anyone, anywhere and anytime, it provides an opportunity for a lifelong and continuous learning without limitation of time and context (Nagella & Govindarajulu, 2009). Research shows that M-learning is a trend of mobile technology usage and it can support learners in many ways; with functionality ranging from the simple use of SMS texting, providing private availability of on-demand study support to the more advanced use of mobile and smartphones for pedagogy. However, the literacy gaps between M-learning technologies and organisational issue associated with technological application in different learning contexts provide challenges to adoption of this trend. There are a few literature reviews providing the theoretical frameworks to bridge the gap, such as Community of Inquiry (CoI) framework (Garrison et al., 2000) for blended learning. The framework emphasizes social, cognitive and teaching presences when combining face-to-face and online learning. It argues that by finding the appropriate balance between the three presences, students' engagement could be fostered by empowering them to become active learners and producers of knowledge, rather than just passive consumers of information. However, the usage of this framework is mostly focused in Higher Education instead of other organisational learning environment.

This chapter aims to undertake comparative case study to address the challenge and fill the gap described above by investigating two sharp-contrasted learning environments, i.e. Higher Education institution (HEI) in the UK and State-Owned Enterprise (SOE) in China, through the lens of CoI framework. The following two questions were addressed by the research:

- What are the implications for the implementation of a stable community of inquiry in the M-learning context? (by theoretical research)
- How will culture and context impact on the mobile technology application in organisational learning? (by practical research)

In this chapter, 'Background' section will focus on M-learning and CoI framework, together with discussion on different aspects of CoI framework in the context of M-learning. This review will lead to the discussion on how mobile technologies have been used in UK HEIs and China SOEs to encourage successful establishment of community of inquiry. Practical research is followed after literature review in 'Comparative Case Study' section to discuss mobile technologies used in Salford Business School in the UK and China Unicom in China. The two case studies will be compared in terms of characteristics of learner groups, motivation mechanism of their learning journey, mobile functions, contents and learning culture to inform the culture and context impact on mobile technology utilisation in organisational learning. The discussion will be summarised and good practice framework will be proposed in 'Solutions and Recommendations' section. The chapter will be concluded in final section based on both theoretical and practical research discussed in previous sections. Limitation of this research will be addressed and recommendation for future research will be proposed as well.

BACKGROUND

Nowadays, the trend of teaching and learning has been shifted from passive lecture approach to student-engaged one. The concept of 'student engagement' has been researched for several decades, such as Pace's (1980) "quality of effort" concept, Astin's (1999) theory of student involvement, Chickering and Gamson's (1999) principles of good practice in undergraduate education, and Pascarella and Terenzini's (2005) causal model of learning and cognitive development. In 2001, Edgerton also highlighted the need for students to engage in tasks in order to really understand the concepts of the discipline. Because of significance of student engagement, blended learning has become a buzzword in pedagogic research.

Blended learning, sometimes used interchangeably with the term 'Hybrid learning', refers to the learning model that combines traditional classroom teaching activities with online and out-of-class coursework (Bleed, 2001; Williams, 2002). It intends to combine the best practice of both face-to-face and eLearning to support student engagement. Vaughan (2010) argues "at the heart of blended learning redesign is the goal to engage students in critical discourse and reflection". This goal can only be achieved based on the clear understanding of the nature of learning process and intended learning outcomes. And Garrison and Vaughan (2008) believe at the core of this process is "the community of inquiry that supports connection and collaboration between learners". In this Section, the existing research on M-learning and Garrison and Vaughan's 'Community of Inquiry' framework will be reviewed and then how mobile technologies are used to facilitate integration of social, cognitive and teaching presences in learning environment will be discussed.

Community of Inquiry Framework (CoI)

The CoI framework was created by Garrison et al. (2000) to guide the research and practice of computer-mediated communication in learning environment. The framework is defined in Garrison's later book (2008) as *"Community, on one hand, recognises the social nature of education the role interaction, communication and discourse play in constructing knowledge. Inquiry, on the other hand, reflects the process of constructing meaning through personal responsibility and choice. A Community of Inquiry is a cohesive and interactive community of learners whose purpose is to critically analyse, construct and confirm worthwhile knowledge."* The framework identifies three elements that are crucial prerequisites for a successful educational experience, i.e. *Social presence*, *Cognitive presence* and *Teaching presence.* These three elements and their interrelationship are outlined in Figure 1 below:

Successful education experience requires the discipline to interact academically and respectably with member of community. *Social presence* is the primary concern and foundation of creating community of inquiry. The community should be able to provide a trustful and purposeful inquiry environment for learners, so that they would feel secure to reveal their private thoughts, interact with others and open them to critique.

Inquiry is a systematic process of defining problem, seeking information and applying solution, which includes both personal reflective and social interactive processes. *Cognitive presence* is basic to inquiry process. It is a recursive process that combines information gathering, idea creation, solution testing so that understanding and knowledge sharing could be enriched. The Community need to be cognitively rich which supports continuous cognitive development of individuals.

Figure 1. Community of Inquiry Framework
(Garrison, et al., 2000)

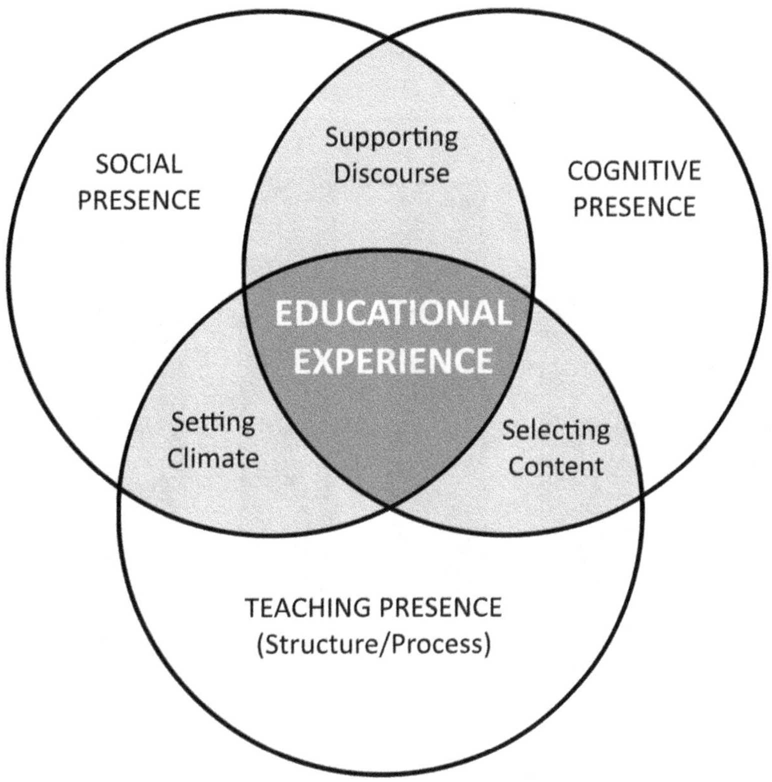

In educational context, *teaching presence* is essential to bring all the elements together to ensure the community is productive. It decides on teaching approach and adopt of tools. It also moderates and guides learners' reflection, interaction and collaboration. It is important to use teaching presence to bring social and cognitive presences together in an effective and efficient manner to support community of inquiry.

There is evidence of construct validity of CoI framework in Higher Education institutions to explore the synergies of face-to-face and online learning (Arbaugh, 2007; Lapointe & Gunawardena, 2004). But what does this mean to mobile tech application in learning in terms of social and teaching presences? How to establish and maintain cognitive presence in other organisational learning environments? What kind of mobile techs can enhance teaching presence? The rest of this chapter will investigate these questions based on both primary and secondary research.

M-Learning

Evolution of Mobile Technologies

Smartphone and devices have been playing a ubiquitous role in different areas of today's society.

As shown in Figure 2, in 2019 the number of mobile phone users is forecast to reach 4.68 billion.

Along with the quickly increased number of mobile holders, the mobile technology has also evolved through several generations with increased capabilities:

Figure 2. Number of mobile phone users worldwide from 2015 to 2020
(Statista, 2018)

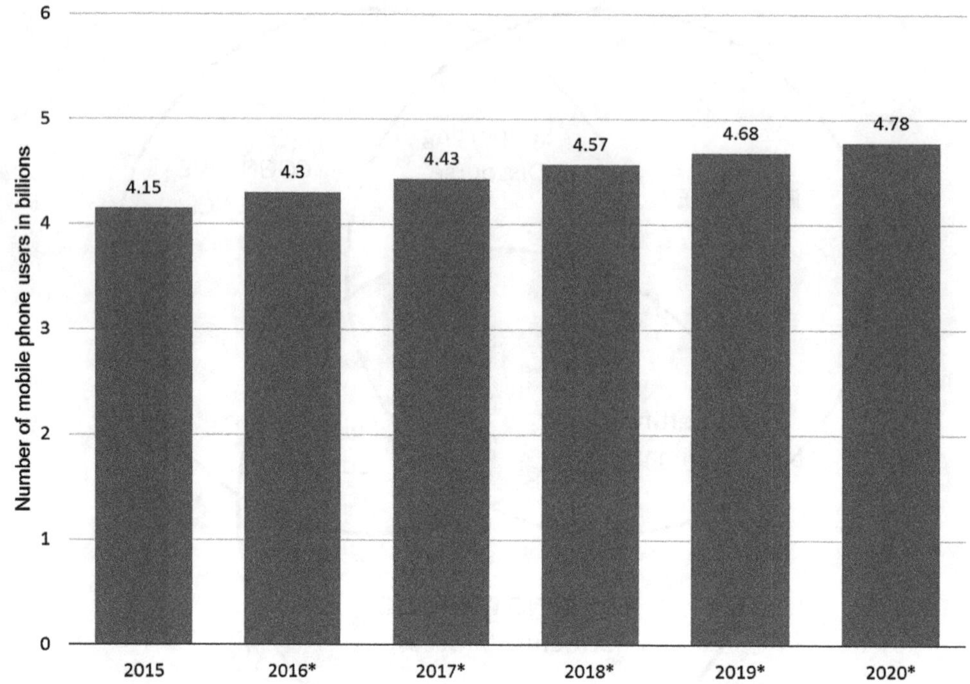

In *Mobile 1.0,* 2G (known as GSM: Global System for Mobile communication) used digital signals and integrated a modem to call dial-up Internet service numbers provided by the phone carrier (Roundtree, 2011). In those earlier days, mobile devices were mainly for communication via phoning or texting people, with few mobile games appearing on mobile devices, such as Nokia's Snake game.

In the next generation of mobile technology, *Mobile 2.0*, the modern networks such as 3G/4G provide very fast data access on phones, including multimedia elements such as audio/video content. Mobile 2.0 technologies enable mobile devices as a tool to complete the computer tasks. They are supported by an operating system and local storage. Users can add and store information, send and receive e-mail, and install programs to the phone. Most of online services are mobilised as mobile applications for convenient use, such as trading, ticket purchasing, event booking and online banking etc.

Mobile technology is now moving to its third generation. *Mobile 3.0* aims to enhance user experiences of using mobile devices by personalisation and contextualisation. E.g. by using Location-Based Service (LBS), Miami International Airport launched iBeacon to benefit passengers in 2014. Miami Airport Beacon Zones allow way-finding and location-related info-receiving to be offered on passenger's phones. It is e used for staff notifications and to beam operational information, such as temperature, noise levels and vibrations, etc., to allow efficient operational management (Airlines & Retailers, 2014).

Research on M-Learning

Widespread usage, accessibility and quickly- developed technologies of mobile devices also extend the educational world. Sung et al. (2016) claim that mobile devices are useful tools to stimulate motivation, strengthen engagements and deliver content. Extended from e-learning concept, using mobile devices in learning is sometimes called M-Learning or Mobile learning by researchers, but with different explanation. Crompton (2013) defined M-learning as '*learning across multiple contexts, through social and content interactions, suing personal electronic devices*'; Traxler and Kukulska-Hulme (2005) claim M-learning as " *a personal, unobtrusive, spontaneous, anytime, anywhere way to learn and to access education tools and materials that enlarges access to education for all.*' As a new enabler, M-learning is considered as a potential learning approach due to its accessibility, flexibility, nature of communication and interaction, and ease of assessment and feedback (Gikas and Grant, 2013; Han & Shin 2016). In research, there has been a growing interest in investigation on m-learning technologies at one hand, and the factors that affect its acceptance on the other hand, such as investigation on novelty effect of different learning tools and how these tools can help with learners' engagement and increase learners' motivation. (Cárdenas-Robledo and Peña-Ayala, 2018; Derounian, 2017; Jeno et al., 2019); empirical study on m-learning adoption in terms of learners' performance, perception and learning contexts (Bakhsh et al., 2017; Garcia-Cabot et al., 2015; Sabah, 2016); practical research on the impact of M-learning technology on students and educators (Abachi and Muhammad, 2014; Althunibat, 2015; Al-Emran et al., 2018a).

Every coin has two sides. Some research also highlight the challenges and downside of M-learning approach, which include technical standardization and comparability, privacy and security issues, delivery contents and learning activities on small screens. (Cheon et al. 2012; Elias, 2011, Parsons, 2009) In order to address these challenges, design and implementation of m-learning settings need carefully scrutiny from both technical and pedagogical perspectives (Alioon and Delialioglu, 2015). Garcia-Cabot et al. (2015) also highlight that context of mobile technologies application in learning is not fixed and should be well understood and defined for good practice.

M-Learning in CoI Framework

Alioon, & Delialioglu. (2015) highlighted that pedagogy, technology and learners play important roles in terms of M-learning based on the analysis of 30 M-learning projects between 20004 and 2012 in various nations. Different conceptual frameworks have been investigated on these roles to facilitate M-learning adoption in order to achieve good practice. Al-Emran et al. (2018b) systematically reviewed and synthesized the studies related to adopting Technology Acceptance Model (TAM) (Davis et al., 1989) in M-learning and provide a comprehensive analysis of 87 research articles from 2006 to 2018. The main findings suggest that although most of the TAM studies involving M-learning focused on extending the TAM with external variables, many other impact factors from other theories/models still need to be examined and investigated for M-learning acceptance and adoption (Al-Emran et al., 2018); Self-determination theory (SDT) (Ryan and Deci, 2017) provides a framework for understanding the factors that promote motivation and healthy psychological and behavioral functioning. Jeno et al. (2018) adopted SDT to investigate the effects of mobile application tools on biology students' achievement and well-being and suggest the mobile application had positive affect through autonomy, competence and intrinsic motivation; The Framework for Rational Analysis of Mobile Education model (FRAME) (Koole & Ally 2006) has also been applied as an analytic framework to explain the dynamics of mobile learning environment and guide the implementation of m-learning content management systems by Asiimwe et al. (2017). The FRAME model includes three interconnected dimensions, the *device aspect*, the *learner aspect,* and the *social aspect,* which are revised concepts of CoI framework three presences discussed in previous session, but focusing on M-learning paradigm. The study verifies that CoI framework has the potential to contribute to M-learning by addressing the nexus of technology and learners' needs. Based on existing research, the authors aim to explore the framework further to investigate each presence more specifically related to practical aspect of mobile technology application in organisational learning settings and extend it with consideration of culture and context impact. The three presences are interpreted as below:

The *social presence* refers to a socially constructed environment that allows interactions between people. It concerns factors that affect social interactions, collaborations, and conversations and how a social, interactive and engaging environment can be created for mobile learners (Assimwe et al., 2017). In M-learning settings, the key decision to make is the selection of mobile platforms, e.g. social media, educational apps or web-based systems?

The *cognitive presence* is closely related to learners. It addresses how the learner can be technologically and pedagogically facilitated and motivated. The delivery contents and facilitation of information exchanging through M-learning are impact factors to mobile learners. As a result, the authors believe contents and usability need to be addressed to ensure successful M-learning adoption in terms of this presence.

The *teaching presence* describes aspects of technologies and infrastructure used in M-learning environment to deliver teaching contents and activities. The goal is to eliminate the barriers but take the advantages of existing and emerging mobile technologies. As discussed in previous session, mobile technologies have been evolved from 1.0 to 3.0. What kind of mobile native functions and how these technologies can be embedded in the m-learning environment to support social and cognitive presences need to be considered for stable establishment of CoI.

Through the lens of the three presences discussed above, the rest of this section will explore how current mobile techs can help with learner engagement and inquiry in UK HEIs and China SOEs.

M-Learning in UK HEIs and China SOEs Through the Lens of CoI Framework

In terms of learning culture and context, HEIs and SOEs are extremely different. Consider study environment as HEI, students are required to stay connected with their academic community. In the community of UK HEIs, learning is learners' priority and learning is highly motivated by assignments, examinations and future employability. As a student, there is an expectation to keep track of exam dates, assignment deadlines, course schedules and so on. Compared with HEIs in the UK, SOEs in China are very different in terms of organisational learning. China has a larger number of SOEs than any other major economies in the world. Employees in these organisations are normally well-educated and comparatively stable. Professional accomplishment is the most desirable target for them. They are also keen to learn something that they really like or useful for their personal life. In order to obtain better understanding of the differences, the three presences of CoI framework are used for investigation.

Social Presence

As discussed in previous section, *Social Presence* is about how to develop a greater sense of community for learners. UK HEIs intend to provide all they can to keep students deeply engaged in the learning environment. The pedagogical functionality of these mobile technologies includes content delivery (e.g. Learning materials on Blackboard), searching for academic support (e.g. M-Library) (Ally & Needham, 2012); and task collaboration and planning (e.g. virtual office & virtual classroom). Furthermore, it offers possibilities to enhance educational setting, such as supporting communication and increasing learner-to-learner interaction (e.g. discussion board and wiki on Blackboard). The capabilities of current mobile technologies support use of multimedia elements, which meets requirements of current and future generation of learners. It also allows both synchronous and asynchronous communication which develops a greater sense of community.

In China, research shows that mobile technologies have been integrated into employee training schedule in organisations. Many enterprises bought professional developed educational apps (Liang, 2017). Some large SOEs have developed Online University to conduct more efficient employee training, e.g. China Telecom Online University (CTOU) and SINOPEC i-ACADEMY. CTOU was founded by China Telecom and aims to be personal assistant employees' work and learning. It allows the employees to download and log in mobile version by scanning QR code using their social media account. With multiple functions, the online university can help the staff to arrange their schedules, sharing documents and discussing within groups during the learning. SINOPEC i-ACADEMY is a traditional IE-based online resource and it serves as 'complex platform for staff development' (Sinopec i-Academic, 2018). Besides huge amount of learning courses, announcement and policies of the company are also shown online. Staffs can figure out popular courses by the number of visits or find something they are interested. They can also see each other's learning progress and be rewarded by various kinds of encouragements (e.g. high marks) by having longest on-line learning hours.

Cognitive Presence

Cognitive Presence has close relationship with learning motivation in certain learning contexts. Organisation instructor can reinforce and encourage this motivation by emphasising desired contents and designing activities that are especially suited to personal requirements. In UK HEIs, bitable-size study materials related to learning tasks are provided by most of library services with email alerts, social media links and Blackboard announcements, e.g. 'Skills for Learning" service provided by the University of Salford Library (Skills for Learning, 2018). This kind of manageable study approach could encourage students' continuous professional development to support their course learning and assignments. In addition, teachers can reinforce and emphasise aspects of the course contents by designing activities that are especially suited to in-person discussion and collaboration for student engagement and activities. Teaching best practice suggests that this method can enhance cognitive aspect of learning for different types of students.

On the other hand, trainings of Chinese SOEs' employees mainly focus on operating-related information and staff's personal development. The former can be regarded as an obligatory and passive learning, and the latter is highly self-motivated. Cognitive presence in this kind of learning environments means to encourage employees to learn something without short-term rewarding but actually will benefit in long run for both staff and the firm. For example, many SOE employees share English learning achievement on social media platform (see Figure 3), but most of them rarely use English at work. Mobile technologies can be used to motivate SOE employees' continuous professional development and learning by the way of personalisation and contextualisation.

Figure 3. Moment sharing of English study on WeChat

Teaching Presence

Teaching Presence decides the tools and methods adopted to moderate and guide learner's reflection, interaction and collaboration. Mobile technologies open the door to engage learners more creatively, e.g. using Software as a Service (SaaS) cloud computing on mobile devices. According to Jordan (2011), cloud computing allows for increased interoperation with all types of systems through shared applications programming interfaces (APIs) and massively aggregated data. This mobile tech has been used widely in the classroom in the UK HEIs to engage students. E.g. using Kahoot (https://kahoot.it/) for in-class quiz and using Slido (https://www.sli.do/) for workshop.

Although Chinese SOEs as a whole pay much more attention on employee training than other kind of business in China, the advantage of Mobile 2.0 and 3.0 technologies have not be maximised to support collaborative learning and encourage idea discussion and activity engagement. There are relatively few research conducted for the application of M-learning in organisations by academic communities or organisations themselves. On the website of National Philosophy and Social Sciences Documentation Centre (www.ncpssd.org), there are only 3 articles related to organisational learning out of 309 search results using 'M-learning' as keyword. The same situation also appears on the official website of the SASAC (www.sasac.gov.cn), which contains news from SOEs at all levels. 9626 search results can be found using 'training' as keywords and the number is down to 2252 if using 'M-learning' as key words, most of which are mainly about 'document learning' instead of 'mobile learning'. This highlights the fact that in the context of increased attention of organisational learning, the degree of research, application and promotion of M-Learning in enterprises are obviously insufficient.

Garrison, et al. argues (2000) CoI should be built on properly-balanced three presences of its core elements. The literature review refers that social and cognitive processes have been well-developed in UK HEIs to provide truthful and continuous support to students because of its nature. The authors believe by applying proper emerging Mobile 3.0 technologies properly, it will provide great opportunities to enhance teaching presence so that a better educational experience will be achieved. On the other hand, Chinese SOEs should focus more on strengthening social and cognitive presence by providing stanalised and moderated learning environment. Teaching presence should be considered later as the bridge to bring those two presences together. In order to obtain a clearer view on this argument, practical research following this hypothesis has been undertaken by investigating two case studies, which will be discussed in the following section.

COMPARATIVE CASE STUDIES

Introduction to Research Approach

The research project described in this chapter was conducted in the interpretive paradigm. Different as positivist paradigm which is mainly used in the natural sciences, interpretivism is to study phenomena under consideration in their nature setting. Its assumption is that knowledge is interpreted and constructed by people (Williamson et al., 2000). Methodologies associated with the interpretive paradigm are mainly qualitative rather than quantitative and often invokes field work. Multiple case studies allows comparison of findings especially for ''the investigation of particular phenomenon in diverse settings'' (Darke et al., 1998). Because the objective of this study is to investigate culture and context impact, the multiple case study approach focusing on comparable organisations in different countries is appropriate.

Selection of Case Study Organisations

In this research, in order to investigate the relationship between the phenomenon (i.e. M-learning) and its nature cultural context, the field work consisted of organisational case studies were conducted in both UK and China. To ensure a successful multiple case study strategy, selected organisation types need to be representative. The education sector presents a rich area for investigation in terms of M-learning because the mobile technologies have been involved in the heart of a University's activities. In other sectors, significant differences lie in the need for alternative methods to set up CoI and the diverse nature of the learner body, which have an effect on the nature and type of communication channels between training staffs and learners and learners to learners.

By way of contrast, two organisations from two different geographical regions and with different cultural characteristics were selected for study, i.e. Salford Business School in the UK and China Unicom in China. The selection of these two case studies was based on authors' existing knowledge, ability to access information via the dominant language and familiarity with particular cultures. And authors believe these two case studies with completely different learner groups can obtain the most significant practice value.

The University of Salford was established in 1967 and is located just 2km from Manchester city centre. It was one of the first UK HEIs to pioneer 'sandwich' courses that gave students practical, work-based experience alongside traditional academic learning. As part of the University of Salford, Salford Business School (SBS) offers Business, Finance and Law subjects-related programmes and its courses are accredited by 14 internationally-recognised professional organisations including the Chartered Institute of Purchasing & Supply (CIPS), Chartered Institute of Personnel Development (CIPD), the Chartered Institute of Marketing (CIM) and the Association of MBAs (AMBA). In order to provide a lively, diverse and supportive learning community, SBS has undertaken several research projects on utilising technologies to facilitate teaching and learning. Salford Location-Based Service (LBS) App project is one of these projects, aiming at move from Mobile 2.0 to Mobile 3.0 to enhance Teaching Presence and support sustainable CoI.

There are three criteria to select SOE case for this research: 1) the organisation should be a representative SOE directly supervised by the SASAC. This kind of organisations can be found on SASAC official website, where includes 96 SOEs in China; 2) the organisation should belongs to communications industry and is directly related to mobile and mobile applications; 3) the organisation should have public information on employee training/learning on line. There are three SOEs which occupy the full share of China's communications markets and directly supervised by the SASAC, namely China Telecom, China Unicom and China Mobile. By reading the social responsibility reports of the three companies in 2015, 16 and 17 years, There is no specific article on employee training for China Telecom; China Mobile's one mentions M-learning, but does not introduce applied methods; China Unicom has continuous report on its in-house mobile training app "Wo-school" (originally named as 'Wo-class' in 2015) every year since 2015 (China Unicom, 2015; China Unicom, 2016; China Unicom, 2017). As a result, the authors believe China Unicom is the most suitable case study for this research. China Unicom was listed on the Shanghai Stock Exchange, New York Stock Exchange and Stock Exchange of Hong Kong Limited. The Company was one of the "Fortune Global 500" companies for consecutive years and ranked 241st in "Fortune Global 500" for the year 2017. It was also voted as "Asia's No.1 Most Honored Telecom Company" for second consecutive year by Institutional Investor. According to its latest Annual Report (CSR Annual Report, 2017), China Unicom employed 252,522 persons and has subsidiaries across China and worldwide.

Data Collection and Analysis

A variety of data collection methods were adopted in this research to ensure a degree of data triangulation. First, documentary sources (such as annual report, journal paper and websites) were used to provide background information and an understanding of state-of-the-art of mobile tech applications in UK HEIs and China SOEs (refer to Section 2). Secondly, meeting, questionnaire and focus group are used as primary means of gathering data because they are well suited for exploration of people's understanding of meaning and phenomena in their societal context, in this case the organisational environment (Kvale, 1996, p. 105).

In the UK case study, the development of Salford LBS app including requirements gathering, wireframing and prototype development was described and questionnaire results for prototype testing was discussed, together with observation from following deployment meetings. While data for the investigation of China Unicom were gathered from three different resources of key players (i.e. Wo-school app development team, trainees and trainers) providing supplementary sources of evidence. Meetings with app development team were employed to understand app developers' viewpoint of the aim, design and deployment of Unicom Wo-school app; Questionnaires were distributed to 150 employees of Unicom to find out their thoughts on user experiences of Wo-school app and perceptions on M-learning; Focus group was used to investigate trainers' perspective of using Wo-school app as a teaching platform. In keeping with the need for exploration, focus group was semi-structured. Prepared lists of questions were used as a prompt to remind authors of topic areas to be covered, but data collection was conducted as far as possible on a conversational basis, with subjects encouraged to expand and explain as much as possible rather than restricting answers to a factual basis only. The respondents were assured that the data will be use for the research purposes only and confidentiality was emphasised.

Case Study One: Salford Business School Location-Based Service App

Wedemeyer (2009) argues that educational apps must go beyond the hardware and equipment and place software above hardware. To extend the benefits of the mobile technologies, mobile 3.0 play a vital role in enhancing positive learning outcomes. There are varieties of apps that help students and academic staffs manage different aspects of their lives. Mobile 3.0 as the new trend of mobile technologies development, enables context-aware applications which are applications that can adjust its behavior and functionality based on the context input around it. Beacon is one of these technologies. It is a Bluetooth Low Energy (BLE) device that broadcasts a unique identifier, repeatedly and mobile applications can be created to listen to these identifiers and perform location or context aware activities when the identifier is detected. This device has the capability to track items, calculate positions or push notifications; therefore, it will be beneficial for various types of tasks in HEI environment. The prediction of using this tech in learning will make learning process more efficient and productive (Vinu et al. 2011).

As one of the M-learning research projects undertaken in Salford Business School, a Location Based Service (LBS) App (i.e. *Salford LBS*) has been developed using Mobile 3.0 Beacon technology. It intends to provide an innovative platform to deliver digital based solutions to increase efficiency, user engagement and drive significant cost reduction. Underpinned by Human Using Computer (HUC) theories, the development was adopting *Agile Methodology* to implement the requirements according to priorities. The project has been progressed in five steps so far as discussion below.

- **Step 1:** App Specification. The app is designed based on two types of user journeys, i.e. Students and Lecturers. The key functions are:
    - **Digital Registry:**
        - *Students* can register their attendance automatically with their phones in their pockets when they walk into the classroom.
        - *Lecturers* can check student attendance and send messages via the app to those that attended.
    - **Marketing Digital Promotions:** Automatically push notifications with an enticing offer to university students or staff when he/she happens to walk by a café that is in the University campus.
    - **Navigation Assistance:** Beacons are perfect to aid navigation within buildings as well as outside. Salford University for example can use their app and Beacon networks to guide students to classrooms and even help them navigate around campus.
- **Step 2:** App Wireframing. Based on the results of informal user experiences test on Creative Entrepreneur event held in Mendacity in November 2016, app wireframe was design to for both student's and Lecturer's journeys (Figure 4). The full presentation of app wirframing can be seen at the link: https://goo.gl/5hx2eA
- **Step 3:** Proof of Concept Development. This stage focuses on the technology development of Salford Business School LBS app. The key technologies such as applied services pack, 3rd party utilities and functionality incorporated in the functionality and the app and Contents Management System (CMS), Web services/API type are discussed implemented.

Mobile app used by Beacon users talks to API and updates the SQL database accordingly. Figure 5 shows its high level technical architecture and Beacon owners can then use CMS to collect and manage the data. This prototype is only available in iOS and can be downloaded in Apple Store by searching '*Salford LBS*'.

Figure 4. Wireframe of Salford LBS App (online version at:)

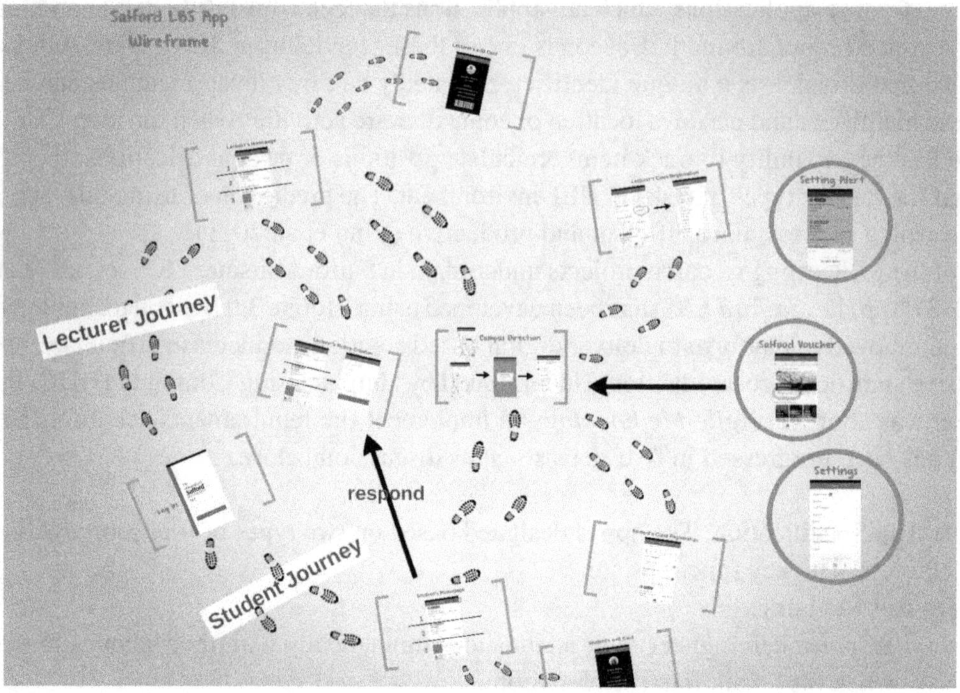

Figure 5. High-level technical architecture of Salford LBS App

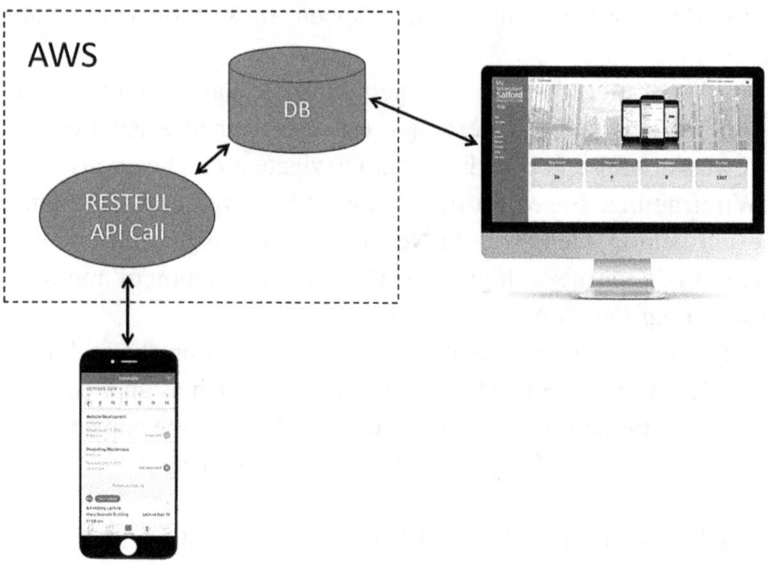

- **Step 4:** Beacon Configuration. As mentioned in previous section, the CMS is a web interface where you add the configured Beacons to the system and then create and assign messages and offers to those Beacons. But in order for the Beacons to be detected by users they will need configuring and adding to the CMS system. This can be done by a 3rd party app called iBeaconSET app (iOS). Each Individual Beacon can be identified by unique *Beacon ID* with major/minor value for management purpose. *Major values* are intended to identify and distinguish a sub group of Beacons– for example all Beacons in on a certain floor or room in your venue could be assigned a unique major value. While *minor values* are intended to identify and distinguish individual Beacons within a group of Beacons assigned with the same major value. *Transmission power level* affects the maximum range of the broadcast signal. This signal also depends on environmental factors. Signals can suffer interference, be diffracted or absorbed. The default transmission power is set to 7 but can be adjusted depending on the proximity required for triggering any messages or offers. See Table 1 for approximate distances for each transmission power setting.

Beacon configuration plan is set up as shown in Table 2. And seven Beacons are all in place in SBS Lady Hale building for pilot test.

- **Step 5:** Prototype testing. This first app prototype has been tested in Salford campus in 2017, which focused on students' experience evaluation and geo-location functions, such as automatic class registration, campus navigation, academic support, push notification for class reminder and lecturer announcements. 14 postgraduate students (6 female and 8 male) participated the test and 13 of them are full-time students. Majority of participants (65%) have the experiences of using the in-house supportive app developed by the University of Salford (i.e. iSalford). The test was hold in a one-day lecture of Operational and Information Management module and the students were provided with a scenario to use Salford LBS app for all its implemented functions. At the end of the day, the students were required to complete an online questionnaire which can be accessed at https://goo.gl/iBKDLN.

Table 1. Beacon transmission power level

Transmission Power Level	Proximity (Meters)
Lowest (-30dBm)	<=2
1 (-20dBm)	7
2 (-16dBm)	10
3 (-12dBm)	15
4 (-8dBm)	22
5 (-4dBm)	27
6 (0dBm)	50
7 (4dBm)	90

Table 2. Salford LBS Beacon configuration plan

No.	Area (Major)	Beacon ID	Major / Minor Value	Location	Trans Pwr (Distance)	Message	Re-Broadcast Period
1	Exterior	MiniBeacon_03217	1 / 10	Plant Pot	7 (4dBm) 90 meters	Welcome, please join the OIM course in Lady Hale Building!	2 Hours
2	Reception Floor (G)	MiniBeacon_03117	2 / 10	Reception	1 (-20dBm) 7 meters	Classroom is PG on G-Floor. For Queries-SBS recpt on 1st Floor.	2 Hours
3	Reception Floor (G)	MiniBeacon_03464	2 / 20	Class Room	2 (-16dBm) 10 metres	You're registered!Duration in classroom recorded automatically	2 Hours
4	Reception Floor (G)	MiniBeacon_03420	2 / 30	Vending Machine	1 (-20dBm) 7 meters	Vending Machine can be found to right of reception (G-Floor)	2 Hours
5	Reception Floor (G)	MiniBeacon_03131	2 / 40	Class Room	Lowest (-30dBm) 2 Meters	n/a	2 Hours
6	Business School (1st Floor)	MiniBeacon_03210	3 /10	Reception	1 (-20dBm) 7 meters)	Welcome to SalfordBus. SchoolReception. Press bell for attention	2 Hours
7	Cafe	MiniBeacon_03299	4 /10	Entrance	5 (-4dBm) 27 meters	Welcome to Salfood Café & Shop. See inside & app for deals!	2 Hours

In addition, the development of Salford LBS app has also been used as a case study of Bsc Digital Business Level 5 Mobile App Development module in Academic year 2017/18 for discussion on mobile technologies usage in HEIs. Meetings were also conducted with Computer Science Engineering Department in the University of Salford for app deployment in July 2018. In class discussion and meetings, concerns were raised regarding personal data protection, ethics and app training cost issues. For example, one staff mentioned "it is uncomfortable to be tracked all time through the campus" and posted question on how the app could protect "huge amount of captured personal data"; another staff questioned on "the money, time and efforts" which needs to devote to train the users. A solution is needed to overcome these challenges before the project is preceded further.

Salford LBS App project so far has verified the power of Mobile 3.0 technologies, but also indicated the difficulties in technology deployment.

Case Study Two: China Unicom Wo-School App

China Unicom, as one of the largest SOEs in China, has been the pioneer to adopt mobile technologies in the employee training and organisational learning. It this research, its in-house developed mobile app of online learning platform, Wo-school App, was investigated to deeply understand the impact of mobile technologies in non-HEI learning environment.

Before practical data collection, initial meeting was arranged between the authors and the app development team in order to understand the background of app design, development and deployment. In the meeting, the authors discovered the aim of app is not just to supplement Unicom's online training so that the employees can study more efficiently by using their fragmentary time, but also to reduce the training cost, collect training feedback and promote the organisition culture. The app has key learning-support functional modules such as registration, learning materials download, discussion board, virtual classroom and quiz. The contents include both compulsory learning blocks (e.g. organization policies, marketing strategy and work flow etc.) and optional ones for individual professional development (e.g. English study, career plan and emerging techs etc.). The app is now at its 3rd version, moving from native app to hybrid app and is internally used inside China Unicom facing all employees. It uses real-name registration system and big data technologies to understand preference of individual's interesting learning areas so that push notification can be sent personally. The app has been continuously modified on its mobile functions and learning contents based on the feedback of users since its initial usage in Unicom in 2016. Until now, Wo-school app has 3753 micro classes with over 1,500,000 clicks and 1,300,000 hours learning duration. The development team indicated a positive believe on continuously using the kind of app in organisitional learning environment.

Following the meeting, the practical research was conducted based on focus group and questionnaire of its two user groups, i.e. trainees and trainers. All data were collected in Chinese because the mother language of both authors are Chinese. The first author who is the lecturer in UK HEI re-examined all items in order to ensure the accuracy of the data. The data collection process is described as below:

1. Questionnaire of Trainees

By applying a purposive sampling technique, online questionnaires were distributed among 150 employees in China Unicom via a Chinese version online form, called WenjuanWang (https://www.wenjuan.com/). The target population of this investigation is new full-time staffs in Unicom because Wo-school app, as an emerging platform, is a compulsory channel for new staff training. In response, 107 usable questionnaires were returned (response rate 73%), consisting of 29% female and 71% male. 105 out of 107 participants are under 30 years old and 102 out of 107 of them have this job as their first job in their career. 67% of them are in junior position.

18 questions are included in the questionnaire with three main aims: 1) to understand whether Wo-school app is a good support to organisitional learning; 2) to investigate feedback of Wo-school app in terms of its functions/contents; 3) to understand personal perception on m-learning trend. The questionnaire was structured in six blocks, i.e.

- Participants' personal information (4 questions) - *Questions about participants' age group, gender, career background and position level in Unicom*

- Participants' M-learning experience (3 questions) - *Questions about platforms and technologies participants have used on their smart devices for learning purpose*
- Participants' feedback on Wo-school functions (2 questions) - *Questions about participants' liked/ disliked Wo-school functional modules and its reasons*
- Participants' feedback on Wo-school contents (3 questions) - *Questions about most/least valuable contents in Wo-school app participants feel*
- Participants' general user experiences of Wo-school app (4 questions) - *Questions to investigate to what extend the usage of Wo-school app facilitate organizational learning, its usability and capability with individual m-learning experience*
- Participants' personal perception on m-learning trend (2 questions) - *Questions to investigate participants' viewpoints on the future of m-learning in general*

2. Focus Group of Trainers

The explanatory focus group was conducted with Unicom trainers who have rich experiences of using Wo-school app as training platform. A total of 5 trainers participated in the focus group interviews to collect their feedback on Wo-school app usage for teaching and learning activities. They are responsible for organising virtual class, micro-class and online quiz. We asked the interviewees 'What are the most popular functions and contents of Wo-school app in employee training?'; The authors also asked their perspectives on learning engagement and sharing related to Wo-school usage in general. The data were then analysed based on grounded theory, i.e., letting the data speak to us rather than making any assumption about the data (Glaser & Strauss, 1967; Locke, 2001). While two interviewees acknowledged that Wo-school was not as functional as PC version of online learning platform, all reported that it was a useful tool. For example, one interviewee said 'Wo-school is useful in fostering staffs' learning habit by providing portable and 24/7 accessible learning materials'. Another reported 'The way of online communication by texts rather than face-to-face talk can encourage learners' discussion and study sharing.' They also mentioned Wo-school app could integrate Chinese popular social media platform, such as WeChat and emerging technologies, such as Location Based Service (LBS) to further support learner participation.

Observations from these two case studies have further revealed the influences of context and culture on smooth learning user experiences of using mobile technologies. Comparative discussion based on data analysis is depicted in next section.

Discussion

Both of these two case studies use mobile application as the platform and enable users to engage in learning activities 24/7. In addition, both of these two apps target on a specific user group (i.e. HEI students or SOE staffs) and are controlled by certain organisation. However, in terms of mobile app functions, the UK app focuses on Mobile 3.0 tech, such as using LBS for in-campus navigation, auto class registration and push notification; while the Chinese one aims to provide standalised learning contents and support functions, such as virtual classroom, discussion board and information centre. According to results of both learner questionnaires, high satisfaction on app functionality and usability is observed. E.g. 86% of participants think Wo-school app has significantly enhanced learning environment in China Unicom (Figure 6); and 85% them believe using app as learning platform in SOEs is a promising trend (Figure 7).

Do you think the company has a better learning atmosphere with Wo-school app than without this app?

According to your learning experience and feelings, do you think that App learning will gradually replace or be parallel to other methods, such as online search, face-to-face training, etc. in the future to be the dominant one?

In terms of Salford LBS App, Mobile 3.0-enabled functions are highly accepted by students. Questions on app functionality and usability were asked in Likert Scale format to test the level of 'agreement' and 'disagreement' from '0 – completely disagree' to '5 – completely agree'. They like emerging LBS functions such as auto-registration, in-building navigation and push notification, although there is 1 participant dislikes geo-related shopping voucher to be sent over in campus (Figure 8, Figure 9, Figure 10). They indicated a high acceptance of all locate-enabled functions (Figure 11) and 13 of them think they would like to use the app if it is fully implemented (Figure 12).

This kind of practical result further explains that the social and cognitive presences have added an involving mechanism in non-HEI learning environment, providing the support to learning using frag-

Figure 6. Questionnaire result on "Do you think the company has a better learning atmosphere with Wo-school app than without this app?"

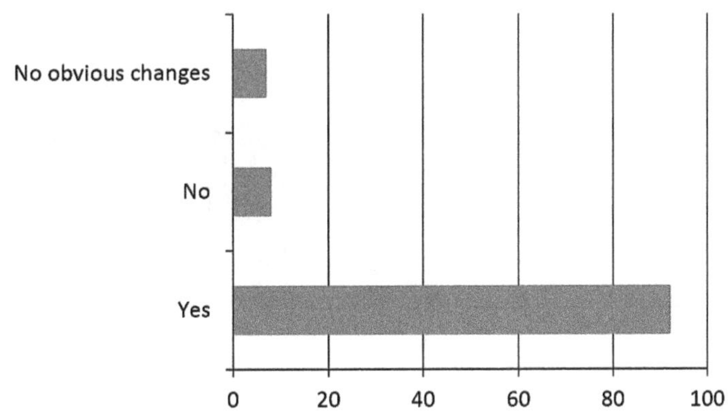

Figure 7. Questionnaire result on "According to your learning experience and feelings, do you think that App learning will gradually replace or be parallel to other methods, such as online search, face-to-face training, etc. in the future to be the dominant one?"

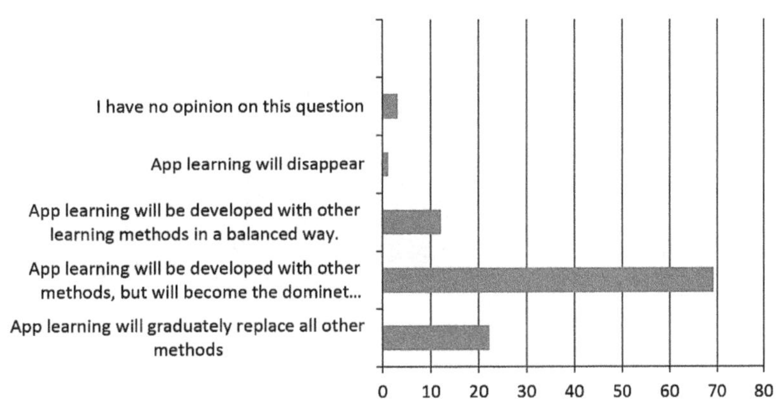

Figure 8. Questionnaire result on "Push Notification" functionality

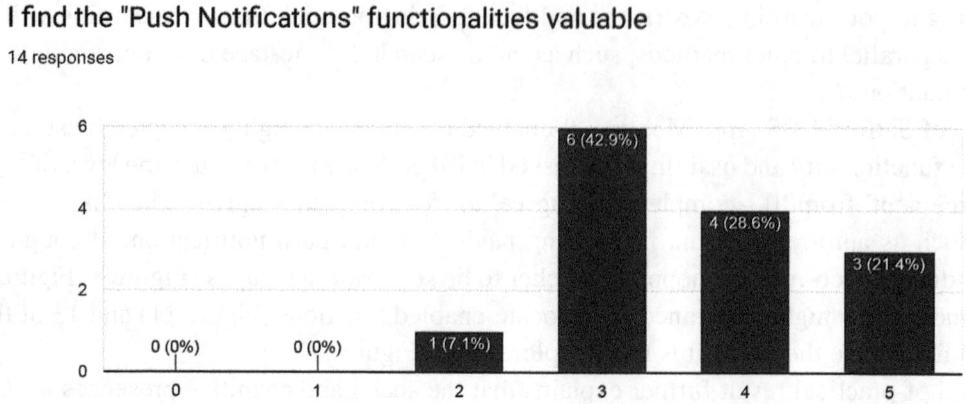

Figure 9. Questionnaire result on "In-Building Navigation" functionality

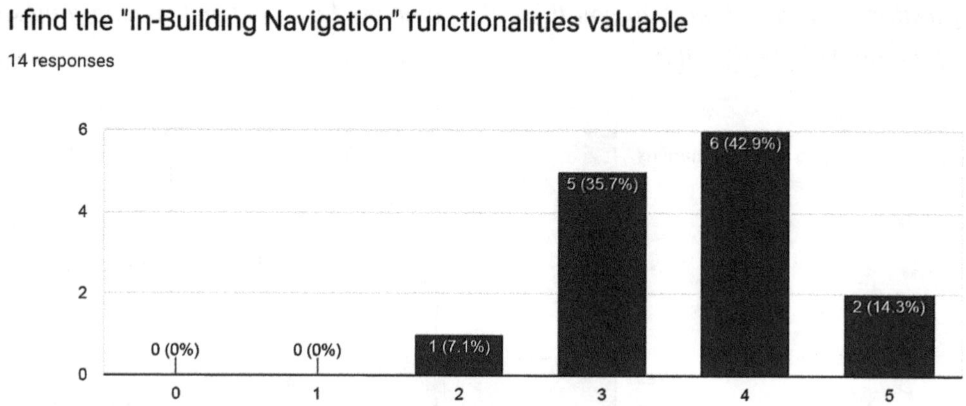

Figure 10. Questionnaire result on "Shopping Voucher" functionality

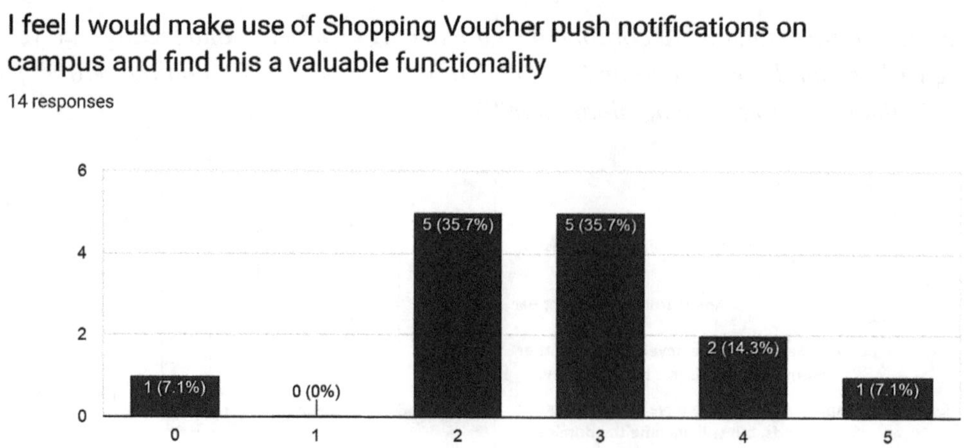

Figure 11. Questionnaire result on functionality in general

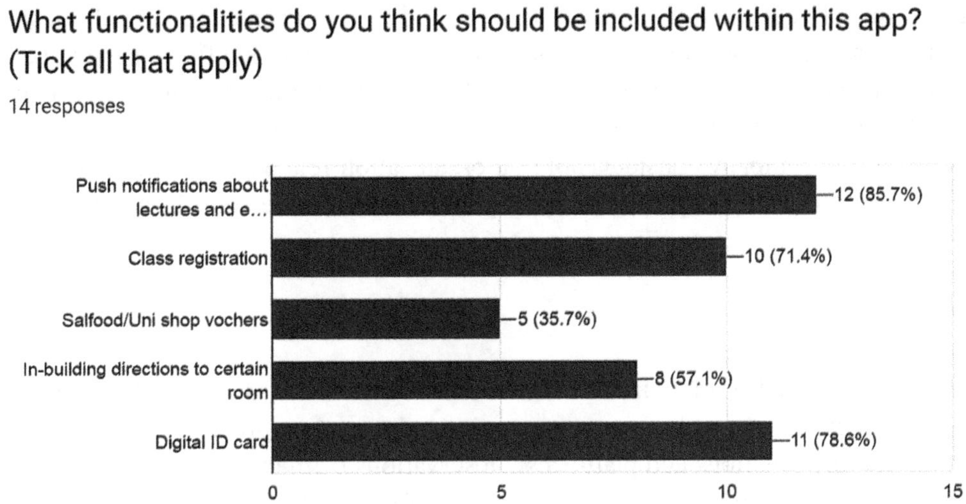

Figure 12. Questionnaire result on Salford LBS app deployment

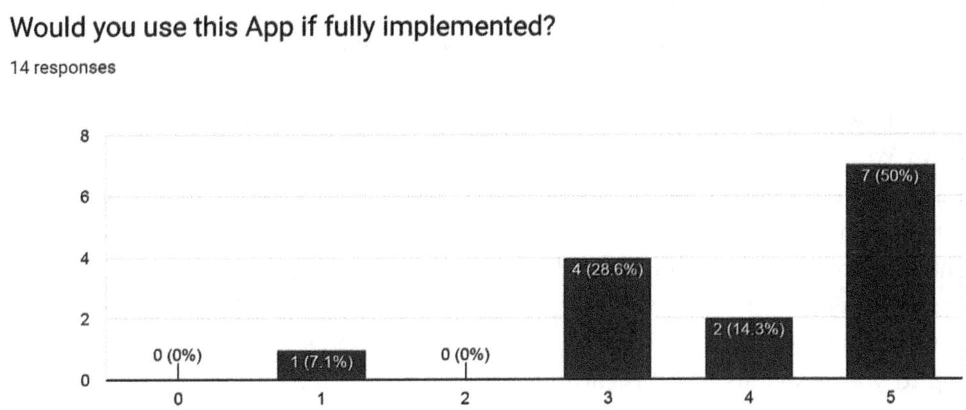

mentary time; while teaching presence could be further enhanced by using mobile emerging technologies in HEI learning environment. Thus the study highlights the relative importance of choosing appropriate mobile technologies, platforms, delivery contents and methods in specific learning environment. In addition, culture and context impact can be observed in practical case studies as discussed below.

Context Impact

The context refers to the situation in which something happens. In this research, context is investigated as an impact factor for M-learning strategy and especially refers to the learners' demographics and motivations.

Different from the simplification of learner groups in UK HEIs, the employees in Chinese SOEs are made of people who have a wide range of age groups, levels of knowledge and learning motivations. The complexity of learner groups leads to various learning motivation, social interaction and study performance. These are necessary precondition for learners to engage in learning activities and factors influencing learning effectiveness ((Malone & Lepper, 1987). As indicated in Wo-school questionnaire results, the participants are interested in various learning subjects, ranging from 'personal career development', 'emerging technologies', 'English study' to 'nourishing of life' although they are in the similar age group and the number of years in China Unicom. The data also indicates the employees who are in higher positions in the company consider more on organisation-related information than junior staffs. The characteristics of learner group need to be fully understood before considering proper contents for learning engagement.

Culture Impact

Given the increasing impact of globalisation, the authors believe in order to enable leaners to achieve the best educational experience in learning community, culture issues should be fully considered when selecting proper mobile functions and platforms. Observations from these two case studies in terms of personal perception and social behavior indicate the significance of culture impact.

1. Personal Perception

As discussed above, students in Salford Business School had a relatively high level of satisfaction with the Salford LBS app functions and usability. But in the following deployment meetings, another user group, teachers show relatively low perceived confidence on personal data collection and security. It generates challenges of applying mobile 3.0 technologies in HEI environment.

For Unicom Wo-school app, 85% of trainers prefer the future development of the app to link with their personal social media platform, such as WeChat, and 16% of them do not really care (Figure 13). In focus group, one of trainers also suggested that 'the integration of Wo-school App with WeChat could encourage interaction between employees and increase their learning motivation'. Comparing with perception of western people regarding social media platforms (i.e. strong indication in personal use and professional use), Chinese people have much more blurred boundary between personal life and career.

According to your previous past M-learning experience and the current experiences of using Wo-school app, are you willing to implement open communication and interaction with social media platform such as WeChat?

2. Social Behaviour

As mentioned by one of trainers in focus group of China Unicom, "discussion forum is not a popular functional module because few people like to post questions/discussion in front public'. The preferred communication way is one-to-one by texts because trainers cannot 'directly observe trainees' facial expressions which can make them feel relaxed". As a result, although the feedback on Wo-school app usage indicates that mobile tech facilitated learning methods enable employees to quickly and easily acquire necessary skills, the methods have not fundamentally changed the characteristics of the traditional training method of one-way communication because of the specific social behavior.

Figure 13. Questionnaire result on "According to your previous past M-learning experience and the current experiences of using Wo-school app, are you willing to implement open communication and interaction with social media platform such as WeChat?"

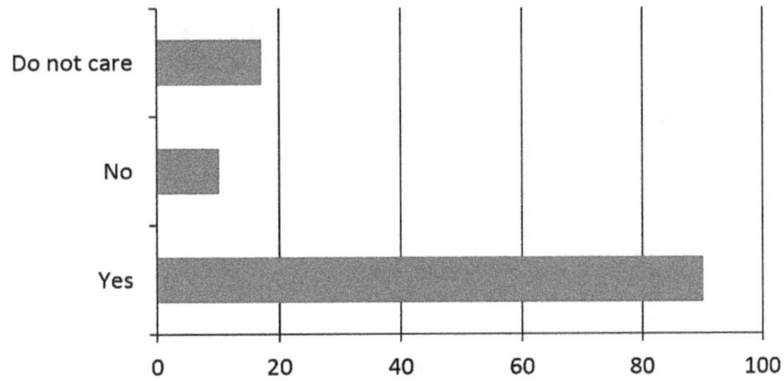

SOLUTIONS AND RECOMMENDATIONS

The rapid development of mobile technologies offers huge potential to help organisations establish community of inquiry to cope various learning requirements. It encourages blended learning by increasing learners' motivation to learn from others, to discuss ideas, to negotiate meaning, and to engage in activities. There are various theories and models explaining M-learning forms. This chapter contributes to such existing research by broadening the understanding of fundamental aspects of a mobile learning environment using CoI framework. The findings derived from this study imply that appropriate usage of mobile technology has a positive impact on organisational learning which leads to greater learning engagement. Implementation of M-learning in different organisations can be guided by properly-balanced three presences of CoI with consideration on mobile platform, delivery contents, usability and native functions. In addition, a range of context and cultural influences should be taken into account in framing implementation strategies of mobile technology application in organisational learning. These findings are transferable and can help inform similar analyses in other organisations. As a result, based on the in-depth understandings of theoretical review to research question one in 'Background' section with reference to practical research in 'Comparative Case Study' section, a framework of good practice is proposed as indicated in Figure 14 below.

The framework suggests that a sustainable CoI should be built on a balanced social, cognitive and teaching system with consideration of culture and context impact. These considerable factors include social behavior, personal perception, learner groups, and their learning motivation. They also decides how to choose proper mobile platform for social presence, adopt relevant mobile 2.0/3.0 functions for teaching presence, enhance mobile usability and include M-learning contents for cognitive presence.

The research also suggests that adoption of M-learning is easier to be implemented in HEIs by using proper strategy and emerging mobile technologies (e.g. Mobile 3.0 as described in case study) because of the nature of its learning environment and learners. Non-HEIs (e.g. China SOEs) would take longer time to develop M-learning environment. Compromising three presences is crucial for sustaining the efforts and keeping up the pace through the process.

Figure 14. Conceptual Framework of mobile technologies-enabled CoI

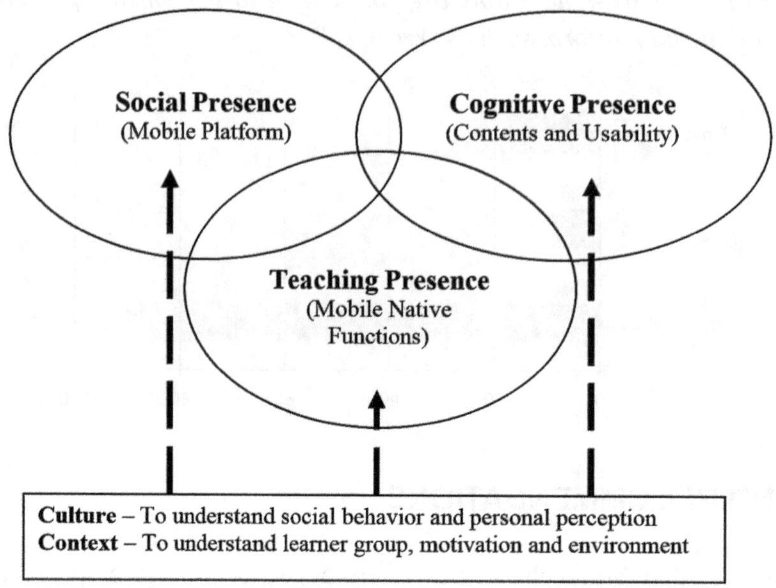

CONCLUSION AND FUTURE RESEARCH DIRECTIONS

The emergence of mobile technology and the introduction of small portable devices have offered much flexibility and convenience for organizational learners. This chapter has reviewed how mobile technologies have been used in UK HEIs and China SOEs to enhance learning experiences and has investigated culture and context impact in organisational learning based on two case studies. In this research, the theoretical framework CoI was served as a powerful tool to frame the design and scope of the research. It also established a basis for comparison to identify differences and similarity using a systematic and authoritative approach.

This framework of good practice proposed by this research (Figure 14) can assist in identifying key factors to use mobile technologies to facilitate establishment of sustainable learning community. It also can inform M-learning strategy in other situations and organisations. But the authors acknowledge the limitations of this study - this study examined the mobile technologies usage only in two specific contexts (i.e. UK HEIs and China SOEs). Therefore, the findings from this study and proposed good practice framework need to be tested in other contexts or different organisations in same place, such as UK SMEs and HEIs in China. Mobile 3.0 should be investigated further to include other facilities in the HEIs, e.g. Library services, for a more comprehensive view of using mobile technology in organisational learning. Also, individuals' experiences may differ depending on the level of mobile technology familiarity in a given context (Day et al., 2010). The impact from individuals, such as level of technology acceptance and resistance to change, are worth to investigate for mobile usage in a certain organisational learning context. In addition, it is equally important to assess issues that affect other stakeholders involved in M-learning environment such as system administrators and their consolidation with learners and teachers.

REFERENCES

Abachi, H. R., & Muhammad, G. (2014). The impact of m-learning technology on students and educators. *Computers in Human Behavior, 30*, 491–496. doi:10.1016/j.chb.2013.06.018

Airlines & Retailers. (2014). *Miami International Airport covered with Beacons.* Available at: https://www.sita.aero/pressroom/news-releases/passengers-airlines-and-retailers-to-benefit-as-miami-international-and-sita-cover-the-airport-with-Beacons

Al-Emran, M., Mezhuyev, V., & Kamaludin, A. (2018a). Students' perceptions towards the integration of knowledge management processes in m-learning systems: A preliminary study. *International Journal of Engineering Education, 34*(2), 371–380.

Al-Emran, M., Mezhuyev, V., & Kamaludin, A. (2018b). Technology Acceptance Model in M-learning context: A systematic review. *Computers & Education, 125*, 389–412. doi:10.1016/j.compedu.2018.06.008

Albrecht, B. (2006). Enriching student experience through blended learning. *Research Bulletin*, 12.

Alioon & Delialioglu. (2015). A Frame for the Literature on M-learning. *Procedia: Social and Behavioral Sciences, 182*(C), 127–135.

Ally, M., & Needham, G. (2012). *M-libraries 3: Transforming libraries with mobile technology (M-Libraries series).* London: Facet Pub. doi:10.29085/9781856049184

Althunibat, A. (2015). Determining the factors influencing students' intention to use m-learning in Jordan higher education. *Computers in Human Behavior, 52*, 65–71. doi:10.1016/j.chb.2015.05.046

Anohah, E., Oyelere, S. S., & Suhonmen, J. (2017). Trend of mobile learning in computing education from 2006 to 2014: A systematic review of research publications. International Journal of Mobile and Blended Learning, 18.

Arbaugh, J., (2007). Does the community of inquiry rameowrk predict outcomes in online MBA courses? *Academy of Management Proceedings, 2007*(1), 1-6.

Asiimwe, E., Grönlund, Y., & Hatakka, M. (2017). Practices and Challenges in an Emerging M-Learning Environment. *International Journal of Education and Development Using Information and Communication Technology, 13*(1), 103–122.

Astin, A. (1999). Student involvement: A developmental theory for higher education. *Journal of College Student Development, 49*(5), 518–529.

Bakhsh, M., Mahmood, A., & Sangi, N. (2017). Examination of factors influencing students and faculty behavior towards m-learning acceptance. *The International Journal of Information and Learning Technology, 34*(3), 166–188. doi:10.1108/IJILT-08-2016-0028

Bleed, R. (2001). A hybrid campus for a new millennium. *EDUCAUSE Review, 36*(1), 16–24.

Cárdenas-Robledo, L. A., & Peña-Ayala, A. (2018). Ubiquitous learning: A systematic review. *Telematics and Informatics, 35*(5), 2097–1132. doi:10.1016/j.tele.2018.01.009

Cheon, J., Lee, S., Crooks, M. S., & Song, J. (2012). An investigation of mobile learning readiness in higher education based on the theory of planned behavior. *Computers & Education*, *59*(3), 1054–1064. doi:10.1016/j.compedu.2012.04.015

Chickering, A. W., & Gamson, Z. F. (1999). Development and adaptations of the seven principles for good practice in undergraduate education. *New Directions for Teaching and Learning*, *80*(80), 75–82. doi:10.1002/tl.8006

Crant, J. (2000). Proactive behavior in organizations. *Journal of Management*, *26*(3), 435–462. doi:10.1177/014920630002600304

Crompton, H. (2013). A Historical Overview of M-learning: Toward Learner-Centred Education. In Handbook of Mobile Learning. Routledge.

Crompton, H. (2013). Mobile learning: New approach, new theory. In Z. L. Berge & L. Y. Muilenburg (Eds.), Handbook of mobile learning (pp. 47–57). Florence, KY: Academic Press.

Crompton, H., & Burke, D. (2018). The use of mobile learning in higher education: A systematic review. *Computers & Education*, *123*, 53–64. doi:10.1016/j.compedu.2018.04.007

Darke, P., Shanks, G., & Broadbent, M. (1998). Successfully Completing Case Study Research: Combining Rigour, Relevance and Pragmatism. *Information Systems Journal*, *8*(4), 273–289. doi:10.1046/j.1365-2575.1998.00040.x

Davis, F., Bagozzi, D., Richard, P., & Warshaw, P. R. (1989). User acceptance of computer technology: A comparison of two theoretical models. *Management Science*, *35*(8), 982–1003. doi:10.1287/mnsc.35.8.982

Day, A., Scott, N., & Kelloway, E. K. (2010). Research in Occupational Stress and Well-being: Vol. 8. Information and communication technology: Implications for job stress and employee well-being. Academic Press.

Derounian, J. G. (2017). Mobiles in class? *Active Learning in Higher Education*, 1–12.

Edgerton, R. (2001). *Education White Paper*. Report prepared for the Pew Charitable Trusts, Pew Forum on Undergraduate Learning.

Elias, T. (2011). Universal Instructional Design Principles for Mobile Learning. *International Review of Research in Open and Distance Learning*, *12*(2), 143–156. doi:10.19173/irrodl.v12i2.965

Garcia-Cabot, A., De-Marcos, L., & Garcia-Lopez, E. (2015). An empirical study on m-learning adaptation: Learning performance and learning contexts. *Computers & Education*, *82*, 450–459. doi:10.1016/j.compedu.2014.12.007

Garrison, A., Anderson, T., & Archer, W. (1999). Critical Inquiry in a Text-Based Environment: Computer Conferencing in Higher Education. *The Internet and Higher Education*, *2*(2), 87–105. doi:10.1016/S1096-7516(00)00016-6

Garrison, D. R., Vaughan, N., & Norman, D. (2008). *Blended learning in higher education*. San Francisco: Jossey-Bass.

Gikas, J., & Grant, M.M. (2013). Mobile computing devices in higher education: Student perspectives on learning with cellphones, smartphones & social media. *Internet and Higher Education, 19*, 18–26.

Glaser, B., & Strauss, A. (1967). *The discovery grounded theory: Strategies for qualitative inquiry.* Chicago, IL: Aldine.

Han, I., & Shin, S. W. (2016). The use of a mobile learning management system and academic achievement of online students. *Computers & Education, 102*, 79–89. doi:10.1016/j.compedu.2016.07.003

Jeno, L. M., Adachi, P. J. C., Grytnes, J., Vandvik, V., & Deci, E. L. (2018). The Effects of M-learning on Motivation, Achievement and Well-being: A Self-Determination Theory Approach. *British Journal of Educational Technology*. doi:10.1111/bjet.12657

Jeno, L. M., Vandvik, V., Eliassen, S., & Grytnes, J. (2019). Testing the novelty effect of an m-learning tool on internalization and achievement: A Self-Determination Theory approach. *Computers & Education, 128*, 398–413. doi:10.1016/j.compedu.2018.10.008

Jordan, J. (2011). Climbing out of the Box and into the Cloud: Building web-scale for libraries. *Journal of Library Administration, 51*(1), 3–17. doi:10.1080/01930826.2011.531637

Koole, M., & Ally, M. 2006, Framework for the Rational Analysis of Mobile Education (FRAME) Model: Revising the ABCs of Educational Practices. In *The International Conference on Networking, International Conference on Systems and International Conference on Mobile Communications and Learning Technologies* (pp. 216-216). IEEE. 10.1109/ICNICONSMCL.2006.103

Kvale, S. (1996). *InterViews: An introduction to Qualitative Research Interviewing*. Thousand Oaks, CA: Sage.

Lapointe, D., & Gunawardena, C. (2004). Developing, testing and refining of a model to understand the relationship between peer interaction and learning outcomes in computer-mediated conferencing. *Distance Education, 25*(1), 83–106. doi:10.1080/0158791042000212477

Liang, J. (2017). *M-learning development for enterprise Online University based on gamification.* Chinese National Centre for Philosophy and Social Science Documentation. Available at: www.ncpssd.org/literatureforregisteredusers

Locke, K. (2001). *Grounded theory in management research.* London: Sage.

Malone, T. W., & Lepper, M. R. (1987). Making learning fun: A taxonomy of intrinsic motivations for learning. In R. E. Snow & M. J. Farr (Eds.), *Aptitude, learning, and instruction: III. Conative and affective process analyses* (pp. 223–253). Hillsdale, NJ: Erlbaum.

Nagella, U. B., & Govindarajulu, D. (2009). Adaptive approaches to context aware mobile learning application. International Journal of Computer Science and Security.

Pace, C. (1980). Measuring the quality of student effort. *Current Issues in Higher Education, 2*, 10–16.

Parsons, G. (2010). Information provision for HE distance learners using mobile devices. *The Electronic Library, 28*(2), 231–244. doi:10.1108/02640471011033594

Pascarella, E., & Terenzini, P. (2005). *How college affects students: A third decade of research* (2nd ed.). San Francisco: Jossey-Bass.

Pask, G. (1976). Styles and strategies of learning. *The British Journal of Educational Psychology, 46*(2), 128–148. doi:10.1111/j.2044-8279.1976.tb02305.x

Roundtree, R. (2011). Article. *Mobile Technology*, 919-921.

Ryan, R. M., & Deci, E. L. (2017). *Self-Determination Theory. Basic psychological needs in motivation, development, and wellness*. The Guilford Press.

Sabah, N. (2016). Exploring students' awareness and perceptions: Influencing factors and individual differences driving m-learning adoption. *Computers in Human Behavior, 65*, 522–533.

Sinopec i-Academic. (2018). Available at: https://sia.sinopec.com/learn/

Skills for Learning. (2018). The University of Salford. Available at: https://www.salford.ac.uk/skills-for-learning

Statista. (2018). *Number of mobile phone users worldwide from 2015 to 2020*. Available at: https://www.statista.com/statistics/274774/forecast-of-mobile-phone-users-worldwide/

Sung, Y.-T., Chang, K.-E., & Liu, T.-C. (2016). The effects of integrating mobile devices with teaching and learning on students' learning performance: A meta-analysis and research synthesis. *Computers & Education, 94*, 252–275. doi:10.1016/j.compedu.2015.11.008

Suresh, S., & Venkatammal, P. (2010). Antecedents of organizational citizenship behavior. *Journal of the Indian Academy of Applied Psychology, 36*(2), 276–286.

Traxler, J., & Kukulska-Hulme, A. (2005). Mobile Learning in Developing Countries. Commonwealth of Learning.

Unicom, C. (2015). *Social Responsibility Report of 2015*. Available at: https://bit.ly/2BWFfl1

Unicom, C. (2016). *Social Responsibility Report of 2016*. Available at: https://www.unglobalcompact.org/participation/report/cop/create-and-submit/active/324231

Unicom, C. (2017). *Social Responsibility Report of 2017*. Available at: https://www.chinaunicom.com.hk/en/csr/reports/csr2017.pdf

Vaughan, N. D. (2010). A blended community of inquiry approach: Linking student engagement and course redesign. *Internet and Higher Education, 13*(1-2), 60–65. doi:10.1016/j.iheduc.2009.10.007

Vinu, P., Sherimon, P., & Reshmy, K. (2011). Towards Pervasive Mobile Learning – the vision of 21st century. *Procedia - Social and Behavioral Science.*

Wedemeyer, C. A. (2009). *Learning at the back door: Reflection on non-traditional learning in the lifespan*. Charlotte, NY: Information Age Publishing.

Williams, C. (2002). Learning on-line: A review of recent literature in a rapidly expanding field. *Journal of Further and Higher Education, 26*(3), 263–272. doi:10.1080/03098770220149620

ADDITIONAL READING

Bersin, J. (2004). *The blended learning book: Best practices, proven methodologies, and lessons learned.* San Francisco, CA: Pfeiffer.

Chung, L., Lee, K. Y., & Choi, J. (2015). Exploring digital creativity in the workspace: The role of enterprise mobile applications on perceived job performance and creativity. *Computers in Human Behavior, 42*, 93–109. doi:10.1016/j.chb.2014.03.055

Denton, J. (1998). *Organisational learning and effectiveness.* London: Routledge.

Garrison, D. (2017). E-learning in the 21st century: A community of inquiry framework for research and practice (Third ed.).

Ishtaiwa, F. (2014). Mobile technology as a learning tool: Use and effects. *10*(4), 1.

Matthews, R., Maccarthy, B., & Braziotis, C. (2017). Organisational learning in SMEs: A process improvement perspective. *International Journal of Operations & Production Management, 37*(7), 970–1006. doi:10.1108/IJOPM-09-2015-0580

Stenbom, S. (2018). A systematic review of the Community of Inquiry survey. *The Internet and Higher Education, 39*, 22–32. doi:10.1016/j.iheduc.2018.06.001

Yueh, L., Lu, M.-H., & Lin, W. (2016). Employees' acceptance of mobile technology in a workplace: An empirical study using SEM and fsQCA. *Journal of Business Research, 69*(6), 2318–2324. doi:10.1016/j.jbusres.2015.12.048

KEY TERMS AND DEFINITIONS

Blended Learning: A pedagogic method that combines online technologies and traditional classroom teaching ways. Sometimes it also refers to "hybrid learning."

Chinese State-Owned Enterprises (SOE): It is a business enterprise where the state has significant control through full, majority, or significant minority ownership.

Community of Inquiry Framework: A conceptual framework which was created by Garrison et al. in 2000 to guide the research and practice of computer-mediated communication in learning environment.

Location-Based Service: The software-level service using location-based technologies and data, such as iBeacong and Wi-Fi access points, to provide geo-related information and functions.

M-Learning: Abbreviation of "mobile learning." It refers to the learning across various online platforms using personal electronic devices.

Mobile 3.0: The third generation of mobile technologies. It aims to enhance user experiences of using mobile devices by personalization and contextualization. iBeacon is one example.

UK Higher Education Institutions: Educational institutes defined as "university," "an institution conducted by a higher education corporation," or "an institution designated as eligible to receive support from funds administered by the Higher Education Funding Council for England (HEFCE)" in the UK.

Chapter 9
Contextualized MALL in Target and Non-Target Countries:
Mobile Activity Evaluation

Anat Cohen
https://orcid.org/0000-0002-8373-3511
Tel Aviv University, Israel

Orit Ezra
https://orcid.org/0000-0001-6596-1345
Tel Aviv University, Israel

ABSTRACT

Mobile-assisted language learning (MALL) is known to be conducive to contextualized language learning. However, the literature lacks an understanding of context. Nor is there a quantitative evaluation tool. The chapter's dependent contextualized variables were the following: real-world context level (one's place), real-life context level (one's life), and device mobility (place diversity). The independent variables were target/non-target country (Taiwan/Israel) and language learning orientation (dedicated/generic). In the target country, the spoken language is the studied language. Dedicated/generic language learning orientation represents the existence/non-existence of pedagogy within activities. Fifty-three Chinese L2 students involved in 296 activities and 519 events were interviewed. Using device mobility and an evaluation index developed to measure real-world/real-life context levels, MALL was found more contextualized in Taiwan only in generic activities. The findings refine our understanding of the benefits of studying in the target country. The index can be used in future studies.

DOI: 10.4018/978-1-5225-8106-2.ch009

INTRODUCTION

In recent decades, the orientation of language learning theories has changed from behaviourist and cognitive approaches to constructivist ones (Comas-Quinn, Mardomingo, & Valentine, 2009; Wong, Chin, Tan, & Liu, 2010). One noticeable shift is toward social constructivist learning which suggests the importance of contextualised, situated language learning as a means to achieve goals, such as more meaningful learning (Comas-Quinn et al., 2009), or context-dependent vocabulary learning (Wong et al., 2010), among others.

With the emergence of mobile devices, it became apparent that Mobile Assisted Language Learning (MALL) might be highly significant in contextualised learning. Mobile device portability enables learners to engage with contexts they find interesting, leading to a more personalised, meaningful, and deeper learning experience (Comas-Quinn et al., 2009). The distinction between MALL and previous Computer Assisted Language Learning (CALL) technologies is evident in MALL's ability to support contextualised learning. Kukulska-Hulme (2012) suggests that mobile device portability offers specific advantages that allow interaction across different contexts. Specific examples of MALL's contextualised learning affordances include location-specific language materials sent using radio-frequency identification (RFID) and Global Positioning System (GPS) technology, among others.

Given the advantages of contextualised language learning and the ability of mobile devices to support it, the importance of contextualised MALL is apparent. However, a review of the implementation case studies (Burston, 2013) found that the lack of consistent approaches to contextualised MALL and the lack of operative and quantitative ways to analyse, design and evaluate it are ongoing problems. Thus, a quantitative evaluation tool is required.

The study investigated contextualised learning in the mobile Chinese learning activities of students in two countries – Taiwan, where the Chinese language (learners' L2) is the spoken language, and Israel, which is predominantly a Hebrew (learners' L1) speaking environment. The study described here examined contextualised learning in MALL activities among students of Chinese using three main variables: real world context levels pertaining to one's place, real life context levels pertaining to one's life, and device mobility pertaining to place diversity. It used two additional variables that could be related to the application of contextualised MALL activities: target country (Taiwan/Israel) and language learning orientation (generic/dedicated). In the target country, Taiwan, where the Chinese language is spoken (target language), context opportunities are abundant, but mobile devices may provide alternative context opportunities in the non-target country where it is not (Comas-Quinn et al., 2009). Generic mobile materials such as social networks (Reinders & Pegrum, 2016) may provide learners with active platforms to create their learning content (Comas-Quinn et al., 2009; Reinders & Pegrum, 2016), related to the contexts they come across (Pegrum, 2014), Thus, these materials might be conducive to contextualised learning. On the one hand, it is questionable whether learners exploit their various available contexts as learning contexts; on the other hand, pedagogically dedicated, pre-designed learning materials (Reinders & Pegrum, 2016) are restricted in their ability to include situations (Reinders & White, 2010).

The purpose of this work was twofold. First, it aimed to devise a consistent approach to contextualised MALL and create an operative evaluation tool for the quantitative analysis of real world and real life contextualised MALL. Second, it aimed to explore target country and language learning orientation variables possible relation with contextualised MALL.

BACKGROUND

Contextualised Mobile Assisted Language Learning (MALL)

An array of studies has noted MALL's contribution to contextualised language learning. For example, one study tries to demonstrate how formal learning could be extended by informal learning, examined Dutch learners of English walking through a zoo; they were equipped with a mobile GPS and were able to interact and play in English based on the animals they saw. The study shows that the vocabulary learning of learners who walked through the zoo and also had extra time practicing at home using their device outperformed non-mobile learners or mobile-learners who were not permitted to take the phone to their homes (Sandberg, Maris, & de Geus, 2011). Several studies in Taiwan (e.g., Hwang, Chen, & Chen, 2011; Hwang & Chen, 2013; Wu, Sung, Huang, Yang, & Yang, 2011) show superior performance in a contextualised English learning system compared to non-contextualised learning. Another study finds positive potential for English learners exchanging information based on QR scanned codes (Rivers, 2009).

Generally, the literature is positive about the outcomes of contextualised MALL and the smartphone's ability to deliver it. However, the above mentioned studies and many more, as described in Burston's review of the MALL field (2013), have a lack of consistency in their understanding of what constitutes contextualised MALL. There is also a lack of operative and quantitative analysis. Finally, most MALL-related studies do no report objectively quantifiable learning outcomes, as presented in Burston`s meta-analysis of the MALL field (2015). Clearer context definitions and a quantitative evaluation tool that may assist in the analysis, design, evaluation and assessment are required.

Context Definitions

Device Mobility

Reinders and Pegrum (2016) suggest various criteria to evaluate MALL resources, such as applications, websites and their accompanying activities. The primary criterion is mobility, wherein the highest scores are allocated to a learning experience altered by the environment. They give an example of students making annotated multimedia recordings of their surroundings and sharing them with peers via online networks. Pegrum (2014) defines the components of mobility: at the most basic level, devices should be mobile and used in varying places, not just one or two locations such as a residence. To allow students to use appropriate language in various situations (Cai, 2008), diversity of context is imperative. Device mobility permits this diversity.

Real World Context

More importantly, the primary constituent of mobility, according to Pegrum (2014), is learning mobility. Here, real world contexts become learning contexts which may turn into learning content and may be informed by or sent to other peers; in other words, a contextualised MALL consists of varied real world places (device mobility) whose potential transformation into learning contexts (real world contexts) happens automatically. A learner who is walking around real world places and getting information from an electronic dictionary is using device mobility to learn, and real world contexts are becoming learning contexts. Looking up words related to coffee while seated on a bus is not real world, but searching for words related to transportation is. Put otherwise, the real world pertains to the place where the learners are located.

Real Life Context

Real life context is also frequently brought up as a constituent of authentic contextualised learning since during daily life actual language is used (Reinders & White, 2010). Vocabulary learning can be embedded in the context of normal, everyday (real life) rather than in abstract definitions. MALL is able to blend learners' learning environments with their real-life contexts (Wong et al., 2010).

Klopfer (2011) refers to mobile learning as taking place in the context of some other activity that may be related or unrelated to the learning. This other activity may be related to one's real life. For instance, Pegrum (2014) refers to learning vocabulary while riding the bus, where the vocabulary is not related to the activity of riding the bus. According to Pegrum, since activity pertains to the learner being active, it seems that learning vocabulary should be considered real life only if there is an actual relationship between the vocabulary and the learner's active performance of the activity, such as purchasing a bus ticket. Otherwise, for example, simply studying bus related words at the bus results in a mobile leaning experience which remains only real world related and not real life; thus, real life pertains to learners' life as reflected in their other activities.

This study examined contextualised MALL using three variables. First, device mobility represents the number of places and is contiguous to the real world context. With more places, there are more potential contexts for learning. Second, real world context level represents places in the world that may turn into learning contexts. Third, real life context level may be manifested through the related activity and may not necessarily represent real world contexts but are attached to the learner's authentic and daily life. To these, it added two variables which may relate to contextualised MALL: learning in the target/non-target country and mobile materials.

Learning in the Target/Non-Target Country

The target language learnt by students who are native speakers of a different language (L1) is referred to as L2 and involves foreign or second language learning (Burston, 2013). Scholarly work agrees on the advantages of learning the L2 language in the country where the target language is spoken. This may be represented as the target culture (Comas-Quinn et al., 2009), or target country. Students traveling to the target country encounter the language used in daily life and participate in real activities in the context of the environment, constructing a new understanding of the language (Anderson, Hwang, & Hsieh, 2008). Petersen, Divitini, and Chabert (2008) say students can interact with local communities and practice the language with native speakers. The cultural context is embedded in this learning process, helping learners understand the language. Contextualised learning becomes highly available in the target country.

As noted by Comas-Quinn et al. (2009), situated, contextual language learning is ideally provided in the country where the language is spoken; put otherwise, the target country offers rich experiences of context. However, they also say mobile learning may provide alternative context-rich experiences in non-target countries.

Accordingly, contextualised MALL case studies cover learning in both target and non-target countries (e.g., Chen & Li, 2010; Ogata et al., 2008). The examples mentioned above are drawn from formal learning programs where contextualised learning is typically designed by program facilitators. A study of students' existing mobile activities, whether implemented by their school or by themselves, could complement these previous findings. Therefore, this study looked at contextualised mobile learning in terms of device mobility, real world context and real life context in target and non-target countries' MALL.

Mobile Materials

Mobile materials include dedicated or generic web services and apps (Reinders & Pegrum, 2016). Dedicated materials are designed for the purpose of learning, so they already have language learning content and a pedagogic method. Dedicated materials in apps are mostly based on behaviourist teacher-centred transmission of information rather than a constructivist, collaborative learner-centred mode of instruction (Burston, 2014). Nevertheless, dedicated pedagogically pre-designed content (likely non-learner created) might be restricted in the scope of its context, as the creators will not know in advance about the contexts in which it will be used, including the participants and the actual situations (Reinders & White, 2010) and without a clear indication of what the contexts will constitute.

Generic materials, such as social networks and video sharing, are not pedagogically pre-designed for learning but may be used to complement the app/web sphere by offering active and collaborative learning (Reinders & Pegrum, 2016). Learners may turn real world/real life contexts into learning contexts and content and share them with peers. Students may produce written texts in blogs and e-books or record audio texts in podcasts and then share these with peers. However, this requires newer hardware, better connectivity and faster software, making it harder to implement (Pegrum, 2014). This, in turn, may hinder the ability of learners to turn the contexts into learning contexts. Generic materials are occasionally created by learners and thus are potentially related to context; again, without a clear indication of what constitutes the context, there is a question of whether learners actually turn it into a learning context.

Given these issues, this study compared generic and dedicated materials in contextualised MALL which consists of device mobility, real world context and real life context.

Mobile Mandarin Chinese Learning

Chinese is difficult to learn, especially for westerners whose European language background is very different from the complicated characters or logographic writing system and the pronunciation or tonal nature of the phonological system of Chinese (Moser, 1991; Tseng, Lu, & Hsu, 2007; Wang, Koda, & Perfetti, 2003). A student's level might not be sufficient for him or her to take advantage of contextual opportunities (Al-Mekhlafi, Hu, & Zheng, 2009). Therefore, it might be harder to transfer knowledge and skills to the real world or real life contexts, but using these contexts to learn can promote this type of transfer (Klopfer, 2008), in fact it is essential to reach language automatization (Marsden, Mitchell, & Myles, 2013). For learners of Chinese, contextualised learning is particularly meaningful and requires particular attention. Thus, this study attempted to answer the contextualised MALL questions by exploring Chinese learning activities.

THE STUDY

Research Aims and Questions

Given the importance of contextualised MALL, this study examined students' contextualised Chinese MALL activities. The research aim was twofold. First, the study sought to devise a consistent approach to the components of contextualised MALL and supply a set of operative and quantitative instructions on how to analyse, design and evaluate mobile activities. To this end, it explored real world and real life contexts in mobile learning in the target and the non-target countries, as well as in dedicated and generic mobile materials or the broader design of their activities, which may also have been generic or dedicated (Reinders & Pegrum, 2016). It also aimed to develop a quantitative evaluation tool based the exploration of contextualised MALL activities of Chinese language students. The second purpose was to explore variables with a possible relation with contextualised MALL: the target/non-target country and the mobile material type or the design of their broader activities. The latter variable represents the materials or activity pedagogically pre-planned for language learning. Henceforth, the mobile material type or its broader activity design is called language learning orientation. To achieve the second goal, the study examined the differences in contextualised language learning (as defined by device mobility, real world context level and real life context level) for the various mobile activities of students of Chinese. Specifically, it asked the following questions:

1. What are the differences between target and non-target countries in contextualised MALL?
2. What are the differences between generic and dedicated language learning orientation in contextualised MALL?
3. Are the differences between target and non-target countries in contextualised MALL a function of generic and dedicated language learning orientation?

Dependent Context Variables

- **Device Mobility:** This variable represented the number of different places where the activity was taking place, measured on a scale from 1 to 5.
- **Real World Context Level:** The degree of the mobile learning activity in relation to the place, the objects of the place, or the situation at the place was measured on a scale from 0 to 5. It was based on Schilit's (1995) definition of a context as including the place and objects and on Dey and Abowd (1999) who suggest that the primary context of a situation should include activities. In this research, activities in the real world are referred to as situations.
- **Real Life Context Level:** This referred to the degree of the mobile learning activity in relation to the learner's real life (Wong et al., 2010). The variable represented the degree of the mobile learning activity in relation to the learner's other activities, whose purpose was not learning and where the learner was active. It was measured on a scale from 0 to 5, and was based on the above definition of other related activity (Klopfer, 2011; Pegrum, 2014).

Figure 1 is a comprehensive illustration of the definition of context dependent variables in this research.

Figure 1. Research context dependent variables

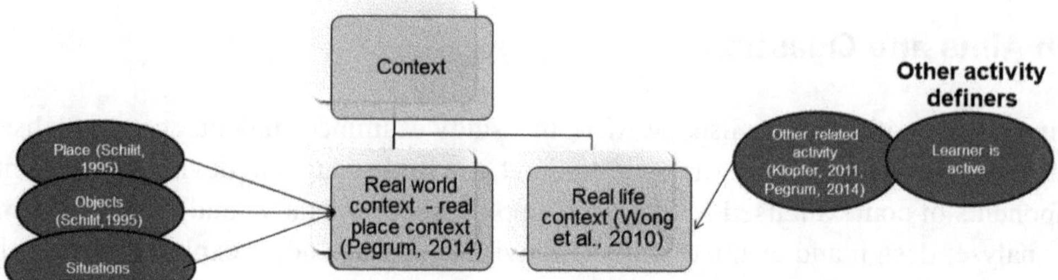

Independent Variables Which May Be Related With Context Learning

- **Target Country:** Taiwan is the target country (1) and Israel is the non-target country (0).
- **Language Learning Orientation:** A two-value variable, generic or dedicated, indicating the content or pedagogic language learning design of the mobile materials or the broader learning activity usage.

METHODOLOGY

Participants and Design

Fifty-three beginner-intermediate students of Chinese as a second or foreign language (L2) at two universities in Israel and one university in Taiwan were interviewed using a fully structured interview format. Data for 296 activity types (such as social communication and electronic dictionary activities), and 519 events (each learner activity may have had a few events) were collected from these fifty-three respondents and analysed.

Procedure

During the interviews, the learners reported on their mobile Chinese learning activities that occurred during the previous month and outside their home. They then answered questions on each activity for each of the dependent context variables. Learners numerated the places at which the activity occurred in the past month. The learners were then asked to describe specific events of the activity which occurred in the past month. They were required to state the place where the event took place, describe it in detail (again, for relevant hybrid activities which may have involved usage of other languages, as well, they were also asked to verify that at least some Chinese existed), express the level of real world and real life context as perceived by them according to the questions described below (0-5) and state their arguments for these context level selections. The real world and real life questions were defined before data collection based on existing definitions of real world and real life context in the literature (Dey & Abowd, 1999; Klopfer, 2011; Pegrum, 2014; Schilit, 1995; Wong et al., 2010).

In line with the research context variables as charted in Figure 1 above, the learners addressed real world/real life context questions for each described event:

- The real world question was: "To what degree was the activity/were the activity materials such as the application, website or pure content related to your physical location at the time, to the objects at the place which you came/did not come in contact with, or to the situation at the place where the activity occurred"?
- The real life question was presented in a three-step process which is illustrated in Figure 2. In Step 1, the question regarding other activities was asked (at the place – the same event, i.e., not on a different day, where the learner was active, which may have been abstract). Merged activities, such as conversing with friends in Chinese via social media tools, were embedded within the mobile activity. Abstract activities were manifested, for example, in searching online information. In Step 2, aiming to locate real life activities with no direct learning purposes, the learners were asked for the main purpose of the other activities. If the main purpose of the other activities was not to learn Chinese or at least partially not to learn Chinese, it was kept; otherwise, it was removed. If the answer was not clear, then learners were asked if they would have done it in English as well, if they would then it was kept; otherwise it was removed. Finally, in Step 3, learners were asked: "Was the (other) activity related to the Chinese mobile learning activity or its applications, websites or content materials?" If it was not related, then the other activity was discarded. If it was related, then learners were asked to grade the other activity by answering the following question: "To what degree was the Chinese learning/practicing/using activity or the activity materials – applications, websites or content, related to the other activity?"

Method

The research employed a mixed method strategy. To address the research questions, the researchers reviewed the students' mobile learning activities and events and analysed them qualitatively. At the end of the qualitative analysis, they developed a quantitative operative evaluation index. Questions 1–3 on the real world and real life contexts drew on the developed quantitative index.

Qualitative Stage

During this stage, each activity and event was individually analysed based on learners' stories and their rankings of the dependent context variables (device mobility, real world context level, real life context level). An inter-judge reliability process was used to establish agreement between the researchers on these three context variables, the independent variable language learning orientation classification, and the definitions of activities and events. A combined top-down and bottom-up approach was used to empirically develop and adjust an index based on the initial literature about real world/real life context definitions and the learners' stories. The development of the evaluation index was based on events analysis (and accompanied by the inter-judge reliability process as stated above).

Real World Index

Three scores were found for each activity's event (Table 1), place score, object score and situation score, based on the amount (partial or full) of content related to the place (P), the non-typical (NTO) or typical objects (TO) of the place, and the typical situations (TS) at the place, with partial content relation or

Figure 2. Overview process for determining other real life activity

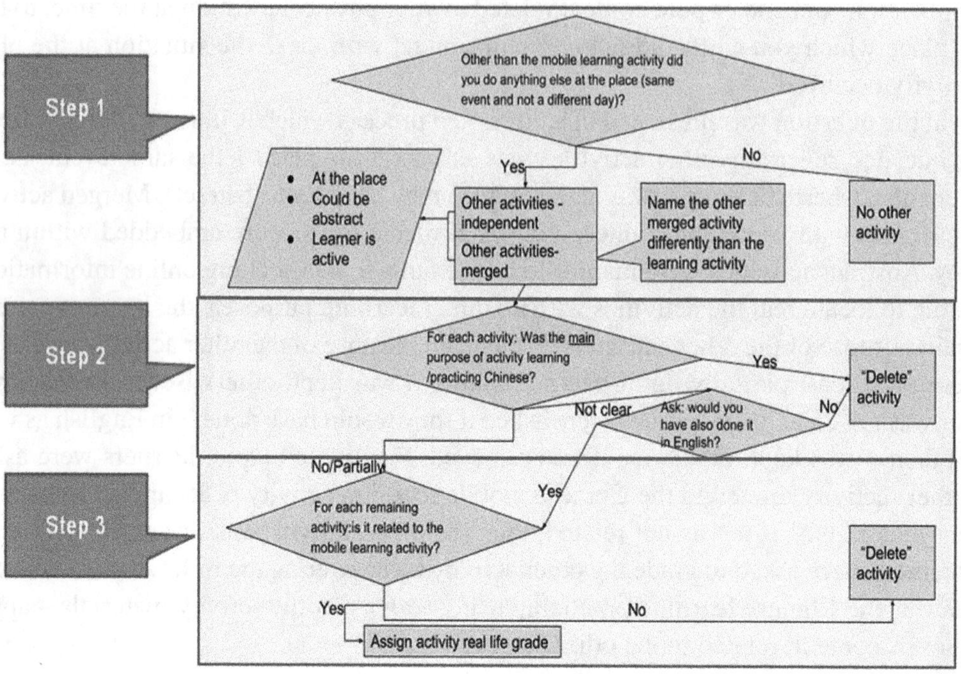

relation to non-typical objects reducing the score in each case. The place score in ascending order could be 0, 3 or 4; the object score in ascending order could be 0, 1, 2, 3 or 4; the situation score in ascending order could be 0, 3 or 4. The object score had two more values, when some or all content was related to non-typical objects of the place. Thus, two additional lower levels, 1 and 2, were added to the higher score levels of 3 and 4, as in the place and situation scores.

Table 2 shows some preliminary examples of activities' events which demonstrate the application of the real world context index. They are ordered by the level of their real world context.

- **Online Conversation Case:** This case scored lowest across the board, receiving 0 for place, object and situation. Learners corresponded with another party (for instance, through social media) in various places and for various purposes, such as scheduling an appointment or just holding a conventional conversation. Words were completely unrelated to the place or its typical objects.

Table 1. Real world context level index

Factors		2. Place Factor			
		Place (P)	Non-Typical Object (NTO)	Typical Object (TO)	Typical Situation (TS)
1. Content factor	No content relation	0	0	0	0
	Some content relation	3	1	3	3
	All content relation	4	2	4	4

Table 2. Real world context level index – examples

Case Name	No Relation	NTO		P	TO	TS	P	TO	TS
		Some Content	All Content	Some Content			All Content		
	0	1	2	3	3	3	4	4	4
Online conversation case	X								
Tattoo case			X						
Typical conversation case				X	X	X			
Place online conversation case					X	X	X		
Sign case							X	X	
Restaurant case							X	X	X

Moreover, since the conversation was not even held at the place (but was online), it did not qualify as a typical situation at the place.

- **Tattoo Case:** The object score for this case was 2. Learners searched their electronic dictionaries for words related to tattoos belonging to someone in their vicinity. The tattoo represented a person who was not necessarily a typical object of the place. However, as mere curiosity triggered these incidents, they did not seem to be typical situations and had no content relation to the place.
- **Typical Conversation Case:** This case received a score of 3 for place, object and situation. Learners used their electronic dictionaries to assist in their conversations about the coffee place they were at, but the conversation went beyond coffee, so only some words they checked were related to the coffee place, the coffee (typical object) or to the typical situation which, in this case, was the actual conversation.
- **Place Online Conversation Case:** This received 3 for the object and situation scores, and 4 for the place score. Learners held online conversations, but the content was related to the place. Nevertheless, content may have been partially related to typical objects and partially to typical situations. For example, a learner at a bar wrote about what she was having (typical objects) and what she was doing (typical situations). This also illustrates the ability of events to have a mixed score (3, 3, 4).
- **Sign Case:** The place and object were both scored 4 in this case. Various manifestations appeared (such as parking lot signs). Learners searched their electronic dictionaries for words completely related to the place which the sign represented and to typical objects of the place – namely, the sign attached to the place. However, mere curiosity triggered these incidents, and there don't seem to have been any typical situations.
- **Restaurant Case:** This case scored the highest, 4 for all three. Various manifestations appeared across consumer venues such as restaurants, supermarkets and stores. Learners searched their electronic dictionaries for words completely related to the place, to its typical objects, such as menu or supermarket and store products, and to typical situations, such as checking menus or searching products.

Real World Context Level Categories: When the real world score was different than 0, an indication of how the content was related to the place, its objects or situations was found. Three categories emerged. The first was content (C) in which only the content was related to the real world context but the learner did not note the relationship and seemed unaware of it. For example, learner 49 in Israel described using Google Translate to practice a lesson book's character handwriting during a bus ride: "Maybe we stood in the traffic jam, and there are traffic jam words [堵车], but at that moment I did not think about it". The second category was learner (L) in which the content was not inherently related to the real world context but the learner made a connection. For example, learner 28 in Israel was practicing the previous lesson's flashcards on the train. Some grammatical structures included the "if" structure, such as 如果, and the learner was able to relate it to the train context: "A little relation because I mentally in my head (not loudly) did a practice … regarding the 'if' structure… train sentence like if I had money I would not ride the train, I would drive in the car". The final category was content and learner (CL) in which the content was related to the real world context and the learner made the connection. The above-mentioned cases (restaurant, sign, about the place online conversation, typical conversation, tattoo) were all manifestations of this category. For example, at a restaurant, a learner was checking the menu while being assisted by an electronic dictionary; in this case, the content (menu products) was related and the learner recognised the connection.

Real Life Index

One score was found for each activity's event, based on the level of assistance/support/contribution of the mobile Chinese learning activity to the other (real life) activity whose purpose was not learning. The other activity could draw support from the learning activity either without any other supporting tools (WOT), such as English or human help, or alongside other supporting tools (AOT). The latter received a lower score. The other activity examined was the smallest activity to which the mobile learning activity contributed; this was referred to as the core activity.

Table 3 demonstrates the scoring process: events in which the mobile learning activity did not support any real life activity received 0, events in which other (real life) activities were supported by the mobile learning activity and other tools received 1, and events in which other (real life) activities were solely supported by the Chinese mobile learning activity received 2. Thus, each activity's event received one of 3 possible scores: 0, 1 or 2.

Table 4 presents preliminary examples of activities' events which demonstrate the application of the real life context index. They are ordered by the degree of their real life context.

- **Sign Case:** This case scored the lowest, 0, for the real life context. Various manifestations appeared across signs (such as parking lot signs). Learners searched their electronic dictionaries for sign words. However, mere curiosity triggered these incidents, and there seemed to be no other core activity at the place (whose purpose was not learning).
- **English Menu Restaurant Case:** This case scored 1 for real life context, as the menu was available in Chinese and English.

Table 3. Real life context level index

No Support	Support Alongside Other Tools (AOT)	Support Without Other Tools (WOT)
0	1	2

Table 4. Real life context level index - examples

Examples and Comments	Other Core Activity	Other Supporting Tools	No Support	AOT	WOT
			0	1	2
Sign case			X		
English menu restaurant case	Check menu	English		X	
Restaurant case	Check menu/ product details				X
Online scheduling appointment case	Scheduling appointment				X

- **Restaurant Case:** This case scored the highest, 2, and was repeated in places such as restaurants and supermarkets. Learners searched their electronic dictionaries when checking menus/products offered only in Chinese.
- **Online Scheduling Appointment Case:** This case also scored the highest for real life context, receiving 2. Learners corresponded with another party through social media for real life purposes. Conversations were held in Chinese only.

The online scheduling appointment case, a manifestation of an online conversation, was an interesting example of something having no relation to real world contexts; however, it did involve a real life context with or without another supporting language. The sign case was an interesting example of a partial relation to real world contexts without involving real life contexts.

The types of supporting tools which emerged, among others, included: English or any other non-Chinese language, for instance, on menus or in online conversations; human help, for instance, a service person assisting in finding a product at a store alongside checking an electronic dictionary for the product name; and parking lot or street signs while using Google Maps (a hybrid blend of English and Chinese) for navigation.

Quantitative Stage

The real world and real life analysis relied on the index developed in the qualitative stage. A linear mixed-effects model was used to analyse activities because the data were not independent: each student may have had a few activities, so activities were grouped within students. Mixed model analysis was conducted for each dependent variable: device mobility, real world context level and real life context level. It is important to note that the real world and real life activity scores were calculated by averaging the respective (index-based) scores of each event while the device mobility score was based on the number of places reported by the learner per activity (1-5).

Further details of the methodology are available in Cohen and Ezra (2018).

RESULTS

Data for 296 activity types (such as social communication and electronic dictionary activities), and 519 events (each learner activity may have had a few events) were collected from the 53 respondents. A detailed list is given in Table 5 for the target and non-target countries.

Descriptive Statistics

Of the 296 activity types, 148 were in Taiwan and 148 were in Israel. Classification of language learning orientation was the following: 132 activities were found to be generic and, in both countries, mostly comprised social communication activities or stand-alone electronic dictionary activities; 164 activities were found to be dedicated and electronic dictionary assistance in studying based on materials such as book or class handouts was common in both countries. Other prevalent activities were relatively balanced between Taiwan and Israel: listening materials, podcasts or learning videos, flashcard activities often related with class book word lists or other formal vocabulary sources such as HSK, and educational applications and games. The real world and real life context level scores were based on the index described above.

Standardising Real World Context Level Scores: As the place and the situation scores had 3 possible values (0, 3, 4) and the object scores had 5 possible values (0, 1, 2, 3, 4), they were standardised and combined into one real world context level joint score. A total standardised score, the *total real world context*, was computed based on the average of the 3 standardised scores.

Differences Between Target and Non-Target Country (Q1)

The results (Table 6) show that the mean scores of device mobility, total real world and real life context levels were higher in Taiwan than in Israel, but still relatively low when compared to the possible range. It is important to clarify the negative results in the table. As total real world context level scores were standardised, negative numbers appeared and some averages were negative, for example, the average in Israel (M = -0.24, SD=0.51).

The Linear Mixed Model was conducted for the three dependent variables. Differences in the target country and the non-target country (Taiwan and Israel) were not found to be significant in device mobility but they were significant for all other variables (Table 7). The target country means in Table 6 show that Taiwan was higher than Israel in total real world context level and real life context level. Thus, the conclusion was that Taiwan was significantly higher than Israel in total real world context level and real life context level. Table 7 presents the main effects of the three dependent variables with regards to the target country.

Table 5. Learners, activities and events figures

Country	*N* Learners	*N* Activity Types	*N* Events	Min-Max Activities per Learner	Event Range per Activity
Target: Taiwan	27	148	255	1–12	1–4
Non-target: Israel	26	148	264	2–18	1–6
Total	53	296	519	1–18	1–6

Table 6. Contextualised MALL variables (Target country)

Context Variables	Values Range	Target Country			
		Taiwan		Israel	
		Mean Score	Std.	Mean Score	Std.
1. Device mobility	1–5	2.3	1.32	2.03	1.08
2. Total real world context level (standardised score)	-3-3	0.24	1.1	-0.24	0.51
3. Real life context level	0–2	0.80	0.83	0.50	0.69

Table 7. Contextualised MALL – main effects (target country)

	Device Mobility	Total RW Context Level	Real Life Context Level
Target country	F = 2.17 P = 0.148 df = 39.92	F = 25.53*** p = 0.000 df = 45.6	F = 6.63* p = 0.015 df = 28.27

$*p < 0.05$ $**p < 0.01$ $***p < 0.001$

Differences Between Generic and Dedicated Language Learning Orientation in Contextualised MALL (Q2)

The results (Table 8) show that the mean score of generic activities was higher than the dedicated score; however, even the generic activities, particularly for device mobility and real world context level, were low and did not reach their potential. It is important to note that the mean score of dedicated activities was especially low (compared to target country) with respect to its potential. It is also important to clarify the negative results in the table. As total real world context level scores were standardised, negative numbers appeared and some averages were negative, for example, the average in dedicated activities (M = -0.33, SD=0.48).

The Linear Mixed Model was conducted for the three dependent variables. Differences in the language learning orientation (generic and dedicated) were found to be significant in all three context variables (Table 9). The language learning orientation means in Table 8 indicate that generic was higher than dedicated learning in device mobility, total real world context level and real life context level. Thus, the conclusion was that generic learning was significantly higher than dedicated learning in device mobility, total real world context level and real life context level. Table 9 presents the main effects of the three dependent variables.

Differences Between Target and Non-Target Countries in Contextualised MALL as a Function of Generic and Dedicated Language Learning Orientation (Q3)

The Linear Mixed Model was also used to calculate the interactions of the two independent variables. Table 10 shows the significant interaction effects of the target country and language learning orientation for all three dependent variables.

Although no main effect was found in the target country for device mobility, follow-up Bonferroni simple effects comparisons found that generic activities in Taiwan had a significantly higher mean score

Table 8. Contextualised MALL variables (language learning orientation)

Context Variables	Values Range	Language Learning Orientation			
		Generic		Dedicated	
		Mean Score	Std.	Mean Score	Std.
1. Device mobility	1–5	2.46	1.31	1.93	1.07
2. Total real world context level (standardised score)	-3-3	0.41	1.1	-0.33	0.48
3. Real life context level	0–2	1.32	0.61	0.11	0.37

Table 9. Contextualised MALL – main effects (language learning orientation)

	Device Mobility	Total RW Context Level	Real Life Context Level
Language learning orientation	F = 13.9*** p = 0.000 df = 279.66	F = 56.41*** p = 0.000 df = 287.8	F = 436.59*** p = 0.000 df = 278.47

$*p < 0.05$ $**p < 0.01$ $***p < 0.001$

Table 10. Contextualised MALL – interaction effects

	Device Mobility	Total RW Context Level	Real Life Context Level
Target country*Language learning orientation	F = 6.49* p = 0.011 df = 279.66	F = 25.87 *** p = 0.000 df = 288.27	F = 13.41*** p = 0.000 df = 279.43

$*p < 0.05$ $**p < 0.01$ $***p < 0.001$

than generic activities in Israel (p=0.005), but dedicated activities did not (p=0.52). Figure 3 illustrates the simple effect of device mobility in the target country; as the figure shows, a significant effect was found only for generic activities. The device's maximum mobility value was 5.

Follow-up Bonferroni simple effects comparisons (with the number of real world events entered as a covariate variable) of total real world context level found that the main effect in the target country was due to a significant simple effect in generic activities, as generic activities in Taiwan had a significantly higher mean score than generic activities in Israel (p=0.000), unlike dedicated activities (p=0.97). Figure 4 illustrates this. As total real world context level was a standardised variable, it had negative scores. The maximum (standardised) value was 3.

Follow-up Bonferroni simple effects comparisons (with the number of real life events entered as a covariate variable) of real life context level found that the main effect in the target country was due to a significant simple effect in generic activities, as generic activities in Taiwan had a significantly higher mean score than generic activities in Israel (p=0.000), unlike dedicated activities (p=0.49). Figure 5 illustrates this. The maximum value of real life context level was 2.

Overall, the differences between Taiwan and Israel in contextualised MALL were manifested only in generic activities. Specifically, total real world context level and real life context level were significantly higher in Taiwan but only in generic activities. Device mobility, which was not significantly different

Figure 3. Device mobility – simple effect
$^{**}p < 0.01$

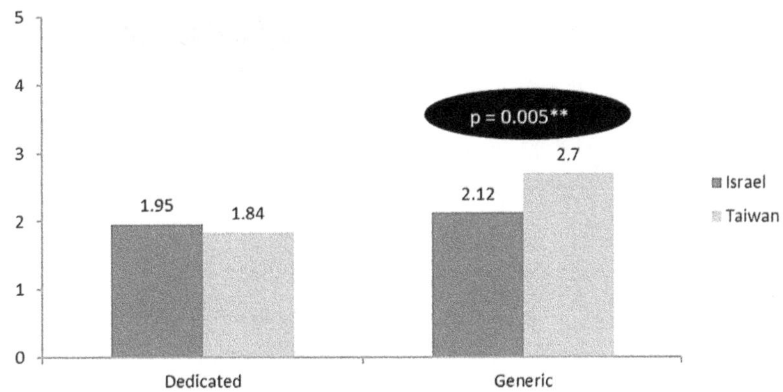

Figure 4. Total real world context level – simple effect
$^{***}p < 0.001$

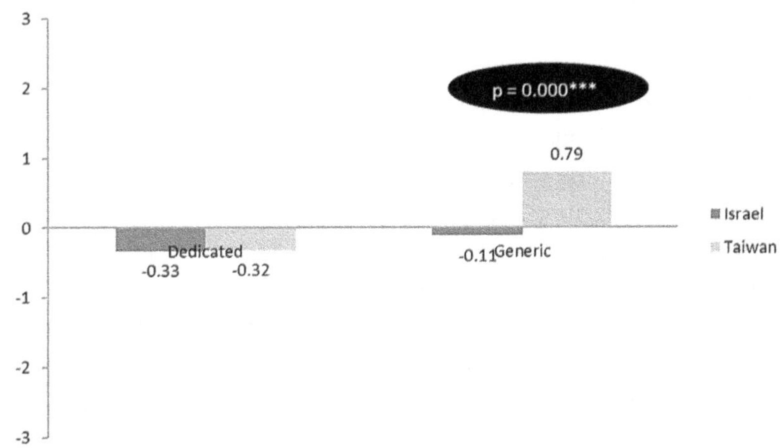

between Taiwan and Israel in main effect, was found to be significantly higher in Taiwan when analysis was restricted to the generic language learning orientation. Both countries were found to be similarly low in dedicated activities.

CONCLUSION AND DISCUSSION

The contextualised MALL evaluation index developed in this study establishes a clearer operative demarcation between real world and real life contexts. The real world context involves the place; simply stated, everything is examined in terms of its relevance to the place. The real life context involves activity which does not necessarily relate to the place but has a context in a person's life; thus, an online conversation, such as online scheduling, is not related to the real world place, but is related to a person's real life. Looking at signs out of curiosity involves real world but not real life. Whereas in real life, learners are very active, in the real world, even when they are not active, they may still perform contextual learn-

Figure 5. Real life context level – simple effect
****p < 0.001*

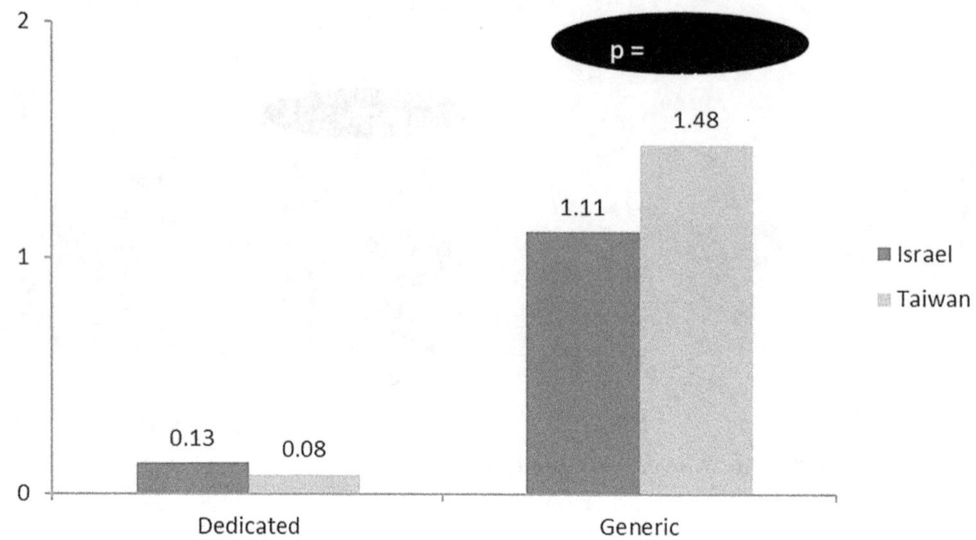

ing. Both content and learner (CL) and learner (L) events, particularly the latter, suggest learners' high *mobile awareness, a finding* which adds to previous findings on the skills desired for mobile learning, such as mobile literacy, a macro literacy underpinning digital literacies which, in the mobile era, is vital in both learning and daily life (Pegrum, 2014).

The finding that the target and non-target countries' contextualised learning was significantly different but only in generic language learning orientation may be attributed to the nature of the study's more common generic activities – social, or stand-alone electronic dictionary not accompanying study materials. For instance, in Israel, monolingual language learners with a similar L1 (Tomlinson, 2016) may have been less prone to speak Chinese on social media, thus reducing their real life context level. Or they may have had less daily need to employ the studied language and less need for electronic dictionaries in restaurants or supermarkets, compared to learners in Taiwan, thus reducing their real world and real life context level. The dedicated activities were contextually relatively lower than they could have been and were lower than generic activities; they were similarly contextually low in the target and the non-target country; this might be attributed to the characteristics of the common dedicated activities found in this study, as described below.

Restricted content (Reinders & White, 2010) may result from the use of similarly restricted school curriculums, such as flashcards based on book materials. If content is restricted, it is less likely to be connected to the rich contextual opportunities that will exist. The study's most common dedicated prevalent activities included dictionaries, listening materials and flashcards; all seem related to a restricted school curriculum. *Lengthy content* might make it more difficult for learners to use the context (Shu, 2008b); the dictionary, flashcards and listening materials used in the study frequently had lengthy content chunks. *Content presentation* is predominantly presented according to a syllabus and not likely to be relevant to the learner's life (Masuhara, 2016). Finally, without *content affective engagement*, knowledge will be unmemorable and will not transfer well to the real world (Immordino-Yang & Damasio, 2007). Learners' ability to connect to the real world or real life contexts may be inhibited by a lack of affective engagement. Further elaboration of target and non-target country results and their interpretation is available in Ezra and Cohen (2018a).

CONTRIBUTION

The conclusions of this study may assist practitioners and researchers alike. Teachers could enhance their knowledge about students' MALL practices in target and non-target countries and be encouraged to further their usage of mobile technology (Van Praag & Sanchez, 2015). They are not there to be replaced; quite the opposite. They should be taking a more advanced role, guiding students in outside activities while not worrying about having to spend extra time doing so (Shu, 2008a). Responding to the important finding of this study on the gap between potential and actual contextualised MALL, some practical suggestions for teachers on how to increase contextualised MALL by taking advantage of their students' existing activities are given next.

All teachers should have device mobility, real world and real life contexts in mind, but depending on the locus of instruction, there are some differences. In a target country, teachers may want to focus on the dedicated activities, while drawing support from the students' generic activities (which will become dedicated as teacher pedagogy becomes involved). In a non-target country, teachers should tackle both generic and dedicated activities. For instance, in generic activities, they may draw support from existing generic usages of the dictionary and social media, as available even in the non-target country and as shown in this study.

Teachers can guide their students to deal with the content factors mentioned above when they are performing dedicated (or generic) activities. More importantly, in both target and non-target countries, teachers may draw upon this study's findings, especially the real world/real life context evaluation index to guide students towards the increasing use of contextualised MALL in generic or dedicated activities. Teachers can help their students diversify the locations of their dedicated and generic activities to increase their device mobility. Their students can learn to see the content relations between dedicated or generic MALL activities and a place, the objects of the place (especially typical objects) and typical situations at the place to increase the real world context level of their learning. Finally, teachers can guide their students to use dedicated or generic MALL activities to support other real life core activities whose purpose is not learning. Students should be recommended to use as few supporting tools as possible, especially if they feel they may be redundant, thus avoiding some tools mentioned in this study, such as: English assistance on menus or in online conversations, or human help in stores looking for product information. Further practical recommendations are available in Ezra and Cohen (2018b).

The newly developed index builds on objective contextual definitions and may be of use in the challenging task of analysing, designing, evaluating and assessing contextualised MALL activities. This has been difficult because of learners' subjectivity (Comas-Quinn et al., 2009) and because of the personal contextual noise which is problematic in definition and evaluation (Traxler, 2007).

The study has value for researchers as well. Other scholars have noted the benefits of studying an L2 language in the target country (Anderson et al., 2008; Comas-Quinn et al., 2009; Petersen et al., 2008). The study quantitatively shows this assertion applies to contextualised MALL, including device mobility, real world contexts and real life contexts, and has vast qualitative data to support it. It also refines this claim by showing that MALL in the target country is more contextualised, although only in generic activities. Finally, the study's demarcation of real world and real life contexts may enable more consistent research of contextualised MALL, particularly as many implementation studies (Burston, 2013) seem to be blurring the real world and real life.

FUTURE RESEARCH

Future studies should validate the real world/real life context index in more target and non-target country settings, looking at existing curriculums and validating quantitative learning outcomes, and perhaps even comparing real world and real life contextual learning, given their differences as shown in this study. Furthermore, the context index may be expanded by investigating more contextual parameters as suggested in the literature; the time context (Dey & Abowd, 1999) might be a good candidate. Attention should be given to other factors that may have an influence on the level of contextualised mobile language learning. Thus, it might be interesting for future research to use the suggested real world/real life index and examine more factors, such as students' language proficiency level or *mobile awareness*.

This research received no specific grant from any funding agency in the public, commercial, or not-for-profit sectors.

REFERENCES

Al-Mekhlafi, K., Hu, X., & Zheng, Z. (2009). (2009). An approach to context-aware mobile Chinese language learning for foreign students. *2009 Eighth International Conference on Mobile Business*, 340-346.10.1109/ICMB.2009.65

Anderson, T. A., Hwang, W., & Hsieh, C. (2008). (2008). A study of a mobile collaborative learning system for Chinese language learning. *Proceedings of International Conference on Computers in Education*, 217-222.

Burston, J. (2013). Mobile-assisted language learning: A selected annotated bibliography of implementation studies 1994-2012. *Language Learning & Technology*, *17*(3), 157–225.

Burston, J. (2014). MALL: The pedagogical challenges. *Computer Assisted Language Learning*, *27*(4), 344–357. doi:10.1080/09588221.2014.914539

Burston, J. (2015). Twenty years of MALL project implementation: A meta-analysis of learning outcomes. *ReCALL*, *27*(1), 4–20. doi:10.1017/S0958344014000159

Cai, Y. X. (2008). Chinese language teaching material editing writing and introduction. In Y. X. Cai, Z. M. Shu, L. F. Chen, X. Y. Zhang, S. Z. He, & M. D. Lai (Eds.), *Introduction to teaching Chinese as a foreign language* (pp. 113–132). Taipei, Taiwan: Sanmin.

Chen, C. M., & Li, Y. L. (2010). Personalised context-aware ubiquitous learning system for supporting effective English vocabulary learning. *Interactive Learning Environments*, *18*(4), 341–364. doi:10.1080/10494820802602329

Cohen, A., & Ezra, O. (2018). Development of a contextualised MALL research framework based on L2 Chinese empirical study. *Journal of Computer Assisted Language Learning, 31*(7), 764–789. doi:10.1080/09588221.2018.1449756

Comas-Quinn, A., Mardomingo, R., & Valentine, C. (2009). Mobile blogs in language learning: Making the most of informal and situated learning opportunities. *ReCALL, 21*(01), 96–112. doi:10.1017/S0958344009000032

Dey, A. K., & Abowd, G. (1999). (1999). Towards a better understanding of context and context-awareness. *Workshop on the what, Who, Where, when, and how of Context-Awareness, 1707*, 304-307.

Ezra, O., & Cohen, A. (2018a). Contextualised MALL: L2 Chinese students in target and non-target country. *Computers & Education, 125*, 158–174. doi:10.1016/j.compedu.2018.06.011

Ezra, O., & Cohen, A. (2018b). *How to increase Contextualised MALL of L2 Chinese adult students in target and non-target countries.* Paper for the 10th International Conference and Workshops on Technology and Chinese Language Teaching (TCLT), National Taiwan Normal University, Taipei, Taiwan.

Hwang, W. Y., Chen, C. Y., & Chen, H. S. (2011). (2011). Facilitating EFL writing of elementary school students in familiar situated contexts with mobile devices. *10th World Conference on Mobile and Contextual Learning: mLearn2011 Conference Proceedings, Beijing, China*, 15-23.

Hwang, W. Y., & Chen, H. S. (2013). Users' familiar situational contexts facilitate the practice of EFL in elementary schools with mobile devices. *Computer Assisted Language Learning, 26*(2), 101–125. doi:10.1080/09588221.2011.639783

Immordino-Yang, M. H., & Damasio, A. (2007). We feel, therefore we learn: The relevance of affective and social neuroscience to education. *Mind, Brain and Education: the Official Journal of the International Mind, Brain, and Education Society, 1*(1), 3–10. doi:10.1111/j.1751-228X.2007.00004.x

Klopfer, E. (2008). *Augmented learning: Research and design of mobile educational games.* Cambridge, MA: MIT Press. doi:10.7551/mitpress/9780262113151.001.0001

Klopfer, E. (2011). Forward/preface. In S. Dikkers, J. Martin, & B. Coulter (Eds.), *Mobile media learning: Amazing uses of mobile devices for learning.* Pittsburgh, PA: ETC Press.

Kukulska-Hulme, A. (2012). Mobile-Assisted language learning. In C. Chapelle (Ed.), *The encyclopedia of applied linguistics* (pp. 3701–3709). New York, NY: Wiley Online Library; doi:10.1002/9781405198431.wbeal0768

Marsden, E., Mitchell, R., & Myles, F. (2013). *Second language learning theories.* London: Routledge.

Masuhara, H. (2016). Brain studies and materials for language learning. In B. Tomlinson (Ed.), *SLA research and materials development for language learning* (pp. 23–32). New York, NY: Routledge.

Moser, D. (1991). Why Chinese is so damn hard. In V. H. Mair (Ed.), *Schriftfestschrift: Essays on writing and language in honor of john DeFrancis on his eightieth birthday, Sino-platonic papers* (pp. 59–70). Philadelphia, PA: University of Pennsylvania.

Ogata, H., Hui, G. L., Yin, C., Ueda, T., Oishi, Y., & Yano, Y. (2008). LOCH: Supporting mobile language learning outside classrooms. *International Journal of Mobile Learning and Organisation, 2*(3), 271–282. doi:10.1504/IJMLO.2008.020319

Pegrum, M. (2014). *Mobile learning: Languages, literacies and cultures*. Basingstoke, UK: Palgrave Macmillan. doi:10.1057/9781137309815

Petersen, S. A., Divitini, M., & Chabert, G. (2008). Identity, sense of community and connectedness in a community of mobile language learners. *ReCALL, 20*(3), 361–379. doi:10.1017/S0958344008000839

Reinders, H., & Pegrum, M. (2016). Supporting language learning on the move. an evaluative framework for mobile language learning resources. In B. Tomlinson (Ed.), *SLA research and materials development for language learning* (pp. 116–141). New York, NY: Routledge.

Reinders, H., & White, C. (2010). The theory and practice of technology in materials development and task design. In N. Harwood (Ed.), *English language teaching materials: Theory and practice* (pp. 58–80). New York, NY: Cambridge University Press.

Rivers, D. J. (2009). Utilizing the quick response (QR) code within a Japanese EFL environment. *JALT CALL Journal, 5*(2), 15–26.

Sandberg, J., Maris, M., & de Geus, K. (2011). Mobile English learning: An evidence-based study with fifth graders. *Computers & Education, 57*(1), 1334–1347. doi:10.1016/j.compedu.2011.01.015

Schilit, W. N. (1995). *A system architecture for context-aware mobile computing* (Doctoral dissertation). Columbia University, New York, NY.

Shu, Z. M. (2008a). Chinese language multimedia and digital teaching. In Y. X. Cai, Z. M. Shu, L. F. Chen, X. Y. Zhang, S. Z. He, & M. D. Lai (Eds.), *Introduction to teaching Chinese as a foreign language* (pp. 281–299). Taipei, Taiwan: Sanmin.

Shu, Z. M. (2008b). Chinese language training and teaching material for listening and speaking. In Y. X. Cai, Z. M. Shu, L. F. Chen, X. Y. Zhang, S. Z. He, & M. D. Lai (Eds.), *Introduction to teaching Chinese as a foreign language* (pp. 231–257). Taipei, Taiwan: Sanmin.

Tomlinson, B. (2016). Applying SLA principles to whole-class activities. In B. Tomlinson (Ed.), *SLA research and materials development for language learning* (pp. 33–49). New York, NY: Routledge. doi:10.4324/9781315749082

Traxler, J. (2007). Defining, discussing and evaluating mobile learning: The moving finger writes and having writ..... *The International Review of Research in Open and Distributed Learning, 8*(2), 1–12. doi:10.19173/irrodl.v8i2.346

Tseng, C. C., Lu, C. H., & Hsu, W. L. (2007). A mobile environment for Chinese language learning. In M. Smith & G. Salvendy (Eds.), Human interface 2007 (vol. 4558, pp. 485-489). Berlin, Germany: Springer. doi:10.1007/978-3-540-73354-6_53

Van Praag, B., & Sanchez, H. S. (2015). Mobile technology in second language classrooms: Insights into its uses, pedagogical implications, and teacher beliefs. *ReCALL, 27*(3), 288–303. doi:10.1017/S0958344015000075

Wang, M., Koda, K., & Perfetti, C. A. (2003). Alphabetic and nonalphabetic L1 effects in English word identification: A comparison of Korean and Chinese English L2 learners. *Cognition*, *87*(2), 129–149. doi:10.10160010-0277(02)00232-9 PMID:12590041

Wong, L. H., Chin, C. K., Tan, C. L., & Liu, M. (2010). Students' personal and social meaning making in a chinese idiom mobile learning environment. *Journal of Educational Technology & Society*, *13*(4), 15–26.

Wu, T. T., Sung, T. W., Huang, Y. M., Yang, C. S., & Yang, J. T. (2011). Ubiquitous English learning system with dynamic personalized guidance of learning portfolio. *Journal of Educational Technology & Society*, *14*(4), 164–180.

KEY TERMS AND DEFINITIONS

Device Mobility: One of the three dependent contextualized variables in the study; it represents the diversity of the places where the mobile learning activity occurs.

L2 Language: Language studies by a native speaker of a different language (L1).

Mobile Awareness: A term coined to represent the skills required by mobile learners. In the context of this paper, it pertains to their ability to relate content to the real-world context especially when there is no inherent relationship.

Mobile-Assisted Language Learning (MALL): The use of mobile technology in language learning; it mostly pertains to using mobile devices such as smart phones (which may be opened in point A, B, and in between) unlike portable devices such as laptops (which are usually opened in point A and B and closed down in between).

Non-Target Country: The country where the studied language (L2) is not the main spoken language.

Real-Life Context Level: One of the three dependent contextualized variables in the study; it represents the degree of the mobile learning activity in relation to the learner's real life.

Real-World Context Level: One of the three dependent contextualized variables in the study; it represents the degree of the mobile learning activity in relation to the place where it occurs.

Target Country: The country where the studied language (L2) is the main spoken language.

Chapter 10
The Use of Mobile Technologies by Immigrant Adolescents in Coping With the New Language and With Their Formal Studies

Gila Cohen Zilka
https://orcid.org/0000-0001-5463-7822
Bar-Ilan University, Israel & Achva Academic College, Israel

ABSTRACT

In light of the extensive use of mobile technologies by adolescents, the chapter examined the usability of mobile technologies for new immigrant adolescents in supporting their absorption process in the new country, acquiring the new language, and coping with their studies. The study also sought to determine what the adolescents' most meaningful experiences were that mobile technologies made possible in the course of their absorption in the new country. Participating in the study were 125 new immigrant adolescents, who have resided in Israel between 6 and 12 months. This is a qualitative study. The study included (1) interviews before completing an open questionnaire, (2) completion of an open-question questionnaire, and (3) personal interviews that took place after the questionnaire was completed. Participants reported that thanks to mobile technologies they were able to integrate into the daily life of their new environment. Translation software, databases, various applications, and social network groups have helped satisfy their needs, assisted in real time with difficulties they encountered, and helped create a sense of social connection and belonging.

DOI: 10.4018/978-1-5225-8106-2.ch010

INTRODUCTION

In light of the extensive use of mobile technologies by adolescents, the present study examined the use of mobile technologies by new immigrant adolescents to support their absorption process in the new country, the acquisition of the new language, and their studies. The study also sought to determine what were the adolescents' most meaningful experiences that mobile technologies made possible in the course of their absorption in the new country.

BACKGROUND

Adolescents and the Digital Era

Researchers have argued that the new media is part of our living environment (Comstock & Scharrer, 2007; Livingstone, 2008, 2007; Millwood-Hargrave, 2007; Millwood-Hargrave & Livingstone, 2006, 2009). People who grow in the digital era develop and shape their personality, attitudes, opinions, and beliefs in relation to the environment. Their daily life experience includes direct, face-to-face experiences as well as new media experiences (Coleman, 2001; Eisen & Lillard, 2016; Millwood-Hargrave & Livingstone, 2009).

Researchers have found that the most popular means for using the new media and its applications is the smartphone (Goggin & Hjorth, 2014; Ito, Baumer, Bittanti, Boyd, Cody, HerrStephenson, Tripp et al., 2010). They use the term "m-learning," which is defined as any learning or learning-support activity that is conducted through mobile technologies (Jan, Ullah, Ali & Khan, 2016; Sung, Chang & Liu, 2016). The use of mobile technologies in learning increases the students' motivation in social and educational engagement, stimulates curiosity, and exposes students to diverse and engaging environments. Mobile technologies help students understand the educational content through images, animation, simulations, and videos, (Choi, 2012; Foti & Mendez, 2012; Hanafi & Samsudin, 2012; Rossie, Miller, Cecil & Stamper, 2012; Schugar, Smith & Schugar, 2013; Warschauer, 2011). Mobile technologies also assist in the transfer of materials between the teacher and the students as well as between the students themselves. Mobile technologies are becoming more and more available, with increased functionality and a wide range of applications that allow students access to learning materials and assistance anytime, anywhere (Cohen, Vincent, Adhikari et al., 2015; Dahlstrom, 2015; Zilka, 2018).

Researchers who have examined the integration of mobile technologies into the learning process found that when it is available anywhere, anytime, it helps make learning a part of the students' daily activities, and fosters student involvement and collaboration(Jan, Ullah, Ali & Khan, 2016; Johnson, Smith, Willis, Levine & Haywood, 2011; Kukulska-Hulme & Shield, 2008; Traxler, 2009). It also creates equal opportunities for students with special needs, with accessibility problems, and more (Ardito, Costabile, Marsico et al., 2006; Gikas & Grant, 2013). At the same time, researchers found a variety of problems, including the fact that m-learning leads to distractions during study; small screens limit the information displayed; storage capacity is inadequate; there is application overload; the small keyboard causes typing fatigue and slower typing than is possible on other devices; short battery life leads to relatively quick battery depletion (Althunibat, 2015; Bidin & Ziden, 2013; Lowenthal, 2010; Park, 2011; Sarrab, Elbasir & Alnaeli, 2016).

In the present study we examined the use of mobile technologies by new immigrant adolescents in support of their integration in the new country, helping them in the acquisition of the new language, and in their studies.

Immigrant Adolescents

Immigration is relocation from one country to another. Many immigrants encounter hardships, such as language problems, issues of housing, financial distress, and difficulties in their cultural and social adjustment. Often there is tension between the immigrants' cultural outlook and their daily experiences in their adoptive society. These tensions and difficulties exacerbate the distress of immigrant children (Choi, 2008; Maldonado-Molina, Reingle, Wesley, Jennings& Prado, 2011; Rossiter & Rossiter, 2009; Spallek, Zeeb & Razum, 2010; Yearwood, Crawford, Kelly & Moreno, 2007). Researchers found that migration to a new country may affect the wellbeing of adolescents. Their ability to study and acquire skills, and their sense of self-worth and social connectedness may be impaired. All these can cause dysfunction in various areas, passivity or aggression, and even social deviance. Therefore, it is important to create an environment that allows a sense of belonging, protection, growth, interest, competence, challenge, success, and feedback an environment that enables meaningful interaction, positive experiences, and success. It is important to provide adolescents with a sense of self-worth following experiences that create feelings of being needed and of making a significant contribution to the environment, and to provide them with an opportunity to exercise skills and receive appreciation and feedback from the environment. It is equally important to foster a sense of social connection through experiences of belonging, close friendships, and social acceptance that invite meaningful interactions, leading to involvement and meaningful roles in the home, school, and community (Birch & Ladd, 1997, 1998; Romi, Savicki, Grupper & Caspi, 2007; Zilka, 2015).

Transition to a new society requires immigrants to adapt to biculturalism and bilingualism. Indeed, immigrant youths are required to recast their identity, which in adolescence is already in the process of being formed. In the new country, knowledge of the spoken language may facilitate the absorption process, as it is one of the main components of the social and cultural integration of new immigrants. Knowledge of the new language contributes to the success of the absorption process in the new country and makes possible better coping with social interactions.

Lack of linguistic fluency is defined as the most prominent difficulty in the functioning of new immigrants (Budría, Colino & Martínez de Ibarreta, 2018; McCarty, 2018; Puigdevall, Walsh, Amorrortu & Ortega, 2018; Sevinç & Dewaele, 2018; Spörlein & Kristen, 2018; Zorlu & Hartog, 2018). The process of acquiring a new language can be characterized by several stages. In the first stage, immigrants try not to speak, and when they start speaking, they use familiar expressions and standard patterns such as "how are you." In the second stage, immigrants create expressions that contain linguistic patterns of the new language, but base these on structures borrowed from their mother tongue, simplifying in the process the system of rules of the new language, in accordance with familiar linguistic norms derived from their mother tongue. The resulting text, however, does not belong to either the mother tongue or the new language. In the third stage, after the acquisition of grammatical mechanisms, the immigrants' ability to express themselves improves. Another difficulty has to do with control of pragmatic aspects of the socio-cultural context, such as irony, humor, rhetorical questions, etc. Therefore, social-cultural mediation in oral and written discourse is of great importance. As the mastery of the language progresses, immigrants learn the discourse patterns of the new language and internalize its pragmatic-rhetorical aspects (Nesterko, Ulitsa, & Friedrich, 2018; Puigdevall, Walsh, Amorrortu, & Ortega, 2018; Repke & Benet-Martínez, 2018; Sevinç & Dewaele, 2018; Spörlein & Kristen, 2018).

This study examined the use of mobile technologies by new immigrant adolescents in support of their absorption process in the new country, in acquiring the new language, and in their studies. The study also sought to identify meaningful experiences that mobile technologies made possible for them in the course of their absorption in the new country.

Research Questions

1. What are the characteristics of the use of mobile technologies by new immigrant adolescents?
2. How does the use of mobile technologies help adolescents in their absorption process in the new country?
3. What are the most meaningful experiences that the use of mobile technologies has created for adolescents in their studies inside and outside school?

METHOD

This qualitative study included (a) interviews before completing an open questionnaire, (b) completion of an open-question questionnaire, and (c) personal interviews that took place after the questionnaire was completed. The content analysis was based on Adler and Adler (2008), Atkinson and Delamont (2006), and Hammersley (2008).

The students studied in classes consisting of both new immigrants and native-born Israelis, and participated in lessons taught by a tutor, whose role was to guide them throughout their absorption process at the school and in the new country.

Wherever a "teacher" is mentioned, it refers to a teacher who teaches a class consisting of both new immigrants and native-born Israelis, and "tutor" refers to an instructor guiding new immigrant students.

Sample

The sample included 125 new immigrant adolescents who had resided in Israel between 6 and 12 months, 15 to 17 years old, 48% boys. Sixty percent of adolescents emigrated from Russia, Ukraine, and other FSU countries, 25% from France, 10% from the US, and 5% from Argentina and Peru.

The sample also included 20 teachers who work in classes that include both immigrants and native-born Israelis, and 20 tutors, that is, graduates who work with new immigrants after school hours. The teachers and tutors we interviewed work with the new immigrant adolescents who participated in the study. They teach them the language and help them with their studies and the process of absorption in Israel.

The purpose of the interviews with the teachers and tutors was to understand in depth the research findings that emerged from the interviews with the adolescents. Therefore only the most relevant findings from the interviews with the teachers are presented.

Research Tools

1. Personal information questionnaire for adolescents, including demographic items such as gender, age, country of origin, year of immigration, marital status of the parents, father's occupation, mother's occupation, number of siblings, residential area, type of residence, number of rooms in the house, sector, economic situation.
2. A questionnaire for adolescents, with ten open questions, such as: How do you use mobile technologies to cope with the new language and with your formal studies? Are mobile technologies formally integrated into your school studies? If yes, is it by your own initiative or by that of your teachers, built into your study program? What apps are used, how, and why? Are you a member of a learning community? If yes, is it by your own and your friends' initiative, or by the initiative of your teachers, built into your study program? Are you using mobile technologies to help you in your absorption process in the new country, and if yes, how? What are the most meaningful experiences that the use of mobile technologies has made possible for you inside and outside of school?
3. Interviews with adolescents. Interviews were conducted with 40 adolescents who completed the questionnaires. Each interview lasted between 20 and 25 minutes. The semi-structured interviews addressed all the research topics, but the order of the questions depended on the dynamics of the interview. The interviewees wanted to share their experiences, both positive and negative, and some of them provided detailed descriptions.
4. Interviews with 20 teachers and 20 tutors. The purpose of the interviews with the teachers and tutors was to understand the findings in depth and from the point of view of the teachers and tutors. Each interview lasted between 20 and 25 minutes.

FINDINGS

In this section, the findings regarding the use of mobile technologies by immigrant adolescents in coping with the new language and with their formal studies, are presented according to the main themes indicated by the adolescents: translations, databases, recordings, scanning, photography, the use of various applications, social networks, and experiences derived from using the telephone during the absorption process in the new country.

Translation Applications

Respondents noted that they search on the phone for information about the material learned in the classroom in their native language, then translate it into Hebrew. The following statements were repeated by a large number of immigrants:

I translate words that I don't understand and by doing so I learn them. I translate words from Hebrew into Russian and vice versa.

When we start learning new material, I search for material in my own language, and then translate it into Hebrew, and that's how I feel ready for class.

Table 1. Main themes in the adolescents' accounts

Key Themes Emerging From the Accounts of New Immigrant Adolescents	Percentage of Respondents	
Translation applications	100%	Extensive use of applications for translating words and explaining their meaning; translators allow displaying and voicing text. All respondents noted that they search on the phone for information about the material learned in the classroom in their native language, then translate it into Hebrew. They also use translation apps in the street, while walking, shopping, etc. They practice pronouncing words according to the way they are pronounced in the translation application. The translation applications quickly illustrate how to pronounce words correctly. They record Hebrew words and use automatic translation of the recorded word, because at times they find it difficult to spell out the words.
	92%	Participants stated that the translation applications help them integrate into society and academically.
Databases	100%	Students search the Internet for in-depth explanations, comments, and elaborations on subjects they learned but did not fully understand in class, or on topics discussed between the students, etc.
	100%	They watch videos on topics learned in class or brought up in conversations; look for materials in Wikipedia, presentations, photos, and movies on topics learned and discussed in the classroom; search for materials on online learning sites, especially sites for new immigrants that have learning materials taught in school in their native language.
Photography, scanning, and recording	100%	Recording the teacher during class, recording of classmates' speech, shopping, conversations in the street, etc.
	100%	Pictures of the board in the classroom
	92%	Photographing/scanning content from other students' notebooks.
	82%	Scanning the worksheets in the classroom, to work on them with the tutor.
Applications	100%	Calculator; schedule application on the smartphone; recording reminders and homework; exam schedule; clock/stopwatch; various applications used to practice the new language, including language games; practicing science through lab simulators; quizzes; applications used to prepare for exams, and more.
Social networks	100%	A group of immigrant students and tutors assists with the language and with subjects requiring high proficiency.
	94%	New immigrants open groups in their native language with students who immigrated from their country of origin.
	87%	Groups in mathematics, physics, with friends who remained abroad.
	93%	Correspondence in the class group, with the help of the translation application, makes them feel that they belong in the class and participate in matters of interest.
Experiences derived from using of the telephone during the absorption process in the new country	92%	The most significant experience participants mentioned was that the use of the smartphone helped in many ways deal with the difficulties they experienced in the process of absorption in the new country. These included the translation applications, the databases, various other applications, the groups on the social networks that responded to their needs and to difficulties they encountered, as well as groups with friends and relatives who remained abroad.

I have a lot of new immigrant friends, we translate entire texts, and that's how we learn the language. We can hear the translation and so we learn to say the words correctly.

I'm mainly looking for the meaning of words I don't understand, during a conversation or lesson.

There are things that are told and everybody understands them except for us, the new immigrants, so we have to translate and try to understand the meaning through the examples that the translation app gives us.

Databases

The new immigrant adolescents search the Internet for in-depth explanations, comments, and elaborations on subjects they learned but did not fully understand in class, or on topics discussed between the students, etc. The following statements were repeated by a large number of immigrants:

I'm searching on the smartphone for summaries of what I'm learning in class, in my native language, made by my native language speakers.

I watch educational presentations on the phone, over and over again, while traveling or during breaks between classes.

I search for information, for example, through Google, Wikipedia, and various other sources, see clips related to the subject of the study, listen to lectures on topics learned in the classroom. Sometimes there's a lesson and I don't understand anything, and then there is a continuation lesson, so during recess I surf and find relevant material, and I no longer feel lost.

I'm looking for online exercises to practice my homework beyond what I'm assigned at school, especially in technological subjects such as computers, math, science.

I search Google images and movies to understand the material I learned in class, which I found difficult to understand.

I'm looking for answers on the Internet to questions on the material learned in the classroom, or topics that came up in conversations with other students, over the Internet."

You can find information on almost any topic in almost any language, especially in the languages of immigrants that are widely spoken, like Russian and Spanish.

Photography, Scanning, and Recording

The new immigrant adolescents record the teacher during class, record classmates' speech, etc. The following statements were made by a large number of immigrants:

I scan notebooks of "good" students to continue working on the subject with the tutor, because my handwriting is not readable and the tutor can't figure out what exactly we were learning.

I record the teacher assigning homework or teaching a new subject, so that after class I can accurately ask for help from the tutor.

I photograph the board when I didn't manage to write down the material, I usually don't have enough time.

Applications

Respondents make extensive use of Calculator, schedule application on the smartphone, recording reminders and homework, exam schedule, practicing science through lab simulators, applications used to prepare for exams, and more. The following statements were made by a large number of immigrants:

I use learning applications, such as applications to learn the new language or applications for practicing and preparing for tests in various subjects.

Sometimes the tutor gives us quizzes through the smartphone to practice in a fun manner the subjects we have learned in class.

We practice with an application that has games teaching new words.

We engage in all sorts of word play related to learning.

We do "barcode" riddle games, we walk around the school with the smartphone and look for barcodes, and each barcode reveals a clue to decipher a riddle.

The teachers prepare joint games for us, for new immigrants and native-born Israelis together, on a smartphone, and it makes me feel like I belong in the class. That's how I learn the language and make friends.

I make extensive use of applications during English lessons. My smartphone rescues me in English classes, otherwise I don't understand a thing.

Social Networks

Participation in social networks helps the new immigrant adolescents feel a sense of belonging. The following statements were made by a large number of immigrants:

I translate words in real time, send messages on the social network, mainly to understand the expressions that are clear to the native-born Israelis but not to newcomers, and free translation is not appropriate.

Students photograph the board in the classroom, showing what the teacher wrote, then upload the picture to the class group.

We post on social networks materials to practice for tests, board snapshots, illustrations, knowledge expansion links, pre-test summaries, homework solutions. If someone asks how a particular word is pronounced, someone sends a recording with that word, and many more things.

We answer questions on homework or complex assignments together, within the learners' community, usually on WhatsApp or Facebook, and everyone participates to achieve a complete answer, explaining to each other, interpreting, explaining expressions or words with multiple meanings.

Experiences Derived From Using the Telephone During the Absorption Process in the New Country

The use of the smartphone helped in many ways deal with the difficulties they experienced in the process of absorption in the new country. The following statements were made by a large number of immigrants:

I don't know how a new immigrant can get by without a smartphone, it's impossible. I constantly use my phone, mainly to translate and hear words. This is how I learn how to pronounce words correctly.

The smartphone enables me to fight the feeling that I'm imprisoned and not free because of language limitations. I feel that with it I will always be able to find solutions in real time to the difficulties that arise from language limitations.

Some of my school teachers encourage me to use the phone during class, so I would not "get lost," and some disagree. They say I'd better listen to the lesson and not occupy myself with the phone. One teacher said to me: "If you don't understand something and you are searching for related material on the phone, you're dealing with the phone and missing even more material, and the material you may find will only confuse you further. So you'd better listen to the lesson and then ask for help.

The tutor who works with us prepares us for the new subjects that the teacher will teach at school, mainly by sending us links to material from the Internet such as pictures, movies, instructional videos, funny videos, animations, content in our mother tongue.

I listen to music when it's hard for me to concentrate. Music helps me concentrate when I learn.

The smartphone helps us communicate with friends who stayed abroad and keep in touch daily through phone calls and social networks.

I play games with friends who stayed in France. We play games that are on my phone and on theirs, and I talk to my relatives and friends who stayed in France. Communication with my friends is mainly through social networks and network games by phone.

Significant experiences thanks to the smartphone are mainly the social networks. They give me the feeling that I'm not alone and that with their help I fight feelings of alienation and of not belonging.

When I just came to Israel, I didn't know Hebrew. Without a smartphone I couldn't express myself, translate words.

We are members of our school distribution list, and we feel that we belong and are part of Israeli society. This is how we learn the Hebrew slang.

With the help of the smartphone, we translated words from French into Hebrew, and it helped teachers and students understand me better, and this way I made friends.

I learn a lot of new words through the Internet, and so my vocabulary expands every month more and more, so I integrate into Israeli society more and more every month.

Translate and understand songs in Hebrew: that way I learn new words and get to know the new culture and also understand it better.

The use of the smartphone really helped me during the absorption process in Israel. Using the translation app helped me learn words faster and start talking, to find friends and understand what they were saying.

Sometimes I don't succeed in saying something, so I use the phone to search for information that will help me express myself.

The use of the smartphone helps me very much in the absorption process. It "cushions" my "landing" in the new country, especially with problems of communication with others, and helps me cope every day.

DISCUSSION

Given the extensive use of mobile technologies by adolescents, the present study examined their usage by new immigrant adolescents to support the process of their absorption in the new country, the acquisition of the new language and their studies, and the most meaningful experiences that mobile technologies made possible in the course of their absorption in the new country. The research questions were: What are the characteristics of use of mobile technologies by new immigrant adolescents? Does the use of mobile technologies help adolescents in their absorption process in the new country, and if so, how? What are the most meaningful experiences that the use of mobile technologies has created for adolescents in their studies inside and outside school?

We found that all the adolescents who participated in the study used mobile technologies with a very high frequency and throughout the day. They use translation apps, databases, various applications, and social networks. They believe that thanks to mobile technologies, they have successfully integrated into their new environment, and many have noted that mobile technologies were responsible to a great extent for their successful absorption in the new country. Other studies also reported extensive use of smartphones (Goggin & Hjorth, 2014; Ito, Baumer, Bittanti, Boyd, Cody, HerrStephenson, Tripp et al., 2010; Livingstone, 2009; Livingstone, Haddon, Vincent, Mascheroni & Ólafsson, 2014; Mascheroni & Ólafsson, 2014).

We also found that the most common and most important use that the respondents mentioned was translation apps, which can quickly instructs them how to pronounce words correctly. Respondents stated that the correspondence over the social networks, with the assistance of translation apps, helped them feel that they belonged in the class and that they were involved. The smartphone gives them the feeling that they are not "imprisoned" by their language limitations.

We found that the smartphone is an integral part of the adolescents' natural environment. Similar findings were reported in previous research (Livingstone, 2008, 2007; Millwood-Hargrave, 2007; Millwood-Hargrave & Livingstone, 2006, 2009). We also revealed that some students find it difficult to study in the classroom and need autonomous learning to understand and internalize the material, therefore photographing worksheets and recording teachers helps them cope with learning difficulties. For example, as a 16-year-old adolescent stated: "At the lesson I have no sense that I understood the material, I feel confused, so I record the teacher, photograph the board, scan the worksheets that remain in the classroom. I don't know what I would do if I couldn't take pictures, scan, and record materials. I wouldn't be able to deal with my studies."

The most significant experience indicated by the adolescents was that use of the smartphone has in many ways facilitated dealing with the difficulties of their absorption process in the new country. The smartphone provides access to translation apps, databases, numerous applications, social network groups that satisfy their needs and help them face the difficulties they encounter. Respondents cited participation in groups with friends in the new country, as well as with friends and relatives who remained abroad.

Using Mobile Technologies During Lessons

Researchers found that mobile technologies increased students' motivation and their social and educational engagement, and helped them understand the learning material (Choi, 2012; Foti & Mendez, 2012; Hanafi & Samsudin, 2012; Rossie, Miller, Cecil & Stamper, 2012; Schugar, Smith & Schugar, 2013; Warschauer, 2011). Mobile technologies are becoming more and more available, offering increasing functionality and a wide range of applications, allowing students access to study aids and learning materials anytime, anywhere (Cohen, Vincent, Adhikari et al., 2015; Dahlstrom, 2015). Mobile technologies also provide equal opportunities for students with special needs (Ardito, Costabile, Marsico et al., 2006; Gikas & Grant, 2013).

This study shows that the use of mobile technologies during lessons has led to increased motivation of new immigrant students, helped them integrate socially and academically, and helped them understand the material learned in class. All the tutors (graduates who work with the new immigrants) allow the adolescents to use the smartphones during lessons. In general, the tutors responded positively to the requests by the adolescents, for example, to learn the new language through songs that "everyone in class is singing," etc.

The respondents noted that most of the teachers (82%) working in classes in which immigrants and native-born students study together, in most schools, allow students to use their phones in class. There were significant differences between adolescents who were allowed to use the smartphone during the lessons and those who were not. We found that in the classes in which new immigrants could use their smartphones, they were able to understand the material better than in classes in which they were not allowed to do so. In the latter, the students felt that they were lost, could not understand what was happening in the classroom, and some even mentioned that they used to "detach" themselves from the classroom and experienced a sense of helplessness. Some teachers allowed students to use their smartphones during class, and some did not. The adolescents noted that there was a significant difference in their functioning between lessons where they were allowed to use the smartphones and those where they were not. New immigrant adolescents who could use the smartphones felt that they understood the material better, that they were "in the loop," whereas those who were not allowed to use their smartphones at times felt helpless and unable to keep track of what was happening in the classroom.

In interviews, we asked teachers who did not allow their new immigrant students to use the phone during the lesson, why they did so. The main themes that emerged from the interviews with the teachers who did not allow immigrant students to use the phone during the lesson were: (a) the temptation to use the smartphone during the lesson is very strong and can distract the students from the lesson to surfing social networks, answering messages, etc.; (b) students searched the Internet for materials related to the subject learned in class, which increased their load, causing them to concentrate on their search and the content they found, rather than on the material the teacher was presenting; at times, students encounter concepts that the teacher had not taught yet, causing them to become overwhelmed and confused. Therefore, students ought not to search the Internet while the teacher is explaining, and it is better for them to listen during the lesson and only afterwards search for additional material, not during class. Most teachers indicated that they sent students links relevant to the subject of the lesson only after the class. A similar finding was reported in other studies that found that m-learning leads to distraction during the lesson (Althunibat, 2015; Bidin & Ziden, 2013; Lowenthal, 2010; Park, 2011; Sarrab, Elbasir & Alnaeli, 2016).

The reasons mentioned by teachers are important and accurate. At the same time, because the transition from one country to another in adolescence is complicated, it is advisable to allow students to use mobile technologies at least in their first months in the new country. It is possible to limit the use of smartphones to a certain extent, but it should be remembered that a simple action, such as real-time word translation, may have a positive effect on the continuation of the lesson, whereas the student's feeling that he "does not understand anything" can lead to a sense of helplessness and detachment from the lesson.

Social Networks

Our results show that adolescents felt that the social networks compensate for the difficulty they experienced with the spoken language and the difficulty in blending into the fabric of the class. Through the social networks and the translation apps, they succeed in participating in class discussions over social networks. All the tutors opened groups on social networks and allowed the new immigrants to correspond freely with the individuals with whom they worked. Some teachers (80%) recognized the positive effect of various social network groups on the absorption and adjustment process of immigrants, and therefore encouraged the opening of groups meeting many criteria, such as a group of new immigrants before a math test, a group of speakers of a particular language before submitting an assignment, or groups where students could feel free to ask questions, without fear of getting embarrassing responses.

It appears that immigrant adolescents perceive communication through Internet applications as providing a compensatory space for the physical distance from those who have stayed abroad. All of them mentioned that they were in contact with friends and family members who have stayed abroad, most of them in daily contact through WhatsApp, Facebook, Skype, and other applications. They "talk" about "everything"—studies, movies, television programs, music, soccer, their friends in the new country, and their experiences in the new country. They are interested in what is happening in their place of origin, talk about how they miss those who stayed behind, share stories, receive updates on what happens to everyone wherever they may be, reminisce, etc. In the interviews, the adolescents said that they found an "attentive ear" in their loved ones abroad. The use of social networks makes them happy and facilitates the process of acclimatization in the new country. Many adolescents stated that they were happy to have regular contact with those who remain abroad, and that both in good times and bad, they had someone with whom to share experiences, in their native language. They felt that communication dispels feelings of longing, enables conducting intimate conversations, sharing emotions, releasing anger, and alleviating frustrations. The findings of this study show that immigrant adolescents extensively use social networks and derive great benefits from it. These findings are similar to those of previous studies (Livingstone & Brake, 2009; Moore & McElroy, 2012; Sheldon, Abad & Hinsch, 2011; Wang, Moon, Kwon, Evans & Stefanone, 2010; Zilka, 2014, 2016, 2018) on the widespread use of social networks by adolescents and the benefits they derive from the use of the related applications.

The findings indicate that the adolescents feel that the transition from the country of origin to the new country is not too sharp, but rather one that still allows for contact with those who remained in the country of origin. From the comparison of the results of this study with the difficulties characteristic of immigrants that have been found in numerous other studies (Choi, 2008; Maldonado-Molina, Reingle, Wesley, Jennings & Prado, 2011; Rossiter & Rossiter, 2009; Spallek, Zeeb & Razum, 2010; Yearwood, Crawford, Kelly & Moreno 2007), it can be seen that adolescents feel that mobile technologies compensate for their physical distance from their place of origin.

Researchers found that adolescents' wellbeing may be affected by migration to a new country. Their ability to study and acquire skills may be impaired, and their sense of self-worth and social cohesion may suffer. These situations can lead to dysfunction in various areas, passivity or aggression, and even social perversion (Birch & Ladd, 1997, 1998; Romi, 2001; Romi, Savicki, Grupper & Caspi, 2007; Zilka, 2014, 2015, 2017a, 2017b). The present study shows that young people feel that mobile technologies allow them to participate in positive experiences that give them a sense of belonging in the class in their new country, by participating in discussions over social networks, and by being able to understand and participate in classroom discussions through translation. They also have the option to continue communicating with their friends and loved ones who have stayed behind. Furthermore, searching for materials raises their motivation, curiosity, and ability to deal with the material that is being learned in school. Searching for relevant materials challenges the students, many of whom deal with the question of whether to search for material originating in their old country, or for material translated from Hebrew into their mother tongue, thereby acquiring a sense of personal growth and meaningfulness. Adolescents also receive positive feedback when they share the materials they have found on social networks. Their sense of self-worth increases as a result of these experiences, which creates a feeling of being needed and of contributing to the environment, as well as an opportunity to demonstrate skills and receive appreciation and feedback from the environment. All these create a sense of social belonging as opposed to alienation.

In conclusion, the findings of the present study indicate that adolescents consider mobile technologies as a window onto the global world, wide, open, and full of possibilities and information, which compensates them for the shortcomings and difficulties they encounter in the new country and for the physical distance between them and their friends and loved ones who have remained behind.

SOLUTIONS AND RECOMMENDATIONS

New immigrant youths should be allowed to use mobile technologies in class. A simple action, such as real-time translation, is likely to have a positive effect on the rest of the lesson, whereas a feeling by students that they "don't understand anything" could lead to helplessness and detachment from the lesson. This study shows that the use of mobile technologies during lessons has led to increased motivation of new immigrants, helped them integrate socially and academically, and helped them understand the material being studied.

LIMITATIONS OF THE STUDY

The study sought to clarify issues related to mobile technologies from the perspective of new immigrant youths, and did not examine in depth other factors that could affect the absorption process of new immigrant youths.

REFERENCES

Adler, P. A., & Adler, P. (2008). Of Rhetoric and Representation: The Four Faces of Ethnography. *The Quarterly Sociological*, *49*(1), 1–30. doi:10.1111/j.1533-8525.2007.00104.x

Althunibat, A. (2015). Determining the factors influencing students' intention to use learning in Jordan higher education. *Computers in Human Behavior*, *52*, 65–71. doi:10.1016/j.chb.2015.05.046

Ardito, C., Costabile, M. F., Marsico, M. D., Lanzilotti, R., Levialdi, S., Roselli, T., & Rossano, V. (2006). An approach to usability evaluation of e-learning applications. *Universal Access in the Information Society*, *4*(3), 270–283. doi:10.100710209-005-0008-6

Atkinson, E., & Delamont, S. (2006). In the roiling smoke: Qualitative inquiry and contested fields. *International Journal of Qualitative Studies in Education: QSE*, *19*(6), 747–755. doi:10.1080/09518390600975974

Bidin, S., & Ziden, A. A. (2013). Adoption and application of mobile learning in the education industry. *Procedia: Social and Behavioral Sciences*, *90*, 720–729. doi:10.1016/j.sbspro.2013.07.145

Birch, S. H., & Ladd, G. W. (1997). The teacher-child relationship and children's early school adjustment. *Journal of School Psychology*, *35*(1), 61–79. doi:10.1016/S0022-4405(96)00029-5

Birch, S. H., & Ladd, G. W. (1998). Children's interpersonal behaviors and teacher-child relationships. *Developmental Psychology*, *34*(5), 934–946. doi:10.1037/0012-1649.34.5.934 PMID:9779740

Budría, S., Colino, A. & Martínez de Ibarreta, C. (2018). The impact of host language proficiency on employment outcomes among immigrants in Spain. *Empirica, Journal of European Economics,* 1-28. doi:10.100710663-018-9414-x

Choi, C. Q. (2012). Textbooks come alive. *Scientific American*, *306*(4), 20–21. doi:10.1038cientificamerican0412-20b PMID:22486109

Choi, Y. (2008). Diversity within: Subgroup differences of youth problem behaviors among Asian Pacific Islander American adolescents. *Journal of Community Psychology*, *36*(3), 352–370. doi:10.1002/jcop.20196 PMID:18645632

Cohen, J., Vincent, J. L., Adhikari, N. K., Machado, F. R., Angus, D. C., Calandra, T., & Tracey, K. (2015). Sepsis: A roadmap for future research. *The Lancet Infectious Diseases*, *15*(5), 581–614. doi:10.1016/S1473-3099(15)70112-X PMID:25932591

Coleman, A. M. (2001). *Dictionary of Psychology*. Oxford University Press.

Comstock, G. A., & Scharrer, E. (2007). *Media and the American Child*. Amsterdam: Academic Press.

Dahlstrom, E. (2015). *ECAR Study of Undergraduate Students and Information Technology*. Research report. Louisville, CO: ECAR.

Eisen, S., & Lillard, A. S. (2016). Young children's thinking about touchscreens versus other media in the US. *Journal of Children and Media*, *11*(2), 167–179. doi:10.1080/17482798.2016.1254095

Foti, M. K., & Mendez, J. (2014). Mobile Learning: How Students Use Mobile Devices to Support Learning. *Journal of Literacy and Technology, 15*(3).

Gikas, J., & Grant, M. M. (2013). Mobile computing devices in higher education: Student perspectives on learning with cellphones, smartphones and social media. *The Internet and Higher Education*, *19*, 18–26. doi:10.1016/j.iheduc.2013.06.002

Goggin, G., & Hjorth, L. (2014). *Routledge Companion to Mobile Media*. New York: Routledge. doi:10.4324/9780203434833

Hammersley, M. (2008). *Questioning Qualitative Research: Critical Essays*. London, UK: Sage. doi:10.4135/9780857024565

Hanafi, H.F., & Samsudin, H. (2012). Mobile Learning Environment System (MLES): The Case of Android-based Learning Application on Undergraduates' Learning International. *Journal of Advanced Computer Science and Applications*, *3*(3).

Ito, M., Baumer, S., Bittanti, M., Boyd, D., Cody, R., HerrStephenson, B., … Tripp, L. (2010). Hanging Out, Messing Around, Geeking Out: Kids Living and Learning with New Media. Cambridge, MA: The MIT Press.

Jan, S. R., Ullah, F., Ali, H., & Khan, F. (2016). Enhanced and Effective Learning through Mobile Learning: An Insight into Students Perception of Mobile Learning at University Level. *International Journal of Scientific Research in Science, Engineering and Technology*, *2*(2), 674–681.

Johnson, L., Smith, R., Willis, H., Levine, A., & Haywood, K. (2011). *The 2011 Horizon Report*. Austin, TX: The New Media Consortium.

Kukulska-Hulme, A., & Shield, L. (2008). An overview of mobile assisted language learning: From content delivery to supported collaboration and interaction. *ReCALL*, *20*(3), 271–289. doi:10.1017/S0958344008000335

Livingstone, S. (2007). Evaluating the online risks for children in Europe. *Telos*, *73*, 52–69.

Livingstone, S. (2008). Taking risky opportunities in youthful content creation: Teenagers' use of social networking sites for intimacy, privacy and self-expression. *New Media & Society*, *10*(3), 393–411. doi:10.1177/1461444808089415

Livingstone, S. (2009). *Children and the Internet: Great Expectations, Challenging Realities*. Cambridge, UK: Polity Press.

Livingstone, S., & Bober, M. (2005). *UK children go online: Final report of key project findings*. London: London School of Economics and Political Science.

Livingstone, S., Haddon, L., Vincent, J., Mascheroni, G., & Ólafsson, K. (2014). *Net Children Go Mobile: The UK Report*. London: London School of Economics and Political Science.

Lowenthal, J. N. (2010). Using mobile learning: Determinates impacting behavioral intention. The Amer. *Journal of Distance Education*, *24*(4), 195–206. doi:10.1080/08923647.2010.519947

Maldonado-Molina, M., Reingle, J. M., Wesley, G., Jennings, W. G., & Prado, G. (2011). Drinking and driving among immigrant and US-born Hispanic young adults: Results from a longitudinal and nationally representative study. *Addictive Behaviors*, *36*(4), 381–388. doi:10.1016/j.addbeh.2010.12.017 PMID:21216535

McCarty, T. L. (2018). Comparing 'new speakerhood': Context, positionality, and power in the new sociolinguistic order. *Journal of Multilingual and Multicultural Development*, *39*(5), 470–474. doi:10.1080/01434632.2018.1429457

Millwood-Hargrave, A. (2007). *Issues facing broadcast regulation*. London: Broadcasting Standards Authority.

Millwood-Hargrave, A., & Livingstone, S. (2006). *Harm and Offence in Media Content: A review of the empirical literature*. Bristol, UK: Intellect Press.

Millwood-Hargrave, A., & Livingstone, S. (2009). *Harm and offence in media content: a review of the evidence* (Revised edition). Bristol, UK: Intellect.

Moore, K., & McElroy, J. C. (2012). The influence of personality on Facebook usage, wall postings, and regret. *Computers in Human Behavior*, *28*(1), 267–274. doi:10.1016/j.chb.2011.09.009

Nesterko, N., Ulitsa, N., Friedrich, M., & Glaesmer, H. (2018). Do They Feel the Same Way? Health-Related Quality of Life and Satisfaction with Life in Jewish Immigrants from the Former Soviet Union in Germany and Israel. *Journal of Cross-Cultural Psychology*, *49*(4), 618–634. doi:10.1177/0022022118759223

Park, Y. (2011). A pedagogical framework for mobile learning: Categorizing educational applications of mobile technologies into four types. *International Review of Research in Open and Distance Learning*, *12*(2), 78–102. doi:10.19173/irrodl.v12i2.791

Puigdevall, M., Walsh, J., Amorrortu, E., & Ortega, A. (2018). 'I'll be one of them': Linguistic modes and new speakers in three minority language contexts. *Journal of Multilingual and Multicultural Development*, *39*(5), 445–457. doi:10.1080/01434632.2018.1429453

Repke, L., & Benet-Martínez, V. (2018). The (Diverse) Company You Keep: Content and Structure of Immigrants' Social Networks as a Window Into Intercultural Relations in Catalonia. *Journal of Cross-Cultural Psychology, 49*(6), 924–944. doi:10.1177/0022022117733475

Romi, S. (2001). Child and youth care in Israel: Trends and dilemmas in training and in therapeutic intervention programs. *Journal of Child and Youth Care Work, 15-16*, 171–184.

Romi, S., Savicki, V., Grupper, E., & Caspi, R. (2007). Occupational burnout among child welfare workers: A work-setting comparison. *International Journal of Child and Family Welfare, 10*(3-4), 93–109.

Rossing, J. P., Miller, W., Cecil, A. K., & Stamper, S. E. (2012). iLearning: The future of higher education? Student's perceptions on learning with mobile tablets. *The Journal of Scholarship of Teaching and Learning, 12*(2), 1–26. Retrieved from http://josotl.indiana.edu/ article/view/2023/1985

Rossiter, M. J., & Rossiter, K. R. (2009). *Immigrant youth and crime: Stakeholder perspectives on risk and protective factors.* Working Paper No. WP02-09.

Sarrab, M., Elbasir, M., & Alnaeli, S. (2016). Towards a quality model of technical aspects for mobile learning services: An empirical investigation. *Computers in Human Behavior, 55*, 100–112. doi:10.1016/j.chb.2015.09.003

Schugar, H. R., Smith, C. A., & Schugar, J. T. (2013). Teaching with interactive picture Ebooks in grades K- 6. *The Reading Teacher, 66*(8), 615–624. doi:10.1002/trtr.1168

Sevinç, Y., & Dewaele, J.-M. (2018). Heritage language anxiety and majority language anxiety among Turkish immigrants in the Netherlands. *The International Journal of Bilingualism*. Prepublished August 05, 2016. doi:10.1177/0123456789123456

Sheldon, K. M., Abad, N., & Hinsch, C. (2011). A two-process view of Facebook use and relatedness need-satisfaction: Disconnection drives use, and connection rewards it. *Psychology of Popular Media Culture, 1*(S), 2–15. doi:10.1037/2160-4134.1.S.2 PMID:21280967

Spallek, J., Zeeb, H., & Razum, O. (2010). Prevention among immigrants: The example of Germany. *BMC Public Health, 10*(1), 92. doi:10.1186/1471-2458-10-92 PMID:20181232

Spörlein, C., & Kristen, C. (2018). Educational Selectivity and Language Acquisition among Recently Arrived Immigrants. *The International Migration Review*. doi:10.1177/0197918318798343

Sung, Y. T., Chang, K. E., & Liu, T. C. (2016). The effects of integrating mobile devices with teaching and learning on students' learning performance: A meta-analysis and research synthesis. *Computers & Education, 94*, 252–275. doi:10.1016/j.compedu.2015.11.008

Traxler, J. (2009). Current state of mobile learning. *Mobile Learning: Transforming the Delivery of Education and Training, 1*, 9-24.

Wang, S. S., Moon, S. I., Kwon, K. H., Evans, C. A., & Stefanone, M. A. (2010). Face-off: Implications of visual cues on initiating friendship on Facebook. *Computers in Human Behavior, 26*(2), 226–234. doi:10.1016/j.chb.2009.10.001

Warschauer, M. (2011). Tablet Computers in Education. Eventually Tablets will Facilitate more Personalized and Interactive Learning. *Educational Technology Debate – Exploring Learning in Developing Countries*. Retrieved from http://edutechdebate.org/tablet-computers-in-education/eventually-tablets-will-facilitate-more-personalized-and-interactive-learning/

Yearwood, E. L., Crawford, S., Kelly, M., & Moreno, N. (2007). Immigrant youth at-risk for disorders of mood: Recognizing complex dynamics. *Archives of Psychiatric Nursing, 21*(3), 62–171. doi:10.1016/j.apnu.2007.02.006 PMID:17556109

Zilka, C. G. (2014). *Empowering parents in the social media age — The Three Element Way (Hebrew).* Butan-Galim.

Zilka, C. G. (2015). Social competence of children at risk: Similarities and differences among the various assessors. In E. Grupper & S. Romi (Eds.), Children and adolescents at risk in Israel: The voice of young people and issues faced by child and youth care workers (Vol. 2, pp. 113–185). Tel Aviv, Israel: MOFET Institute. (in Hebrew)

Zilka, C. G. (2016). Do online friendships contribute to the social development of children and teenagers? The bright side of the picture. *Journal of the Humanities and Social Sciences, 6*(8), 102–112.

Zilka, C. G. (2017a). Children in front of screens: Alone or in company? Desktop or hybrid computer? Children's viewing and browsing habits. *Journal of Educational Technology, 14*(3), 1–14. doi:10.26634/jet.14.3.13855

Zilka, C. G. (2017b). The Elements Way: Empowering parents, educators, and mentors in the age of new media. *Issues in Informing Science and Information Technology, 14*, 101–119. doi:10.28945/3702

Zilka, G. C. (2018). Always with them: Smartphone use by children, adolescents, and young adults – characteristics, habits of use, sharing, and satisfaction of needs. *Universal Access in the Information Society*. doi:10.100710209-018-0635-3

Zorlu, A., & Hartog, J. (2018). *The Impact of Language on Socioeconomic Integration of Immigrants.* IZA Discussion Paper No. 11485. Available at SSRN: https://ssrn.com/abstract=3170274

Section 4

Design of Mobile Solutions for People With Learning Difficulties

Chapter 11
Time for a Change:
Designing a Mobile Application to Assist People With Intellectual Disabilities

Rotem Israel Fishelson
Holon Institute of Technology, Israel

ABSTRACT

In recent years, the use of smartphones by people with intellectual disabilities (ID) has increased, leading to the development of applications designed to ease and improve their lives and independence, and help them integrate into society. Designing applications tailored to the needs of these users is complex, particularly in terms of accessibility and usability. This chapter presents a case study of the design and development process of a dedicated application aimed at helping people with mild ID in handling cash transactions. The chapter explores the design principles of mobile interfaces for people with ID. It introduces the use of the design thinking methodology, which focuses on user-centered design and helps gain an in-depth knowledge of the end users and their needs. Based on the performed user research, recommendations for the design of such applications are highlighted.

DOI: 10.4018/978-1-5225-8106-2.ch011

INTRODUCTION

Society's perception of intellectual disabilities (ID) has changed greatly over the years. Nowadays, there is an attempt to embrace and incorporate people with ID. Instead of regarding disabilities as a medical problem that requires intervention, there is an attempt to acknowledge people with ID as integral and equal parts of society. In order to accomplish this inclusion, there is an increasing use of assistive technology to improve the quality of life, the sense of competence and the level of independence of people with ID (Agree, 2014; Borblik, Shabalina, Kultsova, Pidoprigora, & Romanenko, 2015; Sauer, Parks, & Heyn, 2010). Assistive technology helps to minimize the gaps between people with and without disabilities by circumventing the limitation in order to accomplish a task and/or improving task performance (Dekelver et al., 2015).

Smartphones, which have become an integral part of our lives, offer diverse opportunities to ease and improve the lives of people with ID (Darcy, Green, & Maxwell, 2016; Igual, Plaza, Martin, Corbalan, & Medrano, 2013). Smartphones are becoming more user friendly, hence their adoption by people with ID is rapidly growing (Agree, 2014; Darcy et al., 2016). The applications available for smartphones can help in a variety of areas, including personal communication, memory enhancement, health control, navigation, emergency support and more (Morris, Jones, & Sweatman, 2016). Moreover, the technological development of the devices is a fertile ground for the development of innovative applications adapted to the specific needs of people with ID. However, alongside the advantages of these applications, there is a complexity in their development, especially in terms of accessibility and usability. In order for the potential of these applications to be fully realized, these issues must be considered and the necessary adjustments must be made, allowing users to easily use the applications (Agree, 2014; Darcy et al., 2016).

The main goal of this chapter is to examine the design and development processes of a mobile application developed for people with ID, with an emphasis on the underlying principles of accessibility and usability. The chapter focuses on the difficulties of people with ID in managing their money and the technological possibilities to overcome these difficulties.

The chapter presents a case study of designing an application named "Shkalkalim" for people with mild ID in order to help them manage their money and handle daily cash transactions. It will define the problem, discuss the evaluation process carried out, and outline a possible solution. The chapter describes the process from the needs analysis phase to the development of the application. This procedure is based on the Design Thinking methodology, which focuses on user centered-design and the use of creative thinking for problem solving. The principles of the methodology will be presented and described by definition and by the way they were implemented in this process.

In the course of the study, interesting insights emerged regarding the effectiveness and usability of the interface following a real life pilot of the application with the target population. The chapter describes the feedback received by the users and the adjustments made in light of it. Moreover, it highlights recommendations for proper implementation and training of people with ID with such applications. The chapter concludes with future ideas and directions that can be deducted from the "Shkalkalim" application for developing other applications and features that can support the needs of people with ID.

BACKGROUND

Intellectual Disability

According to the World Health Organization, 15% of the world's population are people with disabilities (World Health Organization, 2011). Of these, close to 200 million people have intellectual disabilities (Special Olympics, 2009).

Intellectual disability (ID) is defined as a neurodevelopmental limitation that effects intellectual functioning and adaptive behavior, and appears before age 18. This disability effects mental/conceptual skills (e.g. reading, writing, arithmetic, reasoning), social skills (e.g. communication, understanding of social rules and norms, social judgment and empathy), and practical skills (e.g. day-to-day independence, money management, task management, personal care and work responsibilities). ID is classified into different levels according to the individual's IQ and the degree of the person's dependence on his surrounding: mild disability (IQ 50-69), moderate disability (IQ 35-49), and severe/profound disability (IQ under 34). The more severe the disability, the lower the level of independence and the degree of dependence on the environment is greater (Merrick, Greydanus, & Patel, 2014).

Approximately 85% of people with ID have a mild disability and their mental age ranges from 9 to 11 years. These people have the ability to acquire everyday skills, but will often need support and guidance in carrying out complex activities that are not part of their routine. These people will find it difficult to learn reading, writing, performing arithmetic operations, managing time and money. They will also have difficulty in abstract thinking, in understanding social situations and in self-management (Merrick et al., 2014).

The western society's approach towards people with disabilities in general and intellectual disabilities in particular has changed significantly over the years (Ouellette-Kuntz, Burge, Brownà, & Arsenaultà, 2010). In the past, the attitude toward people with disabilities was based on the "medical model," according to which disability is considered a medical problem of an individual, an abnormal condition that requires intervention. According to this view, the role of society was to treat these individuals so that they could adapt to the existing environmental conditions in society. No adjustments were required from the individuals, and thus they were excluded from society against their will. In contrast to this view, today, the prevailing perception is based on the "social model". According to this model, disability stems from social norms. The current aspiration is that instead of expecting the disabled individuals to "recover" and adapt to society, the society will embrace their variance and adapt itself so that they can be an integral part of it (Shakespeare & Watson, 2001). This change in perception has led to the enactment of laws and changes in policies of Western countries towards people with intellectual disabilities in order to anchor their right to equal and active participation in all the activities and services offered by society, thereby maximizing their social inclusion (World Health Organization, 2007). Despite the fact that the social model is now widely used, it seems that people with ID experience lower quality of life than people without ID. Nevertheless, it is evident that they have the desire to acquire the knowledge and skills necessary to become more independent and take an active part in society (Abbott & McConkey, 2006).

The ability to manage money is a challenging skill for people with ID, and therefore it often becomes an obstacle for them in their quest for independence (Hua, Woods-groves, Kaldenberg, Lucas, & Therrien, 2015). Only a small percentage of people with ID manage their own money, while the rest are frustrated by the need for external help (Bigby, Bould, & Beadle-brown, 2016).

When people with ID live in a special community household, where they are not required to manage their money, they have no opportunity to deal with the issue, which may increase their dependence on those around them (King et al., 2017).

The difficulty in financial management stems from a significant gap in the acquisition of basic mathematical skills (addition, subtraction, multiplication and division). This gap is expressed in the difficulty of translating problems into mathematical equations, difficulty in identifying the relevant information for the problem and difficulty in solving the problem itself. These difficulties impair the ability to carry out activities such as purchasing a product and managing a budget (Jansen, De Lange, & Van der Molen, 2013). Handling money has the potential to instill a sense of independence in people with ID. It is thus important to find a solution that will enable them to easily learn to overcome this difficulty.

Assistive Technology

The use of technology for improving the quality of life is growing, and more and more applications and software are being developed to help people with different disabilities. Assistive technology is a technology aimed at blurring the boundaries and reducing the gaps between people with and without disabilities (Dekelver et al., 2015). It is a technological aid designed to enable people with special needs to circumvent a limitation while performing a specific task, and help them improve their performance. Assistive technology allows people to accomplish tasks that are impossible without it (World Health Organization, 2018). In fact, assistive technology is a means of accessibility which aims at enabling individuals with disabilities to better integrate into society. It allows people with disabilities to be more independent in performing different tasks, thus becoming less dependent on external help. It allows them to make their own decisions about their daily life activities, which increases their sense of competence, improves their quality of life and thus helps them be integrated better into society (Agree, 2014; Borblik et al., 2015; Sauer et al., 2010). There are various forms of assistive technologies which are offered for people with ID.

Mobile Applications for People With Intellectual Disabilities

In the past decade smartphones have become an integral part of our lives, as they facilitate our communication with others and help us to live independently and to become a part of society (Morris, Jones & Sweatman, 2016). As their distribution grows, so does the variety of applications being offered. The development and improvement of technology invite great creativity and provide a fertile ground for the development of innovative ideas. As part of this trend, applications tailored to people with disabilities and their needs are being developed. These applications can provide assistive technology for people with disabilities and thus facilitate and improve their quality of life, inclusion into society (Darcy et al., 2016; Igual et al., 2013), interpersonal communication, emergency preparedness, and other daily skills (Morris et al., 2016).

Studies show that among people with ID the use of smartphones is increasing. According to these studies, most people with mild ID (70% - 80%) in western countries today use a smartphone (Borblik et al., 2015; de Urturi Breton, Z.S., Hernández, Zorrilla, & Zapirain, 2012; Morris et al., 2016). The use of smartphones by people with ID allows them to feel the same sense of belonging and independence that smartphones provide to people without disabilities (Borblik et al., 2015).

However, despite the advantages of assistive technology applications, there is a complexity in their development (Billi et al., 2010). The inherent difficulty in designing such applications, is even more pronounced for people with ID, especially in terms of accessibility and usability.

In order for the potential of these applications to be fully realized, these issues must be considered and the necessary adjustments should be made accordingly, allowing peoples to easily use the applications (Agree, 2014; Darcy et al., 2016). In addition, studies indicate that training and support processes are also important for application use. People with disabilities don't always understand how to use and operate applications. Thus, in order to make the application accessible and enable them to use it independently, appropriate training and support should be provided (Chan, Lambdin, Graham, Fragale, & Davis, 2014; Darcy et al., 2016).

User Interface Design for People With ID

While designing user interfaces in general and especially while designing for people with ID, it is important to follow accessibility and usability principles to facilitate the learning process of the interface and its actual use. Accessibility refers to the design and adaptation of services and devices for people with disabilities so they can use, capture, understand, navigate, and interact with their interfaces (Henry, 2018; Schweibenz, 2006). Usability refers to a collection of measures to evaluate the ease of use of a product: how the interface works, and how people interact with the interface and its various features. Usability has three key measures: effectiveness, efficiency, and satisfaction (Borblik et al., 2015). In the context of designing interfaces for people with ID, the literature identifies the following principles:

- **Intuitive Interactivity:** Interface components and their roles should be as obvious as possible, without having to decipher their purpose. Care should be taken to create large and distinct clickable areas, and in general, distinct boundaries must be created for the various interface sections (Avila, 2015; Borblik et al., 2015; Dekelver et al., 2015).
- **Maintaining Consistency:** People with ID find it difficult to adjust to changes, even when it comes to changes in mobile interfaces. For example, changing the interface design as part of an application's update may confuse them and make it difficult to use the application. Therefore, the interface structure and navigation menus must be consistent throughout the application to avoid confusion (Avila, 2015; Borblik et al., 2015).
- **Readability:** Emphasis should be put on large text, easy-to-read font (sans-serif), and high contrast between the text and the background (Dekelver et al., 2015).
- **Minimum User Input:** It is recommended to avoid requiring the users to enter content on their own, and provide alternatives such as a list selection or recording (Avila, 2015; Borblik et al., 2015).
- **Input Fields Adjusted to Content:** Input fields and the keyboard should match the expected content. For example, if the user is required to enter a phone number, a keyboard with numbers should be displayed. Additionally, it is important to provide tooltips or descriptive labels so that there is a clear indication of the expected content in the field (Avila, 2015).
- **Clear Feedback to User's Actions:** Physical (e.g. vibrations), visual or auditory feedback should be provided in order to clarify to the users that their action has been recorded (de Urturi Breton et al., 2012).

- **Simple Touch Gestures:** Minimize complex touch gestures or provide clear explanation on how to perform them (Borblik et al., 2015; de Urturi Breton et al., 2012).
- **User Control:** Any action that occurs in the interface should be the result of a user's action. Actions should not occur automatically. For example, warnings and feedbacks should remain on the screen until the user responds to them (Borblik et al., 2015; Dekelver et al., 2015).
- **Graphic Representation:** Universal images and symbols should be preferred as an alternative to textual content in order to strengthen the understanding and reduce cognitive load. It is recommended to include them in controls and text fields to clarify their role (Borblik et al., 2015). Nevertheless, a simple text should accompany the visual elements in order to avoid misinterpretation (Avila, 2015).
- **Linguistic Simplification:** Edit and process information to make it simple, clear and easy to understand. It is important to focus on the main ideas of the content and to omit excessive information. In addition, the vocabulary and sentence structure must be adapted to the capabilities of people with ID (Dekelver et al., 2015; Fajardo, Tavares, Ávila, & Ferrer, 2013).
- **Training:** Appropriate training should be provided as people with ID need guidance while using new interfaces. It is important to give clear instructions, using voice or tooltips, in case the user is expected to perform certain actions. It is recommended to include a tutorial that will accompany the users in their first use of the application, with an option to run it at any time (Avila, 2015; de Urturi Breton et al., 2012).

In the current perception of usability, there is a clear preference for creating intuitive, easy-to-understand and easy-to-use interfaces (Agree, 2014). Placing the user experience as a top goal helps reduce cognitive load and direct users to find the information they need without complex navigation requirements. In today's developed applications there is a tendency to focus only on one content area, which helps maintain concentration and reduces distractions (Avila, 2015). The fact that most applications today are developed according to this usability notion serves people with ID as it facilitates their interaction with the interface and increases the likelihood that they will use it. The following section presents a dedicated mobile application designed for people with ID that aims at helping them in monetary transactions. It examines the adoption of the application in a real setting and its compatibility in terms of accessibility and usability.

"SHKALKALIM" APPLICATION

"Shkalkalim" is a portmanteau, coined from a combination of the Hebrew words "shkalim" (the Israeli currency in plural) and "kal" (meaning 'easy'). It is an application designed for people with mild intellectual disability in order to help them handle daily cash transactions. "Shkalkalim" allows users to bypass the limitations associated with basic mathematical skills, thereby increasing their sense of security, independence, and competence. The application accompanies the users throughout the purchasing process of an item and reduces the need for external help. The application provides information regarding budget management, i.e. the decision if the users have enough money for a purchase of a certain item, which bills and coins they should hand out for a given payment, and whether the correct change has been received. The application is bilingual (Hebrew and Arabic) and is accompanied by voice instructions. It consists of three sections:

1. **"Payment":** A section dedicated for the payment process, including price input, payment and receiving change.
2. **"My Wallet":** A section in which the users enter the amount of money in their possession.
3. **"Know the Currency":** A section to get acquainted with the various bills and coins and strengthening the understanding of their comparable value.

At the beginning of the process, the users input the amount of money in their possession, where the bills and coins are visually presented (Figure 1). Upon arrival at the store, the users choose an item they wish to buy and enter its price (Figure 2). If the users have sufficient funds to acquire the chosen item, the monetary composition (bills and coins) is presented. The composition is selected in order to minimize the need to deal with change (Figure 3). Following the payment, the users have to check whether they received the correct change (Figure 4). Upon completion of the process, the available funds are updated in the "My Wallet" section according to the money paid and the change received.

The Design Process

In order to design an application which caters to the needs of people with ID, the Design Thinking methodology was chosen. The design of interactive learning environments is a complex and challenging process as it combines learning theories with product design and engineering (Mor & Winters, 2007). In recent years, various principles and methodologies were developed for the design of such environments. Design Thinking (DT) is such a methodology, which is aimed at problem solving with a Human-Centered Design approach. This methodology is based on empathy and a deep understanding of the physical and emotional needs of the user. It includes an iterative process and focuses on collaboration, creativity development and out-of-the-box thinking (Brown, 2008). In recent decades, various approaches were developed to this methodology (Tschimmel, 2012), among them was IBM Design Thinking, which was used in this research.

IBM Design Thinking (IDT) was developed by IBM for rapid team product planning, which characterizes the demands of its industry. The term "Design" represents the planning process behind a product. The goal of IDT is to define and construct the product by understanding and empathizing with the end users' needs. The process is done using traditional accessories (such as post-it notes) in order to promote creativity and self-expression in an open environment with mutual effort. According to IDT, it is important that the final product will reflect all the insights identified in the process and provide a better user experience (Johnson, 2016).

Three key principles are at the base of IDT implementation, which frame the way to identify problems: (1) A focus on user outcomes – A principle that defines that the quality of a product is not evaluated by the number of developed features or functionalities but rather by the question how well are the end users' needs met. (2) Multidisciplinary teams – The principle assumes that teams originating from diverse backgrounds and experiences develop more varied ideas than those from homogenous teams. In addition, it is important to make sure that the teams include people with the relevant specialization as well as decision makers. Along the process, a representation of the end users should be involved with the teams in order to receive feedback, guidance and match their expectations. The base line is that any interaction with the end users will decrease the existing gap between the teams' assumptions and the users' reality (Johnson, 2016). (3) Restless reinvention – A principle that emphasizes that there is no "final" product and that the team can always further improve the end product for the benefit of the users.

Figure 1. Visual display of coins and bills in "My wallet"

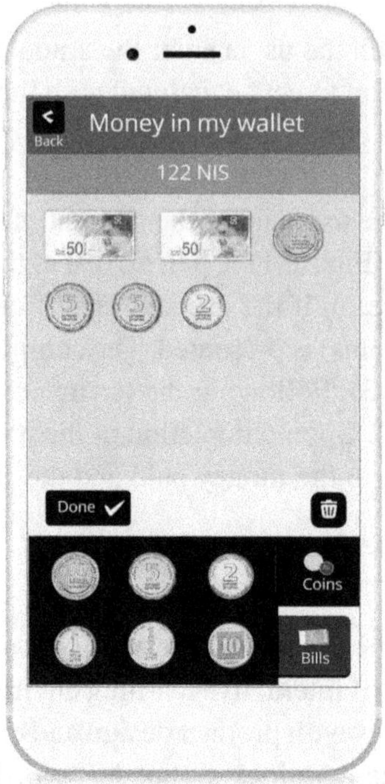

Figure 2. Entering the price of the desired item

Figure 3. The monetary composition required for payment

Figure 4. The feedback for correct change received

Any prototype of a product is another iteration leading to renewed consideration of issues previously considered solved (IBM Corporation, 2017).

The design process of the "Shkalkalim" application was performed using the IDT methodology, as part of a social program called "Kids 4 Kids" meant to help people with ID. The 6-months program was a joint effort of IBM Israel and the Holon Institute of Technology (HIT) and was led and supervised by two IDT advisors from IBM, an academic advisor and two students from the Faculty of Instructional Technologies at HIT. Arab and Jewish junior high-school students applied for participation in the program and 20 were selected by their school's principals according to their academic achievements. The program did not entail any benefit to the students except for their social contribution and desire to help others. The students were introduced to the concept of people with ID but did not have any previous interaction with them. In addition, they were not familiar with the IDT methodology or application design. Only a couple of students in the program had previous basic experience in computer programming. During the program the students learned the basics of the methodology and used it to identify the needs of people with ID. Accordingly, they ideated and designed an application which will assist the end users in their daily lives.

The thought process was led by the end users who acted as the guiding star. In order to create appropriate solutions for them, it was important to understand their point-of-view: the users' perceptions, feelings, needs, and the way they measure success. All of the above were key factors in the process. It was clear that it was impossible to perceive the world from their perspective and thus it was necessary to incorporate the end users in the process as much as possible. During the exploration and acquaintance step, the real needs of the end users arose and among them were the wish to handle their money independently, this need underlies the "Shkalkalim" application.

The process was performed in four stages: 1. User Research, 2. Ideation, 3. Prototyping, 4. Prototype Evaluation & User feedback.

- **User Research:** The first stage was dedicated to familiarizing and deeply understanding the end users. Students were first introduced to people with ID by managers from the AKIM and RAND Institutes for people with ID. The managers debriefed the students and together drafted questions to ask the users from the target group. During the next session, the students met with the target group which included 12 people with mild ID from AKIM and RAND institutes, ages 23-40, who live in joint housing for people with ID. In this meeting the students performed semi-structured interviews in order to understand what are their schedules, levels of independence and technological proficiency. User "D" for example mentioned in the interview: *"I don't know [how to handle] money and don't know math. It's hard for me to know how much money is enough for the clothes I want to buy, if it's a lot of money or a little bit"*. Additionally, she mentioned that: *"I can read the numbers on the bills but if I need to pay a sum which isn't a round sum, I can't do the math"*.

Based on the interviews, the students created an "empathy map", a graphical and textual representation of a potential user (Figure 5). Each quadrant of the empathy map is dedicated to a different aspect of the user, their thoughts, what they say, their feelings and their actions, and those are detailed on the sticky notes. Following the creation of the empathy map, the students discussed daily scenarios of the end users (e.g. acquiring a product in the mall) and their difficulties, among which the handling of money was highlighted.

Figure 5. Empathy maps created by the students representing potential users

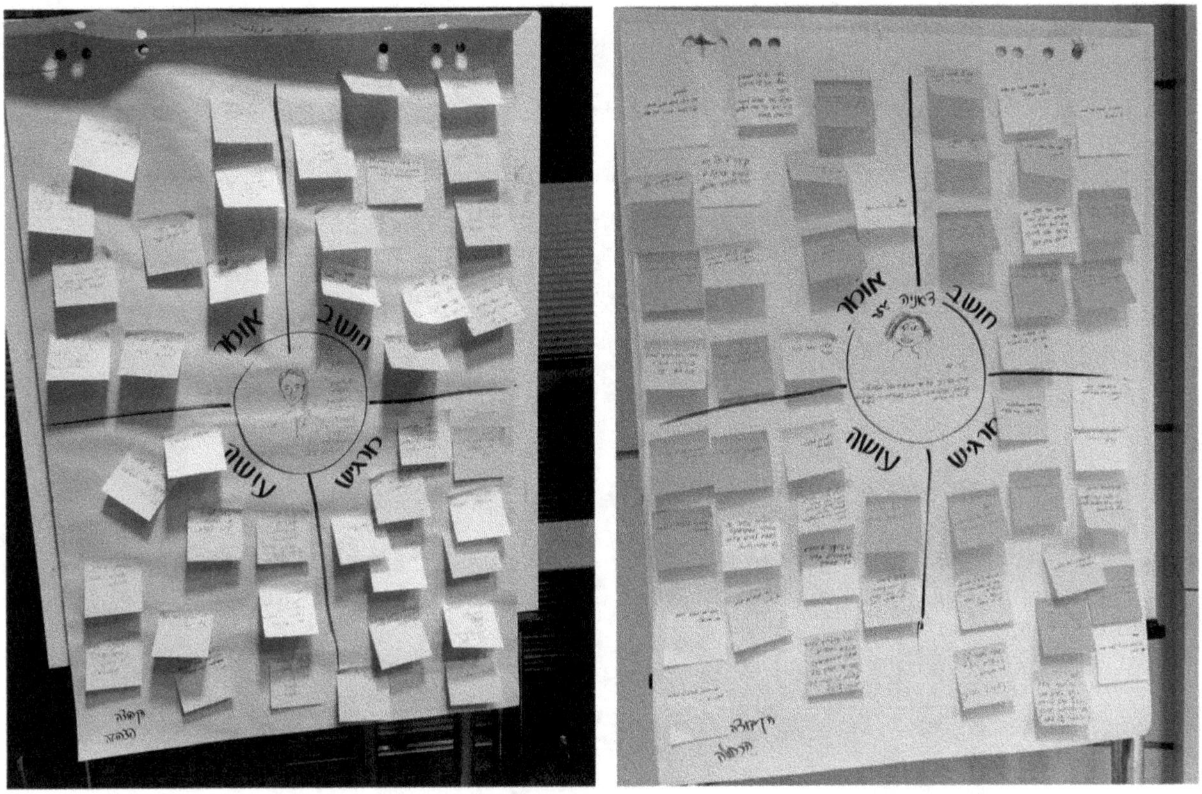

- **Ideation:** In the second stage, the students brainstormed and offered various ideas which will enable people with ID to overcome some of the main difficulties presented previously. The emphasis was put on raising as many ideas as possible (however weird or unrealistic) and avoid judgement to others' ideas. At the end of this stage, the students were asked to rank their ideas on two axis, their importance to the end users (X-axis) and their technological feasibility (Y-axis) (Figure 6).
- **Prototyping:** The third stage focused on creating a prototype for the applications of the selected top ideas. The students created a storyboard (Figure 7) illustrating an idea in a story-like manner using drawings and captions on sticky notes. They then dramatized their ideas to all their peers. This process allowed them to quickly examine various ideas without the "cost" or fear of failure. According to the feedback they received, they created a digital prototype, using the Balsamiq wireframing software, which presented the user interface of the selected ideas.
- **Prototype Evaluation & User Feedback:** In the fourth stage, which focused on assessing the proposed application, the students met again with the target group. They presented their prototypes and allowed the target group to experience them. The students gathered the feedback and insights regarding needed improvements and accordingly corrected and redesigned the user interface.

The overall process was an iterative one and as a result enabled both a deeper acquaintance with the end users and their needs as well as the implementation of the insights in the design of the proposed solutions. After in-depth consideration of the needs and proposed ideas, it was decided that the topic of

Figure 6. Ideas prioritization

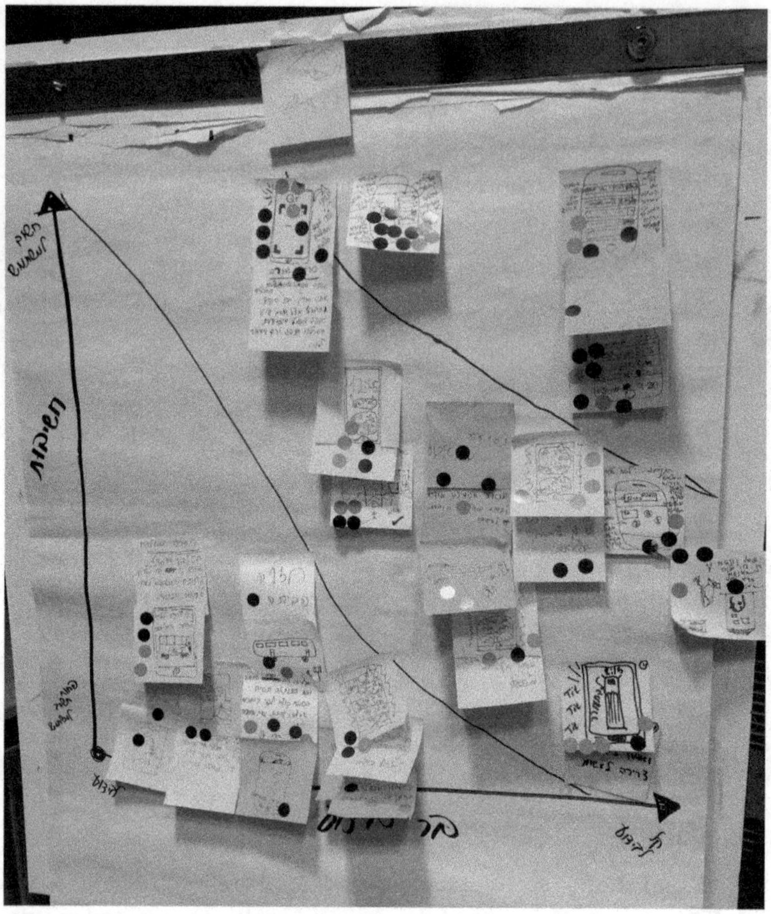

handling money is the most crucial to address and that a mobile application solution would provide a strong sense of independence and impact the lives of people with ID.

Despite the extensive work mentioned above, the prototypes which were created as part of the design thinking process were only preliminary designs since the process mostly focused on the identification of needs and finding suitable solutions for the users. The design of the user interface and experience were only partially considered. Thus, in the next step, the developers/designers of the “Shkalkalim” application (the HIT students who also led the Kids-4-Kids program) performed a deep high fidelity design process. It included the design of mock-ups for all interactions and the needed screens while referring to the design principles and guidelines for people with ID mentioned above.

Implementation of Design Guidelines and Evaluation Process

In the detailed design process, strong emphasis was put on creating a user interface which is easy and intuitive to use and based on the main principles for designing interfaces for people with ID, as detailed above. It was decided that the home screen (Figure 8) will contain large and clear buttons enabling the operation of the three core functions: the payment process, the calculation of change and the input of

Figure 7. Storyboard creation

money to the wallet (Intuitive interactivity and Linguistic simplification principles). In order to distinguish between the functions and in order to maintain a coherent interface, each function received a distinct color which was also the background color in all the screens related to that function (Readability and Maintaining consistency principles). In addition, next to the button's name, a suitable icon was placed as an alternative to the written text (Graphic representation principle). This was done in order to enable suitable adaptability to the end users, some of whom can read simple texts while others cannot read and only understand visual representations.

This decision was also applied to all the feedback given to the users' actions. For example, the feedback "You do not have sufficient money" is accompanied with a sad-face icon, while the feedback "You have sufficient money" (Figure 9) was accompanied with a happy-face icon (Clear feedback to user's actions principle). In order to allow the users to easily define how much money they have in their wallet, a drag-and-drop mechanism was developed. In addition, it was decided to represent all the bills and coins with real images in order to reduce needless mistakes when entering numbers (Input fields adjusted to content principle). This was meant to reduce the cognitive load and sense of confusion. In order to allow for easy navigation between screens, a "back" button was added to all screens (Maintaining consistency principle). In the home screen, another button was added leading to a preferences page meant mainly for the instructors.

Since it was clear that it was not easy to fully understand the end users, it was decided to perform user testing on the design version using the prepared digital high-fidelity mock-ups. The testing was meant to examine the interface's usability for the end users and identify loose ends in the interface before entering the development stage, thus allowing for easier adaptations. The user testing was performed with a group of 12 people with mild ID from the AKIM Institute after receiving authorization from their legal

Figure 8. Home Screen

Figure 9. "Sufficient funds" Feedback screen

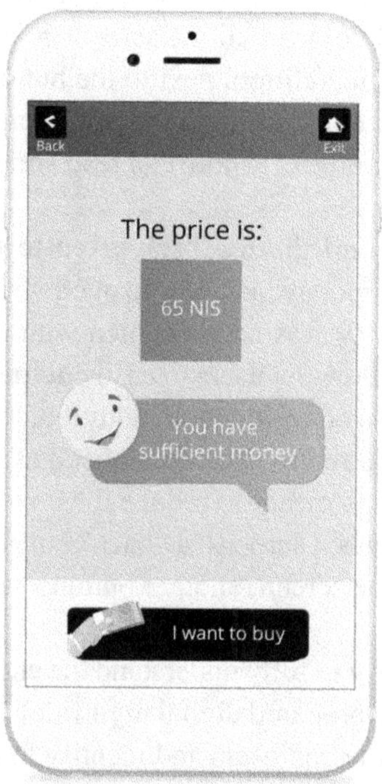

guardians. The testing simulated an acquisition process of an item while using real money and the digital mock-up of the application which was installed on the users' smartphones in order to emulate a scenario as close to reality as possible. The testing took place at "Beit Halomi", an AKIM joint housing, where each user "acquired" an item from a "seller" from the AKIM staff. The interactions were observed by the students from HIT who documented the interactions and insights.

The user testing highlighted several insights regarding the usability of the interface. First, it was found that the scenario of calculating change should not be a standalone feature but rather a part of the payment process. Originally, the scenario was designed to shorten the process for "advanced" users but the user testing showed that the full process is necessary due to the difficulty in remembering and processing all the details. In addition, the user testing proved the importance of immediate feedback to the users' actions as highlighted by de Urturi Breton et al. (2012). When immediate feedback was not presented, users repeated their actions, thought there was a malfunction and became impatient. Additionally, it was found that users should be able to cancel or edit a performed action, especially the input of bills and coins, as they needed to compare the money in their wallet to its representation in the application. The testing also showed that it is important to change to visualization of the bills and coins so they are not stacked but rather presented side-by-side as to ease their perception.

During the testing, the users enjoyed handling the real bills and coins and their ability to identify them in the interface. However, they found it difficult to understand their comparable value and name the selected bills or coins. Thus it was decided to add a "know the Currency" section, which presents the various bills and coins alongside their textual names. The section would help users distinguish between the types of coins and their value, allow them to see both sides of the coin and understand their comparable value to each other. For example, a one shekel coin is worth two half-shekel coins and a 10-shekel coin is worth ten coins of one shekel.

The user testing also raised the need to change and simplify the phrasing in the interface and add voice instructions to the main commands. Due to the differences in skills and abilities among people with ID, it was decided that the voice instructions will not be activated automatically but rather in the preferences page. Furthermore, it was decided to produce a narrated tutorial with a step-by-step explanation on how to use the application. The tutorial is a depiction of the Training principle as was suggested by Avila (2015) and de Urturi Breton et al. (2012).

The above insights led to another iteration of design amendments and only then the development phase began and the "Shkalkalim" application was developed for both Android and iOS operational systems. Following the development, another user testing was conducted, which examined the use of the real application in a real-life situation by five people with mild ID from the AKIM Institute. Each user was accompanied by an instructor from AKIM in the process of acquiring a single item in cash. The use of the application started with viewing the tutorial at home after which they went to a nearby convenience store. The process included inputting money into the application, selecting the item and entering its price, paying at the cashier according to the guidance of the application and confirming that the correct change was received. The users were observed by the HIT students while using the app and their actions were documented using screen recording. At the end of this field experience, a structured interview was performed with each user and the instructors to receive their impression of the application and acquisition process. Below are the tested metrics:

1. Interface usability evaluation

a. **Effectiveness:** Was the user able to perform an independent payment process using the application?
b. **Efficiency:** How much effort was required by the users to achieve their goals?
c. **Satisfaction:** How easy was the use of the application and was it a positive experience?

2. Interface accessibility evaluation
   a. How was the readability of the texts in the interface (size and color)?
   b. How clear was the textual content?

Very positive results came out of the user testing for all metrics. For Effectiveness, all users succeeded in acquiring a product in cash and much excitement was registered from the end user. One user commented: *"I succeeded! At the beginning I had difficulties, but I succeeded eventually"*. During the purchase process, each of the users received a basic support from the instructor. The help was especially necessary in passing through the various steps in the application. According to an instructor: *"I think this is because it is something new for them. This is a population which has difficulty using cash, for N., this [handling money] creates anxiety – whether she is swindled or from the payment process. Now she was suddenly relaxed, content, knew she was doing it the right way"*.

For the efficiency metric, the instructor indicated that the learning curve of the users of "Shkalkalim" was steeper compared to learning other new activities. It was noted that the application is accessible and designed well for the end users. The instructor stated: *"I was really surprised to see how they handled it [the application] on their first time, I expected something worst"*.

Finally, for the satisfaction metric, along the process, some users encountered difficulties but they have all completed the process with a smile on their face. According to an instructor: *"It has been a long time since I saw B. smile like that, he was really pleased. It is a powerful feeling that you can arrive at a place and you know what you need to do and you don't have this confusion in your head"*.

In terms of accessibility, the texts were legible, and the users did not encounter difficulties in understanding them. Most phrasings were clear except for a few which were amended following the testing. Due to reading problems, one user used the voice instructions, which helped him move forward in the process. As one instructor described: *"For the first time it was truly amazing, usually the learning process is much slower. It was very accessible"*.

The user testing of the application raised several conclusions about the interface design and accessibility and adaptations were tailored accordingly. Some of the users lost their concentration while watching the tutorial due to their inactivity while watching. Consequently, it was decided to divide the tutorials into small content segments which would diminish the cognitive burden and facilitate a better understanding of the content. In addition, it was decided to allow navigation between the various tutorials in order to create a stronger sense of control for the users. In the "Know the Currency" section, it was decided to add automatic narration as learning the ratios of bills to coins was complicated enough without the need to add the text understanding layer. Another important point which came up was the need of users to receive guidance prior to their first use. It was found that the leaning process of the end users is slower than that of people without learning disabilities. Therefore, every new activity that they perform requires the support and practice with an instructor. The solution to such difficulties is thus not in the interface but rather in the proper support system. It was decided to create an FAQ section with information on: where and when the users can update the amount in their wallet, how to deal with a product without a price tag, how to manage buying more than one product, etc.

The research showed that the application is adapted to the end users. During the research, users managed to perform tasks they were either not used to or have never done before and experienced a great sense of accomplishment and enjoyment from buying a product independently using the application.

SOLUTIONS AND RECOMMENDATIONS

The design and development of applications for people with intellectual disabilities is a complex process. This case study of the "Shkalkalim" application strengthens the notion that it is crucial to "enter the user's shoes" and interview them carefully to identify their true needs. One should avoid trying to guess their needs and to assume they know and understand them. The end users should take part in the whole process from the needs analysis to the post-development testing. It is important to keep in mind that adaptations according to the users' needs, from the early design onwards, are relatively easier to perform compared to costly alterations post-development (Schweibenz, 2006). Users' experience in a real-world situation can shed light on flaws in the application and on how the users perceive the interface functionality.

There are various user-centric methodologies. This research highlights the strengths of the Design Thinking methodology which enables developers to deal with complex tasks without prior knowledge and with low costs. It enables a deep exploration of the physical and emotional needs of the users, leading to the design of specific and creative solutions to meet these needs. In addition, as was identified by other researchers (Hong & Kwek, 2011; Scheer & Plattner, 2012), this methodology allows the designer to leverage imperative competencies for the 21st century such as problem solving, critical thinking, entrepreneurship and self-efficacy.

The implementation of design principals for people with ID should be used as a scaffold in the development of any assistive technology. Furthermore, the interface should be tested and assessed with the end users in situations as close to reality as possible. Adaptations to the interface should be done according to the feedback received.

The use of mock-ups, as was done in the preliminary user testing, allowed for a quick assessment of the user interface and adaptations were made accordingly. However, it is important to remember that this method requires abstract thinking which might prove difficult for people with ID (Dekelver et al., 2015). The incorporation of tangible elements, such as coins and bills, can help people with ID in grasping and understand the processes and by that produce better conclusions from their testing.

The use of images and symbols, which seemed crucial, was found in the testing to be insufficient and supportive text was sometimes needed. Furthermore, it should be considered that people with ID often interact with communication boards which are not always based on universal symbols and images. This might create misunderstandings for various audiences. This interpretation difference might stem from cultural differences or varying cognitive levels. It is thus important to embed universally and agreed symbols and images and test them with the end users. In addition, accompanying texts should be phrased simply to be clear and indicative.

User testing and past studies, have both emphasized the importance of guidance and training of the users during their first steps with the application in order to promote their independent future use (Chan et al., 2014; Darcy et al., 2016). According to the instructors who supported the process, educating the end users with new activities is usually a slow and repetitive process. Therefore, the instructors need to support the learning process before the latter can perform tasks independently.

People with mild ID are characterized with low self-esteem in relation to learning new skills and might quickly become frustrated (Watson, 2018). Both strong support from the instructors as well as the creation of "success experiences" are needed for these users to take the "risk" of learning a new skill and using the proposed applications. To that end, the "Shkalkalim" application provides support in the configuration of tutorials, voice instructions and an FAQ section, all containing content which helps the instructors assist the end users to overcome the hurdles of using the application.

FUTURE RESEARCH DIRECTIONS

The "Shkalkalim" application is perceived as intuitive and easy to use, however, one of its Achilles' heel is in the need to provide input from the users. Nowadays, there is a growing trend towards the use of image recognition technologies which enable faster and more accurate information gathering and input. The use of such technologies has been found beneficial for various purposes and content areas. Stronger computational power and faster connectivity allow such technologies to be embedded in mobile devices and be incorporated into various applications. Such technologies can be very helpful for people with ID in functions that require users' input. In the "Shkalkalim" application, for example, image recognition can bypass the need of the users to enter the price of the product or the given change. This could streamline the process, prevent mistakes and make the application more user-friendly. While the technologies can offer many new layers of information and opportunities, one must bear in mind whether these are suitable to the end users' needs and abilities.

Today, there are thousands of users in the "Shkalkalim" application. A large scale research is planned in order to test the application's impact on the level of independence of people with ID in their daily monetary transactions. The application also focuses on cash interactions, however, non-cash payment methods are even more popular today. We thus plan to explore whether there is a need to use alternative and more popular payment methods.

In addition, during the iterative IDT process, many additional practical ideas came up to address the needs of people with ID. Technological solutions to address such needs as time management, personal hygiene, and interpersonal interaction all need to be further explored in order to promote the independency and the societal inclusion of people with ID.

CONCLUSION

Mobile technologies have become very popular in recent years. Their penetration rates to every aspect of our lives increases as they ease communication and learning processes (Woodcock, Middleton, & Nortcliffe, 2012). Despite their proliferation, the availability of mobile applications for people with disabilities is still very limited and a deliberate effort should be made to expand the selection (World Health Organization, 2011). Mobile technologies can constitute an accessibility tool enabling people with disabilities to overcome their limitations, decrease their dependence on others and by that become better incorporated in society and improve their quality of life (Agree, 2014; Borblik et al., 2015). Despite the importance of adapting and developing dedicated technologies for people with intellectual disabilities, there are difficulties and complexities in the design and development of such technologies. In the design process, the focus needs to be put on the abilities, rather than the disabilities of people with intellectual disabilities (Dekelver et al., 2015). In addition, these people should be involved in the design and development process as they can provide insights into their own abilities and needs (World Health Organization, 2011). Accesibility and usability principles should be the basis for interface design in order to ensure an optimal user experience.

This chapter described the "Shkalkalim" application as a case study demonstrating an inclusive design and developmental process based on input from the end users along the entire process. The design of the solution began with an in-depth exploration of the way-of-life and needs of people with intellectual disabilities. The ability to use cash independently has been identified as a stumbling block for people with intellectual disabilities, which can become an obstacle on their way to independency (Hua et al., 2015). The difficulty of people with mild intellectual disabilities in performing actions such as purchasing a product or managing a personal budget, stems from a gap in their basic mathematical skills (Jansen et al., 2013).

The "Shkalkalim" application was designed to allow its users to perform the purchase by bypassing these gaps and acts as an assistive technology for their daily monetary interactions. The application enables its users to take their own decisions and to perform daily tasks. Taking advantage of the application improves their self-efficacy, their life quality and incorporation into society (Agree, 2014; Borblik et al., 2015; Sauer et al., 2010). The iterative design and development process was based on the implementation of design principals for people with intellectual disabilities which includes aspects of usability and accessibility. The implementation of such principals and experimentation with real users, allowed optimal adaptation to their needs and abilities. Not without reason was this tailor-made solution adopted by many users.

Today, approximately a year since the application was launched in the Android and iOS application stores, there are over 1,000 active users. Further data is available for some of those users and they log an average use of 2.69 times a week. An estimated total of 50,000 sessions were thus completed so far in the application, supporting hundreds of people with ID in performing regular daily activities.

The implementation of the interface design principles and the recommendations arising from this case study can support and streamline the development of future technological solutions for people with intellectual disabilities. Expanding the range of available mobile applications will support the assimilation of people with intellectual disabilities in society and by that turn it into an inclusive society which aspires to incorporate all its members and shrink the social and digital gaps within it.

ACKNOWLEDGMENT

The author wishes to thank Barak Alon and Yaniv Alon for their support in guiding the "Kids for Kids" project and their assistance in the implementation of the IBM Design Thinking methodology. The author also wishes to thank Noy Misgav and Hadas Rosen for their support in this research and the development of the "Shkalkalim" application.

REFERENCES

Abbott, S., & McConkey, R. (2006). The barriers to social inclusion as percieved by people with intellectual disabilities. *Journal of Intellectual Disabilities, 10*(3), 275–287. doi:10.1177/1744629506067618 PMID:16916851

Agree, E. M. (2014). The potential for technology to enhance independence for those aging with a disability. *Disability and Health Journal, 7*(1), S33–S39. doi:10.1016/j.dhjo.2013.09.004 PMID:24456682

Avila, J. (2015, October). *Designing Mobile Apps for Use by People with Cognitive Disabilities*. Academic Press.

Bigby, C., Bould, E., & Beadle-brown, J. (2016). Conundrums of supported living : The experiences of people with intellectual disability. *Journal of Intellectual & Developmental Disability, 0*(0), 1–11.

Billi, M., Burzagli, L., Catarci, T., Santucci, G., Bertini, E., Gabbanini, F., & Palchetti, E. (2010). A unified methodology for the evaluation of accessibility and usability of mobile applications. *Universal Access in the Information Society, 9*(4), 337–356. doi:10.100710209-009-0180-1

Borblik, J., Shabalina, O., Kultsova, M., Pidoprigora, A., & Romanenko, R. (2015). Assistive technology software for people with intellectual or development disabilities: Design of user interfaces for mobile applications. *Information, Intelligence, Systems and Applications (IISA), 2015 6th International Conference On*, 1–6.

Brown, T. (2008). Design Thinking. *Harvard Business Review*, 85–92. PMID:18605031

Chan, J. M., Lambdin, L., Graham, K., Fragale, C., & Davis, T. (2014). A picture-based activity schedule intervention to teach adults with mild intellectual disability to use an iPad during a leisure activity. *Journal of Behavioral Education, 23*(2), 247–257. doi:10.100710864-014-9194-8

Darcy, S., Green, J., & Maxwell, H. (2016, June). I've got a mobile phone too! Hard and soft assistive technology customization and supportive call centres for people with disability. *Disability and Rehabilitation. Assistive Technology*, 1–11. PMID:27293107

de Urturi Breton, Z.S., Hernández, F. J., Zorrilla, A. M., & Zapirain, B. G. (2012). Mobile communication for Intellectually Challenged people: A proposed set of requirements for interface design on touch screen devices. *Communications in Mobile Computing, 1*(1).

Dekelver, J., Kultsova, M., Shabalina, O., Borblik, J., Pidoprigora, A., & Romanenko, R. (2015). Design of mobile applications for people with intellectual disabilities. In A. Kravets, M. Shcherbakov, M. Kultsova, & O. Shabalina (Eds.), *Creativity in Intelligent Technologies and Data Science. Communications in Computer and Information Science* (Vol. 535). Cham: Springer International Publishing.

Fajardo, I., Tavares, G., Ávila, V., & Ferrer, A. (2013). Towards text simplification for poor readers with intellectual disability: When do connectives enhance text cohesion? *Research in Developmental Disabilities, 34*(4), 1267–1279. doi:10.1016/j.ridd.2013.01.006 PMID:23417132

Henry, S. L. (2018). *Introduction to Web Accessibility*. Retrieved from https://www.w3.org/WAI/fundamentals/accessibility-intro/

Hong, S., & Kwek, D. (2011). Innovation in the classroom: Design thinking for 21st century learning. *Stanford EDU*.

Hua, Y., Woods-groves, S., Kaldenberg, E. R., Lucas, K. G., & Therrien, W. J. (2015).. . *Effects of the TIP Strategy on Problem Solving Skills of Young Adults with Intellectual Disability*, *50*(1), 31–42.

IBM Corporation. (2017). *IBM Design Thinking*. Retrieved April 20, 2017, from https://www.ibm.com/design/thinking/

Igual, R., Plaza, I., Martin, L., Corbalan, M., & Medrano, C. (2013). Guidelines to design smartphone applications for people with intellectual disability: A practical experience. In A. van Berlo, K. Hallenborg, J. M. Corchado Rodriguez, D. I. Tapia, & P. Novais (Eds.), *Ambient Intelligence-Software and Applications* (pp. 65–69). Springer International Publishing. doi:10.1007/978-3-319-00566-9_9

Jansen, B. R. J., De Lange, E., & Van der Molen, M. J. (2013). Math practice and its influence on math skills and executive functions in adolescents with mild to borderline intellectual disability. *Research in Developmental Disabilities*, *34*(5), 1815–1824. doi:10.1016/j.ridd.2013.02.022 PMID:23511346

Johnson, S. (2016). IBM Design Thinking Field Guide. IBM Corporation, 3.4.

King, E., Okodogbe, T., Burke, E., McCarron, M., McCallion, P., & O'Donovan, M. A. (2017). Activities of daily living and transition to community living for adults with intellectual disabilities. *Scandinavian Journal of Occupational Therapy*, *24*(5), 357–365. doi:10.1080/11038128.2016.1227369 PMID:27707082

Merrick, J., Greydanus, D. E., & Patel, D. R. (2014). *Intellectual disability: Some current issues*. New York: Nova Science Publishers, Inc.

Mor, Y., & Winters, N. (2007). Design approaches in technology enhanced learning. *Interactive Learning Environments*, *15*(1), 61–75. doi:10.1080/10494820601044236

Morris, J. T., Jones, M. L., & Sweatman, W. M. (2016). Wireless technology use by people with disabilities: A national survey. In J. Santiago (Ed.), *Annual International Technology and Persons with Disabilities Conference* (Vol. 4, pp. 101–114). Academic Press.

Ouellette-Kuntz, H., Burge, P., Brownà, H. K., & Arsenaultà, E. (2010). Public attitudes towards Individuals with intellectual disabilities as measured by the concept of social distance. *Journal of Applied Research in Intellectual Disabilities*, *23*(2), 132–142. doi:10.1111/j.1468-3148.2009.00514.x

Sauer, A. L., Parks, A., & Heyn, P. C. (2010). Assistive technology effects on the employment outcomes for people with cognitive disabilities: A systematic review. *Disability and Rehabilitation. Assistive Technology*, *5*(6), 377–391. doi:10.3109/17483101003746360 PMID:20392190

Scheer, A., & Plattner, H. (2012). Transforming constructivist learning into action : Design Thinking in education. *Design and Technology Education: An International Journal*, *17*(3), 8–19.

Schweibenz, W. (2006). Accessibility- Why and how to include people with disabilities. *14th International BOBCATSSS Symposium*, 19–30.

Shakespeare, T., & Watson, N. (2001). The social model of disability: An outdated ideology? In Research in Social Science and Disability (Vol. 2, pp. 9–28). Academic Press.

Special Olympics. (2009). *Status and prospects of persons with intellectual disability*. Health, Well-being and Opportunity for People with Intellectual Disability – A Global Imperative.

Tschimmel, K. (2012). Design Thinking as an effective Toolkit for Innovation. In *Proceedings of the XXIII ISPIM Conference: Action for Innovation: Innovating from Experience* (pp. 1–20). Barcelona: Academic Press.

Watson, S. (2018). *How mild intellectual disability is defined*. Retrieved from https://www.thoughtco.com/mild-intellectual-disability-3110889

Woodcock, B., Middleton, A., & Nortcliffe, A. (2012). Considering the Smartphone Learner: Developing innovation to investigate the opportunities for students and their interest. *Student Engagement and Experience Journal, 1*(1), 1–15. doi:10.7190eej.v1i1.38

World Health Organization. (2007). *Atlas: Global Resources for Persons with Intellectual Disabilities*. Geneva: World Health Organization.

World Health Organization. (2011). World report on disability 2011. *American Journal of Physical Medicine Rehabilitation, 91*.

World Health Organization. (2018). *Assistive technology*. Retrieved from http://www.who.int/en/news-room/fact-sheets/detail/assistive-technology

KEY TERMS AND DEFINITIONS

Accessibility: The degree to which devices and services are adapted for people with disabilities, so they can use, understand, and interact with their interfaces.

Assistive Technology: A technological device or service aimed at helping people with disabilities circumvent their limitation while performing a task.

Design Thinking: A human-centered design methodology for problem solving based on empathy and a deep understanding of the physical and emotional needs of the end users.

Human-Centered Design: A design paradigm that places great emphasis on the needs, desires, and limitations of the end user at every stage of the design process.

Intellectual Disability: A significant limitation in mental functioning and adaptive behavior as expressed in conceptual, social, and practical skills.

Mild Intellectual Disability: Refers to people with intellectual disability with an IQ level of 50-69, who need guidance and support in acquiring daily skills.

Usability: A set of measures to examine the ease of use of a product by users in a particular context.

Chapter 12
Design and Development of a Mobile Writing Application for Students With Dysgraphia

Sinan Hopcan
Istanbul University – Cerrahpasa, Turkey

Saniye Tugba Tokel
Middle East Technical University, Turkey

Necdet Karasu
Gazi University, Turkey

Çığıl Aykut
https://orcid.org/0000-0002-0365-9930
Gazi University, Turkey

ABSTRACT

Writing plays a vital role in both daily life and academic life. Students begin learning to write in the first years in school and then they use writing skills their whole life. However, some students, such as students with dysgraphia, experience difficulties in writing. The purpose of the chapter is to develop a mobile writing application for students with dysgraphia and to reveal ideas of special education experts, educational technology experts, and classroom education experts, and a teacher about application. A pilot study was conducted with three students with dysgraphia firstly, and necessary revisions were made. After revisions, experts' views were taken. The results indicated that experts reflected positive ideas about mobile writing application for students with dysgraphia. However, the experts suggested some minor revisions for improving the mobile writing application. To sum up, final improvements were made, and a mobile writing application was developed.

DOI: 10.4018/978-1-5225-8106-2.ch012

INTRODUCTION

Handwriting is one of the most used skills in school and daily life. Students who are attending primary school spend almost half their school days in writing activities (McHale & Cermak, 1992). Students especially with dysgraphia have the additional challenge of having difficulties in writing. They can verbally answer questions, but find it difficult to answer them in writing. Accordingly, frustration may be observed among these students (Zhang, 2000) and the struggles in writing possibly can bring about emotional and social problems. According to Cahill (2009), instead of requesting help or explaining their difficulties, students may start to avoid writing tasks, and failure may begin.

Writing should be considered in two stages. The first stage is called acquisition and the second is called development. Acquisition is about learning and using one's basic knowledge. In other words, it is the stage that is concerned with teaching letters, syllables, words, and sentences more broadly; this stage is about how to use the words and letters in writing. Especially in early grades, writing skills should be given in accordance with the rules and figures (Akyol, 2005). In line with this, if there is a problem in the acquisition stage, it will be challenging to focus on the content. This is where students may first encounter challenges in their writing and their performances can be affected. Students with learning disabilities from the beginning of their school experience show the signs of the struggles in writing in several ways. Typing the letters in wrong directions, having severe organization problems, not being able to correct the mistakes despite all error corrections might be accounted as common problems. Another challenge is that teachers tend to give higher grades for legible handwriting (Graham, Harris & Fink, 2000).

Galanis (2008) states that failures in any of the acquisition stages can have long-term undesirable effects on academic success and self-esteem of students. Considering this, as in the early grades, the knowledge and writing performance of students with writing difficulties can be improved with the use of effective instructional strategies (Harris, Graham, & Mason, 2006). In this context, mobile devices have an essential role for students with writing disabilities. The mobile devices allow students to study at their own pace (Evans, 2008; Kagohara et al., 2013) and in various places (Evans, 2008).

Fernández-López, Rodríguez-Fórtiz, Rodríguez-Almendros, and Martínez-Segura (2013) summarizes the main features and functions of mobile devices such as touch screens, mobility and design, interaction through motion, accessibility, connectivity, and the ease of acquisition. Mobile writing applications have the potential to improve writing skills in students. In this way, the applications may help to close the gap between the practices of both field and literature. Literature reveals that writing requires skills rather than knowledge and skills are acquired by practice (Akyol, 2005; MONE, 2005). Therefore, insufficient number of such applications that have promising potential to provide opportunities for necessary practice is another area that needs to be addressed. To this end, the main purpose of this study is to develop a mobile writing application for students with dysgraphia.

BACKGROUND

Handwriting

Writing is a complex skill with different dimensions, such as cognitive, kinesthetic, and perceptual motor components (Engel-Yeger, Nagauker-Yanuv, & Rosenblum, 2009; Reisman, 1993). In school, students use their writing skills in writing activities most of the time. Writing is a mean that provides students to record what they have learned (Hamstra-Bletz & Blöte, 1993; Phelps, Stempel, & Speck, 1985). Besides, writing is a key event to express feelings and thoughts (Erhardt & Meade, 2005; Rosenblum, Weiss, & Parush, 2003). Writing does not only affect student life, but also it plays an important role in their daily activities (Chicu, Ţicău, & Şoitu, 2014).

Writing is a skill which requires practice (MONE, 2005). Akyol (2005) states that writing instruction can be presented in different ways. For instance, there are different types of handwriting manuscripts, such as cursive, mixed-mostly manuscript, and mixed-mostly cursive (Graham, Weintraub, & Berninger, 1998). The process of teaching how to write should be considered in two stages as acquisition and development. Acquisition is about learning and using the basic knowledge of writing. More broadly, it is concerned with teaching what the letters, syllables, words, and sentences are and how to write them. Especially in early grades, writing skills should be given in accordance with rules and figures. From fourth grade until the end of elementary school, written assignments, exams, and longer written studies start to begin. Namely, it is a transition from learning to write to writing to learn (Reisman, 1993). A primary school student engages in writing for about 50 percent of their school day (McHale & Cermak, 1992). While students need to learn to write in both form and content, some students have difficulty in writing. Writing is one of the relatively common academic problems of the students and between 10-20% of those students later become diagnosed with dysgraphia due to constant and severe writing difficulties.

Definition of the Dysgraphia

Dysgraphia is a writing disability which comes from the Greek words which mean "impaired" and "writing letter by hand" respectively (Chung & Patel, 2015). Galli et al. (2018) defines dysgraphia as a type of learning disability that affects the writing and fine motor skills of individuals. It is a writing difficulty, which manifests poor-quality handwriting (Hamstra-Bletz & Blöte, 1993). Dysgraphia can be defined as being not proficient in writing skills (Kushki, Schwellnus, Ilyas, & Chau, 2011) and characterized by difficulty and disturbance in handwriting (Nicolson & Fawcett, 2011). Children have dysgraphia suffer from difficulties in production of writing related to mechanics of writing (Graham et al., 2000; van Galen & de Jong, 1995). Dysgraphia, writing disability or writing difficulty are terms often used in an interchangeable way (Jena, 2013). APA (2013) uses the term of "impairment in writing" instead of dysgraphia as difficulties with written expression. Students who suffer from dysgraphia can demonstrate slow, illegible, poor and sloppy writing besides spelling errors, and, syntax composition problems (Chung & Patel, 2015).

Galli et al. (2018) describes learning disabilities as an umbrella term which encompasses specific learning disabilities (SLD) such as dysgraphia, dyscalculia and dyslexia. Korkmazlar (2003) defines an individual with SLD as someone without primer psychic illness, apparent brain pathology, and sensory disabilities, who has normal or above-normal level of intelligence (IQ > 85), but suffer from difficulties in writing, reading, arithmetic, listening, and reasoning. Additionally, the individual with SLD, who performs below the age and intelligence level despite standard education, has secondarily problems in self-management, social perception, and interaction. Because SLD affects learners who have normal or above-normal IQ scores (greater than 70), APA (2013) uses unexpected academic under-achievement phase. The diagnostic and statistical manual of mental disorders (DSM-V) defines specific learning disabilities (disorders) as specific deficits in academic skills such as reading, math, or writing (APA, 2013). SLD affect academic achievement as well as daily performance including occupational life which requires writing skills. Writing disability commonly manifests itself at school ages (APA, 2013).

Prevalence

In the literature examining primary school students', in regards to their writing disabilities, researchers report different percentages of such disability as follows: 7%-15% (Katusic, Colligan, Weaver & Barbaresi, 2009), 13% (Karlsdottir & Stefansson, 2002), 34% (Smits-Engelsman, Niemeijer & Van Galen, 2001). In addition, in their review Van Hartingsveldt, De Groot, Aarts and Nijhuis-Van Der Sanden (2011) claim that approximately 16% of school students have writing difficulties. Based on previous studies, prevalence of writing disabilities ranges between 7% and 34%.

Characteristics

Students with SLD exhibit different characteristics. However, they still present some common characteristics. Academic, cognitive, social, and emotional characteristics will be mentioned under this title.

Academic Characteristics

The most prominent characteristic of students with dysgraphia is illegible writing (Alberta Learning, 2002; Chung & Patel, 2015; Galli et al., 2018; Richards, 1998); due to grapho-motor and visual deficiencies, they also have secondary difficulties in tasks such as sentence completion, and choosing words while writing (Chung & Patel, 2015; Richards, 1998). Mixing uppercase and lowercase letters, and making mistakes in writing letters (uncompleted letters, irregular size and/or shape etc.) (Reid, Elbeheri & Everatt, 2015; Richards, 1998). Figure 1 and 2 shows the sample illegible writings of students with dysgraphia. In addition, they also hold the pencil too tightly (Alberta Learning, 2002; Richards, 1998). Due to grapho-motor (and sometimes visual) difficulties, which are expressed in illegibility difficulties, they also have secondary difficulties such as: choosing words, and being unable to complete sentences (Alberta Learning, 2002; Galli et al., 2018; Richards, 1998).

Difficulty in the production of written language is the most exhibited academic characteristics of students with dysgraphia (Alberta Learning and Teaching Branch, 2002; Chung & Patel, 2015; Reid et al., 2015; Richards, 1998; Yiğiter, 2005) Poor handwriting and spelling mistakes (Galli et al., 2018)

Figure 1. Sample writing of a student with dysgraphia
(Crouch & Jakubecy, 2007)

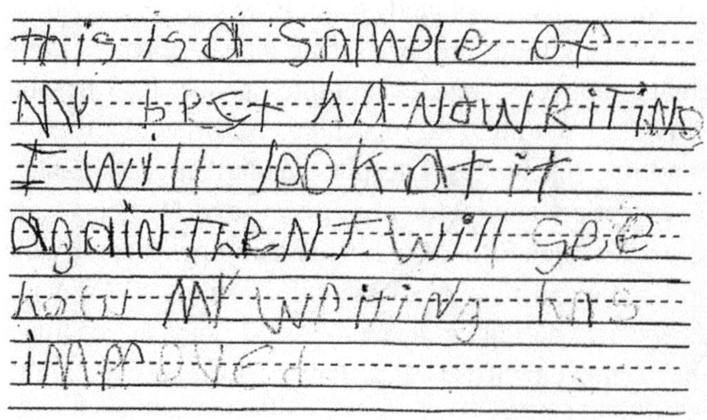

Figure 2. Sample writing of a student with dysgraphia
(Williams, 2014)

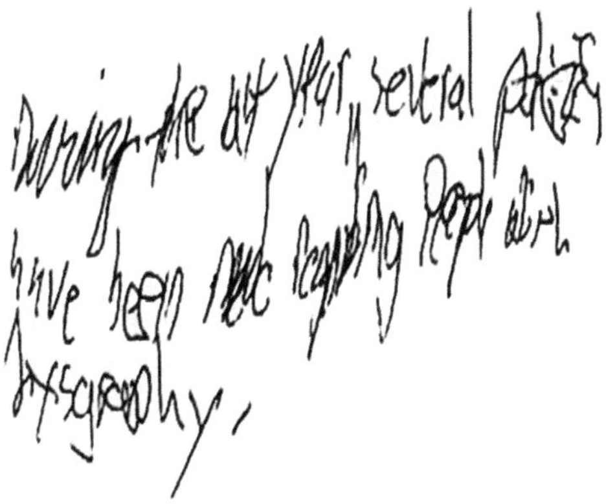

are common characteristics mentioned in the related literature. In addition to their difficulties in basic writing activities, similar difficulties exist in advanced writing activities.

Cognitive Characteristics

Attention deficit disorders may also be comorbid in some of students with SLD. (APA, 2013). Much effort has been exerted to bring attention to important stimuli in the environment (Friend, 2005; McNamara, 2007). Perceptual problems might be skipping the letter or the word despite excellent sight, while reading or misunderstanding the words despite excellent hearing while listening. In addition, there are some difficulties in short term or long-term memory or in both (Friend, 2005; McNamara, 2007). Difficulties that the students encounter are in thinking of their strengths, needs and their learning process. They also have difficulties in selecting and applying new strategies such as metacognitive skills (Alberta Learning and Teaching Branch, 2002; Friend, 2005).

In brief, students with SLD also can have various difficulties in the cognitive domain. Attention and perception problems are frequently mentioned in related literature for these students. Besides, deficiencies in metacognitive skills, memory development are another common point for them.

Social/Emotional Characteristics

Students with SLD may have deficits in social skills; such as having lower self-esteem (Alberta Learning and Teaching Branch, 2002; Friend, 2005; Rowe, 2006; Zhang, 2000) and a lack of recognition by friends (Alberta Learning and Teaching Branch, 2002; Friend, 2005). Furthermore, students with SLD may have low social status compared to their friends and motivational problems because of their academic struggles or social incompetence (Friend, 2005). Difficulty in making statements, understanding thoughts and jokes, participating in discussions, and lower level language skills may also be seen (Alberta Learning and Teaching Branch, 2002).

In brief, students with SLD experience difficulties and problems not only in the academic and cognitive domains but also in the social and emotional skills. The literature indicates that there is a relation between academic skills (success) and social acceptance which is frequently discussed in the literature.

Using Educational Technology to Support Writing Skills for Students With Dysgraphia

Educational technology is a promising solution to meet students with special needs. Under this title, computer-based instruction and mobile devices-based instruction are mentioned.

Computer-Based Instruction

Computer based instruction can offer a wide spectrum of solutions for students who have difficulties in writing (MacArthur, 2009). These solutions are cost-effective, feasible and treatment-effective (Tanimoto, Thompson, Berninger, Nagy, & Abbott, 2015). Zhang (2000) claims that educational technology enables students to organize their ideas and to create sentences and paragraphs by practicing writing. Furthermore, technology allows students to express themselves better when they enjoy writing. Computer based instruction also helps students to focus on writing by reducing attention problems in mainstream classes. Besides this, it allows students to improve their writing skills by allowing them to practice as much as they want at their own pace. Some research are presented below on the use of technology for students with writing disabilities that only used diagnoses tools and then others that used treatment.

Rosenblum, Dvorkin, and Weiss (2006) developed a computerized evaluation tool. Owing to the developed tool, students with dysgraphia can be diagnosed by the educators. By using this tool, they conducted an experimental study in order to examine the handwriting processes of third grade students with and without dysgraphia. 14 students with dysgraphia were assigned to an experimental group and 14 proficient students were assigned to the control group. The results of the study showed that there was a significant difference between two groups in terms of characteristics of handwriting.

Guinet and Kandel (2010) developed a software to understand the handwriting process of both children and adults which was suitable to investigate writing disabilities. Online information was provided by the Ductus software. Ductus was designed to present velocity, duration, pauses, and fluency as different aspects of handwriting, which worked on Windows with Wacom tablet. According to the results, Ductus can facilitate studies that examine handwriting production.

Falk, Tam, Schellnus, and Chau (2011) designed a computer based handwriting assessment tool to diagnose students with writing disabilities in terms of their writing styles of incorrect space and size as well as text alignment and legibility. Out of 35 participants first and second graders, nine of them were identified with handwriting difficulties.

Smits-Engelsman and Van Galen (1997) conducted a longitudinal study with 16 primary school students with dysgraphia. They revealed that the writing speed of children increased after they had used a computer software with a digitizer tablet and a special pen with a pressure-sensing device. They found that the incorrect attempts of children were decreased while they were writing.

Tanimoto et al. (2015) investigated the effectiveness of computerized and mobile device-based writing and reading instruction for students with SLD between the 4th and 9th grade. They conducted a quasi-experimental study. A computerized training (visual motion cue and writing activities on a blank monitor screen) was administered to group A which consisted of 21 students. In addition, an iPad training (sequential, number, arrow cues, and writing activities between lines on iPad) was given to group B which consisted of 11 students. Training which consisted of 18 sessions was continued for 3 months. The findings of the research indicated that group B was significantly more successful than group A.

Azimi and Mousavipour (2014) aimed to investigate effectiveness of an educational multimedia in dictation for second grade students with dysgraphia. Quasi-experimental design was utilized in the study and the control group got traditional educational procedure while experimental group used multimedia dictation. Based on the results of the study, a significant difference was found in favor of the experimental group which had educational multimedia dictation.

Chang and Yu (2014) conducted a pretest posttest design to investigate whether there was a difference among computer-assisted group, sensory motor training group and a control group. The participants of the study were 42 students in the range of 7 to 9 years of age. The study revealed that there was a significant difference among computer-assisted group, sensory motor training group, and a control group. Computer-assisted group showed a promising improvement in writing speed and fluency.

Salih, Abdul-Kahar, Zahari, Khalid, and Rahim (2015) developed a 3D online game for students with writing and reading disabilities (between the ages of 5 to 12 years old) on how to write letters and words. Analysis was conducted after observing games that students played via mobile application. Results showed that games were in great demand for the students in this age group.

Fedora (2015) conducted an exploratory study to investigate experiences of special and primary education teacher candidates about integration dictation software into courses and to seek their future use of technology for students with writing disabilities. 13 pre-service teachers, who took a course related to learning disabilities participated in this study. A survey was administered to seek opinions of teacher candidates about the integration of dictation software into course and the future use of them. The study revealed that teacher candidates had positive attitude towards technology and it is expected that they will use technology in their future classrooms.

Hennion, Gentaz, Gouagout, and Bara (2005) developed a visio-haptic interface (telemaque) which aimed teach writing students with dysgraphia. This interface has a static aspect, which is concerned with the correct shape of letters etc. and a dynamic aspect, which is concerned with the correct order while writing the letters and so further. Four practices were presented by the device and they are as follows: 1) teaching the shape of the letter, 2) teaching correct order while writing the letter 3) teaching retracing the letter 4) teaching writing on a blank space by the visio-haptic interface. The telemaque interface was used by only one student with dysgraphia. The authors reported that further studies would be conducted with students with dysgraphia by using the telemaque interface. Firstly, the telemaque was applied to 42 kindergarten students who were five years of age. The study investigated the effectiveness of telemaque on handwriting fluency of kindergarten students before transition to formal writing instruction. It was an experimental study. Experimental group used telemaque interface and control group used classic methods. Findings of the study showed that there was a significant difference between the two groups. A significant difference was found in favor of the experimental group. However, the authors concluded that this interface should be used for students with dysgraphia to enable them to improve their handwriting skills (Palluel-Germain et al., 2007).

Although there is some research on the use of various technological tools for students with writing disabilities (see MacArthur, 2009), only a limited number of studies have focused on how to teach these students how to write by using technology. There is a still need to help students develop basic writing skills (MacArthur, 2009; Zhang, 2000). In parallel with this, these computer tools such as word processing, word prediction, and speech recognition cannot meet the need for teaching students and developing their handwriting abilities (Giordano & Maiorana, 2014). However, today's technology offers different options for word processing (Zhang, 2000). One example is the computer-based instruction tool such as the interactive educational software, which is designed to improve fine motor skills, hand-eye coordination, and thus improves writing skills as needed (Giordano & Maiorana, 2014). Researchers must recognize the problems and difficulties of students with writing disabilities in order to understand how computer-based instruction helps students to master basic writing skills (Zhang, 2000).

Mobile Devices-Based Instruction

Mobile devices may have a noteworthy potential for students with writing disabilities, first and foremost because mobile phones or tablets are quite common and popular. These devices allow students to study on their own pace (Evans, 2008; Kagohara et al., 2013) and in various places (Evans, 2008).

Fernández-López et al. (2013) state key features of mobile devices are the touch screen, the mobility and design, the interaction through motion, the accessibility, the connectivity, the ease of acquisition.

- **Touch Screen:** There is no need to learn the use of an extra tool such as mouse, keyboard etc. It can be easily used by fingers or stylus pen.
- **Mobility and Design:** The dimensions are small, lightweight, and thus portable. Still, resolution and screen size are adequate to see the objects and symbols clearly. These devices are available everywhere and at all times and their batteries last long enough to use at least for one day.
- **Interaction Through Motion:** Devices give response to rotating and shaking. This feature allows for increased interaction.
- **Accessibility:** Brightness of mobile devices can be adjusted to light and dark environments. Zoom feature in mobile devices enables better visibility of small objects and the symbols are more legible.
- **Connectivity:** These mobile devices connect through Bluetooth and Wi-Fi. They also have USB ports, and connect to both the internet and other devices.
- **Ease of Acquisition:** It is easy to get these devices, to find and download applications. Thus, these devices can help increase the use of instructional material by teachers, students and parents.

Giordano and Maiorana (2014) developed a software that works with tablet or smart phones for students with dysgraphia. The software includes various types of exercises (connecting dots, and writing a given word again a blank space, etc.) and gives feedback to students. The system contributes to individualized learning by processing the data collected from the user in real time. Giordano and Maiorana (2015) developed a mobile web game for students with dysgraphia. They presented design principles as well as preliminary testing with children. Three children participated to the gameplay. Motivation of the children was observed. The students enjoyed the gameplay and gave positive feedback. In addition, it was reported that the writing exercises are sufficiently hidden in the mobile game.

Czyzewski, Odya, Grabkowska, Grabkowski, and Kostek (2009) developed a smart pen to improve writing skills of students with dysgraphia. This pen consists of both hardware and software parts. It allows students to practice with a teacher or a therapist. Results showed that the teacher and students were interested in using the smart pen.

However, there is scarcity any mobile writing applications which may improve a student's writing skill. In this context, this unique study focuses on improving writing skills of students with dysgraphia via a mobile writing application developed by the researchers. Therefore, this study is expected to bring unique insights for both the practitioners and the scholars working in the fields of primary education, special education, and educational technology. The method used in the study is based on an interdisciplinary approach to highlight the needs of students who suffer from the problems of dysgraphia and to develop a mobile application that meets their needs involved in the learning process. In addition, available applications mostly developed in English are not suitable for any other languages. Hence, this is a unique application which addresses not only the needs of students with dysgraphia also introduces the program in the Turkish language.

Development of the Mobile Writing Application

An Android version of an application aims to contribute to acquisition of writing skills (being able to write letters, numbers, syllables, and words) for students with dysgraphia attending 1st-8th grades in primary schools, utilizing Adobe Animate. The application aims to increase students' writing accuracy, writing speed and decreasing their line violation. The application uses a gesture recognition algorithm to recognize a student's hand writing motion during the writing processes. The application was developed for students with dysgraphia that attend first through 8th grade primary schools.

Registration screen is the first screen (Figure 3). After the researcher registers with a user name, a general introduction about the application is presented to students (Figure 4).

Figure 3. Registration screen

Figure 4. Introduction screen

Mobile writing application consists of the following three main parts:

Pretest-Posttest Screens

Pretest-Posttest (Figure 5) aims to compare the writing skills of students before and after using mobile writing application. After the registration and first introduction screens, a pretest is presented to the students while after content screen, posttest is presented. The pretest and posttest includes 29 lowercase letters, 29 uppercase letters, 20 syllabi and words (in Turkish). The content of the pretest and posttest was selected from the course book (Doğan Temur, 2015) and the curriculum (MONE, 2015) considering the views of the classroom teacher and the experts. Pretest and posttest results of students are saved in the tablet memory as jpeg files for the evaluation.

Figure 5. Screenshots from pretest- posttest

Trial Screen

A trial screen aims to introduce the user to the application and helps get them acquainted with the stylus pen. In this screen, there are five different lines on the trial screen and each line has an animal and a food item (Figure 6). If students draw the line, the animal reaches the food.

The Main Part of Application

The main part of application consists of four different contents: 29 lowercase letters, 29 uppercase letters, 10 numbers, 386 syllabi and words in Turkish (Figure 7). The learning objectives were selected from both Turkish MONE (2008) Specific Learning Disabilities Support Education Program (Module: Math, section: natural numbers, learning objective: "writing numbers"; Module: Literacy, the fifth Learning objective: "writing the letters", the seventh learning objective: "writing syllabi" and the ninth learning

Figure 6. Trial screen

objective: "writing words") and MONE (2015) curriculum (Turkish first grade course, T1.3. Writing Section, Learning Objective T1.3.2. Students will be able to write letters, numbers, and arithmetic signs accordance with the technique, Learning Objective T1.3.3. Students will be able to write illegible syllabi and words using cursive writing). The content was selected from course book (Doğan Temur, 2015), considering the views of the classroom teacher and experts. A music player is on the top part of the screen. Students can control it easily. The content is exhibited to students as a dart game. There is a writing line on the screen with a model of the learning objective (the letter/number/syllabi/words) and a guiding gray field to show student where to begin writing. Students with SLD have problems confusing directions (MONE, 2008). The gray field was developed because of this. Students are expected to write the learning objective (a letter, a number, or a word) that is most similar to the model and they need to repeat processes three times correctly. Three levels of clues are exhibited by the application (Figure 8). The first clue shows writing direction (s) of the letter/number/syllabi/words by arrows. The second clue shows how to write the letter/number/syllabi/words by animation. The third clue presents the dotted

Figure 7. Screenshots from the main part of application

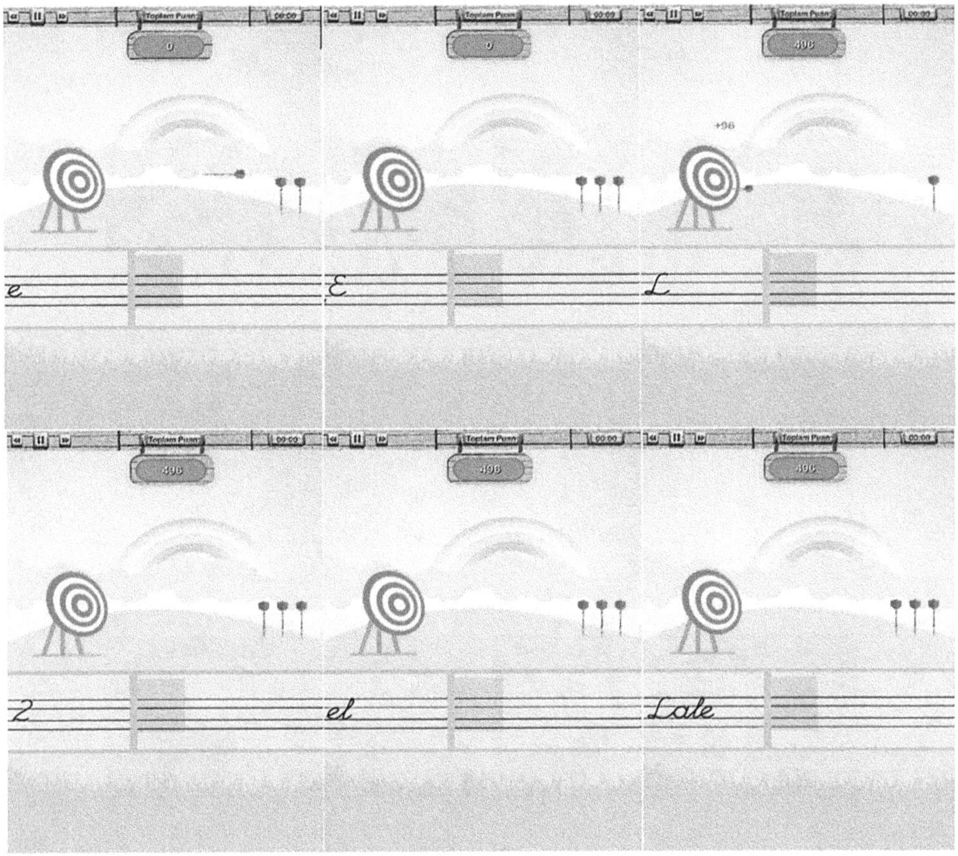

version of the letter/number/syllabi/words in addition to showing how to write the letter/number/syllabi/words by animation. When students write the learning objective three times correctly, the other learning objective is presented. If students make a mistake, the subsequent clue is presented. The clue type changes when students make mistakes two consecutive times. For each learning objective, three arrows are given to students. For each correct correspondence, one arrow is shot and the students gain a score between 88 and 100 in terms of similarity rate to the model of learning objective. When the similarity rate is more than 80%, it is converted to 100 points. If the similarity rate is less than 70%, the writing is not considered to be accurate. The student is given an overall score out of 100. Simultaneously, the student is given a positive or negative sound as feedback. When the students write the learning objective three times correctly, one of the 12 verbal positive reinforcements are given randomly. The total score that the student earned is shown on the scoreboard at the top of the screen.

In addition to correct writing, line control is also checked for each learning objective by the application due to inability of students with dysgraphia to adjust letter size, line and margin (Richards, 1998) considering the needs of these students and special education experts' views. When students can not follow the line and overflow the line 5 pixels, even if they write the learning object correctly, the an-

Figure 8. Screenshots from the three levels of clues

swer is not accepted. Students are given a verbal feedback: “You have to pay attention to the line” and the color of overflowed line (s) change(s) to red (Figure 9). Another novelty of this application that it provides clues (as mentioned above).

Database of the Application

Application database was developed by using SQLite. The time, scores, correct uppercase and lowercase letter, incorrect uppercase and lowercase letter, correct number, incorrect number, correct syllabi, incorrect syllabi, correct word, incorrect words, clues used, percentage of correct writing, number of the lines overflowed, and names of overflowed lines will be kept in the database.

Figure 9. Screenshot of line control feedback

METHODOLOGY

Research Design

The purpose of the study is to develop a mobile writing application for students with dysgraphia taking the views of experts into consideration. The pilot study aimed to evaluate the mobile writing application in terms of instructional design, visual design, and usability through observation. After piloting the study, necessary revisions were made. After the development of the mobile writing application, the views by special education experts, educational technology experts, classroom education experts and the classroom education teacher were taken.

Participants

Two groups participated in the study. The first group of this study consisted of three students, from 3rd grade, who were diagnosed with dysgraphia. The first student was a male, age ten (S1) and other two students were females age nine (S2 and S3). A purposeful sample procedure was employed because the target group of this study was students with dysgraphia. In addition, four special education experts, five educational technology experts, and four classroom education experts/ teacher as a second group participated after the application was developed. Two of the experts had master degrees and the others had doctoral degrees. Their working experiences were over four years. Four of them were female and five of them were male. Table 1 shows the detailed information about the experts who participated.

Instruments

Mobile writing application evaluation form: This form, which included three sub-forms with open-ended questions were developed in order to reveal the ideas of special education experts, educational technology experts, classroom education experts, and the classroom education teacher in terms of perceived ease of use, perceived usefulness, perceived enjoyment, educational content, visual design, multimedia use, and technical features.

The form for special education experts consisted of five questions about perceived ease of use, six questions about perceived usefulness, four questions about perceived enjoyment, five questions about educational content, one question about visual design, and four questions about multimedia use, in total 25 questions. The form for classroom education experts and the classroom education teacher consisted of three questions about perceived ease of use, five questions about perceived usefulness, and seven questions about educational content, in total 15 questions. The form for educational technology experts consisted of five questions about perceived ease of use, five questions about perceived usefulness, four questions about perceived enjoyment, one question about educational content, one question about visual design, four questions about multimedia use and two questions about technical features, in total 22 questions.

Table 1. Information about experts

	Code	Degree	Gender	Experience	Institution
Classroom Education Experts and Classroom Education Teacher	CE1	Ph.D.	Female	13 Years	State University
	CE2	Ph.D.	Female	13 Years	State University
	CE3	MS	Female	6 Years	State University
	CT1	MS	Male	6 Years	Private School
Educational Technology Experts	ET1	Ph.D.	Male	17 Years	State University
	ET2	Ph.D.	Male	16 Years	Private University
	ET3	Ph.D.	Male	9 Years	State University
	ET4	Ph.D.	Female	7 Years	State University
	ET5	Ph.D.	Male	4 Years	Private University
Special Education Experts	SP1	Ph.D.	Male	20 Years	State University
	SP2	Ph.D.	Female	20 Years	State University
	SP3	Ph.D.	Female	20 Years	State University
	SP4	Ph.D.	Female	6 Years	Counseling Center

Procedures and Data Analysis

An application including 29 uppercase and lowercase letters (Turkish alphabet), and numbers was developed for the pilot study. Figure 10 shows a screenshot from pilot version of the application including a progress, control buttons and a writing area. The pilot study was conducted by researchers and was observed by another researcher in the educational technology field.

At the beginning of the pilot study, the parents were informed about the study and parental consent forms were taken from them. Observations and video camera recording were carried out during the pilot study. One session was held for each participant on different dates. The sessions included uppercase letters, lowercase letters and numbers parts. The time for the sessions took approximately 30 minutes for both S2 and S3, and 18 minutes for S1 (Table 2). After piloting the study, necessary revisions were made. They were mentioned in the results section.

The application of the evaluation form took approximately 30 minutes for the experts. The questionnaire was analyzed through descriptive statistics (percentage).

RESULTS

The Pilot Study

The results of the pilot study were discussed with two special education experts and one educational technology expert. Some improvements were made to the mobile writing application according to the observations from the pilot study.

S1 thought his drawing was wrong since his own drawings did not look like the letter displayed on the screen. To overcome this problem, line correction and smoothing algorithm was used in order to prevent the students' writings from looking as if they were incorrect. This way, smoothing lines and pixel skipping problems were resolved during the writing. Due to the lack of palm rejection feature in horizontal axis and toolbar location (bottom) in Samsung Tab 2, some letters could not be written by using stylus pen by S1. For that reason, S1 wanted to write or delete the letters using his finger. Palm rejection feature in vertical axis had been adjusted before, likewise, it was decided to adjust horizontal axis also. Tablet was changed with TAB S2. S1 lost interest in writing each letter three times. Therefore, he passed the letter himself by using the next button. To overcome this problem a dart game was included to the application.

S2 wrote some letters by starting from the opposite direction. To overcome this problem, clues to teach writing direction of the letters were included. Even for several letters she wrote accurately, she could not follow the line. To overcome this problem, line control should be enabled in the application.

Table 2. Completion times of pilot studies

Sections	S1	S2	S3
Lowercase Letters	10 min.	14.5 min.	20 min.
Uppercase Letters	6 min.	17 min.	12 min.
Numbers	2 min.	5 min.	2 min.
Total	18 min.	36.5 min.	34 min.

Figure 10. Screenshots from pilot application

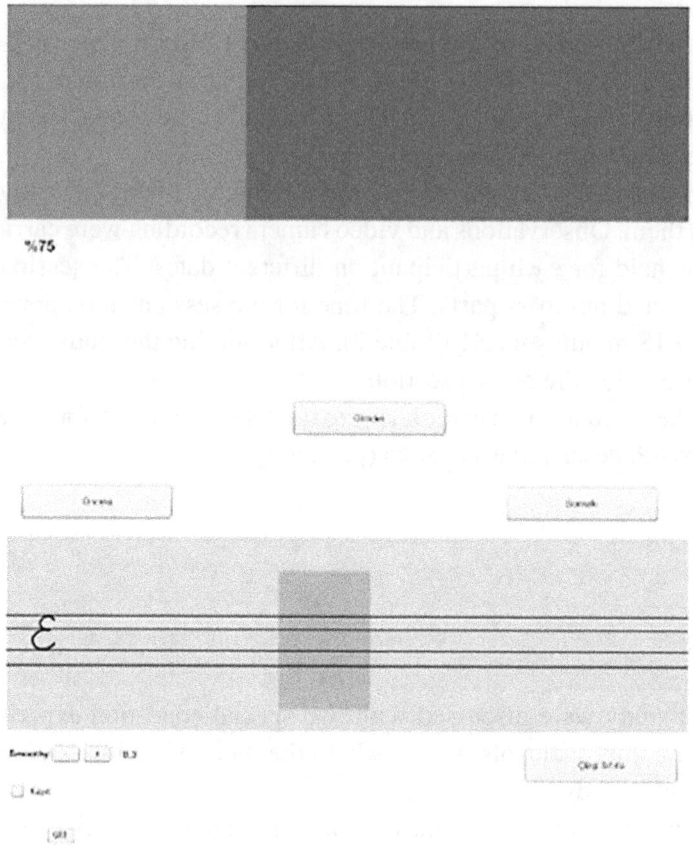

S3 wanted to write close to the sample letter on the screen yet S3 could not write there; as a result, S3 wrote out of the line. A distinguishing space/part should be added between the areas where the student writes and were the sample letter begins. Similar to S2, for several letters even she wrote accurately she could not follow the line. To overcome this problem, line control should be enabled in the application.

The Views of Subject Matter Experts About Mobile Writing Application

The results of the views of experts and the teacher are presented in terms of each category. Table 3 shows the means and standard deviations for the each questions.

Perceived Ease of Use

Educational technology and special education experts found using mobile writing application easy to learn. In addition, they found both the application and stylus pen easy to use. All experts agreed on the need for providing adequate support to start the application by trial screen. Moreover, experts mostly agreed that the application includes a clear, understandable, guiding voice and text instructions for students with dysgraphia.

Perceived Usefulness

All experts perceived mobile writing application as useful for making it easier for students to acquire writing skills, increasing academic performance allowing progressions at their own pace, and providing sufficient practice, and support for writing skills. In general, special education experts perceived mobile writing application as useful for meeting the needs of students with dysgraphia.

Perceived Enjoyment

Educational technology and special education experts thought that the application makes learning enjoyable. Educational technology and special education experts found application attractive, motivating, and fun.

Educational Content

All interviewed experts found the clues, feedback, and reinforcements appropriate. Special education experts, classroom education experts, and classroom teacher found scope, presentation of content, and adequateness to be appropriate in the pretest-posttest. One of classroom education expert suggested that pretest-posttest should include words which contain all letters. It does not have to be all uppercase letters. Uppercase letter should include exceptional letter (like D, N etc.).

Classroom education experts/teacher found adequateness of size of the writing area on the screen to be appropriate. Line spacing was the same as the notebook of students and the font size was 29pt. Three of classroom education experts emphasized that line spacing and font size should be enlarged. One of the special education experts suggested that a lesser amount of learning objective (maximum eight letters) should be at one line and there should be more space between learning objectives in the pretest and posttest.

Visual Design

Educational technology and special education experts found user interface and layout appropriate. In addition, they agreed on appropriateness of colors, characters, buttons, and texts.

Multimedia Use

Educational technology and special education experts found animations, playback feature in animation, sounds, and stop, forward, rewind, playback features for music appropriate. At the same time, most of educational technology experts (n=3) suggested that animations should be slower.

Technical Features

Educational technology experts found installation of application easy. In addition, they agreed that the application is error-free.

Table 3. The views of subject matter experts about mobile writing application

		Educational Technology Experts (n=5)		Special Education Experts (n=4)		Classroom Education Experts / Teacher (n=4)	
		M	Sd	M	Sd	M	Sd
Perceived Ease of Use	Learning to use mobile writing application	5	0	5	0	n/a	n/a
	To use mobile writing application	5	0	4.75	0.5	n/a	n/a
	To provide adequate support to start the application by trial screen	4.6	0.9	5	0	5	0
	Clear, understandable, guiding voice and text instructions for students with dysgraphia	4.8	0.45	4.75	0.5	5	0
	Use of stylus pen	4.2	1.1	5	0	4	1
Perceived Usefulness	Making it easier to acquire writing skill	5	0	5	0	5	0
	Increasing academic performance of students	5	0	5	0	5	0
	Allowing students to progress at their own pace	5	0	5	0	5	0
	Providing a sufficient amount of practice	5	0	4.8	0.5	5	0
	Supporting their writing skill	5	0	5	0	5	0
	Meeting needs of students with dysgraphia in general	n/a	n/a	5	0	n/a	n/a
Educational Content	Scope	n/a	n/a	5	0	5	0
	Presentation of content	n/a	n/a	5	0	5	0
	Clues	5	0	5	0	5	0
	Feedback	5	0	5	0	4.5	0.58
	Reinforcements	5	0	5	0	5	0
	Adequateness of pretest-posttest for evaluation	n/a	n/a	5	0	4.75	0.5
	Adequateness of size of the writing area on the screen.	n/a	n/a	n/a	n/a	4	1
Perceived Enjoyment	Making learning enjoyable	5	0	5	0	n/a	n/a
	Being attractive	4.8	0.45	5	0	n/a	n/a
	Motivating	4.8	0.45	5	0	n/a	n/a
	Having fun	5	0	5	0	n/a	n/a
Visual Design	User Interface	4.8	0.45	5	0	n/a	n/a
	Layout	4.6	0.55	5	0	n/a	n/a
	Colors	5	0	5	0	n/a	n/a
	Characters	5	0	5	0	n/a	n/a
	Buttons	4.8	0.45	5	0	n/a	n/a
	Texts	5	0	4.75	0.5	n/a	n/a
Multimedia Use	Animations	4.8	0.45	5	0	n/a	n/a
	Playback feature in animation	4.8	0.45	5	0	n/a	n/a
	Sounds (in the instructions, background music, feedback and reinforcements)	5	0	5	0	n/a	n/a
	Stop, forward, rewind, playback features for music	4.6	0.9	5	0	n/a	n/a
Technical Features	Easy installation	4.2	0.84	n/a	n/a	n/a	n/a
	Error-free	4	1	n/a	n/a	n/a	n/a

n/a: These questions were not asked to these participants.

All participants, the experts and classroom education teacher, found application adequate for students with dysgraphia to acquire writing skills in terms of "perceived ease of use, perceived usefulness, perceived enjoyment, educational content, multimedia use, and technical features". The findings strongly suggested that the application caused acceptable ratings in all of those. In addition to positive views of the experts, abovementioned, their suggestions were also available.

DISCUSSION AND CONCLUSION

The purpose of the study is to develop a mobile writing application for students with dysgraphia while taking the views of experts into consideration. The pilot study aimed to evaluate the mobile writing application in terms of instructional design, visual design, and usability through observation.

Altogether, the experts and classroom education teacher found application appropriate for target audience to acquire writing skills. Discussion was made in terms of "perceived ease of use, perceived usefulness, perceived enjoyment, educational content, multimedia use, and technical features".

"Perceived Ease of Use" was determined as one of themes that was investigated to analyze the views of experts and the teacher before the utilization of the mobile writing application. All participants perceived mobile writing application as easy to use. As one of the common characteristics of students with dysgraphia, complicated solutions might increase students' problems rather than decreasing. In line with this interpretation, Fernández-López et al. (2013) also highlighted the importance of the ease of use as one of the several principles integrating a technology into special education. Similar to teachers' views, Kagohara et al. (2013) mentioned that the use of these tablet devices are easy. In this study, the researchers received positive remarks with regards to the application but there were some criticisms in relation with the use of stylus pen. The reason for this could be explained with the difference in use and shape when compared the normal pen. This becomes obvious since the tip of the stylus pen is quite different in use. This concern has been revealed in the related scholarship. Annett, Anderson, Bischof, and Gupta (2014), and Helps and Helps (2016) categorized stylus pen as "active and passive" based on the use in capacitive touchscreens. The prices of active stylus pens are higher than the passive ones (Annett et al., 2014; Tanyag & Atienza, 2015). Therefore, the widespread use of stylus pen seems to be limited and constrained in near future due to economic costs. In addition, not all active stylus pen are compatible with all brands/tablet models (Helps & Helps, 2016; Tanyag & Atienza, 2015).

The tips of passive stylus pens vary. Some of them are made from rubber. They are soft and have a thick tip. None of them is able to show the exact written place (Helps & Helps, 2016). This is problematic especially for students who have writing problems. Another type of passive stylus pens are those with hard and transparent tips (Helps & Helps, 2016). This allows the student to see where (s) he writes exactly. In this study, this kind of pen was selected since it was appropriate and inexpensive. In addition to the "perceived ease of use", the "perceived usefulness" was also investigated in the pre-experimentation process.

"Perceived Usefulness" serves as one of the means that leads teachers and students to get involved in the learning process while using a mobile application. Their responses showed that all of them agreed on the usefulness of the application. Their views showed that such applications are needed in learning to write as well as creating a better learning setting both for students and for teachers. The results also showed that students will be able to study on their own and will be able to practice more. In line with this idea, Miller (2002) revealed that students with learning disabilities should practice more than their

normally developing peers. Similarly, Fernández-López et al. (2013) mentioned that mobile devices enable students to learn independently and ubiquitously.

"Perceived Enjoyment" serves as the basis of all the learning processes. When asked about the ways in which experts and the teacher found the application enjoyable, all the participants gave similar responses that highlight the strength of the application as interesting. Their responses indicated the ways in which students' motivation which is common problematic issue for students with SLD could be higher when dealing with such tasks. This view aligns with the idea put forward by Fernández-López et al. (2013) who asserted that students are more willing and able to study longer with such educational activities. Therefore, these views support the idea that the more students get involved and enjoyed the process the more they learn it easily.

The researchers also investigated the views of experts and the teacher with regards to the *"Educational Content"* used in the study in the pre-experimental phase. All participants accepted that educational content was appropriate. In addition, the scope of the application was sufficient to acquire writing skills. Classroom education experts and the teacher mostly agreed on the size of the writing area. Experts noted that it would be easier for the students to write if the line spacing was larger than normal. Considering this fact, the line spacing was increased 1.5 times and the font size increased to 45pt. Similarly, Harley et al. (2013) underlined the importance of expanded line spacing in order to make let students read the text easily. In line with this, Li-Tsang et al. (2013) used triple-line spacing for students with SLD (handwriting problem). Likewise, Romani, Tsouknida, di Betta & Olson (2011) used double-line spacing for students with dysgraphia. All these studies emphasize the possible impact of larger line spacing in literacy.

The researchers also investigated the views of experts and the teacher with regards to *"Visual Design"* in the pre-experimental phase. All the participants found the visual design of the application as appropriate. The colors, which were age-appropriate, were included to increase the motivation of the students. Yılmaz (2008) stated that colors should be used correctly because it affects perception for students with intellectual disabilities. Also, in current study interface design was simple and not distractive. In line with this, Yılmaz (2008) emphasized that especially for students with intellectual disabilities; the interfaces must be in the simplicity that will not distract the student. Besides, too many visual elements were not added for preventing distraction. Similarly, Yılmaz (2008) revealed that irrelevant items should not be used.

In addition to *"Visual Design", "Multimedia Use"* was determined as one of themes that was analyzed from the views of experts and the teacher before the utilization of the mobile writing application. All the participants accepted that multimedia use was appropriate in this study. Thus, it can be inferred from the results that the multimedia used in the application aligns with the content as well as meeting the expectations of the target audience. Two types of animations were used in the application; former aims at making the application enjoyable (the arrow animations for scores) and the latter, facilitating students' learning (second-degree clue). The various use of animations and the positive influence seen on the learning outcomes was also mentioned in the study of Azimi and Mousavipour (2014). Apart from the animations, the background music was also found to be appropriate by the experts. Similarly, Ke and Abras (2013) emphasized that the background music and cartoon characters were grabbed students' attention. In line with this, Takacs, Swart, and Bus (2015) underlined that animations, background music and sound effects which are coherent with the content can facilitate understanding of children with language delay.

The last theme analyzed in the pre-experimentation process was related with the *"Technical Features"* of the mobile application. In this study, the experts analyzed two issues. The first issue was whether it was easy to install. Although their responses fall into the category that represents the ease of use in the downloading process, an expert recommend putting more information for the users about the phases of the download. The researchers informed them that this constraint will be overcome after the application was put on the markets. The second issue was about error-freeness. Experts agreed on the error-freeness of the application. However, they also warned the researchers about the potential risks that could emerge in a long time use in various users. Likewise, Baumgartner and Payr (1996) underlined that educational software should be error-free and user-friendly. So far, the results obtained in the pre-experimentation phase showed that the mobile writing application developed in this study could ready to be used by students with dysgraphia.

CONCLUSION AND FUTURE RESEARCH

The purpose of the present study is to develop a mobile writing application for students with dysgraphia. This study focused specifically on developing writing skills. A writing application was developed for this purpose taking the views of educational technology, classroom education and special education experts into consideration. To sum up, the results of this study revealed that experts' views were very positive about mobile writing application for students with dysgraphia.

The application can help students with dysgraphia not only to write, but to enable them to write legibly and neatly along lines. Students with dysgraphia may face the problem of never being able to write accurately and legibly if their teachers and parents do not give them additional support. In a regular class setting, it is almost impossible to help these students due to the constraints with regards to classroom setting, atmosphere, number of students in a class, teacher competency. These students need to be involved with other students while they are also receiving individual support from their teachers. This becomes extremely difficulty in crowded classrooms, and in most cases, these students seem to get lost in class. Furthermore, having difficulties lead these students to get bored easily with the tendency to avoid writing. This study is expected to bring in an application that will take away some of the barriers that students with dysgraphia encounter in the classrooms and help the teacher to receive additional support for their students. This is because the application developed in this study is expected to allow students to study individually or with little supervision and attract their attention by enriching the learning environments as well as increasing their motivation. In further research, in order to examine effectiveness experimental studies could be administered with students with dysgraphia. Usability studies can be administered as well. New features (audible version of letter, syllabi and words, changeable line space, reminder images of letters and words) could be added to the mobile writing application for future studies. An application can be developed to allow the practitioners to add practices. It can also be used for teaching manuscript handwriting instead of cursive handwriting. The content of mobile writing application can be varied (correct pen grip, line activities, writing sentence, paragraph, adding spelling, and syntax rules). This study brought a perspective on a writing application for student with dysgraphia. In further studies, effectiveness studies on acquisition of writing skills (writing letters, numbers, syllables, and words) could be conducted for students with dysgraphia. These studies could consider on-task behavior, writing speed, correct- incorrect attempts, line violations, and clues in the context of effectiveness. A pen that is more realistic can be developed to use with mobile writing application. This study is expected to bring unique insights for students with dysgraphia and present teachers and experts an alternative method that could be used in the classroom setting. In addition, studies can be conducted that take into account the effectiveness of the application in helping children develop better hand-writing via pen and paper.

Limitations

The limitations of this study could be listed as follows:

- The number of students participated in this study is limited to three in Istanbul.
- The content of mobile application is limited to uppercase and lowercase letters, numbers.
- The duration of the treatment is limited to students' own pace.
- Due to the nature of research method employed, purposeful sampling method was used and it is a limitation for generalization.
- This current study is to develop a mobile writing application for students with dysgraphia.

REFERENCES

Akyol, H. (2005). *Türkçe ilk okuma yazma öğretimi*. Ankara: Pegem A Yayıncılık.

Alberta Learning and Teaching Branch. (2002). *Career and life management guide to implementation.* Edmonton: Alberta Assessment Consortium.

Annett, M., Anderson, F., Bischof, W. F., & Gupta, A. (2014). The pen is mightier: understanding stylus behaviour while inking on tablets. Proceedings of Graphics Interface 2014, 193-200.

APA. (2013). *Diagnostic and Statistical Manual of Mental Disorders* (5th ed.). Washington, DC: APA.

Azimi, E., & Mousavipour, S. (2014). The Effects of Educational Multimedia in Dictation and Its Role in Improving Dysgraphia in Students with Dictation Difficulty. *Contemporary Educational Technology, 5*(4), 331–340.

Baumgartner, P., & Payr, S. (1996). Learning as action: A social science approach to the evaluation of interactive media. *Proceedings of ED-MEDIA 96-World Conference on Educational Multimedia and Hypermedia,* 31-37.

Cahill, S. M. (2009). Where does handwriting fit in? Strategies to support academic achievement. *Intervention in School and Clinic, 44*(4), 223–228. doi:10.1177/1053451208328826

Chang, S. H., & Yu, N. Y. (2014). The effect of computer-assisted therapeutic practice for children with handwriting deficit: A comparison with the effect of the traditional sensorimotor approach. *Research in Developmental Disabilities, 35*(7), 1648–1657. doi:10.1016/j.ridd.2014.03.024 PMID:24770471

Chicu, S. O., Ţicău, A., & Şoitu, L. (2014). Training for new technologies. Handwriting with new technologies. *Procedia: Social and Behavioral Sciences, 142*, 781–785. doi:10.1016/j.sbspro.2014.07.616

Chung, P., & Patel, D. R. (2015). Dysgraphia. *International Journal of Child and Adolescent Health, 8*(1), 27–36.

Crouch, A. L., & Jakubecy, J. J. (2007). Dysgraphia: How It Affects a Student's Performance and What Can Be Done about It. *TEACHING Exceptional Children Plus, 3*(3), n3.

Czyzewski, A., Odya, P., Grabkowska, A., Grabkowski, M., & Kostek, B. (2009, August). *Smart pen: new multimodal computer control tool for dyslexia therapy.* Poster presented at the meeting of SIGGRAPH '09. 10.1145/1599301.1599355

Doğan Temur, Ö. (2015). *İlköğretim okuma yazma öğreniyorum 1 ders kitabi birinci kitap*. Ankara: Yıldırım Yayınları.

Engel-Yeger, B., Nagauker-Yanuv, L., & Rosenblum, S. (2009). Handwriting performance, self-reports, and perceived self-efficacy among children with dysgraphia. *The American Journal of Occupational Therapy*, *63*(2), 182–192. doi:10.5014/ajot.63.2.182 PMID:19432056

Erhardt, R. P., & Meade, V. (2005). Improving handwriting without teaching handwriting: The consultative clinical reasoning process. *Australian Occupational Therapy Journal*, *52*(3), 199–210. doi:10.1111/j.1440-1630.2005.00505.x

Evans, C. (2008). The effectiveness of m-learning in the form of podcast revision lectures in higher education. *Computers & Education*, *50*(2), 491–498. doi:10.1016/j.compedu.2007.09.016

Falk, T. H., Tam, C., Schellnus, H., & Chau, T. (2011). On the development of a computer-based handwriting assessment tool to objectively quantify handwriting proficiency in children. *Computer Methods and Programs in Biomedicine*, *104*(3), e102–e111. doi:10.1016/j.cmpb.2010.12.010 PMID:21376418

Fedora, P. (2015). Integrating technology tools for students struggling with written language. *Kappa Delta Pi Record*, *51*(1), 31–35. doi:10.1080/00228958.2015.988562

Fernández-López, Á., Rodríguez-Fórtiz, M. J., Rodríguez-Almendros, M. L., & Martínez-Segura, M. J. (2013). Mobile learning technology based on iOS devices to support students with special education needs. *Computers & Education*, *61*, 77–90. doi:10.1016/j.compedu.2012.09.014

Friend, M. (2005). *Special Education: Contemporary Perspectives for School Professionals*. Boston: Allyn & Bacon.

Galanis, J. A. (2008). *The effect of the handwriting without tears program on student cursive writing achievement at Central Institute for the Deaf (CID).* Independent studies and capstones, Paper 417. Retrieved from http://digitalcommons.wustl.edu/pacs_capstones/417

Galli, M., Cimolin, V., Stella, G., De Pandis, M. F., Ancillao, A., & Condoluci, C. (2018). Quantitative assessment of drawing tests in children with dyslexia and dysgraphia. *Human Movement Science*. doi:10.1016/j.humov.2018.05.001 PMID:29748041

Giordano, D., & Maiorana, F. (2014). Addressing dysgraphia with a mobile, web-based software with interactive feedback. *Biomedical and Health Informatics (BHI), 2014 IEEE-EMBS International Conference on,* 264-268.10.1109/BHI.2014.6864354

Giordano, D., & Maiorana, F. (2015). A Mobile Web Game Approach for Improving Dysgraphia. In *Proceedings of the 7th International Conference on Computer Supported Education* (vol. 1, pp. 328-333). SCITEPRESS-Science and Technology Publications, Lda. 10.5220/0005449103280333

Graham, S., Harris, K. R., & Fink, B. (2000). Is handwriting causally related to learning to write? Treatment of handwriting problems in beginning writers. *Journal of Educational Psychology*, *92*(4), 620–633. doi:10.1037/0022-0663.92.4.620

Graham, S., Weintraub, N., & Berninger, V. W. (1998). The relationship between handwriting style and speed and legibility. *The Journal of Educational Research*, *91*(5), 290–297. doi:10.1080/00220679809597556

Guinet, E., & Kandel, S. (2010). Ductus: A software package for the study of handwriting production. *Behavior Research Methods*, *42*(1), 326–332. doi:10.3758/BRM.42.1.326 PMID:20160312

Hamstra-Bletz, L., & Blöte, A. W. (1993). A longitudinal study on dysgraphic handwriting in primary school. *Journal of Learning Disabilities*, *26*(10), 689–699. doi:10.1177/002221949302601007 PMID:8151209

Harley, L., Kline, K., Price, C., Jones, A., Mosley, S., Farmer, S., & Fain, B. (2013). A web based voting application study of fonts for voters with dyslexia. *The Information Technology and Innovation Foundation Accessible Voting Technology Initiative Working Paper Series*, 1-30.

Harris, K. R., Graham, S., & Mason, L. H. (2006). Improving the writing, knowledge, and motivation of struggling young writers: Effects of self-regulated strategy development with and without peer support. *American Educational Research Journal*, *43*(2), 295–340. doi:10.3102/00028312043002295

Helps, R., & Helps, C. (2016, September). Measuring stylus and tablet performance for usability in sketching. *Proceedings of the 5th Annual Conference on Research in Information Technology*, 19-24. 10.1145/2978178.2978186

Hennion, B., Gentaz, E., Gouagout, P., & Bara, F. (2005). Telemaque, a new visuo-haptic interface for remediation of dysgraphic children. *Proceedings of WorldHaptic 2005*, 410-419. 10.1109/WHC.2005.125

Jena, S. P. K. (2013). Learning Disability: Theory to Practice. Sage Publication India Pvt. Ltd.

Kagohara, D. M., Meer, L., Ramdoss, S., O'Reilly, M. F., Lancioni, G. E., Davis, T. N., & Sigafoos, J. (2013). Using iPods® and iPads® in teaching programs for individuals with developmental disabilities: A systematic review. *Research in Developmental Disabilities*, *34*(1), 147–156. doi:10.1016/j.ridd.2012.07.027 PMID:22940168

Karlsdottir, R., & Stefansson, T. (2002). Problems in developing functional handwriting. *Perceptual and Motor Skills*, *94*(2), 623–662. doi:10.2466/pms.2002.94.2.623 PMID:12027360

Katusic, S. K., Colligan, R. C., Weaver, A. L., & Barbaresi, W. J. (2009). The forgotten learning disability: Epidemiology of written-language disorder in a population-based birth cohort (1976–1982), Rochester, Minnesota. *Pediatrics*, *123*(5), 1306–1313. doi:10.1542/peds.2008-2098 PMID:19403496

Ke, F., & Abras, T. (2013). Games for engaged learning of middle school children with special learning needs. *British Journal of Educational Technology*, *44*(2), 225–242. doi:10.1111/j.1467-8535.2012.01326.x

Korkmazlar, Ü. (2003). Özel Öğrenme Bozukluğu. In A. Kulaksızoğlu (Ed.), *Farklı Gelişen Çocuklar*. İstanbul: Remzi Kitabevi.

Kushki, A., Schwellnus, H., Ilyas, F., & Chau, T. (2011). Changes in kinetics and kinematics of handwriting during a prolonged writing task in children with and without dysgraphia. *Research in Developmental Disabilities*, *32*(3), 1058–1064. doi:10.1016/j.ridd.2011.01.026 PMID:21315553

Li-Tsang, C. W., Wong, A. S., Leung, H. W., Cheng, J. S., Chiu, B. H., Linda, F. L., & Chung, R. C. (2013). Validation of the Chinese Handwriting Analysis System (CHAS) for primary school students in Hong Kong. *Research in Developmental Disabilities*, *34*(9), 2872–2883. doi:10.1016/j.ridd.2013.05.048 PMID:23816625

MacArthur, C. A. (2009). Reflections on research on writing and technology for struggling writers. *Learning Disabilities Research & Practice*, *24*(2), 93–103. doi:10.1111/j.1540-5826.2009.00283.x

McHale, K., & Cermak, S. A. (1992). Fine motor activities in elementary school: Preliminary findings and provisional implications for children with fine motor problems. *The American Journal of Occupational Therapy*, *46*(10), 898–903. doi:10.5014/ajot.46.10.898 PMID:1463061

McNamara, B. E. (2007). *Learning disabilities: Bridging the gap between research and classroom practice*. Upper Saddle River, NJ: Pearson Merrill Prentice Hall.

Miller, M. (2002). Resilience elements in students with learning disabilities. *Journal of Clinical Psychology*, *58*(3), 291–298. doi:10.1002/jclp.10018 PMID:11836710

MONE. (2005). *Ilköğretim Türkçe dersi öğretim programi ve kilavuzu*. Ankara: Devlet Kitapları Basımevi.

MONE. (2008). *Özel eğitim ve rehabilitasyon merkezi özel öğrenme güçlüğü destek eğitim programı*. Ankara: T.C. Milli Egitim Bakanligi Özel Öğretim Kurumları Genel Müdürlüğü.

MONE. (2015). Türkçe dersi (1-8. Sınıflar) öğretim programı. Ankara: T.C. Milli Egitim Bakanligi, Talim ve Terbiye Kurulu Başkanlığı.

Nicolson, R., & Fawcett, A. (2011). Dyslexia, dysgraphia, procedural learning and the cerebellum. *Cortex*, *47*(1), 117–127. doi:10.1016/j.cortex.2009.08.016 PMID:19818437

Palluel-Germain, R., Bara, F., Hillairet de Boisferon, A., Hennion, B., Gouagout, P., & Gentaz, E. (2007). A Visuo-haptic device – Telemaque - increases kindergarten children's handwriting acquisition. *Proceedings of IEEE World Haptics 2007*, 72–77. doi: 10.1109/WHC.2007.13

Phelps, J., Stempel, L., & Speck, G. (1985). The children's handwriting scale: A new diagnostic tool. *The Journal of Educational Research*, *79*(1), 46–50. doi:10.1080/00220671.1985.10885646

Reid, G., Elbeheri, G., & Everatt, J. (2015). *Assessing children with specific learning difficulties: A teacher's practical guide*. New York, NY: Routledge. doi:10.4324/9781315693873

Reisman, J. E. (1993). Development and reliability of the research version of the Minessota Handwriting Test. *Physical & Occupational Therapy in Pediatrics*, *13*(2), 41–55. doi:10.1080/J006v13n02_03

Richards, R. G. (1998). *The writing dilemma: Understanding dysgraphia*. Riverside, CA: Richards Educational Therapy Center.

Romani, C., Tsouknida, E., di Betta, A. M., & Olson, A. (2011). Reduced attentional capacity, but normal processing speed and shifting of attention in developmental dyslexia: Evidence from a serial task. *Cortex, 47*(6), 715–733. doi:10.1016/j.cortex.2010.05.008 PMID:20655514

Rosenblum, S., Dvorkin, A. Y., & Weiss, P. L. (2006). Automatic segmentation as a tool for examining the handwriting process of children with dysgraphic and proficient handwriting. *Human Movement Science, 25*(4), 608–621. doi:10.1016/j.humov.2006.07.005 PMID:17011656

Rosenblum, S., Weiss, P. L., & Parush, S. (2003). Product and process evaluation of handwriting difficulties. *Educational Psychology Review, 15*(1), 41–81. doi:10.1023/A:1021371425220

Rowe, K. J. (2006). *Effective teaching practices for students with and without learning difficulties: Constructivism as a legitimate theory of learning and of teaching.* Background paper to keynote address presented at the NSW DET Office of Schools Portfolio Forum, Wilkins Gallery, Sydney, Australia. Retrieved from: http://research.acer.edu.au/learning_processes/10

Salih, M. A. M., Abdul-Kahar, R., Zahari, W. A. M. W., Khalid, H. M., & Rahim, K. A. (2015). Strengthening Jawi writing for dyslexia students through online games-analysis of e-Jawi games online in Malaysia. *Journal of Education and Human Development, 4*(2), 147–151. doi:10.15640/jehd.v4n2a17

Smits-Engelsman, B. C., Niemeijer, A. S., & Van Galen, G. P. (2001). Fine motor deficiencies in children diagnosed as DCD based on poor grapho-motor ability. *Human Movement Science, 20*(1), 161–182. doi:10.1016/S0167-9457(01)00033-1 PMID:11471395

Smits-Engelsman, B. C., & Van Galen, G. P. (1997). Dysgraphia in children: Lasting psychomotor deficiency or transient developmental delay? *Journal of Experimental Child Psychology, 67*(2), 164–184. doi:10.1006/jecp.1997.2400 PMID:9388804

Takacs, Z. K., Swart, E. K., & Bus, A. G. (2015). Benefits and pitfalls of multimedia and interactive features in technology-enhanced storybooks: A meta-analysis. *Review of Educational Research, 85*(4), 698–739. doi:10.3102/0034654314566989 PMID:26640299

Tanimoto, S., Thompson, R., Berninger, V. W., Nagy, W., & Abbott, R. D. (2015). Computerized writing and reading instruction for students in grades 4–9 with specific learning disabilities affecting written language. *Journal of Computer Assisted Learning, 31*(6), 671–689. doi:10.1111/jcal.12110 PMID:26858470

Tanyag, R. P., & Atienza, R. O. (2015, September). Implicit palm rejection using real-time hand model filters on tablet devices. *Proceedings of Next Generation Mobile Applications, Services and Technologies, 2015 9th International Conference on,* 347-352. 10.1109/NGMAST.2015.45

van Galen, G. P., & de Jong, W. P. (1995). Fitts' law as the outcome of a dynamic noise filtering model of motor control. *Human Movement Science, 14*(4-5), 539–571. doi:10.1016/0167-9457(95)00027-3

Van Hartingsveldt, M. J., De Groot, I. J. M., Aarts, P. B. M., & Nijhuis-Van Der Sanden, M. W. G. (2011). Standardized tests of handwriting readiness: A systematic review of the literature. *Developmental Medicine and Child Neurology, 53*(6), 506–515. doi:10.1111/j.1469-8749.2010.03895.x PMID:21309763

Williams, C. (2014). *Dysgraphia.* Retrieved from https://wiki.rit.edu/display/msse704012135/Dysgraphia

Yiğiter, S. (2005). *Sınıf öğretmenlerinin özel öğrenme güçlüğüne ilişkin bilgi düzeyleri ile özel öğrenme güçlüğü olan çocuklarin kaynaştirilmasina yönelik tutumlari arasindaki ilişkinin incelenmesi* (Unpublished master's thesis). Marmara University, İstanbul, Turkey.

Yılmaz, S. (2008). *Zihinsel engelli öğrencilerin kavram öğretiminde kullanılan etkileşimli eğitim CD'lerinin görsel tasarım ilkelerine ve içeriğe uyumu* (Unpublished master's thesis). Ondokuzmayıs University, Samsun, Turkey.

Zhang, Y. (2000). Technology and the writing skills of students with learning disabilities. *Journal of Research on Computing in Education*, *32*(4), 467–478. doi:10.1080/08886504.2000.10782292

ADDITIONAL READING

Arndt, P. A. (2016). Computer usage for learning how to read and write in primary school. *Trends in Neuroscience and Education*, *5*(3), 90–98. doi:10.1016/j.tine.2016.07.003

Bara, F., & Gentaz, E. (2011). Haptics in teaching handwriting: The role of perceptual and visuo-motor skills. *Human Movement Science*, *30*(4), 745–759. doi:10.1016/j.humov.2010.05.015 PMID:21272948

Danna, J., & Velay, J. L. (2015). Basic and supplementary sensory feedback in handwriting. *Frontiers in Psychology*, *6*, 169. doi:10.3389/fpsyg.2015.00169 PMID:25750633

Giordano, D., & Maiorana, F. (2014, June). Addressing dysgraphia with a mobile, web-based software with interactive feedback. In *Biomedical and Health Informatics (BHI), 2014 IEEE-EMBS International Conference on* (pp. 264-268). IEEE. 10.1109/BHI.2014.6864354

Kim, Y. S., Collins, M., Bulmer, W., Sharma, S., & Mayrose, J. (2013, April). Haptics Assisted Training (HAT) System for children's handwriting. In *World Haptics Conference (WHC)*, 2013 (pp. 559-564). IEEE.

Mann, A. M., Hinrichs, U., & Quigley, A. (2015). Digital pen technology's suitability to support handwriting learning. In *The Impact of Pen and Touch Technology on Education* (pp. 7–22). Cham: Springer. doi:10.1007/978-3-319-15594-4_2

Mansour, M., Eid, M., & Saddik, A. E. (2007). A multimedia handwriting learning and evaluation tool. *Proc. Intelligent Interactive Learning Object Repositories (I2LOR).*

Palluel-Germain, R., Bara, F., De Boisferon, A. H., Hennion, B., Gouagout, P., & Gentaz, E. (2007, March). A visuo-haptic device-telemaque-increases kindergarten children's handwriting acquisition. In *EuroHaptics Conference, 2007 and Symposium on Haptic Interfaces for Virtual Environment and Teleoperator Systems. World Haptics 2007. Second Joint* (pp. 72-77). IEEE.

Patchan, M. M., & Puranik, C. S. (2016). Using tablet computers to teach preschool children to write letters: Exploring the impact of extrinsic and intrinsic feedback. *Computers & Education*, *102*, 128–137. doi:10.1016/j.compedu.2016.07.007

Thompson, R., Tanimoto, S., Berninger, V., & Nagy, W. (2016). Design Studies for Stylus and Finger-Based Interaction in Writing Instruction on Tablets. In *Revolutionizing education with digital ink* (pp. 51–69). Cham: Springer. doi:10.1007/978-3-319-31193-7_4

KEY TERMS AND DEFINITIONS

Dysgraphia: A disorder of written expression. It affects writing skills of individuals.

Gesture Recognition Algorithm: An algorithm recognize handwriting based on individuals' gestures.

Line Violation: Overflowing a line or not being able to follow a line while writing.

Mobile Writing Application: A mobile application which provides students writing activities.

Registration Screen: Screen where students enter their personal information and register to a system.

Specific Learning Disability (SLD): A difficulty in a specific area such as arithmetic, writing, and reading.

Trial Screen: A screen where students practice to become familiar with mobile application.

Chapter 13
Using Tablet Applications as Assistive Tools in Teaching English as a Foreign Language

Tami Seifert
Kibutzim College of Education, Israel

Veronika Zimon
Kibbutzim College of Education, Israel

ABSTRACT

A case study examined the contribution of tablet applications for English language acquisition and improvement of reading and reading comprehension skills among pupils diagnosed with learning disabilities and those not diagnosed with learning disabilities. Participants were Grade 4 pupils from a public school in central Israel in two groups. Group 1 included two boys and two girls diagnosed with learning disabilities; Group 2 included two boys and two girls not diagnosed with learning disabilities. English learning applications were used with the intent of improving the abilities of letter recognition, reading, reading comprehension, listening comprehension, and oral expression in English in a fun way relevant to the pupils' lives. Findings indicated a more pronounced improvement in English reading and reading comprehension skills among pupils who were diagnosed with learning disabilities. The research also confirms that the use of tablet applications resulted in increased motivation for learning.

DOI: 10.4018/978-1-5225-8106-2.ch013

INTRODUCTION

Those who teach English, often encounter pupils that find it difficult to cope with the acquisition of this language and with the development of English comprehension skills. This is especially evident in pupils diagnosed with learning disabilities. Because of the large number of these pupils, it is important to discover teaching means for this language that could assist their learning, create a successful experience for pupils, and improve their motivation for learning English. Research that examined the contribution of tablet-assisted interventions found that they were able to provide positive contributions (Seifert, 2015). However comprehensive review in this field reveals that a large proportion of these studies lacked methodological rigor in group design studies (Kim, Park & Coleman, 2017).

The present study examined the contribution of tablet applications for the acquisition of the English language and improvement of reading and comprehension skills in two groups of Grade 4 pupils: a group diagnosed with learning disabilities in comparison with a group who were not diagnosed with such disabilities. In order to examine the effectiveness of the tablet applications, use was made of applications in the area of English language studies. These applications are intended to improve the level of pupils' motivation for learning, and have been proven to influence the pupils' decision-making processes and determine their level of progress, and the extent of guidance and effort that will be invested in the learning activities (Cole, Field & Harris, 2004).

In order to examine the effectiveness of tablet applications, lesson plans were prepared combining applications for recital, recording and listening in English, games to exercise vocabulary etc.

Assistive Technology

Assistive technology (hereinafter: AT) is defined as a wide range of strategies, technologies and equipment that enable pupils with special needs to work in the field of their challenge (Hopkins, 2004). The apparatus for AT includes any item, equipment or system that is purchased, prepared or improved in order to help to improve, enlarge or maintain the functional abilities of individuals with special needs (Bausch & Alt, 2008). AT has been supported by legislation and science and thus has become a firm foundation for the process of learning and education for pupils with disabilities and special needs so that they can take part in and complete tasks which are difficult to do by themselves (Lee & Vega, 2005).

Research has found that AT increases the independence and quality of life for people with disabilities and even contributes much to their inclusion in society since they acquire means for communication, mobility, creativity, self-confidence etc. (Yeunjoo & Vega, 2005). The use of AT enables those with special needs to function more effectively under various conditions, and can be adapted to assist various personal needs, helping people with hearing disorders and restricted mobility abilities. Additionally, the use of AT contributes to pupils since it provides access to information, develops skills and assists pupils' progress, competitiveness and effectiveness, specialization, self-worth and self-reliance, peace of mind and social acceptance (Hopkins, 2004).

Tablets have even been found to assist pupils with sensory disorders and reading and writing difficulties and improve social skills, organizational and interpersonal communication skills among children (Suta & Hayhoe, 2013). Tablet-assisted learning is adapted to the characteristics of the younger generation or "digital natives", who have grown up in a technological environment and they enable mobility and adaptation to the pupils' life style (Naismith, Lonsdale, Vavoula & Sharples, 2006). The integration of this technology enables the education system to provide a relevant and authentic response to the world of youth, who are able to learn according to the language, terminology and perceptions of their own world and their activities within that world (Prensky, 2001). The tablet allows the pupils to personalize applications (Al-Okaily, 2015) and they constitute a means for the empowerment for pupils with learning difficulties, for whom they provide a sense of encouragement and hope. A tablet enables practical experience and active thinking during the construction and acquisition of knowledge. These possibilities provide special needs pupils with a meaningful experience of the "real" world (Akpan & Beard, 2013).

Pupils With Learning Disabilities and Studies of English as a Foreign Language

"Learning disability" is a term that describes a varied group of disorders stemming from neurological brain defects (Siegel, 2018). The intellectual ability of pupils with learning disabilities is usually normal. These pupils mostly endure difficulties in their learning functions such as linguistic, thinking, and perceptive disabilities, difficulties in orientation in time and space, memory, concentration and focus, motor functions, perceptual-movement and organization.

The most common types of learning disabilities include:

1. **Dyslexia:** Is mainly characterized by a difficulty using and manipulating speech sounds and thus causes trouble with language skills.
2. **Dysgraphia:** Causes problems forming letters, writing legibly and organizing thoughts on paper.
3. **Dyscalculia:** A math learning disability that causes difficulty in understanding arithmetic concepts.
4. **Aphasia:** A language disorder referring to problems understanding spoken language (Franz, C., Ascherma, L.& Shaftel. J., 2018)

Learning disabilities influence linguistic skills, so that pupils with learning disabilities, trying to learn English as a Foreign Language (hereinafter: EFL) will encounter difficulties relating to the use and mastery of the English language. Moreover, if a pupil has difficulties with his primary language (L1), these difficulties will also appear when acquiring EFL. Often learning disabilities influence the pupil to a slight degree with regard to his primary language but may have greater influence on the English language. This could occur for several reasons: (1) the spelling in the primary language is simple and very similar to the spoken word, in contrast to English there are strong differences between the ways in which the words are pronounced and written (2) the primary language has a poorer vocabulary than English does, which means that words are repeated more in reading (3) a pupil learns to use techniques to compensate in the familiar environment in which they learn to speak, read and write (4) there are languages that do not demand so much visual attention during reading and spelling as the English language (Schwartz, 2002). Learning disabilities are liable to influence audial and visual memory in the short and long term, the understanding of abstract concepts, the perception of temporal terms, writing, spelling and all other aspects of expression and absorption of information. All these functions are essential for the learning process and the acquisition of a foreign language and a disability in any of these functions can hinder the acquisition of that language (Schwartz, 2002).

Motivation for Learning

Children with learning disabilities are mostly characterized as having low motivation for learning. This is due to the low achievements that these children experience throughout their lives. According to the model of mutual influences an early self-perception influences the child's following achievements and prior achievements influence his following self-perception. Different studies support the fact that prior academic achievements are crucial in an academic self-perception, other studies show that the academic self-perception has a major role in academic outcomes (Hadar-Peker, 2013). Many of the children develop a low sense of importance and lose their motivation and desire to prove their abilities and efficacy (Mohan, 2009). Language learning motivation relates to "the effort that learners put into learning an L2 as a result of their need or desire to learn it" (Ellis, 2008, p. 972) and is a crucial variable in language learning (Churchill & DuFon, 2006; Dörnyei, 2009; Ellis, 2008; Willis Allen, 2010).

Most of the teachers teaching languages believe that motivation is a key factor for success in language learning. Many studies have indicated that there are two types of motivation for learning a language: integrative motivation, defined as an individual's desire to integrate within a target culture and instrumental motivation, defined as the desire to learn a language for a distinct purpose, such as employment. The importance of integrative motivation for the learning of EFL has received international interest and has become a major issue for research. At the end of the 1980s and the beginning of the 1990s its importance wanted and the major attention shifted to differences between pupils studying English as a Second Language (hereinafter ESL) living within the culture of the target language and EFL pupils, living within their own mother culture. Researchers indicate that in the context of EFL pupils, who have not experienced the target community's language, motivational factors such as instrumental motivation need to receive special attention and emphasis in the teaching process (Kimura, Nakata & Okmura, 2008).

Research Questions

1. What is the contribution of the use of tablet applications to the improvement of English language skills in the fields of reading and comprehension among pupils diagnosed with learning disorders in comparison to pupils, who have not been diagnosed with learning disorders?
2. What is the contribution of the use of tablet applications to increasing motivation to the learning of the English language among pupils diagnosed with learning disorders in comparison to pupils, who have not been diagnosed with learning disorders?
3. To what extent have the practices of technological means use for learning purposes by pupils diagnosed with learning disorders altered as a result of their experience of work with a tablet, in comparison to pupils, who have not been diagnosed with learning disorders?

METHODOLOGY

Research Population and Sampling Method

Eight pupils from Grade 4 in a government school in the center of Israel participated in the research. Four of the pupils had been diagnosed with learning disabilities (two male and two female); and four of the pupils had not been diagnosed with learning disabilities (two male and two female). All the children are integrated within mainstream classes and were found suitable to participate in the research within the framework of individual lessons in school. The pupils were selected according to their achievements in mapping tests, as all having lower than expected achievements.

The Research Environment

The sessions took place in the school where the pupils study, in a room specially assigned for individual lessons. The tablets were retained by the researcher and brought by her to each session. Each child in both groups was given his own tablet that accompanied them throughout the research sessions, which they shared with a child from the other group, and personal earplugs.

The content of the sessions was constructed in gradual steps, in order to begin from the most basic learning materials and gradually increase the level of difficulty. The first sessions dealt with reading and learning vocabulary from a basic level up to a more progressive level. The next sessions dealt with reading, comprehension and composition of sentences. The last sessions dealt with reading and comprehension of more complex reading texts composed of several sentences and finally reading short stories. In all, 30 different applications were used in these sessions. The following free relevant applications were chosen for the subject of the research according to pedagogical considerations: Vocabulary Builder 4, Drive & Learn, Kids Word Match HD, Kids Picture Dictionary, Vocabulary Spelling City, EF High Flyers Vocabulary Game, Flashcards+, Sight Words 1, ABC First Phonics - Word Families, Complete The Word, Play Sight Words, ABC Phonics Sight Words, Save the Monkey, Kinderspell, Sight Word Sentences, Sentence Maker, Sentence Builder, Make Sentences, Teach Me Sentences, Sight Words 2, Read and Record- Rhymes, Monster Town, Read Along, Elephant Bath, Dog Story, 1st Grade I Like Writing, Animal Farm Tale - My First Numbers, Ebookbox, Five Little Monkeys Jumping on the Bed, and Scooby Doo.

Research Tools

The research tools were adapted for a case study in order to broaden the knowledge base: (1) the participants' preliminary knowledge in terms of comprehension and reading in English (2) participants' preliminary knowledge of technology (3) characterizing the participants according to their level of motivation for learning (4) the experiential learning process and assimilation of the learning material (5) in each pupil's reflective process concerning their feelings as a result of the session.

The following research tools assisted the data collection in order to examine the above-mentioned components:

1. A test in writing to map the pupils' comprehension abilities using an excerpt of reading text suitable for their age, which was accompanied by comprehension questions.
2. An oral test to gauge the pupil's reading ability. This type of reading test was taken from a store of informal reading tests published by the Florida Department of Education (2003).
3. A questionnaire to test motivation for learning based on a motivation questionnaire developed by Pintrich and De Groot (1990). The statements which relate to exam tension in the questionnaire were composed in an opposite manner and so in order to present the data in a uniform manner it was necessary to reverse the grading order so that although the pupils, who were diagnosed with a learning disorder had a lower grade for exam tension, this testified to a higher exam tension.
4. A questionnaire examining use and mastery of technology (adapted from Seifert, 2015).
5. The pupils kept journals to document their experiences and learning at the end of each lesson.
6. A follow-up chart to document the individual sessions.

Research Procedure

Eight pupils from Grade 4 from a school in central Israel participated in the research. The pupils were divided into two groups. One group included four pupils – two boys and two girls who were diagnosed with learning disabilities. The second group also included four pupils – two boys and two girls who were not diagnosed with learning disabilities. First each participant received an explanation of the purpose and substance of the research and its procedure. Then each participant underwent a "pre-test" in writing to map their abilities in the fields of reading comprehension by reading an age-appropriate text and answering questions, performing an oral reading test, and then filling in a motivation for learning questionnaire and a questionnaire on their use and mastery of technology.

The research was conducted over five months during which each group met with the researcher once a week. All the pupils received comprehensive instructions concerning the way in which to perform the tests and how to fill in the questionnaires. Later, each group had a familiarization lesson with the tablet. In this lesson they were shown the different functions that the tablet offers, the way in which to use it and operate it and they were also shown the applications that they would use during the research sessions. It was also clarified to the pupils that all the sessions would be recorded to enable follow-up and documentation of the research. During the research process itself there were weekly sessions with the pupils which were recorded by the researcher. The pupils were asked to keep a diary in which they recorded their experiences and what they learned. This diary was managed through a "notes" application on the tablet and filled in by the pupils at the end of each session. In addition to this documentation, the research completed an individual follow-up chart for the session at the end of the meeting.

All the sessions included teaching performed by the researcher and the independent work of the pupils with the applications chosen for that session. At the end of the sessions each of the pupils filled in a "post-test" in writing to map their abilities in reading comprehension through the reading of an age-appropriate text and responded to questions to test their comprehension. They also filled in a learning motivation questionnaire and a questionnaire to measure their use and mastery of technology and performed an oral reading test. Children with a history of oral language impairment are more likely to present with reading difficulties than their peers. Some research identified this increased likelihood to be as great as four to five times more likely than their peers. It has been shown that children who struggle with phonemic awareness have difficulties acquiring phonic word-attack strategies. There is also evidence that a child's level of vocabulary improved reading development. A recent study of reading comprehension found that both reading accuracy as well as oral language skills, beyond just vocabulary, predict performance on outcome measures Evidence exists linking oral language to the word recognition aspects of reading and/or the comprehension aspects of the reading model (Brooke, 2017).

All the data collected were recorded with the relevant tools. The rights of the pupils to privacy and anonymity were maintained in accord with the researcher's college's and the school's regulations.

Data Analysis Method

The research was a qualitative case study also including quantitative analysis. Case study is defined in different ways by different scholars, but they all agree that a case study enables the researcher to learn about human behavior and about the processes taking place in the studied case. In a case study the researcher aims to investigate human behavior, which in the main has fixed patterns and from which it is possible to make broader generalizations, beyond the specific case (Yin, 2017). This method emphasizes processes and does not focus on products.

The disadvantage of this method is that it only permits a low level of inference from one case to another, so that broad social generalizations and future predictions are limited. Also, since each study relates to a specific case and its particular conditions, it is difficult to draw generalizations concerning a broader social perspective. Nevertheless, the conclusions from a case study can serve as a starting point for more accurate observations, in order to conduct an argument or social discussion on the studied issue (Yin, 2017).

The data collected from the different tools, were processed according to both qualitative and quantitative analysis. Quantitative analysis was used for the calculation of means in order to process data from the motivation questionnaire, the comprehension test and the oral reading test. Qualitative analysis was used to analyze content of the diaries written by the pupils during the sessions.

FINDINGS

Following the use of the tablet applications, there was an improvement in English language skills, in the fields of reading and comprehension for most of the pupils. This is illustrated in Figure 1.

The data shown in Figure 1 show that 50% of the pupils improved their grade in the reading comprehension test. The relative improvement of the pupils not diagnosed with learning disability is shown in Figure 2.

The data in Figure 2 show that the relative improvement in the grades of pupils not diagnosed with learning disabilities in the reading comprehension test ranged from 0 to 127%, with a mean improve-

Figure 1. Grades for pupils who were not diagnosed with learning difficulties in the reading comprehension test

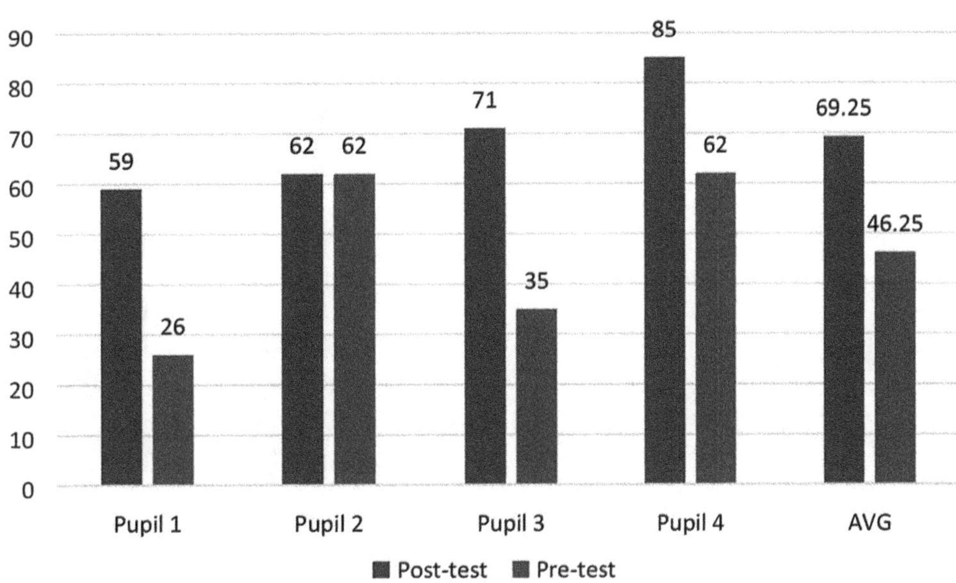

Figure 2. Percentage of improvement for pupils who were diagnosed with learning difficulties in the reading comprehension test

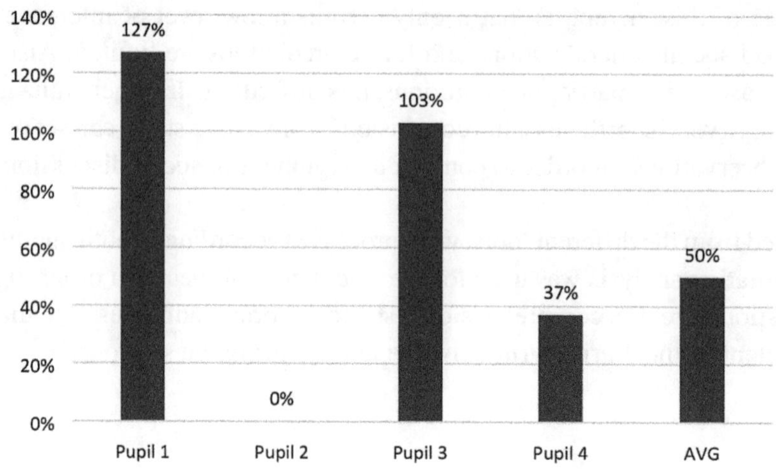

ment of 50%. Also, the number of words read accurately per minute was calculated for these pupils. The results of this test appear in Figure 3.

The data shown in Figure 3 indicate that there was an improvement in the average number of words read accurately per minute. Analysis of the relative mean of words read accurately per minute among the pupils diagnosed with learning disabilities appears in Figure 4.

As can be seen from Figure 4, the data indicate a relative improvement of between 53% for Pupil 4 to 139% for Pupil 3. The mean relative improvement for all the pupils was 85%. Similar tests were conducted for the pupils diagnosed with learning disabilities. The grades for reading comprehension and mean words read accurately per minute for the pupils diagnosed with learning difficulties appear in Figure 5.

Figure 3. Mean words read accurately per minute for pupils not diagnosed with reading disabilities

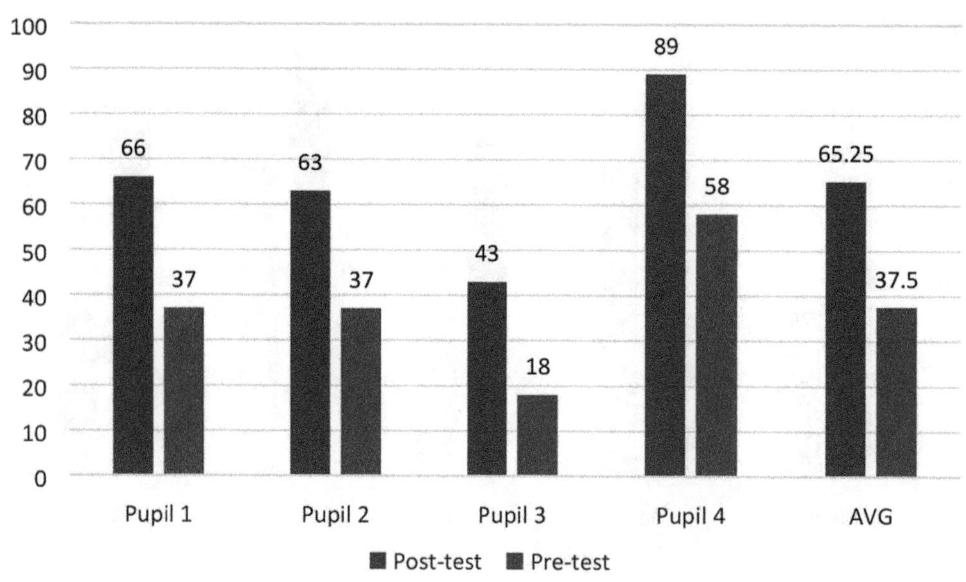

Figure 4. The relative improvement in mean number of words read accurately per minute among pupils not diagnosed with learning disabilities

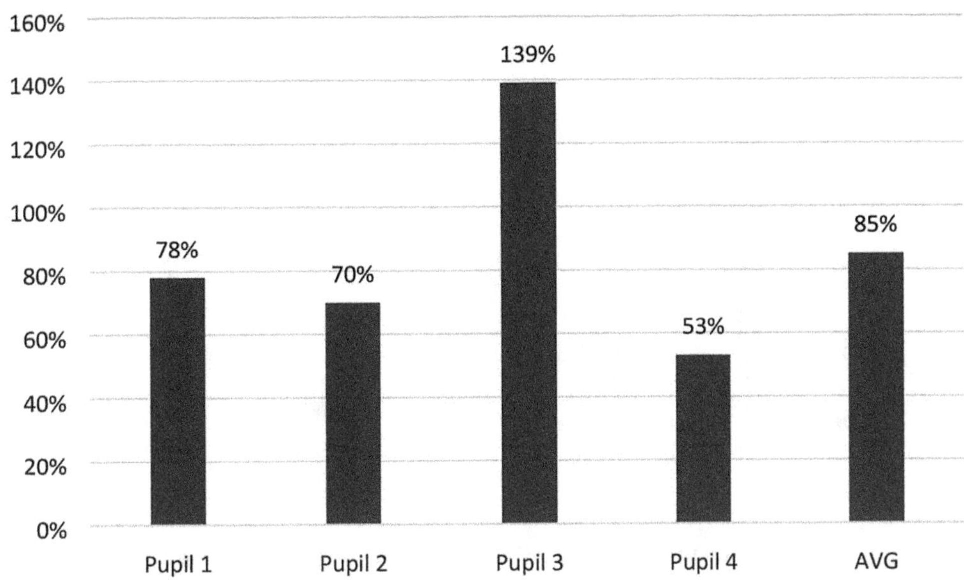

The data shown in Figure 5 indicate that there was an improvement in the post-test in comparison to the pre-test. The relative improvement is shown in Figure 6.

The data shown in Figure 6 indicate that there was a relative improvement in the grades for pupils diagnosed with learning disabilities in the reading comprehension test that ranged between 14% for pupil 8 to 256% for pupil 5. The pupils' mean improvement was 103%. The mean number of words read

Figure 5. Grades for reading comprehension and mean words read accurately per minute for the pupils diagnosed with learning difficulties

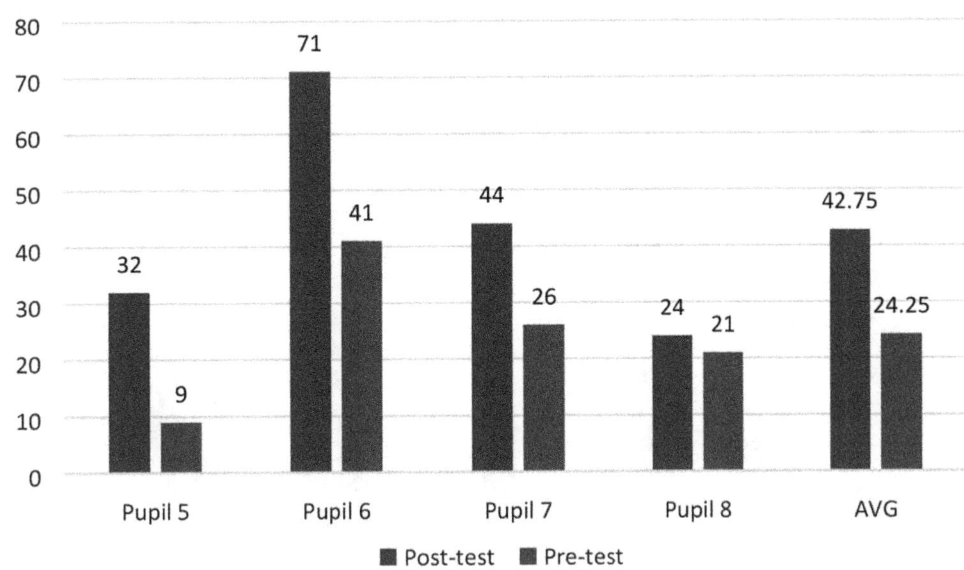

Figure 6. The relative improvement in grades for pupils diagnosed with learning disabilities in the reading comprehension test

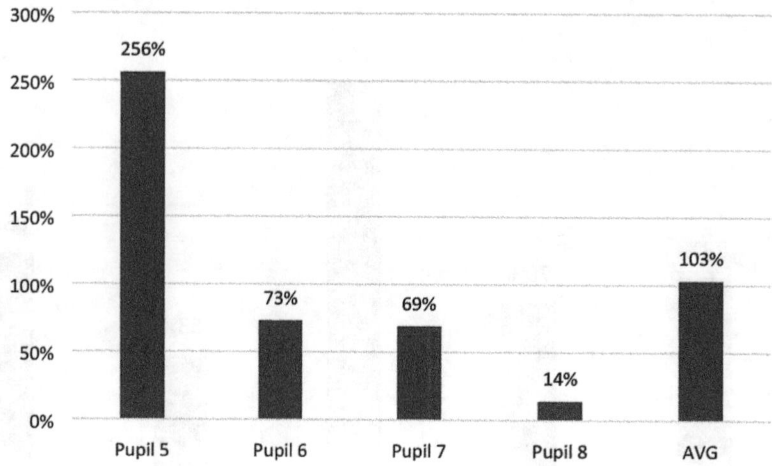

accurately per minute was also tested for pupils diagnosed with learning disabilities. The results appear in Figure 7.

The data shown in Figure 7 indicate that there was an improvement for all the pupils diagnosed with reading disabilities in the post-test in comparison to the pre-test. The relative improvement in the mean words read accurately per minute for these pupils can be seen in Figure 8.

As can be seen from Figure 8, the relative improvement in grades for pupils diagnosed with learning disabilities for mean number of words read accurately per minute ranged between 43% for Pupil 6 to 283% for Pupil 7. In order to investigate the findings from the learning motivation questionnaire between pupils with learning disabilities and pupils without learning disabilities, a comparison was drawn between the two groups, the results appear in Table 1.

Figure 7. Mean number of words read accurately per minute among pupils diagnosed with reading disabilities

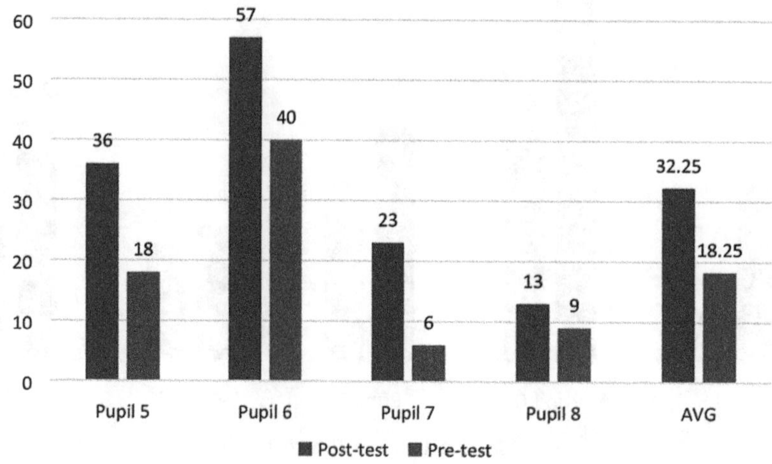

Figure 8. The relative improvement in mean number of words read accurately per minute for pupils diagnosed with learning disabilities

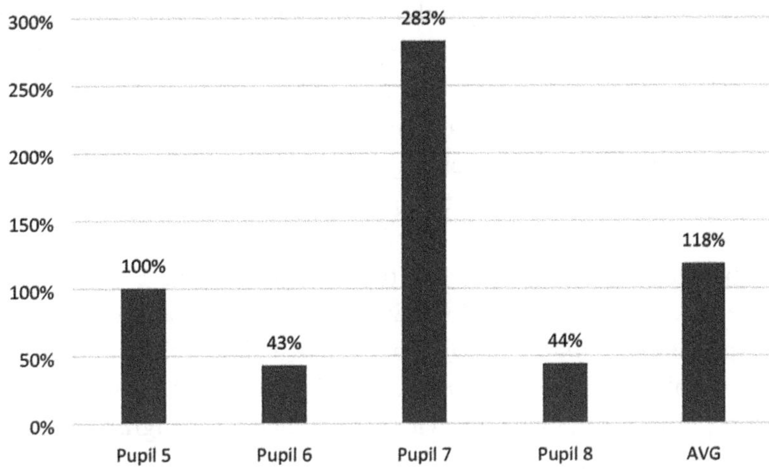

According to the data in Table 1, it is clear that there is a decrease in the sense of self-efficacy among both groups; the group not diagnosed with learning disabilities exhibited a decrease from 4.2 to 3.8, a decrease of 0.4; the group diagnosed with learning disabilities exhibited a decrease from 4.1 to 3.6, a decrease of 0.5. There is also an obvious decrease in intrinsic value for both groups; a decrease of 0.3 -

Table 1. Findings from the motivation for learning questionnaire for the two groups of pupils

Group	Self-Efficacy		Intrinsic Value		Test Anxiety		Total	
	Pre-Test	Post-Test	Pre-Test	Post-Test	Pre-Test	Post-Test	Pre-Test	Post-Test
Pupils not diagnosed with learning disability (M)	4.2	3.8	4.4	4.1	4.3	3.8	4.3	3.9
Extent of change between pre- and post-tests	-0.4		-0.3		-0.5		-0.4	
Pupils diagnosed with learning disability (M)	4.1	3.6	4.2	3.6	4.0	2.1	4.1	3.3
Extent of change between pre- and post-tests	-0.5		-0.6		-1.9		-0.8	

from 4.4 to 4.1 - in the group not diagnosed with learning disabilities and a decrease of 0.6 – from 4.2 to 3.6 - in the group diagnosed with learning disabilities. There was also a decrease in the sense of test anxiety: a decrease of 0.5 – from 4.3 to 3.8 – in the group not diagnosed with learning disabilities, and a decrease of 1.9 – from 4.0 to 2.1 – in the group diagnosed with learning disabilities. There was also a decrease in the sense of general motivation for learning in both groups: 0.4 in the group not diagnosed with learning disabilities and 0.8 in the group diagnosed with learning disabilities.

It is important to note that the findings that emerged from analysis of the pupils' diaries, there appears to be an opposite picture: among the pupils not diagnosed with learning disabilities there were expressions in the summary that was written at the end of the research that perversely indicated an increase in motivation for English language learning: " the tablet made me love English more", " I loved most of the software that we used and it made me love English more", "I want to continue with it next year. There were similar expressions of higher motivation for English learning among the group diagnosed with learning disabilities following the use of the tablet applications: "I want to continue to use it next year", "of all the lessons, I loved it being an individual lesson and the English lesson", "it was really fun to learn to invest an effort and practice in the lessons".

At the last stage of filling in the questionnaire and performance of the tests, the pupils filled in a questionnaire to test their level of use and mastery of technology. In order to produce a picture of the state of their use practices with technology (computer, mobile computer, Smartphone and tablet) and the extent to which they use that technology for learning purposes. The data indicate that half of the children in the group not diagnosed with learning disabilities used their desktop PCs for learning purposes to a reasonable extent, one pupil did not use the PC at all for learning purposes and another did this to a large extent. In contrast in the group diagnosed with learning disabilities, two pupils used their PC to a very large extent, one pupil used it to a large extent and another to a slight extent for learning purposes. Only two of the pupils in the group not diagnosed with learning disabilities possessed a mobile computer, while one of them used it for learning purposes to a large extent, the other did not do so at all. Thus too, in the group diagnosed with learning disabilities, only two of the pupils possessed a mobile computer but they both used it to a very large extent for learning purposes. It was also clear that the group not diagnosed with learning disabilities, three of the pupils had a Smartphone, while one of them did not use it at all for learning and two others used it to a slight extent for learning. In contrast in the group diagnosed with learning disabilities, only two of the pupils had a Smartphone and they used it to a very large extent for learning purposes. Another interesting finding that emerged from the analysis is that none of the pupils in the group not diagnosed with learning disabilities had a tablet and only two of the pupils in the group diagnosed with learning disabilities had a tablet and they used it to a very large extent for learning purposes.

SUMMARY, CONCLUSION AND RECOMMENDATIONS

The unique contribution of this study stems from the fact that it investigated the use of tablet applications among two groups of pupils: pupils diagnosed with learning disabilities and pupils not diagnosed with those disabilities. The research investigated the contribution of use of the tablet applications in two fields: contribution of the tablet applications' use to an improvement in the pupils' English language skills in the areas of reading and reading comprehension and the contribution of these applications to an increase in the pupils' motivation to learn. The research also investigated whether there was any change in the pupils' habits of use of technological means for learning purposes as a result of their work with the tablet applications.

Acquisition of the English language is a complex process for most of the pupils studying EFL, influenced by many variables, including socio-cultural, psychological, cognitive, educational and other background variables such as: age, motivation, socio-economic status, intelligence, teaching methods etc. The process is even more complex and difficult for pupils diagnosed with learning disabilities. Since learning disabilities influence their language skills, the pupil diagnosed with learning disabilities will encounter difficulties in learning, using and mastering the English language. The findings that emerged from the analysis permit several conclusions in the following areas:

1. Contribution to use of tablet applications for the improvement of English language skills in the fields of reading and reading comprehension.

It appears from the findings that the use of tablet applications led to an improvement in English language skills in reading and reading comprehension among pupils diagnosed with learning disabilities in comparison with pupils who had no such diagnosis. The most outstanding improvement was in reading comprehension.

Tablets constitute efficient empowering tools for pupils diagnosed with learning disabilities, allowing them to have practical experience, and active thinking during the process of construction and acquisition of knowledge, and provide them with meaningful experiences from the real world (Beard & Akpan, 2013).

The pupils diagnosed with learning disabilities improved their skills in English reading and reading comprehension. They succeeded in overcoming some of the difficulties that they face due to their learning disabilities, making their achievements even more impressive. Any improvement, however small, is a significant achievement for the child with learning disabilities since it allows them to have a taste of success and progress. Pupil 8 is an example of the importance of such improvement, even if small, for the pupil's experience. He showed the lowest trend to improvement in the reading and reading comprehension tests in comparison to the three other participants in the group diagnosed with learning disabilities: an increase of 3 points in his grade for reading comprehension and an increase of 4 words in the number of words read accurately per minute. From a research point of view this is not seen as a significant improvement, but in the pupil's personal experience, he reported that he felt an improvement and this can be seen in the summary that he wrote in his diary. His words clearly indicate that the he felt a sense of success and improvement in his use of English, although the quantitative data indicated only a slight improvement.

2. Contribution of the use of the tablet to increase motivation to study

The findings indicate that the group diagnosed with learning disabilities had less motivation to study before the research intervention. The pupils diagnosed with learning disabilities have a stronger sense of test anxiety, their sense of self value is lower and as is their sense of self-efficacy in comparison to the group of pupils not diagnosed with learning disabilities. When the data were analyzed, a contradiction was revealed concerning the issue of motivation for studies of pupils in the two groups between the qualitative data collected during the research and the quantitative data collected pre- and post-research. While the qualitative data do not testify to an increase in motivation for studying, the qualitative data indicate an increase in motivation for studying. The reason for this contradiction is apparently pinned in the timing of the administration of the questionnaire: the questionnaire was administered to the pupils one week after the tablets were taken away from them and the individual sessions ended. Thus, because of the timing of the administration of the motivation for studies questionnaire, the quantitative data which were collected after the research were biased. If the questionnaires were administered to the pupils, pre-research, during the research and before its end, before taking away the tables, the quantitative data could have a different appearance. Thus, instead of administering the motivation questionnaires pre- and post-research and after the tablets were no longer available to the pupils, our recommendation for further studies of this kind, is to administer motivation for studying questionnaires at three time points: pre-research, during the research and post-research, but before the participants know about its conclusion. Taking away the tablets and concluding the individual sessions and the return to conventional studies was found to have a substantial influence on the students' motivation to study the English language. Frontal English lessons are substantially different from the individual lessons that the pupils experienced with their tablets. Thus, the pupils' knowledge of the completion of those individual lessons and the end of their work with the tablets apparently led to a sense of "disappointment" and a drop in motivation because the process had ended and they returned to traditional teaching methods in the classroom.

According to the qualitative findings, as they appeared in the pupils' summarizing comments in their diaries, it can be seen that the pupils not diagnosed with learning disabilities had higher motivation for English language studies. A pupil, who performs learning activities and finds them satisfying and pleasurable will according to previous literature be motivated in their studying (Adar, 1969). In light of these findings it appears that in contradiction to the quantitative findings, the qualitative findings do indeed testify that the use of tablet applications led to an increase of motivation for studying for pupils who had not been diagnosed with learning disabilities

Thus too, among the pupils diagnosed with learning disabilities it was possible to see expressions that testified to an increase in motivation for studying following the research intervention. According to the literature, motivational processes play a central role in the creation of pupils' feelings towards school and their studies (Kepel & Ashur, 2001). Thus, a pupil, who had serious difficulties in learning the English language but enjoyed school, and had individual, personalized lessons would certainly have higher motivation for studies. The pupils participating in the research exhibited a will to continue improving their English language skills and to progress. Pupils who set goals for themselves to advance are pupils who want to succeed in their learning, and it can be said that they have strong motivation to learn.

The impression over the research was that there was a more increase in motivation to learn among the pupils diagnosed with learning disabilities, because of the joy and excitement that these pupils expressed each time that they succeeded in completing an English assignment successfully. This was a feeling that was quite foreign for most of them that they had not experienced in frontal English lessons until then. They expressed their satisfaction concerning the progress that they felt they achieved from lesson to lesson and following the entire process, with expressions such as "it was fun for me to progress today in English". These successes are very meaningful for the pupils in this group who have difficulties in acquiring the English longlist and any advance that they make and feel, even the smallest, is a whole world for them. Furthermore, they expressed their desire to continue to use tablets to study English in the coming year. Three out of the four pupils diagnosed with learning disabilities noted this unequivocally in their concluding remarks in their diaries and the participants in this group also told this to their teacher throughout the period of the research. In contrast the group of pupils not diagnosed with learning disabilities expressed less of these desires and they did not express them in their concluding remarks in their diaries. Thus, the research results indicate that the increase in motivation to study was greater for the group of pupils diagnosed with learning disabilities. Most of the language teachers think that motivation is the key factor for success in learning a language (Kimura, Nakata & Okumura, 2008) and so these findings are very important both for teachers teaching in special education and also among teachers teaching in mainstream education and especially for their pupils.

3. Using technological means for learning purposes

Most substantial changes took place in the group diagnosed with learning disabilities through the use of technological means for learning purposes as a result of their work with the tablet; the change was observed in the use of each of the technological means that were tested (computer, mobile computer, Smartphone) apart from the tablet which was used to a very large extent for learning purposes both before the research and during the research.

One possible explanation for this alteration in the use of mobile computers and Smartphones by pupils diagnosed with learning disabilities is the function of mobility close to the body that characterizes these tools. The mobile computer and Smartphone are defined as mobile technology and these tools can accompany the pupils wherever they want to go and connect them to a broad range of information sources. These features enable the pupils to use the tools for learning purposes even when they are outside their homes and they can perform learning assignments, prepare their homework in digital exercise books, read electronic books and search for information on the Internet. These technologies are increasingly popular due to the mobile properties close to the body that have become essential to the lifestyle of the general population and in particular to the pupils for learning purposes (Naismith, Lonsdale, Vavoula & Sharples, 2006). Additionally, the mobile computer, Smartphone and tablet are defined as assistive technologies intended to help pupils with special needs, including those with learning disabilities to take part in or to complete learning assignments that are difficult for them to perform (Lee & Vega, 2005), this is something that most pupils who are diagnosed with learning disabilities cannot do without the support of assistive technologies. Perhaps the improvement and the many experiences of success that the pupils underwent during their work with the tablet enabled them to understand that they can use technological means more for learning purposes and thanks to this use that they may be able to succeed, to enhance their performances and to enjoy experiences of success in other academic disciplines and fields just as they did in their work with the tablet to improve their English language skills. and indeed the professional literature in this field reinforces the claim that assistive technology is effective in helping pupils, who have difficulties with reading and writing (Sultan & Hayhoe, 2013), that are needed in order to succeed in any other discipline studied at school and in life in general.

In the present research findings, it is possible to find two implications for the education system, that deal with the field of education and especially for the pupils: (1) a correlation between the use of tablet applications and improvement of English language skills in the areas of reading and reading comprehensions, while the salient improvement was achieved by pupils diagnosed with learning disorders and (2) the correlation between use of tablet applications and an increase in motivation to study among pupils. Because of the problematic timing in which the motivation questionnaires were administered to the pupils, the data which emerged were biased and so there is a need to rely solely on the qualitative data from the pupils' diaries that illustrated the increase in the pupils' motivation for studying. Consequently, it was not possible to determine in which of the two groups there was a greater increase in motivation to study.

FUTURE RESEARCH DIRECTIONS

It is noted that this study was conducted with a small number of participants. In order to increase the generalizability of the findings, it would be advisable to perform the study with a larger number of participants, in both groups of pupils. Additionally, it is important in research with participants diagnosed with various learning disorders to assist a larger number of them in terms of the contributions of the use of the tablet both for improvement of English language skills and for the improvement of motivation to learn the language and also in terms of the contribution of the use for pupils with specific disabilities.

It is also noted that there was no use in the research of a control group of pupils studying the English language according to the conventional method with the help of regular textbooks and exercise books, so that it was not possible to prove unequivocally that the use of the tablet was the factor which led to the improvement of English language skills in the areas of reading and reading comprehension. However, in a study of this kind it is not possible to transmit the learning materials used in the tablet applications through the means of conventional teaching. In any type of teaching there will be differences between a research group and a control group by the very fact that the methods of teaching differ and so the effectiveness of the use of a control group in research of this type is low. The different tablet applications include a rich and varied vocabulary, different activities to exercise reading, writing, vocabulary and reading comprehension etc., and it is not possible to transmit this variety through teaching methods that do not use English tablet applications. Studying the English lessons in school may also have had some influence on the pupils' progress in reading and reading comprehension and so there would be an influence on the research findings. However, there was no way to conduct the research in the disciplines studied in the school, especially in language learning, in a sterile manner without the influence of the pupils' progress following the lessons transmitted in school, other classes which the children attend, instruction and assistance that they receive in school, exposure to the language on television etc. It is noted that in order to create difference in learning materials studied in frontal lessons and in the individual lessons of the research intervention groups, different learning materials were selected, different stories and games and different activities, but it is still possible that there was some overlap in learning areas, and as noted some of this overlap was inevitable.

Following the research results it is recommended to work with tablets in small groups, and in individual lessons within the school. Additionally, work with the tablets will be effective in special education classes with small number of children. Children in these classes often have difficulties in copying from the blackboard/whiteboard and in correct writing of the letters and numbers. Being able to use a single finger in order to press on icons, letter and number keys and to write numbers and letters on the screen can make it easier for these pupils and help them to overcome these difficulties. Moreover, enjoying the graphics, colors, and interactive applications and learning through games and varied activities, on the tablet can assist students who have difficulties concentrating and paying attention to focus for longer periods.

Another recommendation is to exploit the mobility that the tablet permits. In the present study no meaningful use was made of this feature, apart from transfers from room to room in cases where there was difficulty in availability of rooms for individual lessons in the school. However, the mobile nature of this tool enables the learners to become active learners and to go outside the school spaces to explore, investigate, record and photograph and to share their discoveries with their peers. The tablet has the potential to produce active learning and to encourage pupils to leave their passive learning sitting behind the desk. The tool allows the teachers to develop new lesson interactive systems enabling more meaningful learning relevant to the learners' lives.

If the school or other educational institute faces a choice between use of a mobile computer or the use of tablets within its walls, the use of the tablet is more highly recommended. Today there is no difference in cost between these two tools but there are differences in the comfort of mobility, due to weight and possibilities for use, choice of software, use of the touch screen and relevance to the pupils' lives. The use of tablets of the iPad type is very comfortable. Over the research, there were no technical problems with this tool, it was user-friendly and intuitive for the pupils and the researcher and the selections of free learning applications in the field of the English language was very rich and high-quality and the applications were automatically updated when appropriate. The battery life is very long and there was no need to continually cope with charging the tools and carrying the tool was easy because of its small size and light weight.

Moreover, it was important that the tablets were allotted to the pupils and they could take them home and bring them back to school, allowing them to use the English applications during the lessons and also in their time after school. Thus, the pupils were able to exercise in more depth in their own time, enabling them to increase the improvement of their English language skills in reading and reading comprehension. This ensured professionalization of the pupils' learning (Al-Okaily, 2015), constituting one of the most meaningful features of mobile technologies. In other words, it was possible to adapt the learning with the tablet to the different learning characteristics and organization of the pupils and thus to cope with the diversity of pupils in the group and their different needs.

CONCLUSION

In conclusion, the use of the tablet applications has a potential to assist pupils in learning EFL. The tablet applications can assist both pupils diagnosed or not diagnosed with learning disabilities in improving their English language skills in reading and reading comprehension. This tool is able to increase pupils' motivation for learning the English language, since the use of the tablet applications enables the pupils with dysgraphia (writing difficulties) and dyslexia (reading difficulties) who have serious difficulties acquiring a foreign language to overcome these difficulties, to experience successes and not to feel that they are abnormal in the education system.

Following the use of tablet applications within the school and the use of technological means for learning purposes even beyond school hours, the pupils were ready for the future and were prepared to acquire skills and tools necessary to continue their functioning in society. This readiness is in line with the goal of the education system to adapt itself to the needs of the 21st century through the development and implementation of innovative pedagogy suitable for young pupils, members of the "mouse generation" so that they can become useful citizens in the digital era.

REFERENCES

Akpan, P. J., & Beard, A. L. (2013). Overview of Assistive Technology Possibilities for Teachers to Enhance Academic Outcomes of All Students. *Universal Journal of Educational Research*, *1*(2), 113–118.

Akstein, S. (2006). *Easy to Read.* Raanana: University Publishing Projects.

Alkhawaldeh, A. (2012, November). The EFL Reading Comprehension challenges Faced by Secondary School Students in Jordan. *European Journal of Soil Science, 35*(2), 140–152.

APA. (1994). *DSM-4- Diagnostic and statistical manual of mental disorders* (4th ed.). Washington, DC: American Psychiatric Association.

Bausch, E. M., & Melinda, A. J. (2008). Assistive Technology Implementation Plan: A Tool for Improving Outcomes. *Teaching Exceptional Children, 41*(1), 6–13. doi:10.1177/004005990804100101

Boone, R., & Higinns, K. (2007). The role of instructional design in assistive technology research and development. *Reading Research Quarterly, 42*(1), 135–140. doi:10.1598/RRQ.42.1.5

Brooke, D. E. (2017). *The Critical Role of Oral Language in Reading Instruction and Assessment.* Retrieved from Lexia Learning: https://www.lexialearning.com/resources/white-papers/oral-language

Clarke, B., Svanaes, S., & Zimmermann, S. (2013, July). *One-to-one Tablets in Secondary Schools: An Evaluation study*. Academic Press.

Cole, S. M., Hubert, F. S., & Harris, G. S. (2004). Student Learning Motivation and Psychological Hardiness: Interactive Effects on Students' Reactions to a Management Class. *Academy of Management Learning & Education, 3*(1), 64–85. doi:10.5465/amle.2004.12436819

Florida Department of Education. (2003). Cool Tools: Informal Reading Assessments. Author.

Franz, C., Ascherma, L., & Shaftel, J. (2018). *A Clinician's Guide to Learning Disabilities*. New York: Oxford University Press.

Hadar-Peker, D. D. (2013, December). *The Relationship between the Social-Emotional State and the Scholastic achievements of Schoolchildren.* Retrieved from The Initiative for Applied Research in Education: http://yozma.mpage.co.il/SystemFiles/23051.pdf

Harmon, K. (2013, July 12). *Best iPad® Apps and iPad Strategies to Increase Student Learning (Grades K-6)*. Retrieved from Bureau of Education & Research (BER): http://www.ber.org/bro_pdf/ZDH4F2.pdf

Hasselbring, S. T., & Bausch, E. M. (2005). Assistive Technologies for Reading. *Learning in the Digital Age, 63*(4), 72–75.

Hopkins, J. (2004). School library accessibility: The role of assistive technology. *Teacher Librarian, 31*(3), 15.

Huber, S. (2012). iPads in the Classroom - A Development of a Taxonomy for the Use of Tablets in Schools (M. Ebner, & S. Schön, Eds.). Norderstedt: Books on Demand GmbH.

Kimura, Y., Nakata, Y., & Okumura, T. (2008, October 26). *Language Learning Motivation of EFL Learners in Japan—A Cross-Sectional Analysis of Various Learning Milieus.* Retrieved from Publications of the Japan Association for Language Teaching: http://jalt-publications.org/archive/jj/2001a/art3.pdf

Lan, J., Sung, T., & Chang, E. (2007). *A Mobile-Divice-Supported Peer-Assited Learning System for Collaborative Early EFL Reading.* Retrieved December 15, 2012, from Language Learning & Technology: A refereed for second and foreignlanguage scholars and educators: http://llt.msu.edu/vol11num3/pdf/lansungchang.pdf

Mang, F. C., & Wardley, J. L. (2012). Effective Adoption of Tablets in Post-Secondary Education: Recommendations Based on a Trial of iPads in University Classes. *Journal of Information Technology Education: Innovations in Practice*, *11*, 302–317.

Marés, L. (2012, April). *Tablets in education: Opportunities and challenges in one-to-one programs.* Academic Press.

McClanahan, B. (2012, May/June). A breakthrough for Josh: How Use of an iPad Facilitated. *Reading Improvement.*

Meurant, C. R. (2010, December). iPad Tablet Computing to Foster Korean EFL Digital Literacy. *International Journal of u- and e- Service Science and Technology*, *3*(4), 49–62.

Mock, K. (2004). *Teaching with Tablet PC's.* Anchorage, AK: Academic Press.

Mohan, M. (2009, November). *Designing for Children- With focus on 'Play + Learn'.* Bombay: Academic Press.

Naismith, L., Lonsdale, P., Vavoula, G., & Sharples, M. (2006, September 12). *Report 11: Literature Review in Mobile Technologies and Learning.* Retrieved from Futurelab at NFER: http://www2.futurelab.org.uk/resources/documents/lit_reviews/Mobile_Review.pdf

NJCLD- National Committee on Learning Disabilities. (1994). Learning disabilities issues on definition. In Collective perspectives on issues affecting learning disabilities (pp. 61-66). Austin, TX: PRO-ED.

Ozok, A. A., Dana, B., Chakraborty, J., & Norcio, F. A. (2008). A Comparative Study Between Tablet and Laptop PCs: User Satisfaction and Preferences. *Intl. Journal of Human-Computer Interation*, *24*(3), 329–352. doi:10.1080/10447310801920524

Pintrich, R. P., & De Groot, V. E. (1990). Motivational and Self-Regulated Learning Components of Classroom Academic Performance. *Journal of Educational Psychology*, *82*(1), 33–40. doi:10.1037/0022-0663.82.1.33

Schwarz, L. R. (2002). *LD and the English language learner.* Cambridge, MA: Lesley University.

Stake, E. R. (1995). *The Art of Case Study Research.* Thousand Oaks, CA: Sage Publications.

Sultana, N., & Hayhoe, S. (2013). *Assistive Technology for Students with Special Needs.* Retrieved September 17, 2013, from The Higher College of Technology - Sharjah Colleges: http://shct.hct.ac.ae/events/edtechpd2013/articles/Sultana-Hayhoe.pdf

Thompson, C., & Dekkers, A. (2009). *Accreditation, Video Resources and Interactivity, Tablet PC's and their Advantages.* Retrieved October 2, 2012, from Australasian Association for Engineering Education: http://aaee.com.au/conferences/AAEE2009/PDF/AUTHOR/AE090089.PDF

Thompson, C. A., & Dekkers, A. (2009, December). *Accreditation, Video Resources and Interactivity, Tablet PC's and their Advantages.* Academic Press.

White, E., Wepner, S., & Wetzel, D. (2003). Accessible Education Through Assistive Technology. *Technological Horizons in Education (T.H.E), 30*(7), 24-30.

Wise, C. J., Toto, R., & Lim, K. Y. (2006). *Introducing Tablet PCs: Initial Results From the Classroom.* Retrieved August 17, 2013, from Frontiers in Education Clearing House: http://fieconference.org/fie2006/papers/1683.pdf

Yeunjoo, L., & Vega, L. A. (2005). Assistive Technology: Perceived Knowledge, Attitudes, and Challenges of AT Use in Special Education. *Journal of Special Education Technology*, *20*(2), 60–62. doi:10.1177/016264340502000208

Chapter 14
Real-Time Mobile Assessment of Learning:
Insights From an Experiment With Middle School Students From Remedial, Midstream, and Excellence Tracks

Ina Blau
The Open University of Israel, Israel

ABSTRACT

Real-time mobile assessment (RTMA) can function as assistive technology, since the anonymity of feedback promotes active participation in lessons without being exposed to criticism. This field experiment explored the impact of RTMA, beyond a whole-class technology, on academic achievement, students' perceived learning, academic self-efficacy, learning motivation, and self-esteem among middle school students. The participants were 80 ninth graders in the excellence, mainstream, and remedial tracks being taught by the same experienced language arts teacher. The experiment included nine double language lessons with a random assignment of the technological conditions and the counterbalance procedure of the topics studied. The findings suggested that students in the remedial track benefit from using RTMA in terms of achievement, perceived learning, self-efficacy, and motivation. Students in the excellence track benefit from RTMA in terms of intrinsic motivation, and students in the mainstream track in terms of self-esteem. Implication for educational research and practice are discussed.

DOI: 10.4018/978-1-5225-8106-2.ch014

INTRODUCTION

Active involvement of students in the learning process is important for building solid knowledge and developing competencies (Blau & Shamir-Inbal, 2017; Porat, Blau, & Barak, 2018). Mobile devices have various distinctive features, among others, instant communication and feedback that may be used for enhancing the effects of formative assessment. A meta-analysis and research synthesis study for mobile-integrated education (Sung, Chang, & Liu, 2016) analyzed 110 empirical studies and indicated that very few of them (e.g., Agbatogun, 2014), used mobile devices for formative assessment or quizzes in classrooms or outdoors.

Real-time mobile assessment (RTMA) in education refers to using applications on students' mobile devices to conduct formative assessment in both in-class and out-of-class settings (Hwang & Chang, 2011). Such use has the potential of promoting students' active involvement in learning processes, encourage student voice, thus leading to better learning outcomes (Blau & Shamir-Inbal, 2018). Previous research has demonstrated that using RTMA in classrooms in both schools and higher education leads to increased student involvement in learning processes and higher intrinsic motivation (Camacho-Miñano, & del Campo, 2016; Kowalski et al., 2015).

However, previous research has not provided conclusive data regarding the contribution of RTMA to student achievement (Martyn, 2007; Morgan, 2008). Moreover, to the best of our knowledge, previous studies have not explored the impact of the anonymity of feedback on active participation in lessons, learning motivation, self-esteem in general, and academic self-efficacy in particular, among students in different academic levels, including students in remedial learning tracks. Through RTMA, students become actively involved in the learning process and construct their knowledge (Camacho-Miñano, & del Campo, 2016; Kowalski et al., 2015). In addition, the anonymity of feedback allows students to participate without being exposed to criticism from their teacher or/and classmates, which can be detrimental to their self-efficacy (Blau & Caspi, 2010; Morales, 2011; Weiser, Blau, & Eshet-Alkalai, 2018). Ongoing formative assessment through RTMA is anonymous, since students answers to the assessment questions cannot be traced back to individual students, which may enhance active participation in lessons (Camacho-Miñano, & del Campo, 2016). Previous studies have shown that technology-supported assessments are effective for students with, as well as without, disabilities (for review see: Bryant & Seok, 2017). Thus, it is reasonable to assume that RTMA may function as *assistive technology*, which enhances the learning potential of students in remedial learning tracks (Schnorr, Freeman-Green, & Test, 2016) who are characterized by low levels of achievement, academic self-efficacy, learning motivation, and self-esteem.

In order to explore the added value of ongoing mobile assessment, this field experiment compared the academic achievement, perceived learning, academic self-efficacy, learning motivation, and self-esteem of middle-school students at different academic levels: students from excellence, mainstream, and remedial learning tracks. All of the students learned in three technological conditions: (1) with RTMA and whole-class technology - Interactive Whiteboards, (2) with whole-class technology alone, and (3) without technology.

LITERATURE REVIEW

Brunner's (2001) constructivist approach to learning describes learning as an active process of constructing new knowledge by linking new information to prior knowledge. In order to construct knowledge, students must be actively involved in the learning process (Blau, Peled, & Nussan, 2016; Shamir-Inbal & Blau, 2016; Weiser et al., 2018). RTMA is one of the ways in which technology can be used to facilitate active learning and ongoing assessment. Studies have found that, compared to traditional teaching, using RTMA in classrooms leads to increased student involvement and intrinsic motivation for learning (BoCrews et al., 2011; Martyn, 2007; Stowell, 2015). In addition to higher motivation, a mobile self-assessment procedure in a secondary school Physics class resulted in a significant increase in students' achievement, especially for low-achieving students (Nikou & Economides, 2016). The anonymity that characterizes the use of RTMA encourages students to partake in lessons and respond to the teacher's questions in front of the teacher and other classmates without fear of damaging their self-efficacy beliefs (Blau & Caspi, 2010; Morales, 2011).

However, previous research leaves the question of whether RTMA does (Martyn, 2007) or does not contribute to academic achievement (Morgan, 2008) unresolved. Moreover, some authors have even found that RTMA has a negative impact on students' metacognition (Brady, Seli, & Rosenthal, 2013). A recent study that focused on interactive highlighting for just-in-time formative assessment during whole-class instruction (Ponce, Mayer, Figueroa, & Lopez, 2018) explored the effects of such ongoing assessment on learners' vocabulary learning and reading comprehension in language learning. The technology projected on the class screen words that most of the students did not understand and the instructor based her explanation on this information. After a single session, the experiment group scored significantly higher on a vocabulary test than the control group, which studied without interactive highlighting. This effect was replicated in a "real" learning environment: following participation in a 10-week program, high-school students in the experimental group, who learned with RTMA, scored significantly higher on a standardized reading comprehension test in comparison with the control group. Although these results are promising and suggest that real-time formative assessment results in better learning outcomes, it should be noted that instead of providing feedback to individual students, the experiment actually used RTMA to provide feedback to the instructor and focus her teaching on the average learner's needs.

Whole-class technologies provide opportunities for multimedia learning by combining visual and verbal representations (Mayer, 2001), and for non-linear learning - by linking slides to external learning materials (Blau, 2011). Both multimedia and non-linear learning contribute to the learning process and increase motivation for learning (Blau, Grinberg, & Shamir-Inbal, 2018; Shamir-Inbal & Blau, 2017). However, compared to RTMA, active participation might be lower in classes with whole-class technologies, because, as mentioned earlier, anonymity allows students to participate without being exposed to the potential criticism of their teacher and peers. In this sense, RTMA can function as an assistive technology that helps remedial students who are characterized by low self-esteem, self-confidence, motivation and achievement, to overcome at least some of their learning-related weaknesses. Assistive technology is a broad concept which, when used in an educational context, includes a range of devices and services to increase, maintain, or improve the functional capabilities of students (Ahmed, 2018). Similarly, a recent study conducted in academia (He, Holton, & Farkas, 2018) demonstrated that students with lower high-school GPAs benefited more from an active learning approach in a partially flipped classroom, while student with higher GPAs were not affected by this pedagogical approach. The authors concluded that pedagogy based on the active involvement of students in the learning process has the potential of bridging the achievement gap over time. Moreover, for low-performing students in He et al.'s study (2018), low learning motivation was found to be a limiting factor for achievement stronger than academic ability, and eventually impacted students' course grades.

Research Goals and Hypotheses

The unique feature of technology in general, and mobile devices in particular, is that they can increase school-related communication (Blau & Hameiri, 2017; Caspi & Blau, 2011b) and enhance the essential functionalities of certain teaching methods (for review see: Sung et al., 2016). Sung and colleagues argued that because each student has a mobile device, it has become more accessible for usage in educational practice. Previous research has demonstrated the positive effect of mobile assessment on the achievement and motivation of secondary students studying physics, which was found to be different for high- and low-achievers (Nikou & Economides, 2016). However, we could not find studies exploring the impact of RTMA on students' self-efficacy or on the learning process as perceived by the students themselves, i.e., on different aspects of perceived learning. Sung and colleagues (2016) criticized the fact that most of the studies in their review focused on academic achievement of content knowledge. The authors called for expanding measurements of dependent variables in studies on the educational use of mobile devices. Moreover, previous studies have not explored the added value of anonymous feedback through RTMA on the learning processes and outcomes of students at different academic levels.

Thus, the current study examined the effect of students' academic level (students from remedial, mainstream, and excellence learning tracks) and the extent of students' involvement in active technology-enhanced learning (students who learned with RTMA, with whole-class technology, and without whole-class technology) on the comprehensive set of dependent variables relevant for the learning processes and outcomes. These include academic achievement, cognitive, emotional and social aspects of perceived learning, academic self-efficacy, intrinsic and extrinsic learning motivation, and self-esteem. The cognitive aspect of perceived learning refers to acquiring new understanding; the emotional aspect includes the extent to which students find the learning process to be interesting and enjoyable; and the social aspect refers to enjoyment of interpersonal interactions with teachers and peers (Blau & Caspi, 2008, 2010; Caspi & Blau, 2008, 2011).

It was hypothesized that students' academic achievement, perceived learning, self-efficacy, learning motivation, and self-esteem would change after studying new content from a language arts curriculum. The variance would be a factor of: (1) The *level of students' activity* in technology-enhanced learning (RTMA, whole-class technology, and without technology). (2) Students' *academic level* (students studying in the excellence, mainstream, and remedial learning tracks). (3) The *interaction effect* between the technology characteristics and academic level of students. Namely, it was hypothesized that active learning with anonymous feedback through RTMA would empower students studying in remedial classes, while relatively passive presentations using whole-class technology, which contain multimedia and link to external resources, would provide better affordances for knowledge acquisition of students in the excellence class.

METHOD

Participants

The experimental group included 80 9th-graders (14-15 years old), 51% of them female, and their language arts teacher. Conducting a field experiment in the natural environment of the participants strengthens the ecological validity of the findings (Blau & Barak, 2012; Eden, 2017). All of the students studied in three classes in a large secondary school situated in Northern Israel. Conducting the experiment with all the students within one stratum in a school in which students are segregated into different learning tracks (rather than into different schools) according to their academic performance, ensured a diverse sample of students coming from different academic levels, social-economic background and ethnic origin (Blau & Eshet-Alkalai, 2017). Twenty of the participants were students who learned in the remedial learning track, 36 participants studied in the mainstream track, and 24 learned in the excellence track. To avoid measuring the "novelty effect", this experiment took place in a school where both students and teachers were familiar with these technologies and had experience of more than one year teaching and learning with both RTMA and whole-class technology. The students in the sample did not differ in any systematic way from the general population of Israeli secondary school students.

Instruments

Achievement was measured by testing students' knowledge and comprehension of the new content learned in each lesson – three measurements in total. Each test included 10 questions: three questions were multiple-choice and seven were open-ended questions. To ensure comprehensive measurement of the knowledge constructed during the intervention, half of the questions measured basic understanding, while another half were inference questions (Blau, Weiser, & Eshet-Alkalai, 2018). Lessons from the language arts curriculum for 7th grade dealt with the lexicography and the use of different types of dictionaries. An example of a multiple-choice question from the test is: "A multiple-words dictionary uses the language style: (a) slang, (b) highly-formal style, (c) moderately-formal style, (d) informal style. An example of an open-ended question measuring basic understanding is: "What are the grammatical items that each dictionary displays?" An example of an inference question is: "Compare the following quotations (Figure 1) from two types of dictionaries on the criteria of (a) order of meaning display, (b) the source of the examples".

The content validity of the tests was approved by two Hebrew (native language) literacy specialists familiar with the language arts curriculum for middle schools.

Table 1 presents descriptive statistics for the achievement tests.

The other dependent variables were based on students' self-report.

Students' *perceived learning* (Perceived Learning Scale; Blau & Caspi, 2008, 2010; Caspi & Blau, 2008, 2011a) was measured as a three-dimensional variable including cognitive, emotional, and social aspects of perceived learning, based on a confirmatory factor analysis with Varimax rotation, and in accordance with the results of previous studies. Five items representing the *cognitive aspect* of perceived learning were averaged: After this lesson… (1) "I know more things", (2) "I expect to remember what I have learned", (3) "my prior knowledge expanded", (4) "I clearly understand the topic", (5) "I feel that I learned new things", Cronbach's α=.88-.94 in different lessons.

Figure 1. Quotations from two dictionaries used in the research

מילון אבן שושן (מהדורת 2003)

נָתִיב ז׳ [אונריתית: נתב; ראו גם נתב] 1. שְׁבִיל, דֶּרֶךְ: "טָמוּן בָּאָרֶץ חַבְלוֹ וּמַלְכֻּדְתּוֹ עֲלֵי נָתִיב" (איוב יח י). "נָתִיב לֹא־יְדָעוֹ עָיִט וְלֹא שְׁזָפַתּוּ עֵין אַיָּה" (איוב כח ז). 2. [בהשאלה] כִּוּוּן, אֹרַח מוּסָר: "הַדְרִיכֵנִי בִּנְתִיב מִצְוֹתֶיךָ" (תהילים קיט לה). "וּבִנְתִיב יִרְאָתְךָ נַחֵנִי" (יהודה הלוי, אלוהים שיחותיך). "וְסָלְלָה נְתִיב חַיִּים עַד בֵּית־הַיְשִׁיבָה" (ביאליק, המתמיד). 3ת״. מִנְהָג, נֹהַג, חֹק הַקָּבוּעַ בַּטֶּבַע: "נָתִיב הוּא מִשֵּׁשֶׁת יְמֵי בְרֵאשִׁית, רַבִּים שָׁתוּ – רַבִּים יִשְׁתּוּ, כְּמִשְׁתֶּה רִאשׁוֹנִים כָּךְ מִשְׁתֶּה אַחֲרוֹנִים" (כתובות ח:). 4ת״. קַו תַּחְבּוּרָה שֶׁנִּקְבַּע לִתְנוּעָה בַּיַּבָּשָׁה, בַּיָּם אוֹ בָּאֲוִיר: נְתִיבֵי אֲוִיר יִשְׂרָאֵל-הוֹלַנְד. נְתִיב רַכֶּבֶת. נְתִיב אוֹפַנַּיִם בְּצִדֵּי הַכְּבִישׁ. 5ת״. [בכבישים] רְצוּעָה בִּכְבִישׁ הַמְסֻמֶּנֶת בְּקַו לָבָן מַרְסָק וְהַמְיֻעֶדֶת לִתְנוּעַת טוּר אֶחָד שֶׁל מְכוֹנִיּוֹת. בִּכְבִישׁ רָחָב מְסַמְּנִים בְּקַוֵּי אֹרֶךְ מְרַסְּקִים נְתִיבִים אֲחָדִים זֶה בְּצַד זֶה. הַנַּהָגִים נִדְרָשִׁים לִנְסֹעַ בְּדֶרֶךְ כְּלָל בַּנָּתִיב הַיְמָנִי, וּלְהִשְׁתַּמֵּשׁ בַּנָּתִיב הַשְּׂמָאלִי לְצֹרֶךְ עֲקִיפָה בִּלְבַד. – [נְתִיב־, נְתִיבִים, נְתִיבֵי־]

מילון רב מילים

נתיב נָתִיב *שם ז׳*

1. דרך, שביל, משעול או מסלול, ובייחוד כאלה שנכבשו ברגלי הצועדים; קו נסיעה, הליכה, מעוף וכד׳ שהותווה בדרך כלשהי במהלך התנועה של מישהו או של משהו. *הגשש עלה על נתיב הבריחה של המחבלים.* • *הצפרים עקבו בדריכות אחר נתיב מעופו של הנשר.* **2. כל אחד מן המסלולים המקבילים המסומנים בכביש והמיועדים לתנועתם של כלי רכב בכיוון אחד.** *הנהגים מתבקשים לנסוע בנתיב הימני של הכביש ולהשתמש בנתיב השמאלי רק לצורך עקיפה או פנייה שמאלה.* **3. קו או מסלול שנקבע והוגדר היטב, המיועד לתנועה של כלי תחבורה או של כוח צבאי באוויר, בים וביבשה בדרכם ליעד מסוים.** *בעקבות חידוש היחסים עם פולין נקבע נתיב תעופה לתנועת מטוסים בין שתי המדינות.* **4. אורח חיים, דרך התנהגות; כיוון או מסלול של חיים, של פעילות וכד׳.** *הוא צועד בנתיב האמונה בעקבות הוריו האדוקים.* [נְתִיבִים; נְתִיב־; **נ.ת.ב**]

The *emotional aspect* of perceived learning was measured by the average on four items: During this lesson… (1) "It was easy to learn", (2) "It was easy to stay focused", (3) "I enjoyed learning", (4) "Learning was interesting", Cronbach's α=.73-.81 in different lessons.

The *social aspect* of perceived learning was assessed by the average on three items: During this lesson… (1) "I enjoyed interacting with the teacher", (2) "I enjoyed interacting with other students", (3) "I enjoyed the interactions of the teacher with students", Cronbach's α=.76-.82 in different lessons. Students' answers on the cognitive, emotional and social aspects of perceived learning were ranked on a six point Likert scale ranging from 1 – "not at all true of me" to 6 – "very true of me".

Academic self-efficacy for learning and performance (Motivated Strategies for Learning Questionnaire - MSLQ; Pintrich, et al., 1991) is a self-appraisal of one's ability to master a task and confidence in one's skills to perform the task. Academic self-efficacy was assessed by eight items from the MSLQ adapted to the language arts context, for example, "Considering the difficulty of this subject, the teacher, and my language skills, I think I will do well in the language arts lessons", Cronbach's α=.91-.95.

Table 1. Achievement tests - descriptive statistics

	Test 1	Test 2	Test 3
Mean	77.75	69.6	74.99
Median	85	83	80
SD	28.56	33.45	23.10
Skewness	-1.42	-1.15	-1.26
Minimum	0	0	0
Maximum	100	100	100

Motivation (Motivated Strategies for Learning Questionnaire; Pintrich, et al., 1991) was measured as intrinsic and extrinsic goal orientation and the items were adapted for the language arts context. *Intrinsic goal orientation* refers to the degree to which students perceive themselves as learning for reasons such as challenge, curiosity, or mastery, for example, "In language lessons I prefer learning content that arouses my curiosity, even if it is difficult to learn". *Extrinsic goal orientation* refers to the degree to which students perceive themselves as learning for reasons such as grades, rewards, performance, competition, or evaluation by others, for example, "My main concern in language arts lessons is getting a good grade". The validity and reliability of the self-efficacy and motivation scales has been shown in numerous previous studies. In this study, confirmatory factor analysis with Varimax rotation revealed, as expected, the two components of motivation. Examination of internal consistency indicated that Cronbach's α=.85-.88 for four items of intrinsic motivation and α=.88-.90 for four items of extrinsic motivation.

Students' responses in relation to academic self-efficacy, as well as intrinsic and extrinsic motivation, were rated on a six point Likert scale ranging from 1 – "not at all true of me" to 7 – "very true of me".

Self-esteem (Self-Esteem Scale; Rosenberg, 1979) was measured by ten items, originally designed to measure the self-esteem of high-school students. An example of a self-esteem item is: "I feel that I'm a person of worth". The validity and reliability of this scale has been shown in numerous previous studies. The internal consistency in this study was very high, Cronbach's α=.93-.94 in different lessons. Responses to these items were rated on a 4-point Likert scale ranging from "strongly disagree" to "strongly agree".

Procedure

The experiment was approved by the Institutional Ethics Committee. Students received explanations that their participation is anonymous and cannot impact their grades in language arts. They were asked to openly express their views and to do their best on the language arts tests, since the results will help the school to optimize their learning processes and therefore, can indirectly affect their future learning effectiveness and achievement.

The three learning tracks (remedial, mainstream, and excellence) were randomly assigned to the technological conditions (RTMA, whole-class technology, and without technology). In the RTMA condition, the teacher projected students' responses in real time on the screen, commented on the appropriateness of their responses, provided further explanations and conducted class discussions. In the whole-technology condition, the teacher used an Interactive Whiteboard and in the without-technology condition – a board to support her explanations and class discussions. Each class studied three double language arts lessons in topics included in the standard Hebrew language arts curriculum for the 9th grade. In total, nine double lessons (18 academic hours) were conducted. To insure that the lessons' topics do not vary in difficulty or in any other systematic way, and would not affect the results, the lesson topics were counterbalanced. In order to control the potential impact of the teachers' personal characteristics and professional skills on the findings, the same experienced language arts teacher, who was unaware of the research hypotheses, taught all of the lessons using similar pedagogical approaches.

Lessons plans were planned with the research team prior to the experiment, in order to ensure that the teaching methods used by the teacher did not differ between the technological conditions or the lessons' topics. The following example illustrates a teaching-learning activity that was used in the different experimental conditions. In the condition without technology, the teacher distributes an assessment that includes the dictionary entry from two different dictionary types. After the students read the definitions, the teacher conducts a class discussion about the differences between them. The teacher writes the students' answers on the board. At the end of the discussion, after the types of dictionary definitions have come up, the teacher writes the types of dictionary definitions and examples of each type of definition on the board. The students copy the explanation into their notebooks. After the explanation, students practice by writing on a worksheet that the teacher distributes. After the practice, a number of student-volunteers read and explain their answers.

In the whole-class technology condition, the teacher writes the students' answers on the interactive whiteboard, presents pre-prepared definitions and examples, highlights the words showing the difference between the two types of dictionaries, and promises to share the file with the students, so that they do not need to copy. It is easier for students to follow the explanation while seeing the highlighted text and more time is spent on listening to students' comments and answering their questions. The practice and assessment stages are identical to the condition without technology.

The RTMA condition is identical to the whole-class technology condition, except for the practice stage, which is combined with instant assessment through students' smartphones, followed by discussion and additional explanations by peer-volunteers or by the teachers, when needed. All students participate, the assessment is anonymous, and more time is spent on clarifying issues in relation to which a significant number of students provided incorrect answers.

At the end of each lesson, the students' comprehension of the new content learned in that lesson was assessed by the knowledge test. In addition, students reported the levels of perceived learning, academic self-efficacy, motivation and self-esteem, as described above. Since students completed the same self-report questionnaires three times, the experimental lessons were conducted with intervals of two weeks between them.

All the knowledge tests were assessed by an experienced middle-school language arts teacher unaware of the research hypotheses, of the technological conditions of each test or of the students' identity. Twenty five percent of the tests were re-assessed by another middle-school language arts teacher and the inter-rater reliability was Cohen's kappa $\kappa=.88$, indicating a satisfactory level of agreement (Landis & Koch, 1977).

Results

In order to explore the effect of student's academic level and their involvement in lessons through different technologies on the dependent variables, a series of analyses of variance - Repeated Measures ANOVAs – were conducted, with achievement as the dependent variable, student academic level as the between-subjects factor, and technology as the within-subjects factor.

Student Achievement

Tables 2-3 present descriptive statistics and the analysis of variance for the effect of student academic level and active involvement in learning through different technological conditions on student grades on knowledge tests.

Tables 2-3 show a significant main effect with a large effect size of students' academic level on their achievement. Bonferroni post-hoc tests revealed that, as expected, students' grades on knowledge tests were significantly lower in the remedial track compared to the mainstream and excellence tracks (p's $<.001$). The main effect of type of technology was not statistically significant. A significant interaction effect with a medium effect size was found, revealing that, unlike the mainstream and excellence tracks,

Table 2. Achievement – descriptive statistics

Student Academic Level	Technological Condition	Mean	SD
Remedial class (n=20)	Without technology	60.23	27.99
	Whole-class technology	54.71	36.46
	RTMA	67.38	23.46
	Total	60.63	29.69
Mainstream class (n=36)	Without technology	89.39	11.94
	Whole-class technology	79.78	17.97
	RTMA	74.23	24.66
	Total	81.01	19.91
Excellence class (n=24)	Without technology	88.32	13.13
	Whole-class technology	89.05	16.05
	RTMA	81.25	19.09
	Total	86.23	16.29
Total (n=80)	Without technology	83.47	19.80
	Whole-class technology	77.09	25.58
	RTMA	74.99	23.10
	Total	78.54	23.08

Table 3. Achievement – analysis of variance

Factor	F	df	p	Partial η^2
Student academic level	19.32	2, 238	<.001	.17
Technological condition	1.08	2, 238	=.34	.01
Interaction effect	2.91	2, 238	<.05	.04

students' grades in the remedial track were significantly higher after active learning with RTMA compared to learning with whole-class technology only (p<.01) or learning without technology (p<.05).

Perceived Learning

Tables 4-5 present descriptive statistics and the analysis of variance for the effect of student academic level and their active involvement in learning through different technological conditions on the cognitive, emotional and social aspects of perceived learning.

As can be seen from Tables 4-5, the emotional and social aspects of perceived learning were affected, with a medium effect size, by the academic level of students, but the cognitive aspect of perceived learning was not. Bonferroni post-hoc tests showed that learning in the remedial track was perceived as more pleasant in comparison to the mainstream and excellence tracks (p's <.05). An interaction effect between student academic level and the technological condition was also statistically significant for the emotional and social aspects of perceived learning, both with a medium effect size, but not for the

Table 4. Perceived learning – descriptive statistics

Student Academic Level	Technological Condition	Cognitive Aspect Mean	SD	Emotional Aspect Mean	SD	Social Aspect Mean	SD
Remedial class (n=20)	Without technology	3.51	1.43	3.88	1.10	3.58	0.10
	Whole-class technology	3.63	1.45	4.14	1.03	3.75	1.16
	RTMA	3.58	0.93	4.24	0.83	3.54	1.18
	Total	3.57	1.26	4.13	1.30	3.63	1.09
Mainstream class (n=36)	Without technology	3.59	0.96	3.88	1.15	3.67	0.75
	Whole-class technology	3.12	0.95	3.14	1.35	3.08	1.05
	RTMA	3.31	1.33	3.40	1.28	3.16	1.33
	Total	3.34	1.11	3.48	1.29	3.30	1.09
Excellence class (n=24)	Without technology	3.71	1.18	3.89	1.37	3.55	1.10
	Whole-class technology	3.76	1.23	4.079	1.52	3.97	1.23
	RTMA	3.08	1.13	3.70	1.31	3.50	1.01
	Total	3.52	1.20	3.89	1.39	3.66	1.12
Total (n=80)	Without technology	3.61	1.12	3.88	1.30	3.61	0.91
	Whole-class technology	3.41	1.18	3.63	1.48	3.48	1.18
	RTMA	3.29	1.20	3.71	1.22	3.33	1.21
	Total	3.44	1.17	3.76	1.33	3.48	1.11

Table 5. Perceived learning – the analysis of variance

Factor	F	df	p	Partial η^2
Cognitive Aspect				
Student academic level	0.77	2, 198	=.46	.01
Technological condition	0.84	2, 198	=.43	.01
Interaction effect	1.13	4, 196	=.35	.02
Emotional Aspect				
Student academic level	3.29	2, 198	<.05	.04
Technological condition	0.25	2, 198	=.78	.00
Interaction effect	2.16	4, 196	<.05	.03
Social Aspect				
Student academic level	3.62	2, 198	<.05	.04
Technological condition	0.65	2, 198	=.53	.00
Interaction effect	2.52	4, 196	<.05	.03

cognitive aspect. Regarding the emotional aspect of perceived learning, students in the remedial track perceived learning with RTMA as more pleasant than learning with whole-class technology only or learning without technology (p's<.05). In contrast, learning with whole-class technology was the most pleasant for students in the excellence track, while learning without technology was the most enjoyable for students in the mainstream track (*p*'s<.05). Concerning the social aspect of perceived learning, no significant difference between the conditions was found in the remedial track, while for the excellence track learning with whole-class technology, and for the mainstream track learning without technology, were perceived as the most supportive learning-related social interactions (*p*'s<.05).

Academic Self-Efficacy

Tables 6-7 present descriptive statistics and the analysis of variance for the effect of student academic level and their active involvement in learning through different technological conditions on students' academic self-efficacy.

As can be seen from Tables 6-7, academic self-efficacy was, as expected, significantly higher, with a medium effect size, in the excellence track compared to the mainstream and remedial tracks (*p*'s<.01). The main effect of the technological condition was not significant. A significant interaction effect, with a medium effect size, was found, suggesting that in the remedial track, studying with RTMA and without

Table 6. Self-efficacy – descriptive statistics

Student Academic Level	Technological Condition	Mean	SD
Remedial class (n=20)	Without technology	5.28	0.74
	Whole-class technology	4.70	1.02
	RTMA	5.05	1.07
	Total	4.99	0.98
Mainstream class (n=36)	Without technology	5.32	1.22
	Whole-class technology	4.89	1.12
	RTMA	4.94	1.43
	Total	5.05	1.27
Excellence class (n=24)	Without technology	5.58	1.03
	Whole-class technology	5.94	0.92
	RTMA	5.40	1.25
	Total	5.63	1.08
Total (n=80)	Without technology	5.40	1.08
	Whole-class technology	5.16	1.23
	RTMA	5.08	1.34
	Total	5.22	1.22

Table 7. Self-efficacy – the analysis of variance

Factor	*F*	*df*	*p*	Partial η^2
Student academic level	5.55	2, 198	<.001	.06
Technological condition	0.97	2, 198	=.38	.01
Interaction effect	2.06	4, 196	<.05	.03

technology resulted in significantly higher academic self-efficacy compared to learning with whole-class technology. Among students in the excellence track, the highest level of academic self-efficacy was found among those who studied with whole-class technology, and among students in the mainstream track, among those who studied without technology (all p's <.05).

Motivation

Tables 8-9 present descriptive statistics and the analysis of variance for the effect of student academic level and their active involvement in learning through different technological conditions on intrinsic and extrinsic learning motivation of students.

As Tables 8-9 show, the analysis of variance revealed a significant main effect, with a medium effect size, of student academic level on both intrinsic and extrinsic goals. Post-hoc comparisons showed that in the remedial and excellence tracks, the two types of motivation were significantly higher compared to students in the mainstream track (p's < .05). *Intrinsic motivation* in the remedial track was even higher than in the excellence class (p< .05). The main effect of type of technology on the two types of motivation was not statistically significant. The interaction effect between student academic level and technology was significant, with a medium effect size, for both intrinsic and extrinsic motivation types. As for intrinsic motivation, in the remedial track, it was in the following declining order: whole-class technology – RTMA – without technology (p's < .05). For the excellence track, intrinsic motivation was higher among students who studied with RTMA compared to students in the other technological

Table 8. Motivation – descriptive statistics

Student Academic Level	Technological Condition	Intrinsic Goals Mean SD		Extrinsic Goals Mean SD	
Remedial class (n=20)	Without technology	4.31	1.06	4.60	1.09
	Whole-class technology	5.02	1.02	4.87	1.06
	RTMA	4.75	1.02	5.18	1.05
	Total	4.69	1.04	4.91	1.06
Mainstream class (n=36)	Without technology	3.96	1.34	4.11	1.57
	Whole-class technology	4.13	1.41	4.69	1.38
	RTMA	3.81	1.42	4.50	1.56
	Total	3.96	1.38	4.43	1.51
Excellence class (n=24)	Without technology	4.31	1.03	3.99	1.15
	Whole-class technology	4.28	1.01	3.79	1.58
	RTMA	4.78	1.06	3.83	1.56
	Total	4.45	1.04	3.87	1.41
Total (n=80)	Without technology	4.14	1.42	4.17	1.46
	Whole-class technology	4.36	1.54	4.44	1.51
	RTMA	4.25	1.46	4.41	1.59
	Total	4.25	1.47	4.34	1.52

Table 9. Motivation – the analysis of variance

Factor	F	df	p	Partial η^2
Intrinsic Goals				
Student academic level	4.02	2, 198	<.05	.04
Technological condition	0.61	2, 198	=.55	.01
Interaction effect	2.78	4, 196	<.05	.03
Extrinsic Goals				
Student academic level	5.15	2, 198	<.01	.05
Technological condition	0.34	2, 198	=.66	.00
Interaction effect	2.55	4, 196	<.05	.03

conditions ($p < .05$), while in the mainstream track no significant differences were found in the intrinsic motivation of students as a function of the technological conditions.

Regarding *extrinsic motivation*, in the remedial track it was in the following declining order: RTMA – whole-class technology – without technology (p's $< .05$). In the mainstream track, external motivation was significantly higher in both types of technology as compared to learning without technological integration (p's $< .05$), while in the excellence track there were no significant differences in extrinsic motivation between the technological conditions.

Self-Esteem

Tables 10-11 present descriptive statistics and the analysis of variance for the impact of students' academic level and their active involvement in learning through different technological conditions on students' self-esteem.

As can be seen in Tables 10-11, a significant main effect, with a medium effect size, was found for students' academic level on their self-esteem. Post-hoc comparisons showed, as expected, that the self-esteem of students in the remedial track was significantly lower compared to students in the mainstream and excellence tracks (p's$< .01$). The main effect of technology was not statistically significant; however, a significant interaction effect with a medium size effect was found between students' academic level and technology. It seems that the self-esteem of remedial students is lower in learning with technology – either RTMA or whole-class technology compared to learning without technology (p's$<.05$). The self-esteem of the students in the mainstream track is higher in learning with RTMA compared to learning with whole-class technology or without technology (p's$<.05$), while for the excellence track, the presence or absence of technology was not related to self-esteem.

Discussion

This study examined how academic achievement, students' perceived learning, academic self-efficacy, learning motivation, and self-esteem are affected by their academic level and the extent of their involvement in active technology-enhanced learning. In contrast to the *first research hypothesis* regarding the

Table 10. Self-esteem – descriptive statistics

Student Academic Level	Technological Condition	Mean	SD
Remedial class (n=20)	Without technology	5.02	0.48
	Whole-class technology	4.48	1.08
	RTMA	4.25	1.23
	Total	4.58	1.01
Mainstream class (n=36)	Without technology	5.08	1.16
	Whole-class technology	4.96	1.18
	RTMA	5.23	1.14
	Total	5.10	1.15
Excellence class (n=24)	Without technology	5.13	0.60
	Whole-class technology	5.18	0.71
	RTMA	5.00	0.70
	Total	5.10	0.66
Total (n=80)	Without technology	5.09	0.90
	Whole-class technology	4.93	1.05
	RTMA	4.97	1.10
	Total	5.00	1.02

Table 11. Self-esteem – the analysis of variance

Factor	F	df	p	Partial η^2
Student academic level	4.13	2, 198	<.001	.04
Technological condition	1.05	2, 198	=.35	.01
Interaction effect	2.21	4, 196	<.05	.03

main effect of technology, technological conditions per se (RTMA, whole-class technology, and without technology) did not influence any of the dependent variables. This finding is not consistent with the study assumptions of the benefit of multimedia learning (Mayer, 2001) and non-linear learning with a whole-class technology (Blau, 2011) over learning without technology, nor with the impact of active students' involvement in the learning process through RTMA over learning with a whole-class technology. It is possible that the diversity of the participants in different classes impeded on exploration of the main effect of technology. In future studies we recommend investigating the effect of RTMA in more homogeneous samples.

In accordance with the *second research hypothesis*, student academic level was found to affect the dependent variables. As expected based on previous research (e.g., Meissel, & Rubie-Davies, 2016; Schnorr et al., 2016), students' achievement and academic self-efficacy were the highest in the excellence track and the lowest in the remedial track. Is was encouraging to find that the cognitive aspect of perceived learning, namely, acquiring new understanding, was as high in the remedial track as in the excellence track. Moreover, the emotional aspect of perceived learning was the highest in the remedial track, suggesting that remedial students find the learning processes to be more interesting and enjoyable. Regarding learning motivation, the intrinsic motivation for learning was as high in the remedial track as in the excellence track and extrinsic motivation was the highest in the remedial track and the lowest in the excellence track. Altogether, these findings suggest that, despite the learning difficulties of the students in the remedial classroom, their motivation is very high and the learning processes seem to be effective. However, as expected, and in accordance with previous research (e.g., Alves-Martins, Peixoto, Gouveia-Pereira, Amaral, & Pedro, 2002), self-esteem was the lowest for students in the remedial track.

The most interesting findings in this study relates to the interaction effect between the technological conditions and students' academic level. Consistent with the *third hypothesis*, a significant interaction effect was found between the participants' academic level and technological conditions on all the dependent variables. In the remedial track, achievement and self-efficacy were significantly higher after lessons using RTMA compared to lessons with whole-class technology and learning without technology. These findings are consistent with previous research on using "low-tech" - response cards among remedial students to increase their participation (e.g., Schnorr et al., 2016). Regarding the emotional aspect of perceived learning and extrinsic motivation, consistent with the research hypothesis, remedial students enjoyed learning the most with RTMA and the least when learning without technology. Intrinsic motivation in the remedial track was the highest for whole-class technology and the lowest for learning without technology. Altogether, the findings suggest that students in remedial education benefit more from learning with technology, and especially from RTMA technology. This effect can be explained by both the more active participation of *all* students in the learning process through RTMA and by the benefits of frequent anonymous feedback (Blau & Caspi, 2010). It seems that RTMA functions as an *assistive technology*, which enhances the learning potential of remedial students. The findings of the current study are consistent with previous findings in higher education (He et al., 2018), according to which students with lower high-school GPAs benefited more from an active learning approach in a partially flipped classroom, while student with higher GPA were not affected by this this pedagogical approach. However, the self-esteem of our remedial students was the lowest in leaning with RTMA, which monitored their errors, and the highest in learning without technology. It seems that remedial students still need to build more confidence in relation to learning from mistakes when using formative ongoing technology-enhanced feedback.

For students in the excellence track, the cognitive and emotional aspects of perceived learning and self-efficacy were the highest in the whole-class technology condition. However, intrinsic motivation in the excellence track was the highest when learning with RTMA. For all technological conditions, the achievement and self-esteem of these students were high, while extrinsic motivation was relatively low. It seems that students in the excellence track are confident high-achievers, who do not require external approval to boost their motivation (Alves-Martins et al., 2002). However, even these students benefit in terms of intrinsic motivation from ongoing feedback through RTMA.

Surprisingly, for students in the mainstream track the highest level of achievement, perceived learning, and self-efficacy were found among those learning without technology. The highest extrinsic motivation was for learning with whole-class technology, while intrinsic motivation and self-esteem did not differ significantly among the technological conditions. In general, students in the mainstream track showed relatively low level of intrinsic motivation, and it seems that the lack of motivation impeded upon their benefits from technological integration (Blau & Barzel-Rubin, 2013). However, the self-esteem of students in the mainstream track was higher when learning with RTMA compared to the other conditions, suggesting that this group of students can benefit from externalization of their learning progress through ongoing mobile assessment.

CONCLUSION AND IMPLICATIONS

The results of this study suggest that students at different academic levels benefit from different learning technologies. RTMA seems to be the appropriate assistive technology for remedial students, while whole-class technology enhances the learning of students in excellence tracks. However, students in the excellence track benefit from RTMA in terms of intrinsic motivation, whereas students in the mainstream track – in terms of self-esteem. This study suggests that the benefits of RTMA can be attributed to the anonymous feedback, which makes learning more relaxed and effective for low-achievers and more exciting for high-achievers. Based on these findings, it is recommended to incorporate RTMA in learning processes, especially as assistive technology for low-achievers, as well as anonymous feedback for the general population - when studying complicated subjects or discussing delicate topics.

Although this study employed a field experiment research design, this experiment was conducted in a single secondary school and might be affected by its specific learning culture. Future studies may explore these issues and try to replicate the findings in a more diverse and larger sample, as well as employing other research methods, e.g., classroom and outdoor observations. In addition, in the future, the anonymity effect of RTMA can be more easily isolated, manipulated and studied in a lab experiment.

REFERENCES

Agbatogun, A. O. (2014). Developing learners' second language communicative competence through active learning: Clickers or communicative approach? *Journal of Educational Technology & Society, 17*(2), 257–269.

Ahmed, A. (2018). Perceptions of using assistive technology for students with disabilities in the classroom. *International Journal of Special Education, 33*(1), 129–139.

Alves-Martins, M., Peixoto, F., Gouveia-Pereira, M., Amaral, V., & Pedro, I. (2002). Self-esteem and academic achievement among adolescents. *Educational Psychology, 22*(1), 51–62. doi:10.1080/01443410120101242

Blau, I. (2011). Teachers for "Smart classrooms": The extent of implementing of an Interactive Whiteboard-based professional development program on elementary teachers' instructional practices. *Interdisciplinary Journal of E-Learning and Learning Objects, 7*, 275–289. doi:10.28945/1524

Blau, I., & Barak, A. (2012). How do personality, synchronous media, and discussion topic affect participation? *Journal of Educational Technology & Society, 15*(2), 12–24.

Blau, I., & Barzel-Rubin, A. (2013). Assisting learning disabilities and empowering excellence by "clickers" and interactive whiteboards. In *Proceedings of the 15th Biennial EARLI2013 Conference for Research in Learning and Instruction.* TUM School of Education, Technical University Munich.

Blau, I., & Caspi, A. (2008). Do media richness and visual anonymity influence learning? A comparative study using Skype. In Y. Eshet, A. Caspi, & N. Geri (Eds.), *Learning in the Technological Era* (pp. 18–24). Ra'anana, Israel: Open University of Israel.

Blau, I., & Caspi, A. (2010). Studying invisibly: Media naturalness and learning. In N. Kock (Ed.), Evolutionary psychology and information systems research: A new approach to studying the effects of modern technologies on human behavior (pp. 193-216). New-York: Springer.

Blau, I., & Eshet-Alkalai, Y. (2017). The ethical dissonance in digital and non-digital learning environments: Does technology promotes cheating among middle school students? *Computers in Human Behavior*, *73*, 629–637. doi:10.1016/j.chb.2017.03.074

Blau, I., Grinberg, R., & Shamir-Inbal, T. (2018). Pedagogical perspectives and practices reflected in metaphors of learning and digital learning of ICT leaders. *Computers in the Schools*, *35*(1), 32–44. doi:10.1080/07380569.2018.1427960

Blau, I., & Hameiri, M. (2017). Ubiquitous mobile educational data management by teachers, students and parents: Does technology change school-family communication and parental involvement? *Education and Information Technologies*, *22*(3), 1231–1247. doi:10.100710639-016-9487-8

Blau, I., Peled, Y., & Nusan, A. (2016). Technological, pedagogical and content knowledge (TPACK) in one-to-one classroom: Teachers developing "digital wisdom". *Interactive Learning Environments*, *24*(6), 1215–1230. doi:10.1080/10494820.2014.978792

Blau, I., & Shamir-Inbal, T. (2017). Digital competences and long-term ICT integration in school culture: The perspective of elementary school leaders. *Education and Information Technologies*, *22*(3), 769–787. doi:10.100710639-015-9456-7

Blau, I., & Shamir-Inbal, T. (2018). Digital technologies for promoting "student voice" and co-creating learning experience in an academic course. *Instructional Science*, *46*(2), 315–336. doi:10.100711251-017-9436-y

Blau, I., Weiser, O., & Eshet-Alkalai, Y. (2017). How do medium naturalness and personality traits shape academic achievement and perceived learning? An experimental study of face-to-face and synchronous e-learning. *Research in Learning Technology, 25*. http://repository.alt.ac.uk/2380/

BoCrews, T. B., Ducate, L., Rathel, J. M., Heid, K., & Bishoff, S. T. (2011). Clickers in the classroom: Transforming students into active learners. *EDUCAUSE - ECAR Research Bulletin*, *9*, 1–12.

Brady, M., Seli, H., & Rosenthal, J. (2013). Metacognition and the influence of polling systems: How do clickers compare with low technology systems. *Educational Technology Research and Development*, *61*(6), 885–902. doi:10.100711423-013-9318-1

Bryant, B. R., & Seok, S. (2017). Introduction to the special series. *Technology and Disabilities in Education, 29*(3), 121-122.

Camacho-Miñano, M. D. M., & del Campo, C. (2016). Useful interactive teaching tool for learning: Clickers in higher education. *Interactive Learning Environments*, *24*(4), 706–723. doi:10.1080/10494820.2014.917108

Caspi, A., & Blau, I. (2011a). Collaboration and psychological ownership: How does the tension between the two influence perceived learning? *Social Psychology of Education*, *14*(2), 283–298. doi:10.100711218-010-9141-z

Caspi, A., & Blau, I. (2011b). Media choice for intra-school communication: The role of environment, user, and medium. *Journal of Educational Multimedia and Hypermedia*, *20*(4), 337–360.

Eden, D. (2017). Field experiments in organizations. *Annual Review of Organizational Psychology and Organizational Behavior*, *4*(1), 91–122. doi:10.1146/annurev-orgpsych-041015-062400

He, W., Holton, A., & Farkas, G. (2018). Impact of partially flipped instruction on immediate and subsequent course performance in a large undergraduate chemistry course. *Computers & Education*, *125*, 120–131. doi:10.1016/j.compedu.2018.05.020

Hwang, G. J., & Chang, H. F. (2011). A formative assessment-based mobile learning approach to improving the learning attitudes and achievements of students. *Computers & Education*, *56*(4), 1023–1031. doi:10.1016/j.compedu.2010.12.002

Kowalski, F. V., Kowalski, S. E., Colling, T. J., Cuba, J. G., Gardner, T. Q., Greivel, G., & Ruskell, T. (2015). Using inksurvey with pen-enabled mobile devices for real-time formative assessment: I applications in diverse educational environments. In *The Impact of Pen and Touch Technology on Education* (pp. 297–305). Cham: Springer. doi:10.1007/978-3-319-15594-4_31

Martyn, M. (2007). Clickers in the classroom: An active learning approach. *EDUCAUSE Quarterly*, *2*, 71–74.

Mayer, R. E. (2001). *Multimedia learning*. Cambridge, UK: Cambridge University Press. doi:10.1017/CBO9781139164603

Meissel, K., & Rubie-Davies, C. M. (2016). Cultural invariance of goal orientation and self-efficacy in New Zealand: Relations with achievement. *The British Journal of Educational Psychology*, *86*(1), 92–111. doi:10.1111/bjep.12103 PMID:26856404

Morales, L. (2011). Can the use of clickers or continuous assessment motivate critical thinking? A case study based on corporate finance students. *Higher Learning Research Communications*, *1*(1), 33–42. doi:10.18870/hlrc.v1i1.31

Morgan, R. P. (2008). Exploring the pedagogical effectiveness of clickers. *Scholarly Teaching*, *3*, 31–36.

Nikou, S. A., & Economides, A. A. (2016). The impact of paper-based, computer-based and mobile-based self-assessment on students' science motivation and achievement. *Computers in Human Behavior*, *55*, 1241–1248. doi:10.1016/j.chb.2015.09.025

Pintrich, P. R., Smith, D. A. F., Garcia, T., & McKeachie, W. J. (1991). *A manual for the use of the Motivated Strategies for Learning Questionnaire (MSLQ)*. Ann Arbor, MI: National Center for Research to Improve Postsecondary Teaching and Learning, University of Michigan.

Ponce, H. R., Mayer, R. E., Figueroa, V. A., & López, M. J. (2018). Interactive highlighting for just-in-time formative assessment during whole-class instruction: Effects on vocabulary learning and reading comprehension. *Interactive Learning Environments*, *26*(1), 42–60. doi:10.1080/10494820.2017.1282878

Porat, E., Blau, I., & Barak, A. (2018). Measuring digital literacies: Junior high-school students' perceived competencies versus actual performance. *Computers & Education*, *126*, 23–36. doi:10.1016/j.compedu.2018.06.030

Rosenberg, M. (1965). *Society and the adolescents Self Image*. Princeton, NJ: University Press. doi:10.1515/9781400876136

Schnorr, C. I., Freeman-Green, S., & Test, D. W. (2016). Response cards as a strategy for increasing opportunities to respond: An examination of the evidence. *Remedial and Special Education*, *37*(1), 41–54. doi:10.1177/0741932515575614

Shamir-Inbal, T., & Blau, I. (2016). Developing Digital Wisdom by students and teachers: The impact of integrating tablet computers on learning and pedagogy in an elementary school. *Journal of Educational Computing Research*, *54*(7), 967–996. doi:10.1177/0735633116649375

Shamir-Inbal, T., & Blau, I. (2017). Which pedagogical parameters predict the general quality of ICT integration from the perspective of elementary school leaders? *Computers in the Schools*, *34*(3), 168–191. doi:10.1080/07380569.2017.1347427

Stowell, J. R. (2015). Use of clickers vs. mobile devices for classroom polling. *Computers & Education*, *82*, 329–334. doi:10.1016/j.compedu.2014.12.008

Sung, Y. T., Chang, K. E., & Liu, T. C. (2016). The effects of integrating mobile devices with teaching and learning on students' learning performance: A meta-analysis and research synthesis. *Computers & Education*, *94*, 252–275. doi:10.1016/j.compedu.2015.11.008

Weiser, O., Blau, I., & Eshet-Alkalai, Y. (2018). How do medium naturalness, teaching-learning interactions and Students' personality traits affect participation in synchronous E-learning? *The Internet and Higher Education*, *37*, 40–51. doi:10.1016/j.iheduc.2018.01.001

Chapter 15
Leveraging Mobile Technologies to Support Active Learning for All Students:
Smartphones to Support Learning

Jennifer L. Gallup
Idaho State University, USA

Beverly B. Ray
Idaho State University, USA

Cory A. Bennett
Idaho State University, USA

ABSTRACT

This chapter explores the use of smartphones, a category of mobile learning, as learning devices that align well with tenets of universal design for learning (UDL). The theoretical foundation for this use is explored along with pertinent research. The chapter continues with an exploration of specific apps that reflect the tenets of UDL even as they promote learning, specifically executive function, collaboration, and knowledge acquisition and expression. Guidelines for select and successfully using apps as learning tools are provided along with a discussion of barriers to use. Recommendations for use, including actionable ideas for integration, are provided.

DOI: 10.4018/978-1-5225-8106-2.ch015

INTRODUCTION

As technology continues its rapid evolution, keeping up with that change and integrating technology into teaching and learning remains essential. Tomorrow's high school graduates must be prepared to compete in a digital, global economy that is becoming more mobile every day. Yesterday's desktops and other stationary devices have given way to mobile devices such as laptops, smartphones, and tablets that continue to be integrated into classrooms be it through school-based one-to-one programs or with students using their own devices to use during class time. The devices themselves are only part of this evolution. When the tools are paired with rich but uncomplicated applications (apps) that support content acquisition, engagement, and collaboration, the way in which students learn changes. The tools not only become more accessible, but the learning shifts to mastery-based and individualized approach which is ideal for a larger spectrum of learners including students with disabilities within general education classes.

Smartphones are ubiquitous and used by most teens and adults for communicating, socializing, gaming, conducting business, finding information and identifying solutions to problems, and navigating from point to point. Smartphone ownership and use has become commonplace among teenagers ages 13-17 years old. Eighty-eight percent of teenagers have or have access to a cell phones or smartphones (Lenhart, 2015); 91 percent of teenagers access the Internet on cell phones, tablets and other mobile devices at least occasionally (Lenhart, 2015). This group of young people has been dubbed, now for more than a decade, "Digital Natives" (Prensky, 2005), the "Net Generation" (Tapscott, 1999; Tapscott, 2009) as a result of the impact that digital devices and Internet capabilities have had on all areas of their lives. Access and ownership of smartphones increases the amount of time teens spend online, even in learning situations. We are now in an era that requires meaningful ways to effectively integrate apps and smartphones into the classroom, rather than ban them.

Research continues to be conducted on teen's use of smartphones and the Internet. In a recent report, Lenhart (2015) noted 92% of teens report going online daily with an overwhelming 24% of teens age 13-17 indicating they consistently access the internet while more than half (56%) of teens access the Internet several times a day, and only 12% report once-a-day use. Within learning contexts, accessing information for academic projects, analysis of information, and collaborative interactions through mobile devices supports engagement through readily available and accessible tools and resources. Some educators realize the vital need to integrate technology as a means for the delivery of information, such as with classroom blogs to post relevant information related to course content, participating in conversations outside of class related to the course, and facilitating synchronous group conversations through such platforms as Facetime, Skype, or Messenger.

Through the use of mobile technologies, students engage in more personalized learning, which leads to higher academic achievement for all students, especially, those with disabilities (CAST, 2014). Furthermore, preparing students, especially students with disabilities, to be an active participant in their own learning creates more inclusive learning as all students can better leverage their strengths and interests to support learning (Basham & Marino, 2013). However, the design of lessons greatly determines the overall effectiveness of learning. If lessons are only designed around the "average" student, then others, such as students with disabilities, are often unintentionally left out. For students who are already unintentionally marginalized, such as students with disabilities, barriers to learning will continue to persist. Some barriers for students with disabilities include comprehension, equitable access to curriculum based on reading level, and alternative options for expression of learning. Leveraging research-based frameworks that are purposefully designed around an inclusive lens allows teachers to be more effective in supporting learning for all students in the class. One such framework that continues to be recognized for its efficacy in supporting students with various learning differences is Universal Design for Learning (UDL).

The purpose of this chapter is to explore the use of smartphones and apps as a category of mobile learning, which align well with tenets of UDL. This chapter assists practitioners by understanding the role and tenets of UDL, in brief, and exploring specific apps that promote learning through UDL. Some apps that attend to students' executive function, collaboration, knowledge acquisition, and personal expression of their learning are highlighted. Additionally, guidelines and recommendations for selecting, using, and overcoming barriers are provided. As such, teachers, teacher educators, professional development facilitators, and administrators from international contexts will benefit from this chapter. These potential areas of interest are an outgrowth of a global society fully engaged in identifying methods of integrating technology into everyday learning.

A CONTEXT FOR SMARTPHONE USE

This chapter is grounded in two primary areas of research: 1) supporting inclusive learning environments, for all students, by leveraging the curricular design aspects of UDL and 2) the use of smartphones as viable tools for learning within a constructivist framework because of their ubiquitous and dynamic nature. When considered together, these two areas of research describe a method by which all students have equitable access to learning by leveraging the potential of handheld technologies, like smartphones. Each of these two areas will be addressed below in greater detail followed by a description of smartphone apps that support student learning through UDL.

Universal Design for Learning: An Overview

Universal Design for Learning has been identified as an effective instructional framework for diverse learners across academic contexts. The premise of UDL is based on the understanding that all students can learn, and if one support may be good for a student with a disability, the support could be beneficial for all students. A lesson plan or curriculum designed through the lens of UDL provides multiple paths for each student to achieve the same goal. These standards do not need to be modified for an individual with a disability, the standards need to be met differently, and perhaps with apps. While each student would use these same criteria and standards as benchmarks of achievement, the process, product, and overall path of learning the content and expressing understanding will likely be different.

In essence, a UDL lesson, plans for learning beyond the margins, or the "middle of the road learners," to include more students within inclusive learning experiences at grade-level and aligned to grade-level standards. This allows teachers to reach populations, such as students with disabilities, often organized in core content classes, through purposefully designed choices that leverage students' strengths. This is the premise of UDL, allowing choice through multiple means of representation (MMR) such as: texts, audio, or video representations of content. Multiple means of engagement (MME) by using student choice, varying demand, and by developing purposeful and motivated learning experiences. Finally, by supporting multiple means of action and expression (MMAE). Through a UDL curriculum based on these three principles a teacher can represent information, allow students to engage with the content, and support learning through their actions and expressions of knowledge, in more than one way.

However, designing learning experiences with the UDL framework requires intentional planning to integrate these three principles to create an inclusive educational environment for all students (Gallup & Bennett, 2018). Although UDL does not necessitate the integration of technology, teachers who are able to leverage technology in their instructional practices increase their capabilities to incorporate the principles of UDL for the benefit of all students (Edyburn, 2010). This means that appropriate technology is often a key aspect of effective implementation and integration of learning. Smartphones, commonly brought to classes (e.g., BYOD), when purposefully used, can create equitable opportunities for students to demonstrate understanding through UDL.

Supporting Inclusive Learning Through UDL

Today's classrooms are diverse and require lessons designed to meet the needs of a wide range of student abilities. Students with disabilities make up approximately 30% of all general education classes. This does not include the number of students with possible undiagnosed learning disabilities, students who are struggling with content acquisition, or English Learners. Designing and implementing inclusive learning practices has generated substantial attention over the past two decades and more recently since the revision of the Individuals with Disabilities Education Act (IDEA) in 2004. Universal Design for Learning is a theoretical framework used to help proactively design curricula rather than retrofitting a lesson to meet the needs of individuals with differing learning needs.

- **Multiple Means of Engagement:** Teaching is a complex process built on a series of decisions to support student learning. This process includes delivering content in a manner that supports critical and analytical thinking, discourse, and making complex ideas understandable while helping students develop their own ideas related to the content presented. Engagement or the "why" of learning is predicated on recruitment of interest, sustaining effort and persistence, and self-regulation. First, teachers need to engender interest in the topic presented by supporting individual choice and autonomy by creating lessons that are relevant, of value, and authentic learning experiences (Bennett, 2012). Through engagement, students will sustain effort and persistence by varying the demand (based on student level as needed), fostering collaboration, and increasing mastery based feedback. Feedback supports engagement by providing checkpoints throughout a lesson to redirect a student's learning. Engagement supports accessing the curriculum; when possible, CAST, (2014) recommends that students have as much autonomy over: (1) perceived challenge, (2) reward and recognition, (3) context and/or content used for practicing and assessing skills, (4) informational tools used, (5) design and layout of projects/assignments, and (6) sequence or timing or task and completion of assignment. A smartphone can support engagement by allowing a student to have autonomy over projects, develop projects with peers, and access materials in a context or content of choice.
- **Multiple Means of Representation (MMR):** Students often prefer content represented in multiple or varied methods. Learning is difficult when the information presented is too complex to understand or requires extraordinary effort to dissect word, content, or relationships. To reduce the challenge with presentation of new information all material should be presented in multiple ways. For example the same information should be presented in different modalities, such as reading a text, listening to an audio, or watching a video. By incorporating MMR a teacher can ensure that students of all abilities have equitable access to the content while attending to sensory and

perceptual challenges or disabilities. Information presented using MMR though a smartphone supports language and symbols by clarifying vocabulary, syntax, and structure, and will promote understanding across languages.

- **Multiple Means of Action and Expression (MMAE):** Multiple Means of Action and Expression provides options for physical action, manipulation of content, varying the methods of response and by optimizing access to tools and assistive technologies. This in turn supports the expression by using multimedia, construction, and development of projects and assignments, and by building fluency with scaffolded support. Additionally, MMAE also considers executive functioning such as goal setting, planning and strategy implementation, management of information, and progress monitoring. Consider that students may choose to create their representation of understanding using different media, tools, or materials. They may choose to write out their work by hand, leverage audio-based input devices, use a keyboard to type out their thinking or use other adaptive devices such as a joystick or certain apps on a smartphone. Since it is possible that no two students will choose the same outlet of expression, it is important to have clear and consistent criteria, and ideally well aligned rubrics, so all learners can interact with the content and demonstrate understanding of the standards. A curriculum that is equitable in access is designed with materials to provide a seamless interface with technologies such as a smartphone and other assistive technologies. Often individuals with movement impairments, vision challenges, or cognitive disabilities can navigate and express what they learned better. A smartphone can support MMAE by allowing the student to develop timelines, checkpoints, and reminders all from a calendar app, interact with and collaborate with groups through collaboration and chat options, or create products to demonstrate mastery using smartphone apps as well.

A variety of apps support student's participation individually, in group, or within the context of whole class activity. Smartphones present multiple opportunities for students with disabilities to participate in core-content classes commensurate with their peers. Additionally, smartphones hold the potential to support differentiation across learning preferences and content for all students.

Smartphones as Viable Learning Tools

The theoretical foundation supporting the use of smartphones as educational tools is situated within an understanding of social and intellectual learning theory, wherein learners actively engage knowledge using technology as a catalysts for deep and meaningful learning (Piaget & Inhelder, 1969; Vygotsky, 1999). Specifically, constructivists view learning as an intellectual process that asks the learner to construct new knowledge by combining new ideas with those acquired during previous learning experiences (Schunk, 2015). Depth of understanding is affected by previous knowledge and interests (Bruner, 1993; Piaget & Inhelder, 1969). Likewise, use of smartphones supports additional key tenets of constructivism wherein students are responsible for their own learning. When students use smartphones for educational purpose, they meaningfully interact with peers to demonstrate understanding and clarify misconceptions.

Today the use of smartphones is ubiquitous and by harnessing the power of technology that is commonplace, students have more opportunities to access content on demand. Gone are the days of flipping through encyclopedias or card catalogs in the hopes of finding what is needed. Additionally, the Common Core State Standards (CCSS) discuss preparing students to be competitive on a national and international level. It makes sense to utilize tools that potentially can support broader perspectives and access to information that was once not easily attainable This means that in many contexts, smartphones are literally at the tips of students fingers and can provide immediate access to information beyond the scope of textbook content.

Support for Smartphones and Equitable Access

The Elementary and Secondary Education Reauthorization Blueprint for Reform in 2010 (U.S. Department of Education, 2010a) set forth a motion in nearly all federal education policy initiatives to utilize and increase technology to support the equitable access to content specific to instruction and assessment. For example, The National Education Technology Plan (NETP; U.S. Department of Education, 2010b) states, "the challenge for our educational system is to leverage the learning sciences and modern technology to create engaging, relevant, and personalized learning experiences for all learners that mirror students' daily lives and the reality of their futures" (p. 4).

Smartphones can provide assistive technology (AT), technology that improve a student's functional capabilities, and instructional technology (IT), technology that supports curriculum, instruction, and assessment. Some tools such as advanced planners, voice recognition software, reminders, virtual dictionaries, and an ever-increasing supply of instructional apps can also be considered IT. All of which can be beneficial to students with disabilities. Using technology such as a smartphone gives a child a voice and fosters independence (i.e., autonomy). This allows students to succeed in ways that work for them and their learning preferences, blending, school, home, and community environments. When technology is used efficaciously it opens learning opportunities for students that otherwise may not be possible.

Additionally, adolescents with disabilities, those students who are in grades five to twelve, indicate they prefer virtual learning over traditional learning environments including traditional reading, discussions, and labs (Marino, Israel, Beecher, & Basham, 2013). Many classrooms are still created using reading tasks, writing prompts, and/or worksheets as the primary means for developing skills or content knowledge. However, when learning is designed using the UDL framework, and when students are presented opportunities to use technology in meaningful and applicable ways, equitable access to the content increases (Gallup & Bennett, 2018). While the research on using smartphones is emergent, there is evidence that they can support all learners, including those with disabilities and students who are at risk of failure (Rappolt-Schlitman et al., 2013).

Technology and Education Engagement and Augmentation of Learning

With the rapid increase of technology and apps, the ability to support education has become easier and at the fingertips of teachers and students. Mobile devices offer many benefits, when compared with desktop computers. The benefits include flexibility of learning, access through portable devices and lower cost, and ease of use with timely apps. Therefore, smartphones and other portable devices should be seen as potential tools to facilitate mobile learning, and, specifically, mobile adaptive expression and engagement for all students, especially for those with disabilities.

Tablets, such as iPads and other handheld portable devices, have been incorporated into education to support critical and higher-order thinking (Ritz, 2009), which supports all students. Swine (2012) noted when digital devices like smartphones were used as tools in the classroom, enhanced engagement and creativity during active, passive, and independent learning times increased. In addition to supporting education for students without disabilities, it has been identified that handheld digital devices have enhanced learning and classroom experiences for students with disabilities (Pennington, 2010). Students with disabilities are using iPads to produce visuals of content and abstract concepts as well as support social and communicative language for students with severe disabilities and Autism Spectrum Disorder (Ploog, Bertram, Scharf, Nelson, & Scharf, 2013; Xin & Sutman, 2011). When considering a handheld devices, smartphones are often more readily available and smaller, making their portability easier than the traditional tablet especially for individuals with disabilities.

EXAMPLES OF UDL GROUNDED SMARTPHONE APPS

Despite evidence of use, including potentially effective use, it is not easy for educators to identify quality educational apps. That is because many apps, even when marketed as educational apps, do not meet the needs of all learners (Papadakis, et. al, 2017). Because educators are often overwhelmed by the sheer volume of apps and may not yet know how to use them, the following section identifies a number apps along with practical actions for their use. However, this is not a definitive list but rather a list of apps the authors use when working with developing and experienced teachers. It is recommended that the reader explore these apps on their own prior to implementation with students to determine the extent to which they fit the needs of students within the respective context or learning environment. See the Appendix A for information on some suggested apps, accessing the apps discussed in this chapter, and a possible UDL category for the app.

Multiple Means of Action and Expression, Representation, and/or Collaboration

A variety of apps support student's participation individually, in group, or within the context of whole class activity. A variety of formats exist across these apps to do so. Those include apps that facilitate multiple choice, fill in the blank, and short answer assessment responses. The following apps are a start to understanding a few to support MMR and MMAE. Representation includes videos, text to speech, readings, and navigation of material while expression and collaboration are also part of these apps. For example Google Presentations, as part of the Google Suite,has features that double both as a tool to represent content using video and voice-over, as well as a speech to text function allowing teachers to represent material so they are more accessible and students can utilize the platform for collaborative expression in virtually any content area.

Google Suite

Google Suite is comprised of a tool to develop documents, presentations, spreadsheets, and other tools such as forms/surveys. Other platforms on Google Suits can be equally as versatile in all content areas such as: Google sites, maps, and drawings. One specific example is the Google Docs, a versatile web-based tool that is perfect for synchronous and asynchronous collaboration. It can be accessed from a smartphone through an app (Apple, Android, Windows) or any portable digital device. Common uses for Google Docs are collaborative writing projects where students can work nearby or at a distance on the same document at the same time, or at any time that is convenient for individual group members. Another use for the Google Doc app is collaborative brainstorming. Google Docs has a robust set of drawing tools that can be used to create thinking maps and visual diagrams that can be accessed and edited by all collaborators (See Figure 1 for a sample of a concept map developed using a handheld device and Google Drawing). Google Docs can also be used to create templates, checklists, or rubrics that you can make available to students to use when completing projects which assists in executive functioning. These tools are often used for collaboration and expression; however, the speech-to-text within Google Docs supports students with physical disabilities and an aspect of executive functioning by organizing text in color, with embedded notes, and auto-save.

Lucidchart

Lucidchart is an online app that allows users to connect content by creating a diagram that visually displays connections to major concepts and ideas presented in any content area. Lucidchart allows the user to create simple flowcharts to complex diagrams either as a team or individually. Lucidchart has an option to allow the user to embed live data and automatically build charts. Lucidchart is an app that can help students who benefit from visual connections create a diagram of content both for comprehension and expression of ideas (See Appendix B for a sample of Lucidchart and a chapter dissection).

Padlet

Padlet is a free app that makes sharing ideas and posting information using smartphones easy. This resource can be accessed on any device (computer desktop, tablet, and phone) and is available as a free iOS and Android app. Padlet is a virtual bulletin board that allows individuals to post questions, reviews, opinions, or feedback on a select topic. Padlet users create a board by selecting from a variety of premade templates. Once a board is created, the user can share it with the audience of choice. Contributors have

Figure 1. Sample of a Google Draw concept map

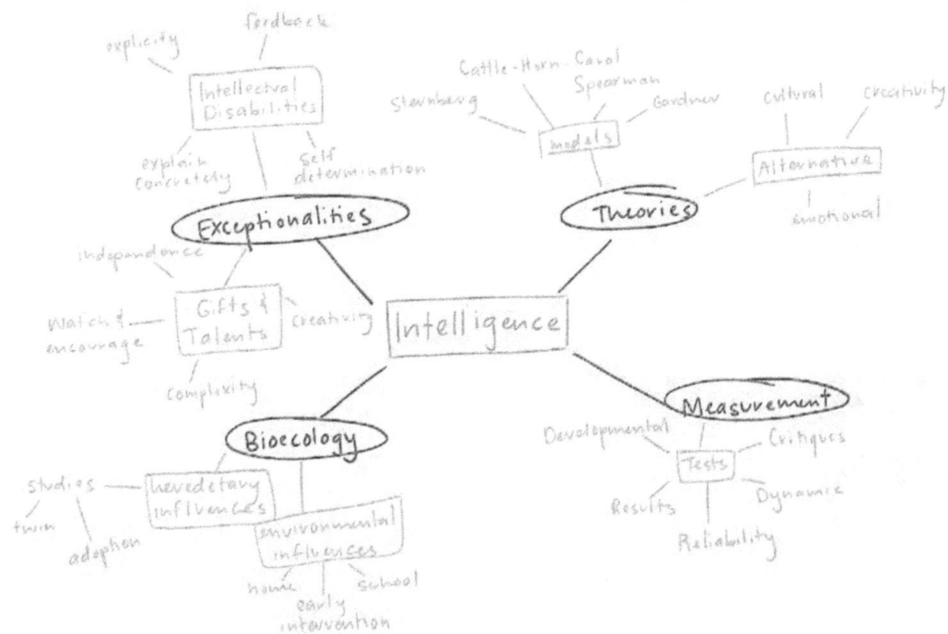

the ability to share notes, links, videos, and more. The versatility of sharing in Padlet supports specific needs of students with disabilities allowing equitable access to group discussions, and content representation through speech-to-text, video, and photos. The settings feature allows teachers to control the privacy of each board, choosing to make the board completely open for public contributions, completely private, or moderated (the creator approves all posts before they show on the board). *Padlet* is also ideal for organizing project based learning and sharing collaborative work with peers (see Figure 2).

Twitter

Twitter is an online social network that allows members to *tweet* messages up to 140 characters in length. Teachers can use Twitter to send out reminders to students, review content, or send out assignments. It can also be used to share information and successes with parents. Students can use the app to collaborate with peers, send and receive homework help, share notes, or keep up with a current event topic.

Multiple Means of Representation

The following apps support MMR by offering alternative ways to receive information through audio, visual, and custom displays. These apps also can support the activation of background knowledge through pre-teaching options and can highlight connections between content. Additionally, the options for representation may help support language acquisition by clarifying vocabulary through contextualization.

LibriVox

Librivox was founded in 2005 and is comprised of volunteers from all over the world who record public domain texts: poetry, short stories, whole books, dramatic works, and in many different languages. All LibriVox recordings are housed in the public domain in the United States of America and available as free downloads on the internet. *LibriVox* supports representation of content in speech format and ultimately equitable access to text-based publications for all individuals, especially those with disabilities.

Figure 2. Sample of a science project

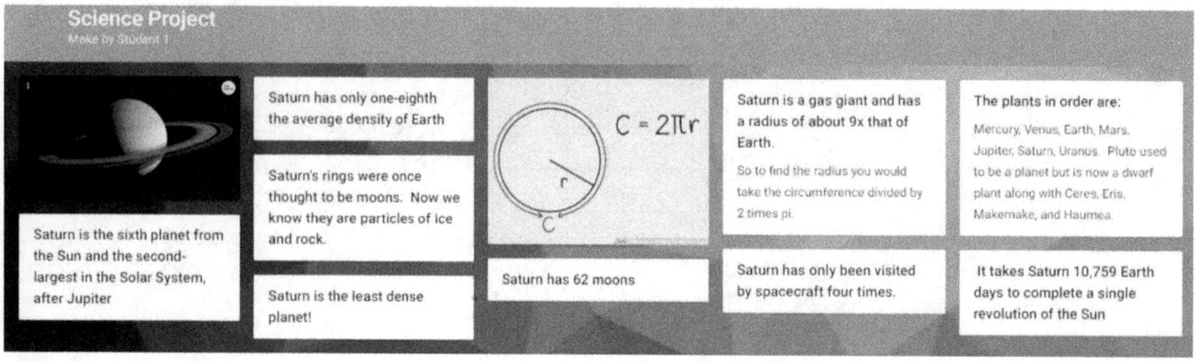

Seesaw

Seesaw is a digital portfolio application teachers can use in their classroom. It can be used for Pre K-12 students. Teachers are able to make assignments, blog, journal, keep up with students goals, and so much more. Students have easy access using their smartphones or other devices and can share their work with family members across the academic year. Use of this app can help teachers use smartphones to build community between school and home.

- **Formative Assessment:** In addition to the broadly focused educational apps outlined above, apps can be used to support specific learning and learners with repetition through content seen as potential tools to facilitate mobile learning, and specifically mobile adaptive formative or summative assessment.

Kahoot!

Kahoot! is a game-based app that helps teachers get students excited about studying a new concept or reviewing for a test. Feedback is instant for both student and teacher. The Kahoot tool is live, simple, and easy to use and effective for any grade level, subject, or language. Additionally, Kahoot supports choral response which enhances student self-esteem, content acquisition, and ultimately student retention of content often helpful for students with disabilities. Another great option for student involvement is creating their own Kahoot games. They can work with others in this endeavor. Students become part of the creation process.

When this app is used with the UDL framework in mind it is possible to attend to MME, MMR, and MMAE. For example, students with disabilities can master content through feedback, support sustained effort and persistence through MME, activate or supply background knowledge, maximize transfer and generalization with opportunities to practice through MMR, and finally MMAE by using multiple tools for construction and composition of work. Like many apps, Kahoot! Is a multipurpose tool that can help clarify information during a lesson and help students develop connections to previous lessons and related topics.

Poll Everywhere

Poll Everywhere is another useful app for sharing information with students and parents. The app allows teachers to create different "polls," or assessments, for students to take using their smartphones. Using the app and smartphones, teachers can check for student understanding quickly while using the app as a quizzes or warm-up. Creating and sharing polls is easy to create and easy to share with students. It works well for any subject area. Additionally, it can be used to kick start discussions, connect content to other real-life situations or other classes, and begin a discussion on sensitive topics by providing a platform for students to make initial comments without revealing their identity. Its anonymous nature encourages reluctant discussants to enter in class discussions giving teachers an opportunity to assess learning and identify student misconceptions.

Plickers

Plickers is a tool that is relatively simple to use and lets teachers collect real-time formative assessment data without the need for student devices. By using Plickers teachers can implement quick formative assessments to gauge student's level of comprehension. Students participate without being self-conscious about providing the incorrect answer as all responses are made using individual symbols and scanned into the teacher's app. To generate student participation, students can develop quizzes based on a topic they were assigned to research and teachers can implement student-led quizzes.

Apps for MMAE and Executive Function

The following apps can support MMAE by creating seamless connections between documents, videos, and reminders to support executive functioning. Executive function is a set of mental processes that helps connect past experience with present action. Executive functioning is required for activities such as planning, organizing, strategizing, paying attention to and remembering details, and managing time and space are performed. Executive functioning effects working memory and can impair a student's ability to function to their highest potential without supports and often one of the biggest challenges for students with disabilities.

Executive function allows learners to engage in a number of higher order thinking skills include: planning for and keeping track of one or more tasks, meaningful transfer of previous learning into current discussions, evaluating the depth and quality of ideas, and using new information to revise or change one's thinking about a topic. It can also be as simply as knowing when to wait to speak when called upon. Innovative apps integrated into instructional planning or provided as an IEP or 504 accommodation can be used to support students discreetly. The apps described below can be used to provide significant support student success in classrooms by increasing engagement, task management, and self-monitoring tasks such as:

- Take step-by-step approaches to work; rely on visual organizational aids.
- Use tools like time organizers, computers or watches with alarms.
- Prepare visual schedules and review them several times a day.
- Ask for written directions with oral instructions whenever possible.
- Plan and structure transition times and shifts in activities.

Additionally, providing students with tools to be successful and teaching them to implement, self-monitor, and learn to be self-aware will support life-long skills and habits that will promote post-high school success as part of the whole curricular design through the lens of UDL.

Remind

Remind is a great UDL appropriate app for all students to use, particularly for students with disabilities, because it helps students use their smartphones to keep on top of assignments and due dates and set priorities for what work needs to be accomplished first. Keeping track of deadlines and responding to collaborators is an executive function skill also connected to soft-skills. Often students with disabilities struggle with time and task management and when and how to respond. Apps such as Remind allows educators to communicate with parents and guardians; when helping students with disabilities, parent interaction is critical to support the development of a skill such as executive functioning.

In particular, special education teachers need fast, reliable ways to reach parents to share information related to behavior, medicines, and home life to know what is working for or against the student on a given day as when students are on-task and engaged learning happens. Parents are valuable to special education teachers because they can carry over learning from school and provide home life information and strategies that can be considered for meeting, adapting and modifying IEP goals which in turn supports academic acquisition and inclusion from a curricular design.

Trello

Trello is a more advanced collaboration tool, best used by older students for multi-step projects. It can organize the individual tasks for each student allowing all to see who is responsible for each and what the progress is on each task. *Trello* can help group members assign tasks, organize materials, and manage information asynchronously as well. This app helps students stay organized and on task, but also requires that students accurately keep the app updated. See Figure 3 for a sample of a *Trello* board.

Thirty/Thirty

The Thirty/Thirty (30/30) app functions as a time management and on task manager recommended for students age 6 and over. It allows students to create timed and categorized tasks with audible or vibrating alerts to keep track of their own progress and manage their break times. Students learn to plan their time while the app helps them to work their plan. This app is available for all handheld platforms.

Figure 3. Science Project Trello board sample

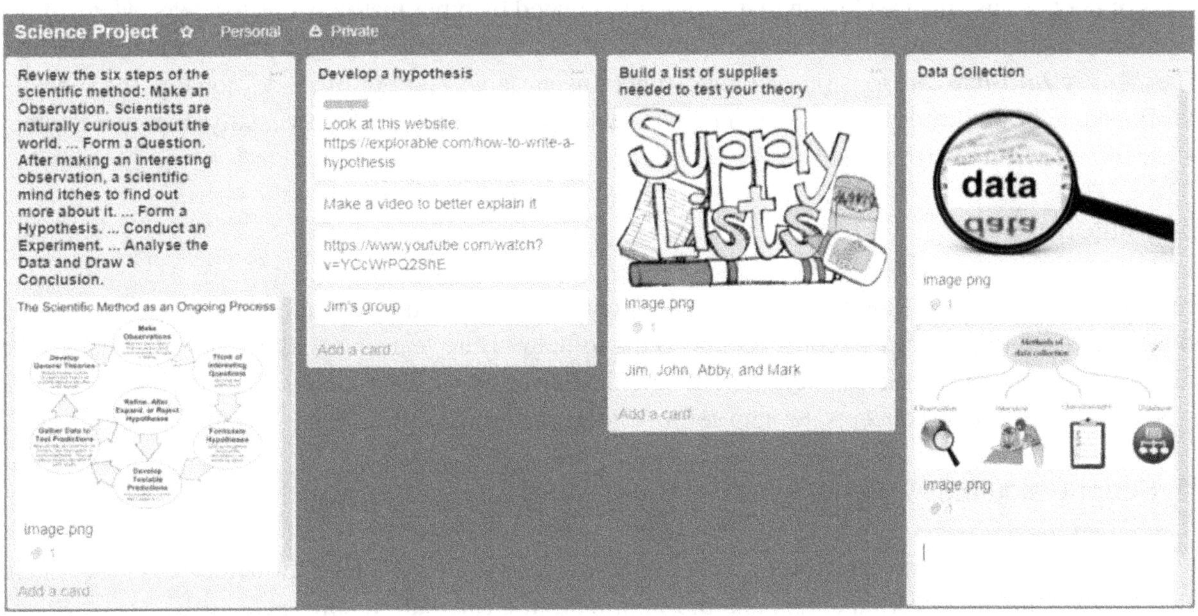

YouNote!

YouNote! is a flexible note taking app available for multiple handheld platforms. Students can take notes by entering text, drawing, or voice recordings. The app supports task management at home, in school, or to bridge the two. New tasks can be added, prioritized, and categorized by adding searchable keywords. The options provided by *YouNote!* support greater self-awareness by catering to the individual's unique needs and strengths. For users that need additional support or have an executive functioning goal can work with the teacher to learn what style works best for them to support their self-monitoring of individual tasks and tailor their personal *YouNote!* library to match their needs.

inClass

This app supports both student and teacher organization. Teachers and students can work together to prioritize, color code, elaborate on tasks, and set timers for engaging in the identified work. This app targets older learners and is well suited for middle and high school students who have access to an iPhone or iPad.

Content Specific Apps: Mathematics

Calculators are a common learning tool in mathematics classrooms and may also serve as an accommodation for some students. There are a number of interesting calculators that provide unique interactions for students. For example:

Desmos

This web and mobile app is a free graphing calculator and is appropriate for middle through high school (and post-secondary learning) classrooms. Teachers can create their own learning experiences for students or they can use the vast lesson plan repository created by other mathematics teachers. Many of the explorations available are highly engaging as they create a game-like feel to learning about functions. The "*Angry Bird-like*" lessons on quadratic functions are always a big hit. Additionally, the recently developed audio indicator for visually impaired students (co-developed by a visually impaired person) is excellent for "showing" the shape of a function.

Brightstorm

This app assists in presenting material in smaller chunks for secondary math and science lessons. The videos are used to present key concepts and skills to help insure learning with over 2500 videos. The videos are supportive of varied audio, visual, and display options. The videos can be used to pre-teach, expose students to new concepts, or help solidify content through alternate representations of material.

MyScript Calculator

MyScript Calculator allows students to use a finger to write their problem on the screen. As they watch, their handwriting is converted to text and the solution to the calculation appears.

My Talking Calculator

This app allows students to enter their problem into the calculator and hear the math sentence read to them as they enter the numbers. Once they push the equal sign the solution appears and is spoken by the app.

Beyond calculators, there are other apps and mobile websites that support personalized learning and provide a wealth of opportunities for both the teaching and learning of mathematics for kindergarten through high school.

Khan Academy

This smartphone supported mobile site and app is a personalized learning solution for differentiation and remediation in math for students of all ages. While Khan now includes a variety of content, its most extensive curriculum is mathematics. The app provides learners with instructional videos, practice exercises, and a personalized dashboard that allows users to track their progress. Teachers can assign work, track student progress, and generate reports on learning outcomes, all from a smartphone or other portable digital device.

MathNEXUS

This site, while no longer being actively updated, is a wealth of resources for teaching and learning mathematics and can be useful for cross curricular explorations in history and literacy. It includes over 5,000 pages of content on such things as interesting mathematical websites, classic mathematical problems to solve, historically important mathematicians, and mathematical humor.

National Council of Teachers of Mathematics

The NCTM has multiple mobile sites that are great for teachers, students, and families. Two of these are *Student Explorations in Mathematics* and *Figure This!* In both of these examples, students have opportunities to develop problem solving skills that are not often found in regular curricular materials. Teachers can use these materials to supplement learning or as a way to differentiate learning experiences for students in mathematics. It should be noted that some of the apps require membership to the NCTM but others are free, but memberships are affordable for individuals and funding may be possible through school or district funds.

National Library of Virtual Manipulatives

This mobile site and app is an accessible site from handheld devices for students at all levels in mathematics. The manipulatives can be used to introduce, reinforce particular mathematical concepts, to support acceleration options. Teachers should review the different activities and select those manipulatives that are most appropriate for different subgroups of students.

Concept Specific Apps: Social Sciences

World Wiki

This app provides useful demographic information for over 250 countries around the world. Data contained here assist educators in the development of students' data literacy and content mastery, particularly in world and cultural studies courses such as geography and world history. It also allows students to look at issue facing their local communities and provide solutions based on their understanding of the data provided within the app. This is a free, easy to use app that helps students learn how to sort through information even as they make decisions about what is accurate and what is important to include in written or oral reports.

iAmerica

iAmerica is a reference book app that contains basic information about the White House and all 45 US Presidents. Information provided in the app assist educators develop students' knowledge and understanding of the historic and current White House along with information that allows research on individual presidents past and present. It is particularly useful in elementary and secondary United States history and civics courses. Quick access can support inquiry about the accuracy of information, whole or small group discussions, and written and oral research projects.

U.S. Constitution

These apps contain the entire United States Constitution in a digital format that allows students to quickly access, search, read, and critique any part of these important founding documents. Documents contained within the app assist educators in the development of students' inquiry skills even as it helps promote content mastery, particularly in elementary and secondary level United States history and civics courses. Cross cultural comparisons can be made between these documents and those international students are studying in their own classrooms, further broadening their understanding of the varied types of governments that exist around the world. Like the previous apps, this one can serve as a launch pad for inquiry, discussion, and written or oral reports about topics of current interests to students and their communities.

Given the sheer volume of educational apps available for smartphones, educators should consider the apps presented here as a starter ideas for their own thinking about integration and use as well as apps preloaded on the smartphone.

Preinstalled Smartphone Apps

While a plethora of apps are available for download, several apps are available for use on phones that are pre-programmed and should not be overlooked. For example all smartphones come preloaded with quality video and still image capturing apps that can be quickly mastered by users. Videos can be used as a way for students to express their content knowledge as well as by teacher's to expand on the representation of content. Photo apps can also support students by providing a quick option to capture handwritten notes or figures on a classroom board or and handout. The photo and video apps are useful because we know that the use of video and photo can support a student's ability to seamlessly integrate imagery into a multimedia presentation or link to existing documents such as Google Documents. Other pertinent written or typed information, such as schedules, can be quickly captured and shared using these devices in the calendar app. This function supports executive functioning through time and task management. Alternatively, students can use the preinstalled voice capture app to take oral notes when listening to directions is preferred, over re-reading directions. Another app often preloaded is a note taking app, this can serve as a tool to help students with executive functioning, task development, and time management. Finally, most phones come with or calendar reminders that can also help students complete and follow schedules, tasks, or set aside time to collaborate.

RESEARCH SUPPORTING THE USE OF SMARTPHONES AND APPS

Recent research on P-12 smartphone use demonstrates an increase global interest in their efficacy as learning tools. Regardless of their acceptance as a learning tool, in formal classroom contexts, smartphones were being used by nearly half of all students to support their learning (Walker, 2013). Smartphones alone, are not as effective; however, when utilized with the standard apps or other available apps, Purposeful educational use makes them highly effective classroom tool.

Smartphones, coupled with the right apps, can serve as important tools to improve learning and teaching, allowing for the exploration of advanced concepts, including those once thought to be too demanding and incompatible for young learners (Neumann & Neumann, 2015). Apps can also be used to support non-academic learning and behaviors, students in kindergarten used apps to develop their sensorimotor skills, emotional wellbeing, and verbal skills (Noorhidawati, Ghalebandi, & Hajar, 2015). Increases in early adoption can further support learning in the middle grades. For example, Monem (2016) noted that when mobile devices were used to support post-lesson reviews and students with disabilities had an increase in content acquisition compared to those not using mobile devices. This means that leveraging mobile devices as a learning tool can positively impact learning for all students especially those with disabilities. Such findings are in direct alignment with the premise of UDL; namely, that all students can learn and make gains when given the right tools and supports.

Incorporating smartphones into classroom instruction in a manner that supports all students, especially those with disabilities can take many forms, have multiple purposes, and can evolve as technology changes. Smartphones provide a medium that supports communication, socialization, and often soft-skills for individuals with disabilities (Gallup, Little, Serianni, & Kocaoz, 2017). Additionally, as noted by Peterson-Karlan and Parette (2005) technology such as smartphones, virtual environments, and related apps are a medium that typically developing individuals are using to develop socialization and communication skills; these skills are often lacking in millennial students (i.e., those who entered school after the year 2000) with disabilities. While the role of apps in supporting standards based learning is emergent within the literature base, some research indicates the efficacy of using apps to meet standards (i.e. Heath-Reynolds & VanWeelden, 2015; Monem, 2016). This highlights the importance of educators exploring apps prior to implementation to ensure the selected app is pedagogically sound with respect to content, design, and methods of play (Judge, Floyd, & Jeffs, 2015).

IMPLEMENTATION BARRIERS AND RECOMMENDATIONS

For many, barriers to implementation can be overwhelming. Therefore, this section is offered to identify examples of some barriers and to suggest ways of overcoming, or mitigating, them. Barriers might include lack of technology for some students, access to the Internet if individuals are in a remote area, effective integration, and an abundance of choice. Well this is not a comprehensive list, these are some barriers that may arise. First, consider the student and the student's needs and abilities and the goal of the lesson(s). Then reflect on what technology is already available as well as access to Internet and with that, parental concerns.

How to Choose and Implement Apps in the Classroom

Integration of apps is an effective way to support learning, increase engagement and self-regulation, and reduce off-task behaviors. Many apps often support individual student interest through a modality that they are interested in. For example if a student must learn about geometry and has an interest in architecture, that student could watch a video on how geometry is used in architecture. This attends to understanding the student and tasks that need to be accomplished.

App Selection

To select an app for a particular lesson or project, teachers must first determine the purpose of the lesson and, how they want students to participate in selecting an app. Additionally, a primary concern for app selection is the purpose of the app. For example, students with Autism struggle in lessons where social reciprocity, high level of communication or collaboration is expected or needed to accomplish a given task, as well as managing multiple tasks or executive functioning skills. Options for communication using technology or virtual collaboration hold the potential to remove extraneous variables and support communication (Gallup, Serianni, Duff, Gallup, 2016). An app that provides support with executive functioning to stay on task or an app that is particularly engaging holds the potential to support collaboration which will increase inclusion in content area classes such as mathematics or history. For example, a teacher or student could select a platform such as *Trello* to help divide up tasks and assign tasks to individuals with reminders.

Considerations for App Selection

Consider for a moment the impact a lack of inclusion and social-skills training has on individuals with Autism and other disabilities; a challenge with social-skills will extend to independent living and chances of obtaining employment. A common problem for content classes such as science, mathematics, and history are reading levels, group work, and persistence when being assessed. Therefore it is critical to understand the students and their individual needs to identify apps that can support individual and group needs. For example, alternate reading levels such as text to speech, virtual collaboration through the Google Suite, or apps to support content acquisition through video such as *Brightstorm.* Most importantly, a teacher needs to ask the question: "What do I need this app to do, to help the student have equitable access to the curriculum.?"

Understanding What Devices Are Present in Classes as a "Bring Your Own Device"

To effectively consider apps in the classroom a teacher must know what devices are available, what students regularly bring to class, as well as capabilities of the devices such as download-ability, memory, and processing speed. Then after considering the needs, abilities, and interests of the students, collaboration with content experts (e.g. general education teacher, community expert), to identify how different apps can support the acquisition of knowledge. This is a process that begins at the inception of the lesson. Consider the standards, goals and objectives, key vocabulary, and procedural knowledge necessary to complete the tasks associated with the lesson. As with any instructional tool or technology paired with a lesson, an app must have measurable objectives to be effective. In essence, how do you, the instructor, know that the app is supporting the student as needed by acquiring information (e.g. text to speech app), taking notes through *Endnote,* or using a graphic organizer such as *Trello* to manage a project and learn new vocabulary or content knowledge. It is important to note that the teacher does not need to be the expert in the app; this can be a collaborative process with the student. To help guide app selection, the following questions can be helpful:

1. Determine the support that students will need and the tasks that will be required during the lesson or task (e.g. collaboration, organization, writing, etc.). What task do the students need to complete and how does that relate back to the navigation of content and purpose of the lesson?
2. What do I want my students to get out of the content? How will the app support assessment and expression of learning?
3. What do I want students to know or do as it relates to real-world experiences and 21st century learning? How will they navigate their learning environment?
4. Review available apps. The app serves as the tool to enhance learning, engagement, and expression, not replace learning.

Understanding Individual Needs

Selecting apps to meet specific classroom needs requires understanding the individual needs of students. Student needs can vary across learning preferences, abilities, and disabilities. Student age, maturity, disability, and learning preferences often require strategies or tools to stay on-task, access instructional content, and complete work to maximize learning gains. Handheld devices can add layers of support that can be used by those who need or want specific supports. Apps can assist in differentiating instruction and providing both class and individualized support.

Barriers to Use

Many of the barriers to instructional use are similar to those for other instructional technologies, including tablets and laptops. First, as demonstrated above many researchers report issues related to finding *quality* educational apps that are relevant to and aligned with to the curriculum. Second, educators must be able to identify appropriate smartphone apps before determining how to align them to standards. Perhaps, just as important in terms of barriers to use, is that little work has been done to align educational app use to instructional technology theory and practice. Certainly, many teachers will need help determining how best to integrate their use into instruction. Motivation to use them, however, may rest on whether or not they perceive that use of smartphones and apps motivates learning. Additional, barriers include perceptions among educators, both real and imagined, about their educational value. Beyond that, issues such as alignment of app use with instructional standards and curricular goals challenge their use, as does potential concerns about how best to assess learning when using smartphones and apps. Further, barriers might include time constraints, budget concerns, lack of support, and lack of access for some learners whose parents either cannot afford or do not want their children to use these devices without their supervision. Because these and other barriers can seem insurmountable, teacher educators and others who recognize the educational value of these devices must work with teachers and other stakeholders even as they advocate for their *appropriate* integration into instructional settings. Finally, students' access to smartphones in the classroom can result in classroom management issues. As Ferriter (2010) reported,

Cell phones can be a real disruption to learning when used improperly. With almost 60 percent of teens reporting that they've sent and received text messages and 25 percent reporting that they've made phone calls while in class, the disruptions are real (p. 85).

Therefore, it is important to remember that many schools still ban their use (Ferriter, 2010) and, that resistance to use, may continue in many settings. Careful focus on purposeful and managed use can help mitigate this and other barriers to use. However, doing so will require discussion, experimentation, and the sharing of results with colleagues and peers locally, nationally, and internationally.

Without a concerted effort aimed at intervention for effective use of smartphones and apps, coupled with advocacy for their appropriate use, many educators may not learn how to discern educationally appropriate apps or may not use them at all. Barriers to effective integration resolve with improvements in teacher attitudes toward technology and presence in the classroom, administrative support, and professional development that is specific to their individual classroom needs (Kalonde, 2017). If these barriers, including inclusion, communication and social skills, are to be addressed a call to support effective integration needs to be answered.

Overcoming Barriers

The most effective way to overcome barriers is to simply identify an area of concern in teaching that could potentially be enhanced by effective integration of cell phones or other handheld devices. We encourage teachers to take a risk, select an app, or have students present some apps that may be interesting to use in class. Try the app out and identify if student are more eager to persist in their learning and become engaged in advancing their understanding of content. The key is to find something that meets the comfort level of the teacher, the needs of students, and identified learning outcomes. If apps and technology are switched too frequently and too quickly, students will be confused, and the teacher will spend too much time teaching the app, setting up apps on new systems thus missing out on valuable active learning time. Some suggested first steps are to:

1. Identify who in the class has a handheld device
2. Collectively explore some apps
3. Have students research and present apps that they are interested in using.
4. Try one app that will support the set goals and improve engagement.
5. Evaluate the app based on a set criteria such as, did the app:

a. Support equitable access to the content (improve reading or connections)?
b. Protect the privacy rights of minors?
c. Help students remain engaged in the topic?
d. Help students find new ways to express their understanding of the content that would still meet the terms of the rubric or grading criteria?
e. Was the app user friendly and support 21st century skills?

CONCLUSION

This chapter examined the use of smartphones and mobile apps as a means of supporting more inclusive, dynamic, and accessible learning that aligns with the tenets of UDL. Mobile technologies are not common place for most current educators and thus recognizing how they can be used to support learning for diverse groups of students may not be immediately apparent. However, teachers need not fully understand how to implement such tools prior to using them with students. The best learning will occur as students and teachers alike collaborate on which tools work the best and for what reasons. Sometimes, the best opportunities for integrating technologies arise because the students had a choice in how to explore and represent their learning (i.e. MMAE, MME). Essentially, by putting the ownership and responsibility for demonstrating learning on the students, the teachers have an opportunity to learn more about which mobile tools were the most useful, beneficial, and easiest to use. Such information builds teachers' pedagogical practices and knowledge for crafting future learning experiences, too.

UDL, is a powerful framework for supporting all learners, and as outlined above, there are numerous apps to help develop students' executive function, collaboration, knowledge acquisition, and personal expression of their learning. Implementing a mobile device through the lens of UDL may seem daunting at first, but, over time, teachers will find the framework greatly assists in designing and delivering learning experiences that help all students better understand their strengths and ways of thinking. As such, the personalized learning and opportunities for self-expression through mobile technologies greatly supports students' skills in interacting and communicating with others in a global society through every day learning. Mobile technologies are here. Perhaps it is time to leverage their potential so all students can better reach their potential, too.

REFERENCES

Basham, J., & Marino, M. (2013). Understanding STEM education and supporting students through universal design for learning. *Teaching Exceptional Children*, *45*(4), 8–15. doi:10.1177/004005991304500401

Bennett, C. A. (2012). Using tiered explorations to promote reasoning. *Mathematics Teaching in the Middle School*, *18*(3), 166–173. doi:10.5951/mathteacmiddscho.18.3.0166

Blackwell, C. K., Lauricella, A. R., & Wartella, E. (2016). The influence of TPACK contextual factors on early childhood educators' tablet computer use. *Computers & Education*, *98*, 57–69. doi:10.1016/j.compedu.2016.02.010

Bricker, L. A., & Bell, P. (2012). “GodMode is his video game name”: Situating learning and identity in structures of social practice. *Cultural Studies of Science Education, 7*(4), 883–902. doi:10.100711422-012-9410-6

Bruner, J. T. (1993). *Schools for thought: A science of learning in the classroom*. Cambridge, MA: MIT Press.

Center for Applied Special Technology. (2011). *Universal design for learning guidelines version 2.0*. Wakefield, MA: CAST.

Ferriter, W. M. (2010). Digitally speaking: Cell phones as teaching tools. *Educational Leadership, 68*(2), 85 - 86. Retrieved from http://www.ascd.org/publications/educational-leadership/oct10/vol68/num02/Cell-Phones-as-Teaching-Tools.aspx

Flewitt, R., Messer, D., & Kucirkova, N. (2015). New directions for early literacy in a digital age: The iPad. *Journal of Early Childhood Literacy, 15*(3), 289–310. doi:10.1177/1468798414533560

Gallup, J., & Bennett, C. A. (2018). Creating effective inclusion practices for students with exceptionalities. In B. Eisenbach & P. Greathouse (Eds.), *The Online Classroom: Resources for Effective Middle Level Virtual Education*. Charlotte, NC: Information Age Publishing.

Gallup, J., Little, M. E., Serianni, B., & Kocaoz, O. (2017). The Potential of Virtual Environments to Support Soft-Skill Acquisition for Individuals with Autism. *Qualitative Report, 22*(9), 2509–2532. Retrieved from http://nsuworks.nova.edu/tqr/vol22/iss9/16

Gardner, H., & Davis, K. (2013). *The app generation*. New Haven, CT: Yale University Press.

Geist, E. A. (2012). A qualitative examination of two year--olds interaction with tablets. *Journal of Instructional Psychology, 39*(1).

Heath-Reynolds, J., & VanWeelden, K. (2015). Integrating apps with the core Arts Standards in the 21st-century elementary music classroom. *General Music Today, 29*(1), 24–27. doi:10.1177/1048371315589127

Hirsh-Pasek, K. H., Zosh, J. M., Golinkoff, R. M., Gray, J. H., Robb, M. B., & Kaufman, J. (2015). Putting education in “Educational” apps: Lessons from the science of learning. *Psychological Science in the Public Interest, 16*(1), 3–34. doi:10.1177/1529100615569721 PMID:25985468

Judge, S., Floyd, K., & Jeffs, T. (2015). Using mobile media devices and apps to promote young children’s learning. In K. L. Heider & M. Renck Jalongo (Eds.), *Young children and families in the information age, educating the young child* (pp. 117–131). New York: Springer. doi:10.1007/978-94-017-9184-7_7

Kabali, H. K., Irigoyen, M. M., Nunez-Davis, R., Budacki, J. G., Mohanty, S. H., Leister, K. P., & Bonner, R. L. (2015). Exposure and use of mobile media devices by young children. *Pediatrics, 136*(6), 2015–2151. doi:10.1542/peds.2015-2151 PMID:26527548

Kucirkova, N. (2016). iRPD--a framework for guiding design-based research for iPad apps. *British Journal of Educational Technology*, 48598–48610.

Lenhart, A., & Page, D. (2015). *Original survey report: Teens, social media & technology overview 2015*. Washington, DC: Pew Research Center. Retrieved from http://www.pewinternet.org/2015/04/09/teens-social-media-technology-2015/

Lunden, I. (2015). *6.1B smartphone users globally by 2020, overtaking basic fixed phone subscriptions*. TechCrunch. Retrieved from https://techcrunch.com/2015/06/02/6-1b-smartphone-users-globally-by-2020-overtaking-basic-fixed-phone-subscriptions/

Monem, R. (2018). The effect of mobile technology as an active student response system on the acquisition of U.S. history content of secondary students with specific learning disabilities. *Dissertation Abstracts International Section A: Humanities and Social Sciences, 77*(1-A)(E).

Neumann, M. M., & Neumann, D. L. (2015). The use of touch-screen tablets at home and pre-school to foster emergent literacy. *Journal of Early Childhood Literacy*, *17*(2), 203–220. doi:10.1177/1468798415619773

Noorhidawati, A., Ghalebandi, S. G., & Siti Hajar, R. (2015). How do young children engage with mobile apps? Cognitive, psychomotor, and affective perspective. *Computers & Education*, *87*, 385–395. doi:10.1016/j.compedu.2015.07.005

Papadakis, S., Kalogiannakis, M., & Zaranis, N. (2017). Designing and creating an educational app rubric for preschool teachers. *Education and Information Technologies*, *22*(6), 3147–3165. doi:10.100710639-017-9579-0

Pennington, L. (2010). Measuring communication outcomes. *Developmental Medicine & Outcomes.*, *7-8*. doi:10.1111/j.1469-8749.2009.03467.x PMID:20419868

Peterson-Karlan, G. R., & Parette, P. (2005). Millennial students with mild disabilities and emerging assistive technology trends. *Journal of Special Education Technology*, *4*(20), 27–38. doi:10.1177/016264340502000404

Piaget, J., & Inhelder, B. (1969). *The psychology of the child*. New York, NY: Basic Books.

Ploog, B. O., Scharf, A., Nelson, D. S., & Brooks, P. J. (2013). Use of computer-assisted technologies (CAT) to enhance social, communicative, and language development in children with autism spectrum disorders. *Journal of Autism and Developmental Disorders*, *43*(2), 301–322. doi:10.100710803-012-1571-3 PMID:22706582

Rappolt-Schlichtmann, G., Daley, S. G., Lim, S., Lapinski, S., Robinson, K. H., & Johnson, M. (2013). Universal Design for Learning and elementary school science: Exploring the efficacy, use, and perceptions of a web-based science notebook. *Journal of Educational Psychology*, *105*(4), 1210–1225. doi:10.1037/a0033217

Ritz, J. (2009). A new generation of goals for technology education. *Journal of Technology Education.*, *2*(20), 50–64.

Schunk, D. H. (2015). *Learning theories: An educational perspective* (7th ed.). Upper Saddle River, NJ: Pearson Merrill/Prentice Hall.

Tapscott, D. (1999a). Educating the net generation. *Educational Leadership*, *56*(5), 611.

Tapscott, D. (2009b). *Grown up digital: How the net generation is changing your world.* New York: McGraw-Hill.

U.S. Department of Education. (2010a). *A blueprint for reform: The reauthorization of the Elementary and Secondary Education Act*. Retrieved from http://www2.ed.gov/policy/elsec/leg/blueprint/blueprint.pdf

U.S. Department of Education. (2010b). *Transforming American education: Learning powered by technology.* National Education Technology Plan 2010. Retrieved from http://www.ed.gov/technology/netp-2010

Vygotsky, L. S. (1978). *Mind in society: The development of higher psychological processes*. Cambridge, MA: Harvard University Press.

Walker, R. (2013). I don't think I would be where I am at now: Pupil perspectives on using mobile devices for learning. *Research in Learning Technology, 21*(2216). doi:10.3402/rlt.v21i0.22116

Yadav, S., & Chakraborty, P. (2017a). Children aged two to four are able to scribble and draw using a smartphone app. *Acta Paediatrica: Nurturing the Child, 106*(6), 991–994. doi:10.1111/apa.13818 PMID:28271551

Yadav, S., & Chakraborty, P. (2017b). Using smartphones with suitable apps can be safe and even useful if they are not misused or overused. *Acta Paediatrica (Oslo, Norway), 107*(3), 384–387. doi:10.1111/apa.14042 PMID:28833547

Yin, K. Y., & Fitzgerald, R. (2015). Pocket learning: A new mobile learning approach for distance learners. *International Journal of Mobile Learning and Organization, 9*(3), 271–283. doi:10.1504/IJMLO.2015.074215

APPENDIX A

Table 1. Apps Supporting UDL

App	Access Information	UDL Category
Brightstorm	https://www.brightstorm.com/	Learning & Collaboration
Desmos	https://www.desmos.com/	Content Specific (Mathematics)
Google Docs	https://docs.google.com/document/u/0/?tgif=d	Learning & Collaboration
inClass	http://learningworksforkids.com/apps/inclass/	MMAE & EXEC Function
iAmerica	http://iamerica.org/	Content Specific (Social Studies)
Kahoot!	http://www.kahoot.com	Learning & Collaboration
Khan Academy	https://www.khanacademy.org/	Content Specific (Varied)
LibriVox	https://itunes.apple.com/us/app/librivox-audio-books/id596159212	Learning & Collaboration
MathNEXUS	http://mathnexus.wwu.edu/	Content Specific (Mathematics)
MyScript Calculator	https://www.myscript.com/calculator/	Content Specific (Mathematics)
My Talking Calculator	https://my-talking-calculator.en.softonic.com/	Content Specific (Mathematics)
National Council of Teachers of Mathematics: Student Explorations in Mathematics	https://www.nctm.org/sem/	Content Specific (Mathematics)
National Library of Virtual Manipulatives	http://nlvm.usu.edu/en/nav/vlibrary.html	Content Specific (Mathematics)
Padlet	Http://www.padlet.com	Learning & Collaboration
Poll Everywhere	https://www.polleverywhere.com/	Learning & Collaboration
Remind	https://www.remind.com/	MMAE & EXEC Function
Seesaw	https://web.seesaw.me/	Learning & Collaboration
Thirty/Thirty	http://3030.binaryhammer.com/	MMAE & EXEC Function
Trello	http://www.trello.com	MMAE & EXEC Function
Twitter	http://www.twitter.com	Learning & Collaboration
YouNote!	http://learningworksforkids.com/apps/younote/	MMAE & EXEC Function
US Constitution		Content Specific (Social Studies)
World Wiki	http://itunes.apple.com/us/app/world-wiki/id289322538?mt=8	Content Specific (Varied) (Social Studies/Media Literacy)

APPENDIX B

Figure 4.

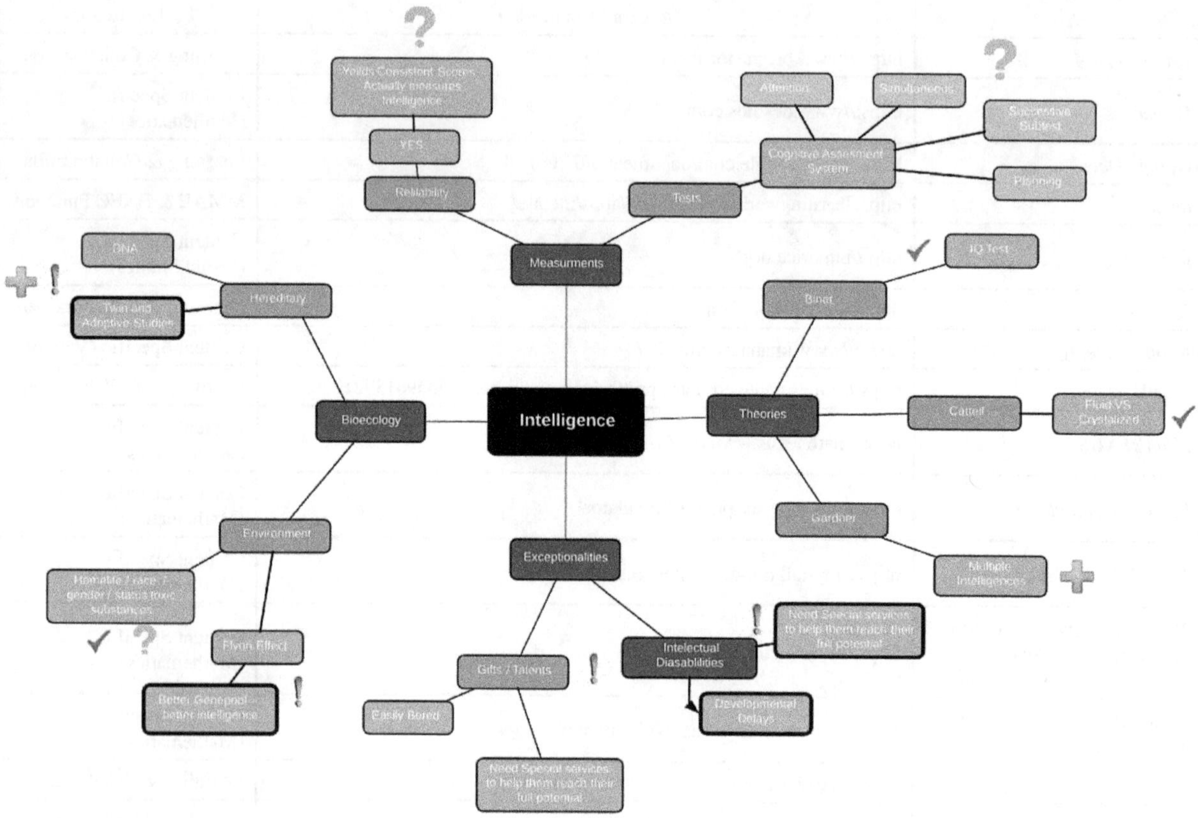

Compilation of References

Abachi, H. R., & Muhammad, G. (2014). The impact of m-learning technology on students and educators. *Computers in Human Behavior*, *30*, 491–496. doi:10.1016/j.chb.2013.06.018

Abbott, S., & McConkey, R. (2006). The barriers to social inclusion as percieved by people with intellectual disabilities. *Journal of Intellectual Disabilities*, *10*(3), 275–287. doi:10.1177/1744629506067618 PMID:16916851

Adams Becker, S., Cummins, M., Davis, A., Freeman, A., Hall Giesinger, C., & Ananthanarayanan, V. (2017). *NMC Horizon Report: 2017 Higher Education Edition*. Austin, TX: The New Media Consortium.

Adler, P. A., & Adler, P. (2008). Of Rhetoric and Representation: The Four Faces of Ethnography. *The Quarterly Sociological*, *49*(1), 1–30. doi:10.1111/j.1533-8525.2007.00104.x

Admiraal, W., van Vugt, F., Kranenburg, F., Koster, B., Smit, B., Weijers, S., & Lockhorst, D. (2017). Preparing pre-service teachers to integrate technology into K-12 instruction: Evaluation of a technology-infused approach. *Technology, Pedagogy and Education*, *26*(1), 105–120. doi:10.1080/1475939X.2016.1163283

Agarwal, P. K., Finley, J. R., Rose, N. S., & Roediger, H. L. III. (2017). Benefits from retrieval practice are greater for students with lower working memory capacity. *Memory (Hove, England)*, *25*(6), 764–771. doi:10.1080/09658211.2016.1220579 PMID:27531308

Agarwal, R., & Karahanna, E. (2000). Time flies when you're having fun: Cognitive absorption and beliefs about information technology usage. *Management Information Systems Quarterly*, *24*(4), 665–694. doi:10.2307/3250951

Agbatogun, A. O. (2014). Developing learners' second language communicative competence through active learning: Clickers or communicative approach? *Journal of Educational Technology & Society*, *17*(2), 257–269.

Agree, E. M. (2014). The potential for technology to enhance independence for those aging with a disability. *Disability and Health Journal*, *7*(1), S33–S39. doi:10.1016/j.dhjo.2013.09.004 PMID:24456682

Aharony, N. (2014). Library and Information Science students' perceptions of m-learning. *Journal of Librarianship and Information Science*, *46*(1), 48–61. doi:10.1177/0961000613518819

Ahmed, A. (2018). Perceptions of using assistive technology for students with disabilities in the classroom. *International Journal of Special Education*, *33*(1), 129–139.

Airlines & Retailers. (2014). *Miami International Airport covered with Beacons*. Available at: https://www.sita.aero/pressroom/news-releases/passengers-airlines-and-retailers-to-benefit-as-miami-international-and-sita-cover-the-airport-with-Beacons

Akçayır, G., Akçayır, M., Pektaş, H. M., & Ocak, M. A. (2016). Augmented reality in science laboratories: The effects of augmented reality on university students' laboratory skills and attitudes toward science laboratories. *Computers in Human Behavior*, *57*, 334–342. doi:10.1016/j.chb.2015.12.054

Akçayır, M., & Akçayır, G. (2017). Advantages and challenges associated with augmented reality for education: A systematic review of the literature. *Educational Research Review*, *20*, 1–11. doi:10.1016/j.edurev.2016.11.002

Akpan, P. J., & Beard, A. L. (2013). Overview of Assistive Technology Possibilities for Teachers to Enhance Academic Outcomes of All Students. *Universal Journal of Educational Research*, *1*(2), 113–118.

Akstein, S. (2006). *Easy to Read.* Raanana: University Publishing Projects.

Akyol, H. (2005). *Türkçe ilk okuma yazma öğretimi.* Ankara: Pegem A Yayıncılık.

Alberta Learning and Teaching Branch. (2002). *Career and life management guide to implementation.* Edmonton: Alberta Assessment Consortium.

Albrecht, B. (2006). Enriching student experience through blended learning. *Research Bulletin*, 12.

Albugami, S. S. (2016). Developing a strategic approach to ICT implementation in Saudi secondary schools. Retrieved from https://search.proquest.com/docview/1885877026?accountid=26642

Al-Emran, M., Mezhuyev, V., & Kamaludin, A. (2018a). Students' perceptions towards the integration of knowledge management processes in m-learning systems: A preliminary study. *International Journal of Engineering Education*, *34*(2), 371–380.

Al-Emran, M., Mezhuyev, V., & Kamaludin, A. (2018b). Technology Acceptance Model in M-learning context: A systematic review. *Computers & Education*, *125*, 389–412. doi:10.1016/j.compedu.2018.06.008

Alexsandrova, M. (2018). Augmented reality in education: The hottest EdTech trend 2018 and how to apply it to your business. *Eastern Peak*. Retrieved from https://easternpeak.com/blog/augmented-reality-in-education-the-hottest-edtech-trend-2018-and-how-to-apply-it-to-your-business/

Alioon & Delialioglu. (2015). A Frame for the Literature on M-learning. *Procedia: Social and Behavioral Sciences*, *182*(C), 127–135.

Alkhawaldeh, A. (2012, November). The EFL Reading Comprehension challenges Faced by Secondary School Students in Jordan. *European Journal of Soil Science*, *35*(2), 140–152.

Ally, M., Grimus, M., & Ebner, M. (2014). Preparing teachers for a mobile world, to improve access to education. *Prospects*, *44*(1), 43–59. doi:10.100711125-014-9293-2

Ally, M., & Needham, G. (2012). *M-libraries 3: Transforming libraries with mobile technology (M-Libraries series).* London: Facet Pub. doi:10.29085/9781856049184

Al-Mekhlafi, K., Hu, X., & Zheng, Z. (2009). (2009). An approach to context-aware mobile Chinese language learning for foreign students. *2009 Eighth International Conference on Mobile Business*, 340-346.10.1109/ICMB.2009.65

Almerich, G., Suárez-Rodríguez, J. M., Belloch, C., & Bo, R. M. (2011). Las necesidades formativas del profesorado en TIC: perfiles formativos y elementos de complejidad. Relieve, 17(2), 1-28. Retrieved from http://www.uv.es/RELIEVE/v17n2/RELIEVEv17n2_1.htm

Althunibat, A. (2015). Determining the factors influencing students' intention to use m-learning in Jordan higher education. *Computers in Human Behavior*, *52*, 65–71. doi:10.1016/j.chb.2015.05.046

Alves-Martins, M., Peixoto, F., Gouveia-Pereira, M., Amaral, V., & Pedro, I. (2002). Self-esteem and academic achievement among adolescents. *Educational Psychology*, *22*(1), 51–62. doi:10.1080/01443410120101242

Ambrose, S. A., Bridges, M. W., DiPietro, M., Lovett, M. C., & Norman, M. K. (2010). *How learning works: Seven research-based principles for smart teaching*. Hoboken, NJ: Wiley.

Anderson, T. A., Hwang, W., & Hsieh, C. (2008). (2008). A study of a mobile collaborative learning system for Chinese language learning. *Proceedings of International Conference on Computers in Education*, 217-222.

Andrade, T. (2005). Inovação e Ciências Sociais: Em Busca de Novos Referenciais. Sistema de Información Científica: Red de Revistas Científicas de América Latina y el Caribe, 20(58), 145-156.

Angelo, T. A., & Cross, K. P. (1993). *Classroom assessment techniques* (2nd ed.). San Francisco: Jossey-Bass.

Annett, M., Anderson, F., Bischof, W. F., & Gupta, A. (2014). The pen is mightier: understanding stylus behaviour while inking on tablets. Proceedings of Graphics Interface 2014, 193-200.

Annetta, L., Burton, E. P., Cheng, R., Chmiel, M., & Frazier, W. (2012). Augmented reality games: Using technology on a budget. *Science Scope*, *36*(3), 54–60.

Anohah, E., Oyelere, S. S., & Suhonmen, J. (2017). Trend of mobile learning in computing education from 2006 to 2014: A systematic review of research publications. International Journal of Mobile and Blended Learning, 18.

Antonioli, M., Blake, C., & Sparks, K. (2014). Augmented reality applications in education. *The Journal of Technology Studies,* 96-107.

APA. (1994). *DSM-4- Diagnostic and statistical manual of mental disorders* (4th ed.). Washington, DC: American Psychiatric Association.

APA. (2013). *Diagnostic and Statistical Manual of Mental Disorders* (5th ed.). Washington, DC: APA.

Apple Store Downloads Statistics. (n.d.). *Statista.* Retrieved from www.statista.com/statistics/263794/number-of-downloads-from-the-apple- app-store/

Arbaugh, J., (2007). Does the community of inquiry rameowrk predict outcomes in online MBA courses? *Academy of Management Proceedings, 2007*(1), 1-6.

Arcavi, A., & Karsenty, R. (n.d.). *Enhancing Mathematics Teachers' Knowledge and Reflection Through Peer-Discussions of Videotaped Lessons: A Pioneer Program in Israel*. Academic Press.

Ardichvili, A., Maurer, M., Li, W., Wentling, T., & Stuedemann, R. (2006). Cultural influences on knowledge sharing through online communities of practice. *Journal of Knowledge Management*, *10*(1), 94–107. doi:10.1108/13673270610650139

Ardito, C., Costabile, M. F., Marsico, M. D., Lanzilotti, R., Levialdi, S., Roselli, T., & Rossano, V. (2006). An approach to usability evaluation of e-learning applications. *Universal Access in the Information Society*, *4*(3), 270–283. doi:10.100710209-005-0008-6

Asiimwe, E., Grönlund, Y., & Hatakka, M. (2017). Practices and Challenges in an Emerging M-Learning Environment. *International Journal of Education and Development Using Information and Communication Technology*, *13*(1), 103–122.

Asongu, S., & Boateng, A. (2018). Introduction to Special Issue: Mobile Technologies and Inclusive Development in Africa. *Journal of African Business*, *19*(3), 297–301. doi:10.1080/15228916.2018.1481307

Astin, A. (1999). Student involvement: A developmental theory for higher education. *Journal of College Student Development*, *49*(5), 518–529.

Atkinson, E., & Delamont, S. (2006). In the roiling smoke: Qualitative inquiry and contested fields. *International Journal of Qualitative Studies in Education: QSE*, *19*(6), 747–755. doi:10.1080/09518390600975974

Ausubel, D. P., Novak, J. D., & Hanesian, H. (1978). *Educational psychology: A cognitive view* (2nd ed.). New York: Holt, Rinehart and Winston.

Avidar, R., Ariel, Y., Malka, V., & Levy, E. C. (2013). Smartphones and young publics: A new challenge for public relations practice and relationship building. *Public Relations Review*, *39*(5), 603–605. doi:10.1016/j.pubrev.2013.09.010

Avidov-Ungar, O., & Forkosh-Baruch, A. (2018). Professional identity of teacher educators in the digital era in light of demands of pedagogical innovation. *Teaching and Teacher Education*, *73*, 183–191. doi:10.1016/j.tate.2018.03.017

Avila, J. (2015, October). *Designing Mobile Apps for Use by People with Cognitive Disabilities*. Academic Press.

Avouris, N. M., & Yiannoutsou, N. (2012). A Review of Mobile Location-based Games for Learning across Physical and Virtual Spaces. *Journal of Universal Computer Science*, *18*, 2120–2142.

Azimi, E., & Mousavipour, S. (2014). The Effects of Educational Multimedia in Dictation and Its Role in Improving Dysgraphia in Students with Dictation Difficulty. *Contemporary Educational Technology*, *5*(4), 331–340.

Azuma, R., Baillot, Y., Behringer, R., Feiner, S., Julier, S., & MacIntyre, B. (2001). Recent advances in augmented reality. *IEEE Computer Graphics and Applications*, *21*(6), 34–47. doi:10.1109/38.963459

Bacca, J., Baldiris, S., Fabregat, R., & Graf, S. (2014). Augmented reality trends in education: A systematic review of research and applications. *Journal of Educational Technology & Society*, *17*(4), 133.

Bacca, J., Baldiris, S., Fabregat, R., & Graf, S. (2015). Mobile augmented reality in vocational education and training. *Procedia Computer Science*, *75*, 49–58. doi:10.1016/j.procs.2015.12.203

Baeten, M., Struyven, K., & Dochy, F. (2013). Student-centered teaching methods: Can they optimize students' approaches to learning in professional higher education? *Studies in Educational Evaluation*, *39*(1), 14–22. doi:10.1016/j.stueduc.2012.11.001

Bailey, L. W. (2019). Virtual and augmented reality: Exploring the potential for instructional application. In L. W. Bailey (Ed.), *Educational technology and the new world of persistent learning* (pp. 12–23). Hershey, PA: IGI Global. doi:10.4018/978-1-5225-6361-7.ch002

Bakhsh, M., Mahmood, A., & Sangi, N. (2017). Examination of factors influencing students and faculty behavior towards m-learning acceptance. *The International Journal of Information and Learning Technology*, *34*(3), 166–188. doi:10.1108/IJILT-08-2016-0028

Bandura, A. (1986). The explanatory and predictive scope of self-efficacy theory. *Journal of Social and Clinical Psychology*, *4*(3), 359–373. doi:10.1521/jscp.1986.4.3.359

Baran, E. (2014). A Review of research on mobile learning in teacher education. *Journal of Educational Technology & Society*, *17*(4), 17–32.

Barber, M., & Mourshed, M. (2007). *How the world's best-performing school systems come out on top*. McKinsey & Company.

Barkhuus, L., & Polichar, V. E. (2011). Empowerment through seamfulness: Smart phones in everyday life. *Personal and Ubiquitous Computing*, *15*(6), 629–639. doi:10.100700779-010-0342-4

Basham, J., & Marino, M. (2013). Understanding STEM education and supporting students through universal design for learning. *Teaching Exceptional Children*, *45*(4), 8–15. doi:10.1177/004005991304500401

Baumgartner, P., & Payr, S. (1996). Learning as action: A social science approach to the evaluation of interactive media. *Proceedings of ED-MEDIA 96-World Conference on Educational Multimedia and Hypermedia,* 31-37.

Bausch, E. M., & Melinda, A. J. (2008). Assistive Technology Implementation Plan: A Tool for Improving Outcomes. *Teaching Exceptional Children, 41*(1), 6–13. doi:10.1177/004005990804100101

Bennett, C. A. (2012). Using tiered explorations to promote reasoning. *Mathematics Teaching in the Middle School, 18*(3), 166–173. doi:10.5951/mathteacmiddscho.18.3.0166

Bers, M. U., Flannery, L., Kazakoff, E. R., & Sullivan, A. (2014). Computational thinking and tinkering: Exploration of an early childhood robotics curriculum. *Computers & Education, 72*, 145–157. doi:10.1016/j.compedu.2013.10.020

Bianchi, A., & Phillips, J. G. (2005). Psychological predictors of problem mobile phone use. *Cyberpsychology & Behavior, 8*(1), 39–51. doi:10.1089/cpb.2005.8.39 PMID:15738692

Bidin, S., & Ziden, A. A. (2013). Adoption and application of mobile learning in the education industry. *Procedia: Social and Behavioral Sciences, 90*, 720–729. doi:10.1016/j.sbspro.2013.07.145

Bigby, C., Bould, E., & Beadle-brown, J. (2016). Conundrums of supported living : The experiences of people with intellectual disability. *Journal of Intellectual & Developmental Disability, 0*(0), 1–11.

Billi, M., Burzagli, L., Catarci, T., Santucci, G., Bertini, E., Gabbanini, F., & Palchetti, E. (2010). A unified methodology for the evaluation of accessibility and usability of mobile applications. *Universal Access in the Information Society, 9*(4), 337–356. doi:10.100710209-009-0180-1

Billinghurst, M., & Dunser, A. (2012). Augmented reality in the classroom. *Computer, 45*(7), 56–63. doi:10.1109/MC.2012.111

Billinghurst, M., Kato, H., & Poupyrev, I. (2001). The magic book: Moving seamlessly between reality and virtuality. *IEEE Computer Graphics and Applications, 21*, 6–8.

Birch, S. H., & Ladd, G. W. (1997). The teacher-child relationship and children's early school adjustment. *Journal of School Psychology, 35*(1), 61–79. doi:10.1016/S0022-4405(96)00029-5

Birch, S. H., & Ladd, G. W. (1998). Children's interpersonal behaviors and teacher-child relationships. *Developmental Psychology, 34*(5), 934–946. doi:10.1037/0012-1649.34.5.934 PMID:9779740

Bíró, G. I. (2014). Didactics 2.0: A pedagogical analysis of gamification theory from a comparative perspective with a special view to the components of learning. *Procedia: Social and Behavioral Sciences, 141*, 148–151. doi:10.1016/j.sbspro.2014.05.027

Bitter, G., & Corral, A. (2014). The pedagogical potential of augmented reality apps. *International Journal of Engineering Science Invention, 3*(10), 13–17.

Blackwell, C. K., Lauricella, A. R., & Wartella, E. (2016). The influence of TPACK contextual factors on early childhood educators' tablet computer use. *Computers & Education, 98*, 57–69. doi:10.1016/j.compedu.2016.02.010

Blau, I., & Barzel-Rubin, A. (2013). Assisting learning disabilities and empowering excellence by "clickers" and interactive whiteboards. In *Proceedings of the 15th Biennial EARLI2013 Conference for Research in Learning and Instruction.* TUM School of Education, Technical University Munich.

Blau, I., & Caspi, A. (2010). Studying invisibly: Media naturalness and learning. In N. Kock (Ed.), Evolutionary psychology and information systems research: A new approach to studying the effects of modern technologies on human behavior (pp. 193-216). New-York: Springer.

Blau, I., Weiser, O., & Eshet-Alkalai, Y. (2017). How do medium naturalness and personality traits shape academic achievement and perceived learning? An experimental study of face-to-face and synchronous e-learning. *Research in Learning Technology, 25*. http://repository.alt.ac.uk/2380/

Blau, I. (2011). Teachers for "Smart classrooms": The extent of implementing of an Interactive Whiteboard-based professional development program on elementary teachers' instructional practices. *Interdisciplinary Journal of E-Learning and Learning Objects*, *7*, 275–289. doi:10.28945/1524

Blau, I., & Barak, A. (2012). How do personality, synchronous media, and discussion topic affect participation? *Journal of Educational Technology & Society*, *15*(2), 12–24.

Blau, I., & Caspi, A. (2008). Do media richness and visual anonymity influence learning? A comparative study using Skype. In Y. Eshet, A. Caspi, & N. Geri (Eds.), *Learning in the Technological Era* (pp. 18–24). Ra'anana, Israel: Open University of Israel.

Blau, I., & Eshet-Alkalai, Y. (2017). The ethical dissonance in digital and non-digital learning environments: Does technology promotes cheating among middle school students? *Computers in Human Behavior*, *73*, 629–637. doi:10.1016/j.chb.2017.03.074

Blau, I., Grinberg, R., & Shamir-Inbal, T. (2018). Pedagogical perspectives and practices reflected in metaphors of learning and digital learning of ICT leaders. *Computers in the Schools*, *35*(1), 32–44. doi:10.1080/07380569.2018.1427960

Blau, I., & Hameiri, M. (2017). Ubiquitous mobile educational data management by teachers, students and parents: Does technology change school-family communication and parental involvement? *Education and Information Technologies*, *22*(3), 1231–1247. doi:10.100710639-016-9487-8

Blau, I., Peled, Y., & Nusan, A. (2016). Technological, pedagogical and content knowledge (TPACK) in one-to-one classroom: Teachers developing "digital wisdom". *Interactive Learning Environments*, *24*(6), 1215–1230. doi:10.1080/10494820.2014.978792

Blau, I., & Shamir-Inbal, T. (2017). Digital competences and long-term ICT integration in school culture: The perspective of elementary school leaders. *Education and Information Technologies*, *22*(3), 769–787. doi:10.100710639-015-9456-7

Blau, I., & Shamir-Inbal, T. (2018). Digital technologies for promoting "student voice" and co-creating learning experience in an academic course. *Instructional Science*, *46*(2), 315–336. doi:10.100711251-017-9436-y

Bleed, R. (2001). A hybrid campus for a new millennium. *EDUCAUSE Review*, *36*(1), 16–24.

Blonder, R. (2018). Student-curated exhibitions: Alternative assessment in chemistry education in Israel. In C. Cox & W. E. Schatzberg (Eds.), *International perspectives on chemistry education research and practice* (Vol. 1293, pp. 39–55). American Chemical Society.

Blonder, R., Jonatan, M., Bar-Dov, Z., Benny, N., Rap, S., & Sakhnini, S. (2013). Can You Tube it? Providing chemistry teachers with technological tools and enhancing their efficacy beliefs. *Chemistry Education Research and Practice*, *14*(3), 269–285. doi:10.1039/C3RP00001J

Blonder, R., & Mamlok-Naaman, R. (2016). Learning about teaching the extracurricular topic of nanotechnology as a vehicle for achieving a sustainable change in science education. *International Journal of Science and Mathematics Education*, *14*(3), 345–372. doi:10.100710763-014-9579-0

Blonder, R., & Sakhnini, S. (2012). Teaching two basic nanotechnology concepts in secondary school by using a variety of teaching methods. *Chemistry Education Research and Practice*, *13*(4), 500–516. doi:10.1039/C2RP20026K

BoCrews, T. B., Ducate, L., Rathel, J. M., Heid, K., & Bishoff, S. T. (2011). Clickers in the classroom: Transforming students into active learners. *EDUCAUSE - ECAR Research Bulletin*, *9*, 1–12.

Bonwell, C. C., & Eison, J. A. (1991). *Active learning: Creating excitement in the classroom*. Washington, DC: The George Washington University, School of Education and Human Development.

Boone, R., & Higinns, K. (2007). The role of instructional design in assistive technology research and development. *Reading Research Quarterly*, *42*(1), 135–140. doi:10.1598/RRQ.42.1.5

Booth, S. E. (2012). Cultivating knowledge sharing and trust in online communities for educators. *Journal of Educational Computing Research*, *47*(1), 1–31. doi:10.2190/EC.47.1.a

Borblik, J., Shabalina, O., Kultsova, M., Pidoprigora, A., & Romanenko, R. (2015). Assistive technology software for people with intellectual or development disabilities: Design of user interfaces for mobile applications. *Information, Intelligence, Systems and Applications (IISA), 2015 6th International Conference On*, 1–6.

Boruff, J. T., & Storie, D. (2014). Mobile devices in medicine: A survey of how medical students, residents, and faculty use smartphones and other mobile devices to find information. *Journal of the Medical Library Association: JMLA*, *102*(1), 22–30. doi:10.3163/1536-5050.102.1.006 PMID:24415916

Bouhnik, D., & Deshen, M. (2014). WhatsApp goes to school: Mobile instant messaging between teachers and students. *Journal of Information Technology Education: Research*, *13*, 217–231. doi:10.28945/2051

Bovill, C., Aitken, G., Hutchison, J., Morrison, F., Roseweir, K., Scott, A., & Sotannde, S. (2010). Experiences of learning through collaborative evaluation from a postgraduate certificate in professional education. *The International Journal for Academic Development*, *15*(2), 143–154. doi:10.1080/13601441003738343

Bovill, C., Cook-Sather, A., & Felten, P. (2011). Students as co-creators of teaching approaches, course design, and curricula: Implications for academic developers. *The International Journal for Academic Development*, *16*(2), 133–145. doi:10.1080/1360144X.2011.568690

Bovill, C., Cook-Sather, A., Felten, P., Millard, L., & Moore-Cherry, N. (2016). Addressing potential challenges in co-creating learning and teaching: Overcoming resistance, navigating institutional norms and ensuring inclusivity in student-staff partnerships. *Higher Education*, *71*(2), 195–208. doi:10.100710734-015-9896-4

Bower, M., Howe, C., McCredie, N., Robinson, A., & Grover, D. (2014). Augmented reality in education – cases, places and potentials. *Educational Media International*, *51*(1), 1–15. doi:10.1080/09523987.2014.889400

Brady, M., Seli, H., & Rosenthal, J. (2013). Metacognition and the influence of polling systems: How do clickers compare with low technology systems. *Educational Technology Research and Development*, *61*(6), 885–902. doi:10.100711423-013-9318-1

Brennan, K., Valverde, A., Prempeh, J., Roque, R., & Chung, M. (2011). More than code: The significance of social interactions in young people's development as interactive media creators. In T. Bastiaens, & M. Ebner (Eds.), *Proceedings of World Conference on Educational Multimedia, Hypermedia and Telecommunications 2011* (pp. 2147-2156). Chesapeake, VA: AACE.

Bricker, L. A., & Bell, P. (2012). "GodMode is his video game name": Situating learning and identity in structures of social practice. *Cultural Studies of Science Education*, *7*(4), 883–902. doi:10.100711422-012-9410-6

Britner, S. L., & Pajares, F. (2006). Sources of science self-efficacy beliefs of middle school students. *Journal of Research in Science Teaching*, *43*(5), 485–499. doi:10.1002/tea.20131

Brooke, D. E. (2017). *The Critical Role of Oral Language in Reading Instruction and Assessment.* Retrieved from Lexia Learning: https://www.lexialearning.com/resources/white-papers/oral-language

Brown, E. (2010). Introduction to location-based mobile learning. In E. Brown, M. Sharples, G. Clough, B. Tangney, J. Wishart, M. Wijers, & M. Specht (Eds.), *Education in the wild: Contextual and location-based mobile learning in action*. Nottingham, UK: Learning Sciences Research Institute, University of Nottingham.

Brown, J. S., Collins, A., & Duguid, S. (1989). Situated cognition and the culture of learning. *Educational Researcher, 18*(1), 32–42. doi:10.3102/0013189X018001032

Brown, T. (2008). Design Thinking. *Harvard Business Review*, 85–92. PMID:18605031

Bruner, J. S. (1977). *The process of education.* Cambridge, MA: Harvard University Press.

Bruner, J. T. (1993). *Schools for thought: A science of learning in the classroom.* Cambridge, MA: MIT Press.

Bryant, B. R., & Seok, S. (2017). Introduction to the special series. *Technology and Disabilities in Education, 29*(3), 121-122.

Budría, S., Colino, A. & Martínez de Ibarreta, C. (2018). The impact of host language proficiency on employment outcomes among immigrants in Spain. *Empirica, Journal of European Economics,* 1-28. doi:10.100710663-018-9414-x

Bujak, K. R., Radu, I., Catrambone, R., MacIntyre, B., Zheng, R., & Golubski, G. (2013). A psychological perspective on augmented reality in the mathematics classroom. *Computers & Education, 68*, 536–544. doi:10.1016/j.compedu.2013.02.017

Burden, K., & Hopkins, P. (2016). Barriers and challenges facing pre-service teachers use of mobile technologies for teaching and learning. *International Journal of Mobile and Blended Learning, 8*(2), 1–20. doi:10.4018/IJMBL.2016040101

Burden, K., & Kearney, M. (2018). Designing an educator toolkit for the mobile learning age. *International Journal of Mobile and Blended Learning, 10*(2), 88–99. doi:10.4018/IJMBL.2018040108

Burns-Sardone, N. (2014). Making the case for BYOD instruction in teacher education. *Issues in Informing Science and Information Technology, 11*, 191-201. Retrieved from http://iisit.org/Vol11/IISITv11p191-201Sardone0505.pdf

Burston, J. (2013). Mobile-assisted language learning: A selected annotated bibliography of implementation studies 1994-2012. *Language Learning & Technology, 17*(3), 157–225.

Burston, J. (2014). MALL: The pedagogical challenges. *Computer Assisted Language Learning, 27*(4), 344–357. doi: 10.1080/09588221.2014.914539

Burston, J. (2015). Twenty years of MALL project implementation: A meta-analysis of learning outcomes. *ReCALL, 27*(1), 4–20. doi:10.1017/S0958344014000159

Butler, D., & Winne, P. (1995). Feedback and self-regulated learning: A theoretical synthesis. *Review of Educational Research, 65*(3), 245–281. doi:10.3102/00346543065003245

Caetano, L. M. D. (2015). Tecnologia e Educação: Quais os desafios? [UFSM]. *Educação, 40*(2), 295–309. doi:10.5902/19846444

Cahill, S. M. (2009). Where does handwriting fit in? Strategies to support academic achievement. *Intervention in School and Clinic, 44*(4), 223–228. doi:10.1177/1053451208328826

Caine, R. N., & Caine, G. (2011). *Natural learning for a connected world: Education, technology, and the human brain.* New York: Teachers College Press.

Cai, S., Wang, X., & Chiang, F. K. (2014). A case study of augmented reality simulation system application in a chemistry course. *Computers in Human Behavior*, *37*, 31–40. doi:10.1016/j.chb.2014.04.018

Cai, Y. X. (2008). Chinese language teaching material editing writing and introduction. In Y. X. Cai, Z. M. Shu, L. F. Chen, X. Y. Zhang, S. Z. He, & M. D. Lai (Eds.), *Introduction to teaching Chinese as a foreign language* (pp. 113–132). Taipei, Taiwan: Sanmin.

Camacho-Miñano, M. D. M., & del Campo, C. (2016). Useful interactive teaching tool for learning: Clickers in higher education. *Interactive Learning Environments*, *24*(4), 706–723. doi:10.1080/10494820.2014.917108

Cansoy, R. (2017). Teachers' professional development: The case of WhatsApp. *Journal of Education and Learning*, *6*(4), 285–293. doi:10.5539/jel.v6n4p285

Carbonell, C. C., & Bermejo, L. A. A. (2017). Augmented reality as a digital teaching environment to develop spatial thinking. *Cartography and Geographic Information Science*, *44*(3), 259–270. doi:10.1080/15230406.2016.1145556

Carbonell, C. C., Saorin, J. L., & De la Torre Cantero, J. (2018). Teaching with AR as a tool for relief visualization: Usability and motivation study. *International Research in Geographical and Environmental Education*, *27*(1), 69–84. doi:10.1080/10382046.2017.1285135

Cárdenas-Robledo, L. A., & Peña-Ayala, A. (2018). Ubiquitous learning: A systematic review. *Telematics and Informatics*, *35*(5), 2097–1132. doi:10.1016/j.tele.2018.01.009

Carter, S., & Yeo, A. C. M. (2017). From hedonism and utilitarianism to anticipated actual behavior: Do UK and Malaysian postgraduate students behave differently to mobile apps? *Economics, Management, and Financial Markets*, *12*(2), 11–34. doi:10.22381/EMFM12220171

Caspi, A., & Blau, I. (2011a). Collaboration and psychological ownership: How does the tension between the two influence perceived learning? *Social Psychology of Education*, *14*(2), 283–298. doi:10.100711218-010-9141-z

Caspi, A., & Blau, I. (2011b). Media choice for intra-school communication: The role of environment, user, and medium. *Journal of Educational Multimedia and Hypermedia*, *20*(4), 337–360.

Cavanaugh, C., Maor, D., & McCarthy, A. (2018). Mobile learning. In K. Kennedy & R. E. Ferdig (Eds.), *Handbook of Research on K-12 Online and Blending Learning* (2nd ed.; pp. 575–591). Pittsburgh, PA: ETC Press. doi:10.1184/R1/6686813.v1

Celuch, K., Bačić, D., Chen, M. W., Maier-Lytle, J., & Smothers, J. (2018). The potential of student co-creation in extracurricular experiences. *Marketing Education Review*, 1–14. doi:10.1080/10528008.2017.1419432

Center for Applied Special Technology. (2011). *Universal design for learning guidelines version 2.0.* Wakefield, MA: CAST.

Chalcraft, D., Hilton, T., & Hughes, T. (2015). Customer, collaborator or co-creator? what is the role of the student in a changing higher education servicescape? *Journal of Marketing for Higher Education*, *25*(1), 1–4. doi:10.1080/08841241.2015.1044790

Chang, G., Morreale, P., & Medicherla, P. (2010). Applications of augmented reality systems in education. In *Society for Information Technology & Teacher Education International Conference,* San Diego, CA (pp. 1380–1385).

Chang, H., Lee, S. W., Liang, J. C., & Wu, H. K. (2013). Current status, opportunities and challenges of augmented reality in education. *Computers & Education*, *62*, 41–49. doi:10.1016/j.compedu.2012.10.024

Chang, S. H., & Yu, N. Y. (2014). The effect of computer-assisted therapeutic practice for children with handwriting deficit: A comparison with the effect of the traditional sensorimotor approach. *Research in Developmental Disabilities*, *35*(7), 1648–1657. doi:10.1016/j.ridd.2014.03.024 PMID:24770471

Chan, J. M., Lambdin, L., Graham, K., Fragale, C., & Davis, T. (2014). A picture-based activity schedule intervention to teach adults with mild intellectual disability to use an iPad during a leisure activity. *Journal of Behavioral Education*, *23*(2), 247–257. doi:10.100710864-014-9194-8

Chan, T. W., Roschelle, J., Hsi, S., Kinshuk, Sharples, M., Brown, ... Hoppe, U. (2006). One-to-one technology-enhanced learning: An opportunity for global research collaboration. *Research and Practice in Technology Enhanced Learning*, *1*(1), 3–29. doi:10.1142/S1793206806000032

Chen, B., & Denoyelles, A. (2013). Exploring students' mobile learning practices in higher education. *EDUCAUSE Review*, 7.

Chen, C. C., & Lin, P. H. (2016). Development and evaluation of a context-aware ubiquitous learning environment for astronomy education. *Interactive Learning Environments*, *24*(3), 644–661. doi:10.1080/10494820.2014.915417

Chen, C. H., Ho, C. H., & Lin, J. B. (2015). The development of an augmented reality game-based learning environment. *Procedia: Social and Behavioral Sciences*, *174*, 216–220. doi:10.1016/j.sbspro.2015.01.649

Chen, C. M., & Li, Y. L. (2010). Personalised context-aware ubiquitous learning system for supporting effective English vocabulary learning. *Interactive Learning Environments*, *18*(4), 341–364. doi:10.1080/10494820802602329

Cheng, E. C. (2013). Enhancing the quality of pre-service teachers' learning in teaching practicum. *Education in Science*, *1*, 6–16.

Cheon, J., Lee, S., Crooks, M. S., & Song, J. (2012). An investigation of mobile learning readiness in higher education based on the theory of planned behavior. *Computers & Education*, *59*(3), 1054–1064. doi:10.1016/j.compedu.2012.04.015

Cheung, W. S., & Hew, K. F. (2009). A review of research methodologies used in studies on mobile handheld devices in K-12 and higher education settings. *Australasian Journal of Educational Technology*, *25*(2). doi:10.14742/ajet.1148

Chiang, T. H., Yang, S. J., & Hwang, G. J. (2014). An augmented reality-based mobile learning system to improve students' learning achievements and motivations in natural science inquiry activities. *Journal of Educational Technology & Society*, *17*(4), 352–365.

Chickering, A. W., & Gamson, Z. F. (1999). Development and adaptations of the seven principles for good practice in undergraduate education. *New Directions for Teaching and Learning*, *80*(80), 75–82. doi:10.1002/tl.8006

Chicu, S. O., Ţicău, A., & Şoitu, L. (2014). Training for new technologies. Handwriting with new technologies. *Procedia: Social and Behavioral Sciences*, *142*, 781–785. doi:10.1016/j.sbspro.2014.07.616

Choi, C. Q. (2012). Textbooks come alive. *Scientific American*, *306*(4), 20–21. doi:10.1038cientificamerican0412-20b PMID:22486109

Choi, Y. (2008). Diversity within: Subgroup differences of youth problem behaviors among Asian Pacific Islander American adolescents. *Journal of Community Psychology*, *36*(3), 352–370. doi:10.1002/jcop.20196 PMID:18645632

Christensen, R., & Knezek, G. (2018). Reprint of Readiness for integrating mobile learning in the classroom: Challenges, preferences and possibilities. *Computers in Human Behavior*, *78*, 379–388. doi:10.1016/j.chb.2017.07.046

Chung, C., Hwang, G., & Lai, C. (2019). A review of experimental mobile learning research in 2010–2016 based on the activity theory framework. *Computers & Education*, *129*, 1–13. doi:10.1016/j.compedu.2018.10.010

Chung, P., & Patel, D. R. (2015). Dysgraphia. *International Journal of Child and Adolescent Health, 8*(1), 27–36.

Church, K., & de Oliveira, R. (2013). *What's up with whatsapp?: comparing mobile instant messaging behaviors with traditional SMS*. Paper presented at the 15th international conference on Human-computer interaction with mobile devices and services, Munich, Germany. 10.1145/2493190.2493225

Cieb. (2017). Nota technical. Retrieved from http://www.cieb.net.br/wp-content/uploads/2018/01/CIEB_NotaTecnica-8_nov2017.pdf

Clarke, B., Svanaes, S., & Zimmermann, S. (2013, July). *One-to-one Tablets in Secondary Schools: An Evaluation study*. Academic Press.

Clifford, P., & Friesen, S. (2001). *The stewardship of the intellect: Classroom life, educational innovation and technology. Issues in the Integration of Technology into Teaching, Learning, and School Culture*. Calgary, AB: Detselig Enterprises.

Cobb, P., Yackel, E., & Wood, T. (1992). Interaction and learning in mathematics classroom situations. *Educational Studies in Mathematics, 23*(1), 99–122. doi:10.1007/BF00302315

Cochrane, T., & Bateman, R. (2010). Smartphones give you wings: Pedagogical affordances of mobile Web 2.0. *Australasian Journal of Educational Technology, 26*(1), 1–14. doi:10.14742/ajet.1098

Cochran-Smith, M., & Lytle, S. L. (1999). Relationships of knowledge and practice: Teacher learning in communities. *Review of Research in Education, 24*, 249–305.

Coffman, T. (2017). *Designing instruction to promote higher level thinking* (3rd ed.). Lanham, MD: Rowman & Littlefield Education.

Coffman, T., & Klinger, M. B. (2014). Collaboration and communication in the online classroom through a brain-based approach. *International Journal of Information Communication Technologies and Human Development, 6*(3), 42–52. doi:10.4018/ijicthd.2014070104

Cohen, A., & Ezra, O. (2018). Development of a contextualised MALL research framework based on L2 Chinese empirical study. *Journal of Computer Assisted Language Learning, 31*(7), 764–789. doi:10.1080/09588221.2018.1449756

Cohen, J., Vincent, J. L., Adhikari, N. K., Machado, F. R., Angus, D. C., Calandra, T., & Tracey, K. (2015). Sepsis: A roadmap for future research. *The Lancet Infectious Diseases, 15*(5), 581–614. doi:10.1016/S1473-3099(15)70112-X PMID:25932591

Cohen, L., Mannion, L., & Morrison, K. (2018). *Research Methods in Education*. London: Routledge.

Coleman, A. M. (2001). *Dictionary of Psychology*. Oxford University Press.

Cole, S. M., Hubert, F. S., & Harris, G. S. (2004). Student Learning Motivation and Psychological Hardiness: Interactive Effects on Students' Reactions to a Management Class. *Academy of Management Learning & Education, 3*(1), 64–85. doi:10.5465/amle.2004.12436819

Comas-Quinn, A., Mardomingo, R., & Valentine, C. (2009). Mobile blogs in language learning: Making the most of informal and situated learning opportunities. *ReCALL, 21*(01), 96–112. doi:10.1017/S0958344009000032

Comstock, G. A., & Scharrer, E. (2007). *Media and the American Child*. Amsterdam: Academic Press.

Connolly, T. M., Boyle, E. A., MacArthur, E., Hainey, T., & Boyle, J. M. (2012). A systematic literature review of empirical evidence on computer games and serious games. *Computers & Education, 59*(2), 661–686. doi:10.1016/j.compedu.2012.03.004

Cook, J., & Pachler, N. (2012). Online people tagging: Social (mobile) network (ing) services and work-based learning. *British Journal of Educational Technology*, *43*(5), 711–725. doi:10.1111/j.1467-8535.2012.01346.x

Cook-Sather, A., & Agu, P. (2013). Students of color and faculty members working together toward culturally sustaining pedagogy. *To Improve the Academy, 32*, 271-285.

Cook-Sather, A., & Abbot, S. (2016). Translating partnerships: How faculty-student collaboration in explorations of teaching and learning can transform perceptions, terms, and selves. *Teaching & Learning Inquiry*, *4*(2), 1–14. doi:10.20343/teachlearninqu.4.2.5

Cook-Sather, A., Bovill, C., & Felten, P. (2014). *Engaging students as partners in learning and teaching: A guide for faculty*. San Francisco, CA: Josey-Bass.

Corrin, L., Lockyer, L., & Bennett, S. (2010). Technological diversity: An investigation of students' technology use in everyday life & academic study2. *Learning, Media and Technology*, *35*(4), 387–401. doi:10.1080/17439884.2010.531024

Costabile, M. F., De Angeli, A., Lanzilotti, R., Ardito, C., Buono, P., & Pederson, T. (2008, April). Explore! Possibilities and challenges of mobile learning. In *Proceedings of the SIGCHI Conference on Human Factors in Computing Systems*, Florence, Italy (pp. 145-154). ACM.

Coutinho, C. P., & Lisbôa, E. S. (2013). Social networks as spaces for informal teacher professional development: Challenges and opportunities. *International Journal of Web Based Communities*, *9*(2), 199–211. doi:10.1504/IJWBC.2013.053244

Coutinho, R. X., Folmer, V., & Puntel, R. L. (2014). Aproximando universidade e escola por meio do uso da produção acadêmica na sala de aula. *Ciência & Educação (Bauru)*, *20*(3), 765–783.

Craig, A. B. (2013). *Understanding augmented reality: Concepts and applications*. London, UK: Elsevier.

Crant, J. (2000). Proactive behavior in organizations. *Journal of Management*, *26*(3), 435–462. doi:10.1177/014920630002600304

Crescente, M. L., & Lee, D. (2011). Critical issues of m-learning: Design models, adoption processes, and future trends. *Journal of the Chinese Institute of Industrial Engineers*, *28*(2), 111–123. doi:10.1080/10170669.2010.548856

Creswell, J. W. (2018). *Research design: Qualitative, quantitative, and mixed methods approaches* (5th ed.). Thousand Oaks, CA: Sage Publications, Inc.

Cristancho, S. M., Moussa, F., & Dubrowski, A. (2011). A framework-based approach to designing simulation-augmented surgical education and training programs. *American Journal of Surgery*, *202*(3), 344–351. doi:10.1016/j.amjsurg.2011.02.011 PMID:21871988

Crompton, H. (2013). A Historical Overview of M-learning: Toward Learner-Centred Education. In Handbook of Mobile Learning. Routledge.

Crompton, H. (2013). Mobile learning: New approach, new theory. In Z. L. Berge & L. Y. Muilenburg (Eds.), Handbook of mobile learning (pp. 47–57). Florence, KY: Academic Press.

Crompton, H. (2013). Mobile learning: New approach, new theory. In Z. L. Berge & L. Muilenburg (Eds.), *Handbook of mobile learning* (pp. 47–57). New York: Routledge.

Crompton, H., & Burke, D. (2018). The use of mobile learning in higher education: A systematic review. *Computers & Education*, *123*, 53–64. doi:10.1016/j.compedu.2018.04.007

Crompton, H., Burke, D., & Gregory, K. H. (2017). The use of mobile learning in PK-12 education: A systematic review. *Computers & Education*, *110*, 51–63. doi:10.1016/j.compedu.2017.03.013

Crompton, H., & Traxler, J. (2019). Learning with mobile devices. In M. Khosrow-Pour (Ed.), *Advanced Methodologies and Technologies in Modern Education Delivery* (pp. 793–808). IGI Global. doi:10.4018/978-1-5225-7365-4.ch062

Crouch, A. L., & Jakubecy, J. J. (2007). Dysgraphia: How It Affects a Student's Performance and What Can Be Done about It. *TEACHING Exceptional Children Plus*, *3*(3), n3.

Cuendet, S., Bonnard, Q., Do-Lenh, S. D., & Dillenbourg, P. (2013). Designing augmented reality for the classroom. *Computers & Education*, *68*, 557–569. doi:10.1016/j.compedu.2013.02.015

Cumming, T. M., Strnadová, I., Knox, M., & Parmenter, T. (2014). Mobile technology in inclusive research: Tools of empowerment. *Disability & Society*, *29*(7), 999–1012. doi:10.1080/09687599.2014.886556

Curcio, I. D. D., Dipace, A., & Norlund, A. (2016). Virtual realities and education. *Research on Education and Media*, *8*(2), 60–68. doi:10.1515/rem-2016-0019

Czyzewski, A., Odya, P., Grabkowska, A., Grabkowski, M., & Kostek, B. (2009, August). *Smart pen: new multimodal computer control tool for dyslexia therapy*. Poster presented at the meeting of SIGGRAPH '09. 10.1145/1599301.1599355

da Silva, I. N., da Rocha, J. S., da Silva, J. B., & Bilessimo, S. M. S. (2018). Uso de dispositivos móveis na disciplina de guarani para estudantes de uma escola multisseriada indígena. Renote, *16*(1).

da Silva, I. N., da Silva, K. C. N., Lotthammer, K. S., & Bilessimo, S. M. S. (2017). Inclusão digital em escolas públicas através de tecnologias inovadoras de baixo custo no ensino de disciplinas stem. *Renote, 15*(2).

da Silva. *I.N.* (2016). *Impacto da aplicação de tecnologia no ensino de empreendedorismo para estudantes concluintes da educação básica na rede pública. 2016. Universidade Federal de Santa Catarina*, Araranguá.

da Silva, O. G., & Navarro, E. C. (2012). A relação professor-aluno no processo ensino-aprendizagem. *Revista Eletrônica Interdisciplinar*, *2*(8), 95–100.

Dahlstrom, E. (2015). *ECAR Study of Undergraduate Students and Information Technology*. Research report. Louisville, CO: ECAR.

Dalal, M., Archambault, L., & Shelton, C. (2017). Professional Development for International Teachers: Examining TPACK and Technology Integration Decision Making. *Journal of Research on Technology in Education*, *49*(3-4), 117–133.

Dann, C., & Allen, B. (2015). Mobile video collection in preservice teacher practicum placements. *Journal of Technology and Teacher Education*, *23*(1), 5–27. Retrieved from https://www.learntechlib.org/primary/p/129814/

Darcy, S., Green, J., & Maxwell, H. (2016, June). I've got a mobile phone too! Hard and soft assistive technology customization and supportive call centres for people with disability. *Disability and Rehabilitation. Assistive Technology*, 1–11. PMID:27293107

Darke, P., Shanks, G., & Broadbent, M. (1998). Successfully Completing Case Study Research: Combining Rigour, Relevance and Pragmatism. *Information Systems Journal*, *8*(4), 273–289. doi:10.1046/j.1365-2575.1998.00040.x

Darling-Hammond, L., Austin, K., Orcutt, S., & Rosso, J. (2001). How people learn: Introduction to learning theories. *The learning classroom: Theory into practice. A Telecourse for Teacher Education and Professional Development.* Retrieved from https://web.stanford.edu/class/ed269/hplintrochapter.pdf

Darling-Hammond, L., & Baratz-Snowden, J. (2007). A good teacher in every classroom: Preparing the highly qualified teachers our children deserve. *Educational Horizons*, *85*(2), 111–132. Retrieved from https://www-jstor-org.ezproxy.canterbury.ac.nz/stable/42926597

Darling-Hammond, L., & Sykes, G. (1999). *Teaching as the learning profession: Handbook of policy and practice*. San Francisco: Jossey-Bass.

Davis, F., Bagozzi, D., Richard, P., & Warshaw, P. R. (1989). User acceptance of computer technology: A comparison of two theoretical models. *Management Science*, *35*(8), 982–1003. doi:10.1287/mnsc.35.8.982

Davis, N. E. (2010). Technology in preservice teacher education. In P. Peterson, E. Baker, & B. McGaw (Eds.), *International Encyclopaedia of Education* (3rd ed.; Vol. 8, pp. 217–221). Oxford: Elsevier. doi:10.1016/B978-0-08-044894-7.00748-X

Day, A., Scott, N., & Kelloway, E. K. (2010). Research in Occupational Stress and Well-being: Vol. 8. Information and communication technology: Implications for job stress and employee well-being. Academic Press.

de Oliveira, L. A., & Junior, M. T. F. (2016). A inserção das tecnologias da informação e comunicação nas escolas. In *EnPED-Simpósio Internacional de Educação a Distância e Encontro de Pesquisadores em Educação a Distância.*

de Oslo, O. M. (2005). *Diretrizes para coleta e interpretação de dados sobre inovação*. OECD Publishing.

de Urturi Breton, Z.S., Hernández, F. J., Zorrilla, A. M., & Zapirain, B. G. (2012). Mobile communication for Intellectually Challenged people: A proposed set of requirements for interface design on touch screen devices. *Communications in Mobile Computing, 1*(1).

Dede, C. (1995). The evolution of constructivist learning environments: Immersion in distributed, virtual worlds. *Educational Technology*, *35*(5), 46–52.

Dede, C. (2011). Emerging technologies, ubiquitous learning, and educational transformation. In C. D. Kloos, D. Gillet, R. M. Crespo García, F. Wild, & M. Wolpers (Eds.), *Towards Ubiquitous Learning* (pp. 1–8). Springer. doi:10.1007/978-3-642-23985-4_1

Dede, C., & Bjerede, M. (2011). *Mobile learning for the 21st century: Insights from the 2010 Wireless EdTech Conference*. San Diego, CA: Qualcomm.

Dekelver, J., Kultsova, M., Shabalina, O., Borblik, J., Pidoprigora, A., & Romanenko, R. (2015). Design of mobile applications for people with intellectual disabilities. In A. Kravets, M. Shcherbakov, M. Kultsova, & O. Shabalina (Eds.), *Creativity in Intelligent Technologies and Data Science. Communications in Computer and Information Science* (Vol. 535). Cham: Springer International Publishing.

DeLucia, A., Francese, R., Passero, I., & Tortoza, G. (2012). A collaborative augmented campus based on location–aware mobile technology. *International Journal of Distance Education Technologies*, *10*(1), 55–71. doi:10.4018/jdet.2012010104

Deng, L., & Tavares, N. J. (2013). From moodle to facebook: Exploring students' motivation and experiences in online communities. *Computers & Education*, *68*, 167–176. doi:10.1016/j.compedu.2013.04.028

Department of Education and Skills. (2015). Digital strategy for schools 2015-2020: Enhancing teaching, learning and assessment. Author.

Derounian, J. G. (2017). Mobiles in class? *Active Learning in Higher Education*, 1–12.

Derry, S. J. (1990). Flexible cognitive tools for problem solving instruction. *Paper presented at the annual meeting of the American Educational Research Association*, Boston, MA.

Dewey, J. (1938). *The experience of education*. New York: Simon & Schuster.

Dey, A. K., & Abowd, G. (1999). (1999). Towards a better understanding of context and context-awareness. *Workshop on the what, Who, Where, when, and how of Context-Awareness, 1707*, 304-307.

Diemer, T. T., Fernandez, E., & Streepey, J. W. (2012). Student perceptions of classroom engagement and learning using iPads. *Journal of Teaching and Learning with Technology, 1*(2), 13–25.

Ding, Y., & Chai, K. H. (2015). Emotions and continued usage of mobile applications. *Industrial Management & Data Systems, 115*(5), 833–852. doi:10.1108/IMDS-11-2014-0338

Doğan Temur, Ö. (2015). *İlköğretim okuma yazma öğreniyorum 1 ders kitabi birinci kitap*. Ankara: Yıldırım Yayınları.

Domingo, M. G., & Gargantè, A. B. (2016). Exploring the use of educational technology in primary education: Teachers' perception of mobile technology learning impacts and applications' use in the classroom. *Computers in Human Behavior, 56*, 21–28. doi:10.1016/j.chb.2015.11.023

Dorsey, E. R., Glidden, A. M., Holloway, M. R., Birbeck, G. L., & Schwamm, L. H. (2018). Teleneurology and mobile technologies: The future of neurological care. *Nature Reviews. Neurology, 14*(5), 285–297. doi:10.1038/nrneurol.2018.31 PMID:29623949

Dunaway, J., Searles, K., Sui, M., & Paul, N. (2018). News attention in a mobile era. *Journal of Computer-Mediated Communication, 23*(2), 107–124. doi:10.1093/jcmc/zmy004

Dunleavy, M. (2014). Design principles for augmented reality learning. *TechTrends, 58*(1), 28–34. doi:10.100711528-013-0717-2

Dunleavy, M., Dede, C., & Mitchell, R. (2009). Affordances and limitations of immersive participatory augmented reality simulations for teaching and learning. *Journal of Science Education and Technology, 18*(1), 7–22. doi:10.100710956-008-9119-1

Dunlosky, J., Rawson, K. A., Marsh, E. J., Nathan, M. J., & Willingham, D. T. (2013). Improving students' learning with effective learning techniques: Promising directions from cognitive and educational psychology. *Psychological Science in the Public Interest, 14*(1), 4–58. doi:10.1177/1529100612453266 PMID:26173288

Eden, D. (2017). Field experiments in organizations. *Annual Review of Organizational Psychology and Organizational Behavior, 4*(1), 91–122. doi:10.1146/annurev-orgpsych-041015-062400

Edgerton, R. (2001). *Education White Paper.* Report prepared for the Pew Charitable Trusts, Pew Forum on Undergraduate Learning.

Edmonds, R., & Smith, S. (2016). Location-based mobile learning games: Motivation for and engagement with the learning process. In *Proceedings of the 15th World Conference on Mobile and Contextual Learning; mLearn 2016, Mobile Learning Futures–Sustaining Quality Research and Practice in Mobile Learning* (pp. 72-76).Edmonds, R., & Smith, S. (2017). From playing to designing: Enhancing educational experiences with location-based mobile learning games. *Australasian Journal of Educational Technology, 33*(6).

Efstathiou, I., Kyza, E., & Georgiou, Y. (2018). An inquiry-based augmented reality mobile learning approach to fostering primary school students' historical reasoning in nonformal settings. *Interactive Learning Environments, 26*(1), 22–41. doi:10.1080/10494820.2016.1276076

Eisen, S., & Lillard, A. S. (2016). Young children's thinking about touchscreens versus other media in the US. *Journal of Children and Media, 11*(2), 167–179. doi:10.1080/17482798.2016.1254095

El-Hussein, M., Osman, M., & Cronje, J. C. (2010). Defining mobile learning in the higher education landscape. *Journal of Educational Technology & Society, 13*(3).

Elias, T. (2011). Universal Instructional Design Principles for Mobile Learning. *International Review of Research in Open and Distance Learning, 12*(2), 143–156. doi:10.19173/irrodl.v12i2.965

Engel-Yeger, B., Nagauker-Yanuv, L., & Rosenblum, S. (2009). Handwriting performance, self-reports, and perceived self-efficacy among children with dysgraphia. *The American Journal of Occupational Therapy*, *63*(2), 182–192. doi:10.5014/ajot.63.2.182 PMID:19432056

Erhardt, R. P., & Meade, V. (2005). Improving handwriting without teaching handwriting: The consultative clinical reasoning process. *Australian Occupational Therapy Journal*, *52*(3), 199–210. doi:10.1111/j.1440-1630.2005.00505.x

Evans, C. (2008). The effectiveness of m-learning in the form of podcast revision lectures in higher education. *Computers & Education*, *50*(2), 491–498. doi:10.1016/j.compedu.2007.09.016

Ezra, O., & Cohen, A. (2018b). *How to increase Contextualised MALL of L2 Chinese adult students in target and non-target countries.* Paper for the 10th International Conference and Workshops on Technology and Chinese Language Teaching (TCLT), National Taiwan Normal University, Taipei, Taiwan.

Ezra, O., & Cohen, A. (2018a). Contextualised MALL: L2 Chinese students in target and non-target country. *Computers & Education*, *125*, 158–174. doi:10.1016/j.compedu.2018.06.011

Facer, K., Joiner, R., Stanton, D., Reid, J., Hull, R., & Kirk, D. (2004). Savannah: Mobile gaming and learning? *Journal of Computer Assisted Learning*, *20*(6), 399–409. doi:10.1111/j.1365-2729.2004.00105.x

Fajardo, I., Tavares, G., Ávila, V., & Ferrer, A. (2013). Towards text simplification for poor readers with intellectual disability: When do connectives enhance text cohesion? *Research in Developmental Disabilities*, *34*(4), 1267–1279. doi:10.1016/j.ridd.2013.01.006 PMID:23417132

Falk, T. H., Tam, C., Schellnus, H., & Chau, T. (2011). On the development of a computer-based handwriting assessment tool to objectively quantify handwriting proficiency in children. *Computer Methods and Programs in Biomedicine*, *104*(3), e102–e111. doi:10.1016/j.cmpb.2010.12.010 PMID:21376418

Farias Filho, M. C., & Arruda Filho, E. J. (2000). *Planejamento da pesquisa científica*. Editora Atlas, SA.

Faulds, D. J., Mangold, W. G., Raju, P. S., & Valsalan, S. (2018). The mobile shopping revolution: Redefining the consumer decision process. *Business Horizons*, *61*(2), 323–338. doi:10.1016/j.bushor.2017.11.012

Fedora, P. (2015). Integrating technology tools for students struggling with written language. *Kappa Delta Pi Record*, *51*(1), 31–35. doi:10.1080/00228958.2015.988562

Fernández-López, Á., Rodríguez-Fórtiz, M. J., Rodríguez-Almendros, M. L., & Martínez-Segura, M. J. (2013). Mobile learning technology based on iOS devices to support students with special education needs. *Computers & Education*, *61*, 77–90. doi:10.1016/j.compedu.2012.09.014

Fernandez, M. (2017). Augmented virtual reality: How to improve education systems. *Higher Learning Research Communications*, *7*(1), 1–15. doi:10.18870/hlrc.v7i1.373

Ferrington, G., & Loge, K. (1992). Virtual reality: A new learning environment. *The Computing Teacher*, *9*(7), 16–19.

Ferriter, W. M. (2010). Digitally speaking: Cell phones as teaching tools. *Educational Leadership*, *68*(2), 85 - 86. Retrieved from http://www.ascd.org/publications/educational-leadership/oct10/vol68/num02/Cell-Phones-as-Teaching-Tools.aspx

Fickel, L., & Mackey, J. (2013). Rapporteurs' report: What should initial teacher education programs for 2022 look like and why? *Waikato Journal of Education*, *18*(1), 117–119. doi:10.15663/wje.v18i1.145

Fink, D. L. (2013). *Creating significant learning experiences: An integrated approach to designing college courses.* San Francisco: Jossey-Bass.

Fischer, A., Greiff, S., & Funke, J. (2012). The process of solving complex problems. *The Journal of Problem Solving*, *4*(1), 19–42. doi:10.7771/1932-6246.1118

FitzGerald, E. (2012), Creating user-generated content for location-based learning: An authoring framework. *Journal of Computer Assisted Learning*, *28*(3), 195–207.

Fleischman, D., Raciti, M., & Lawley, M. (2015). Degrees of co-creation: An exploratory study of perceptions of international students' role in community engagement experiences. *Journal of Marketing for Higher Education*, *25*(1), 85–103. doi:10.1080/08841241.2014.986254

Flewitt, R., Messer, D., & Kucirkova, N. (2015). New directions for early literacy in a digital age: The iPad. *Journal of Early Childhood Literacy*, *15*(3), 289–310. doi:10.1177/1468798414533560

Flint, A. (2016). Moving from the fringe to the mainstream: opportunities for embedding student engagement through partnership. *Student Engagement in Higher Education Journal, 1*(1), 1-6. Retrieved from https://journals.gre.ac.uk/index.php/raise/article/view/382/340

Florida Department of Education. (2003). Cool Tools: Informal Reading Assessments. Author.

Fonseca, J. J. S. 2002(). Metodologia da pesquisa científica. Fortaleza: Apostila.

Forkosh Baruch, A., & Erstad, O. (2018). Upbringing in a Digital World: Opportunities and Possibilities. *Technology. Knowledge and Learning*, *23*(3), 377–390. doi:10.100710758-018-9386-8

Forkosh-Baruch, A., & Hershkovitz, A. (2012). A case study of Israeli higher-education institutes sharing scholarly information with the community via social networks. *The Internet and Higher Education*, *15*(1), 58–68. doi:10.1016/j.iheduc.2011.08.003

Foti, M. K., & Mendez, J. (2014). Mobile Learning: How Students Use Mobile Devices to Support Learning. *Journal of Literacy and Technology, 15*(3).

Fraga, L. M., & Harmon, J. (2014). The flipped classroom model of learning in higher education: An investigation of preservice teachers' perspectives and achievement. *Journal of Digital Learning in Teacher Education*, *31*(1), 18–27. doi:10.1080/21532974.2014.967420

Franz, C., Ascherma, L., & Shaftel, J. (2018). *A Clinician's Guide to Learning Disabilities*. New York: Oxford University Press.

Friend, M. (2005). *Special Education: Contemporary Perspectives for School Professionals*. Boston: Allyn & Bacon.

Gadotti, M. (2000). Perspectivas atuais da educação. *São Paulo em perspectiva, 14*(2), 03-11.

Galanis, J. A. (2008). *The effect of the handwriting without tears program on student cursive writing achievement at Central Institute for the Deaf (CID)*. Independent studies and capstones, Paper 417. Retrieved from http://digitalcommons.wustl.edu/pacs_capstones/417

Galli, M., Cimolin, V., Stella, G., De Pandis, M. F., Ancillao, A., & Condoluci, C. (2018). Quantitative assessment of drawing tests in children with dyslexia and dysgraphia. *Human Movement Science*. doi:10.1016/j.humov.2018.05.001 PMID:29748041

Gallup, J., & Bennett, C. A. (2018). Creating effective inclusion practices for students with exceptionalities. In B. Eisenbach & P. Greathouse (Eds.), *The Online Classroom: Resources for Effective Middle Level Virtual Education*. Charlotte, NC: Information Age Publishing.

Gallup, J., Little, M. E., Serianni, B., & Kocaoz, O. (2017). The Potential of Virtual Environments to Support Soft-Skill Acquisition for Individuals with Autism. *Qualitative Report, 22*(9), 2509–2532. Retrieved from http://nsuworks.nova.edu/tqr/vol22/iss9/16

Gannod, G. C., & Bachman, K. M. (2011). Integrating M-Learning in a broad context: issues and recommendations. In I.A. Sanchez, & P. Isaias (Eds.), *Proceedings of the IADIS International Conference Mobile Learning*, Avita, Spain (pp. 291-294).

Garcia-Cabot, A., De-Marcos, L., & Garcia-Lopez, E. (2015). An empirical study on m-learning adaptation: Learning performance and learning contexts. *Computers & Education, 82*, 450–459. doi:10.1016/j.compedu.2014.12.007

Gardner, H., & Davis, K. (2013). *The app generation*. New Haven, CT: Yale University Press.

Garrison, D. R., Anderson, T., & Archer, W. (1999). Critical inquiry in a text-based environment: Computer conferencing in higher education. *The Internet and Higher Education, 2*(2–3), 87–105. doi:10.1016/S1096-7516(00)00016-6

Garrison, D. R., Vaughan, N., & Norman, D. (2008). *Blended learning in higher education*. San Francisco: Jossey-Bass.

Geer, R., & Sweeney, T. A. (2012). Students voice about learning with technology. *Journal of Social Sciences, 8*(2), 294–303. doi:10.3844/jssp.2012.294.303

Geist, E. A. (2012). A qualitative examination of two year--olds interaction with tablets. *Journal of Instructional Psychology, 39*(1).

Gess-Newsome, J. (2015). A model of teacher professional knowledge and skill including PCK: results of the thinking from the PCK summit. In A. Berry, P. J. Friedrichsen, & J. J. Loughran (Eds.), *Re-examining pedagogical content knowledge in science education* (pp. 28–42). New York, NY: Routledge.

Getenet, S. T. (2016). Adapting technological pedagogical content knowledge framework to teach mathematics. *Education and Information Technologies, 22*(5), 2629–2644.

Giemza, A., Verheyen, P., & Hoppe, H. U. (2012, March). Challenges in scaling mobile learning applications: The example of quizzer. In *2012 IEEE Seventh International Conference on Wireless, Mobile and Ubiquitous Technology in Education* (pp. 287-291). 10.1109/WMUTE.2012.69

Gikas, J., & Grant, M.M. (2013). Mobile computing devices in higher education: Student perspectives on learning with cellphones, smartphones & social media. *Internet and Higher Education, 19*, 18–26.

Gikas, J., & Grant, M. M. (2013). Mobile computing devices in higher education: Student perspectives on learning with cellphones, smartphones and social media. *The Internet and Higher Education, 19*, 18–26. doi:10.1016/j.iheduc.2013.06.002

Gil, A. C. (2007). *Como elaborar projetos de pesquisa*. São Paulo: Atlas.

Giordano, D., & Maiorana, F. (2014). Addressing dysgraphia with a mobile, web-based software with interactive feedback. *Biomedical and Health Informatics (BHI), 2014 IEEE-EMBS International Conference on,* 264-268.10.1109/BHI.2014.6864354

Giordano, D., & Maiorana, F. (2015). A Mobile Web Game Approach for Improving Dysgraphia. In *Proceedings of the 7th International Conference on Computer Supported Education* (vol. 1, pp. 328-333). SCITEPRESS-Science and Technology Publications, Lda. 10.5220/0005449103280333

Glaser, B., & Strauss, A. (1967). *The discovery grounded theory: Strategies for qualitative inquiry*. Chicago, IL: Aldine.

Goggin, G., & Hjorth, L. (2014). *Routledge Companion to Mobile Media*. New York: Routledge. doi:10.4324/9780203434833

Goodlad, J. (1994). *Educational renewal: better teachers, better schools.* San Francisco, CA: Jossey-Bass.

Graham, S., Harris, K. R., & Fink, B. (2000). Is handwriting causally related to learning to write? Treatment of handwriting problems in beginning writers. *Journal of Educational Psychology, 92*(4), 620–633. doi:10.1037/0022-0663.92.4.620

Graham, S., Weintraub, N., & Berninger, V. W. (1998). The relationship between handwriting style and speed and legibility. *The Journal of Educational Research, 91*(5), 290–297. doi:10.1080/00220679809597556

Grossman, P., Wineburg, S., & Woolworth, S. (2001). Toward a theory of teacher community. *Teachers College Record, 103*(6), 942–1012. doi:10.1111/0161-4681.00140

Guinet, E., & Kandel, S. (2010). Ductus: A software package for the study of handwriting production. *Behavior Research Methods, 42*(1), 326–332. doi:10.3758/BRM.42.1.326 PMID:20160312

Guskey, T. R. (2003). What makes professional development effective? *Phi Delta Kappan, 84*(10), 748–750. doi:10.1177/003172170308401007

Gustafson, T. M. (2015). *Teacher education as distance education: Pre-service experience and the uses of technology.* Michigan State University.

Hadar-Peker, D. D. (2013, December). *The Relationship between the Social-Emotional State and the Scholastic achievements of Schoolchildren.* Retrieved from The Initiative for Applied Research in Education: http://yozma.mpage.co.il/SystemFiles/23051.pdf

Hainey, T., Connolly, T, M., Boyle, E, A., Wilson, A., & Razak, A. (2016). A systematic literature review of games-based learning empirical evidence in primary education. *Computers & Education.* doi:10.1016/j.compedu.2016.09.001

Hammersley, M. (2008). *Questioning Qualitative Research: Critical Essays.* London, UK: Sage. doi:10.4135/9780857024565

Hamstra-Bletz, L., & Blöte, A. W. (1993). A longitudinal study on dysgraphic handwriting in primary school. *Journal of Learning Disabilities, 26*(10), 689–699. doi:10.1177/002221949302601007 PMID:8151209

Hanafi, H.F., & Samsudin, H. (2012). Mobile Learning Environment System (MLES): The Case of Android-based Learning Application on Undergraduates' Learning International. *Journal of Advanced Computer Science and Applications, 3*(3).

Han, I., & Shin, S. W. (2016). The use of a mobile learning management system and academic achievement of online students. *Computers & Education, 102*, 79–89. doi:10.1016/j.compedu.2016.07.003

Harley, L., Kline, K., Price, C., Jones, A., Mosley, S., Farmer, S., & Fain, B. (2013). A web based voting application study of fonts for voters with dyslexia. *The Information Technology and Innovation Foundation Accessible Voting Technology Initiative Working Paper Series,* 1-30.

Harley, J. M., Poitras, E. G., Jarrell, A., Duffy, M. C., & Lajoie, S. P. (2016). Comparing virtual and location-based augmented reality mobile learning: Emotions and learning outcomes. *Educational Technology Research and Development, 64*(3), 359–388. doi:10.100711423-015-9420-7

Harmon, K. (2013, July 12). *Best iPad® Apps and iPad Strategies to Increase Student Learning (Grades K-6).* Retrieved from Bureau of Education & Research (BER): http://www.ber.org/bro_pdf/ZDH4F2.pdf

Harris, J. B., & Hofer, M. J. (2011). Technological pedagogical content knowledge (TPACK) in action: A descriptive study of secondary teachers' curriculum-based, technology-related instructional planning. *Journal of Research on Technology in Education, 43*(3), 211–229. doi:10.1080/15391523.2011.10782570

Harris, K. R., Graham, S., & Mason, L. H. (2006). Improving the writing, knowledge, and motivation of struggling young writers: Effects of self-regulated strategy development with and without peer support. *American Educational Research Journal, 43*(2), 295–340. doi:10.3102/00028312043002295

Hasselbring, S. T., & Bausch, E. M. (2005). Assistive Technologies for Reading. *Learning in the Digital Age, 63*(4), 72–75.

Healey, M., Flint, A., & Harrington, K. (2014). *Engagement through partnership: Students as partners in learning and teaching in higher education.* York, UK: Higher Education Academy. Retrieved from https://www.heacademy.ac.uk/engagement-throughpartnership-students-partners-learning-and-teaching-higher-education

Heath-Reynolds, J., & VanWeelden, K. (2015). Integrating apps with the core Arts Standards in the 21st-century elementary music classroom. *General Music Today, 29*(1), 24–27. doi:10.1177/1048371315589127

Helps, R., & Helps, C. (2016, September). Measuring stylus and tablet performance for usability in sketching. *Proceedings of the 5th Annual Conference on Research in Information Technology*, 19-24. 10.1145/2978178.2978186

Hénard, F., & Leprince-Ringuet, S. (2008). *The path to quality teaching in higher education.* OECD. Retrieved from https://www1.oecd.org/edu/imhe/44150246.pdf

Hénard, F., & Roseveare, D. (2012). *Fostering quality teaching in higher education: Policies and practices.* OECD. Retrieved from https://www.oecd.org/edu/imhe/QT%20policies%20and%20practices.pdf

Hennion, B., Gentaz, E., Gouagout, P., & Bara, F. (2005). Telemaque, a new visuo-haptic interface for remediation of dysgraphic children. *Proceedings of WorldHaptic 2005*, 410-419. 10.1109/WHC.2005.125

Henry, S. L. (2018). *Introduction to Web Accessibility.* Retrieved from https://www.w3.org/WAI/fundamentals/accessibility-intro/

Henton, M. C. (2016). Technology integration practices of faculty in the louisiana community and technical college system. Retrieved from https://search.proquest.com/docview/1871305453?accountid=26642

Herrington, J., & Parker, J. (2013). Emerging technologies for authentic learning. *British Journal of Educational Technology, 44*(4), 607–615. doi:10.1111/bjet.12048

He, W., Holton, A., & Farkas, G. (2018). Impact of partially flipped instruction on immediate and subsequent course performance in a large undergraduate chemistry course. *Computers & Education, 125*, 120–131. doi:10.1016/j.compedu.2018.05.020

Hirsh-Pasek, K. H., Zosh, J. M., Golinkoff, R. M., Gray, J. H., Robb, M. B., & Kaufman, J. (2015). Putting education in "Educational" apps: Lessons from the science of learning. *Psychological Science in the Public Interest, 16*(1), 3–34. doi:10.1177/1529100615569721 PMID:25985468

Hoffmann, M., Meisen, T., & Jeschke, S. (2014). Shifting virtual reality education to the next level: Experiencing remote laboratories through mixed reality. In S. Frerich, T. Meisen, & A. Richert et al. (Eds.), *Engineering Education 4.0* (pp. 235–249). New York: Springer.

Hofstein, A., Mamlok, R., & Rosenberg, O. (2006). Varying instructional methods and students assessment methods in high school chemistry. In M. McMahon, P. Simmons, R. Sommers, D. DeBaets, & F. Crawley (Eds.), *Assessment in Science* (pp. 139–148). Arlington, VA: NSTA press.

Hofstein, A., Shore, R., & Kipnis, M. (2004). Providing high school chemistry students with opportunities to develop learning skills in an inquiry-type laboratory: A case study. *International Journal of Science Education, 26*, 47–62. doi:10.1080/0950069032000070342

Holden, C. (2014). Homegrown augmented reality. *TechTrends*, *58*(1), 42–48. doi:10.100711528-013-0719-0

Hong, S., & Kwek, D. (2011). Innovation in the classroom: Design thinking for 21st century learning. *Stanford EDU.*

Hopkins, J. (2004). School library accessibility: The role of assistive technology. *Teacher Librarian*, *31*(3), 15.

Hornecker, E., & Dünser, A. (2009). Of pages and paddles: Children's expectations and mistaken interactions with physical–digital tools. *Interacting with Computers*, *21*(1–2), 95–107. doi:10.1016/j.intcom.2008.10.007

Hosein, A., Ramanau, R., & Jones, C. (2010). Learning & living technologies: A longitudinal study of first-year students' frequency & competence in the use of ICT. *Learning, Media and Technology*, *35*(4), 403–418. doi:10.1080/17439884.2010.529913

Howland, J. L., Jonassen, D. H., & Marra, R. M. (2014). *Meaningful learning with technology*. Melbourne, Australia: Pearson Education Australia.

Hsiao, K. F., Chen, N. S., & Huang, S. Y. (2011). Learning while exercising for science education in augmented reality among adolescents. *Interactive Learning Environments*, *20*(4), 331–349. doi:10.1080/10494820.2010.486682

Hsu, Y. C., & Ching, Y. H. (2013). Mobile app design for teaching and learning: Educators' experiences in an online graduate course. *The International Review of Research in Open and Distributed Learning*, *14*(4). doi:10.19173/irrodl.v14i4.1542

Huang, T. C., Chen, C. C., & Chou, Y. W. (2016). Animating eco-education: To see, feel, and discover in an augmented reality-based experiential learning environment. *Computer Education*, *96*, 72–82. doi:10.1016/j.compedu.2016.02.008

Huang, Y. M., Lin, Y. T., & Cheng, S. C. (2010). Effectiveness of a mobile plant learning system in a science curriculum in Taiwanese elementary education. *Computers & Education*, *54*(1), 47–58. doi:10.1016/j.compedu.2009.07.006

Hua, Y., Woods-groves, S., Kaldenberg, E. R., Lucas, K. G., & Therrien, W. J. (2015).. . *Effects of the TIP Strategy on Problem Solving Skills of Young Adults with Intellectual Disability*, *50*(1), 31–42.

Huber, S. (2012). iPads in the Classroom - A Development of a Taxonomy for the Use of Tablets in Schools (M. Ebner, & S. Schön, Eds.). Norderstedt: Books on Demand GmbH.

Hung, Y. H., Chen, C. H., & Huang, S. W. (2017). Applying augmented reality to enhance learning: A study of different teaching materials. *Journal of Computer Assisted Learning*, *33*(3), 252–266. doi:10.1111/jcal.12173

Husbye, N. E., & Elsener, A. A. (2013). To move forward, we must be mobile: Practical uses of mobile technology in literacy education courses. *Journal of Digital Learning in Teacher Education*, *30*(2), 46–51. doi:10.1080/21532974.2013.10784726

Hwang, G. J., & Chang, H. F. (2011). A formative assessment-based mobile learning approach to improving the learning attitudes and achievements of students. *Computers & Education*, *56*(4), 1023–1031. doi:10.1016/j.compedu.2010.12.002

Hwang, G. J., & Wu, P. H. (2014). Applications, impacts and trends of mobile technology- enhanced learning: A review of 2008–2012 publications in selected SSCI journals. *International Journal of Mobile Learning and Organisation*, *8*(2), 83–95. doi:10.1504/IJMLO.2014.062346

Hwang, W. Y., Chen, C. Y., & Chen, H. S. (2011). (2011). Facilitating EFL writing of elementary school students in familiar situated contexts with mobile devices. *10th World Conference on Mobile and Contextual Learning: mLearn2011 Conference Proceedings, Beijing, China*, 15-23.

Hwang, W. Y., & Chen, H. S. (2013). Users' familiar situational contexts facilitate the practice of EFL in elementary schools with mobile devices. *Computer Assisted Language Learning*, *26*(2), 101–125. doi:10.1080/09588221.2011.639783

IBGE. (n.d.). Estatísticas. Retrieved from http://downloads.ibge.gov.br/downloads_estatisticas.htm

İbili, E., & Şahin, S. (2013). Software design and development of an interactive 3D geometry book using augmented reality: ARGE3D Afyon Kocatepe University. *Journal of Science and Engineering, 13*(1), 1–8.

IBM Corporation. (2017). *IBM Design Thinking*. Retrieved April 20, 2017, from https://www.ibm.com/design/thinking/

ICED. (2014). The Preparation of University Teachers Internationally. Retrieved from http://icedonline.net/wp-content/uploads/2011/08/The-Preparation-of-UniversityTeachers-Internationally.pdf

Igual, R., Plaza, I., Martin, L., Corbalan, M., & Medrano, C. (2013). Guidelines to design smartphone applications for people with intellectual disability: A practical experience. In A. van Berlo, K. Hallenborg, J. M. Corchado Rodriguez, D. I. Tapia, & P. Novais (Eds.), *Ambient Intelligence-Software and Applications* (pp. 65–69). Springer International Publishing. doi:10.1007/978-3-319-00566-9_9

Ihamäki, P. (2014). The potential of treasure hunt games to generate positive emotions in learners: Experiencing local geography and history using GPS devices. *Int. J. Technology Enhanced Learning, 6*(1), 5–20. doi:10.1504/IJTEL.2014.060025

Immordino-Yang, M. H., & Damasio, A. (2007). We feel, therefore we learn: The relevance of affective and social neuroscience to education. *Mind, Brain and Education: the Official Journal of the International Mind, Brain, and Education Society, 1*(1), 3–10. doi:10.1111/j.1751-228X.2007.00004.x

Israel, R., Ronen, M., Kalderon, S., & Shoham, H. (2014) Are we ready for this QRace? In Y. Eshet, I. Blau, A. Caspi et al. (Eds.), *Proceeding of the 10th Chais Conference for Research in Innovation and Learning Technologies,* The Open University of Israel, Raanana (pp. 56-63).

Ito, M., Baumer, S., Bittanti, M., Boyd, D., Cody, R., HerrStephenson, B., … Tripp, L. (2010). Hanging Out, Messing Around, Geeking Out: Kids Living and Learning with New Media. Cambridge, MA: The MIT Press.

Izadinia, M. (2016). Student teachers' and mentor teachers' perceptions and expectations of a mentoring relationship: Do they match or clash? *Professional Development in Education, 42*(3), 387–402. doi:10.1080/19415257.2014.994136

Jacka, L. (2018). *Using virtual worlds in educational settings: Making learning real.* London: Taylor and Francis. doi:10.4324/9780203728901

Jan, S. R., Ullah, F., Ali, H., & Khan, F. (2016). Enhanced and Effective Learning through Mobile Learning: An Insight into Students Perception of Mobile Learning at University Level. *International Journal of Scientific Research in Science, Engineering and Technology, 2*(2), 674–681.

Jansen, B. R. J., De Lange, E., & Van der Molen, M. J. (2013). Math practice and its influence on math skills and executive functions in adolescents with mild to borderline intellectual disability. *Research in Developmental Disabilities, 34*(5), 1815–1824. doi:10.1016/j.ridd.2013.02.022 PMID:23511346

Jena, S. P. K. (2013). Learning Disability: Theory to Practice. Sage Publication India Pvt. Ltd.

Jeno, L. M., Adachi, P. J. C., Grytnes, J., Vandvik, V., & Deci, E. L. (2018). The Effects of M-learning on Motivation, Achievement and Well-being: A Self-Determination Theory Approach. *British Journal of Educational Technology.* doi:10.1111/bjet.12657

Jeno, L. M., Vandvik, V., Eliassen, S., & Grytnes, J. (2019). Testing the novelty effect of an m-learning tool on internalization and achievement: A Self-Determination Theory approach. *Computers & Education, 128*, 398–413. doi:10.1016/j.compedu.2018.10.008

Jerry, T., & Aaron, C. (2010). The impact of augmented reality software with inquiry-based learning on students' learning of kinematics graph. In *2nd International Conference on Education Technology and Computer (ICETC), 2010* (pp. V2-1–V2-5). Shanghai, China: IEEE. 10.1109/ICETC.2010.5529447

Johnson, S. (2016). IBM Design Thinking Field Guide. IBM Corporation, 3.4.

Johnson, D. W., Johnson, R. T., & Smith, K. (1991). *Active learning: Cooperation in the college classroom.* Edina, MN: Interaction.

Johnson, L., Adams Becker, S., Cummins, M., Estrada, V., Freeman, A., & Ludgate, H. (2018). *Technology outlook for community, technical, and junior colleges 2013-2018: An NMC horizon project sector analysis.* Austin, TX: The New Media Consortium.

Johnson, L., Adams Becker, S., Estrada, V., & Freeman, A. (2015). *NMC Horizon Report: 2015 Higher Education Edition.* Austin, TX: The New Media Consortium.

Johnson, L., Adams Becker, S., Estrada, V., Freeman, A., Kampylis, P., Vuorikari, R., & Punie, Y. (2014). *NMC Horizon Report Europe: 2014 Schools Edition.* Austin, TX: The New Media Consortium. Retrieved from https://www.learntechlib.org/p/182011/

Johnson, L., Smith, R., Willis, H., Levine, A., & Haywood, K. (2011). *The 2011 Horizon Report.* Austin, TX: The New Media Consortium.

Jonassen, D. H. (1996). *Computers in the classroom: Mindtools for critical thinking.* Columbus, OH: Merrill/Prentice-Hall.

Jonassen, D. H., Wilson, B. G., Wang, S., & Grabinger, R. S. (1993). Constructivist uses of expert systems to support learning. *Journal of Computer-Based Instruction, 20*(3), 86–94.

Jones, A., Issroff, K., Scanlon, E., Clough, G., & McAndrew, P. (2006). Using mobile devices for learning in informal settings: Is it motivating? In *IADIS International Conference on Mobile Learning,* Dublin (pp. 251–255). IADIS Press.

Jordan, J. (2011). Climbing out of the Box and into the Cloud: Building web-scale for libraries. *Journal of Library Administration, 51*(1), 3–17. doi:10.1080/01930826.2011.531637

Juan, M. C., Furió, D., Alem, L., Ashworth, P., & Cano, J. (2011). ARGreenet and Basic-Greenet: Two mobile games for learning how to recycle. In *Proceedings of the 19th International Conference on Computer Graphics, Visualization and Computer Vision 2011* (pp. 25-32).

Judge, S., Floyd, K., & Jeffs, T. (2015). Using mobile media devices and apps to promote young children's learning. In K. L. Heider & M. Renck Jalongo (Eds.), *Young children and families in the information age, educating the young child* (pp. 117–131). New York: Springer. doi:10.1007/978-94-017-9184-7_7

Kabali, H. K., Irigoyen, M. M., Nunez-Davis, R., Budacki, J. G., Mohanty, S. H., Leister, K. P., & Bonner, R. L. (2015). Exposure and use of mobile media devices by young children. *Pediatrics, 136*(6), 2015–2151. doi:10.1542/peds.2015-2151 PMID:26527548

Kagohara, D. M., Meer, L., Ramdoss, S., O'Reilly, M. F., Lancioni, G. E., Davis, T. N., & Sigafoos, J. (2013). Using iPods® and iPads® in teaching programs for individuals with developmental disabilities: A systematic review. *Research in Developmental Disabilities, 34*(1), 147–156. doi:10.1016/j.ridd.2012.07.027 PMID:22940168

Kamarainen, A., Metcalf, S., Grotzer, T. A., Brimhall, C., & Dede, C. (2016). Atom tracker: Designing a mobile augmented reality experience. *International Journal of Designs for Learning, 7*(2), 111–130. doi:10.14434/ijdl.v7i2.19430

Kao, H. Y., Wei, C. W., Yu, M. C., Liang, T. Y., Wu, W. H., & Wu, Y. J. (2018). Integrating a mobile health applications for self-management to enhance Telecare system. *Telematics and Informatics*, *35*(4), 815–825. doi:10.1016/j.tele.2017.12.011

Karlsdottir, R., & Stefansson, T. (2002). Problems in developing functional handwriting. *Perceptual and Motor Skills*, *94*(2), 623–662. doi:10.2466/pms.2002.94.2.623 PMID:12027360

Karsenty, R., & Arcavi, A. (2017). Mathematics, lenses and videotapes: A framework and a language for developing reflective practices of teaching. *Journal of Mathematics Teacher Education*, *20*(5), 433–455. doi:10.100710857-017-9379-x

Karsenty, R., & Sherin, M. G. (2017). Video as a catalyst for mathematics teachers' professional growth. *Journal of Mathematics Teacher Education*, *20*(5), 409–413. doi:10.100710857-017-9387-x

Katusic, S. K., Colligan, R. C., Weaver, A. L., & Barbaresi, W. J. (2009). The forgotten learning disability: Epidemiology of written-language disorder in a population-based birth cohort (1976–1982), Rochester, Minnesota. *Pediatrics*, *123*(5), 1306–1313. doi:10.1542/peds.2008-2098 PMID:19403496

Kearney, M., Burden, K., & Rai, T. (2015). Investigating teachers' adoption of signature pedagogies. *Computers & Education*, *80*, 48–57. doi:10.1016/j.compedu.2014.08.009

Kearney, M., Schuck, S., Burden, K., & Aubusson, P. (2012). Viewing mobile learning from a pedagogical perspective. *Research in Learning Technology*, *20*(1). doi:10.3402/rlt.v20i0.14406

Ke, F., & Abras, T. (2013). Games for engaged learning of middle school children with special learning needs. *British Journal of Educational Technology*, *44*(2), 225–242. doi:10.1111/j.1467-8535.2012.01326.x

Ke, F., & Hsu, Y. C. (2015). Mobile augmented-reality artifact creation as a component of mobile computer-supported collaborative learning. *The Internet and Higher Education*, *26*, 33–41. doi:10.1016/j.iheduc.2015.04.003

Kennedy, T., Wellman, B., & Clement, K. (2008). Gendering the digital divide. *IT & Society*, *1*(5), 72–96.

Kenny, R. F., Park, C. L., Neste-Kenny, V., Jocelyne, M. C., & Burton, P. A. (2012). *Mobile Self-Efficacy in Canadian Nursing Education Programs-Replication*. Academic Press.

Kerawalla, L., Luckin, R., Seljeflot, S., & Woolard, A. (2006). Making it real: Exploring the potential of augmented reality for teaching primary school science. *Virtual Reality (Waltham Cross)*, *10*(3-4), 163–174. doi:10.100710055-006-0036-4

Kesim, M., & Ozarslan, Y. (2012). Augmented reality in education: Current technologies and the potential for education. *Procedia: Social and Behavioral Sciences*, *47*, 297–302. doi:10.1016/j.sbspro.2012.06.654

Keskin, N. O., & Metcalf, D. (2011). The current perspectives, theories and practices of mobile learning. *Turkish Online Journal of Educational Technology*, *10*(2), 202–208.

Kettle, K. L., & Häubl, G. (2010). Motivation by anticipation: Expecting rapid feedback enhances performance. *Psychological Science*, *21*(4), 545–547. doi:10.1177/0956797610363541 PMID:20424099

Khaddage, F., Müller, W., & Flintoff, K. (2016). Advancing mobile learning in formal and informal settings via mobile app technology: Where to from here, and how? *Journal of Educational Technology & Society*, *19*(3), 16–26.

Kim, J., Ilon, L., & Altmann, J. (2013). Adapting smartphones as learning technology in a Korean university. *Journal of Integrated Design & Process Science*, *17*(1), 5–16.

Kimura, Y., Nakata, Y., & Okumura, T. (2008, October 26). *Language Learning Motivation of EFL Learners in Japan—A Cross-Sectional Analysis of Various Learning Milieus*. Retrieved from Publications of the Japan Association for Language Teaching: http://jalt-publications.org/archive/jj/2001a/art3.pdf

King, E., Okodogbe, T., Burke, E., McCarron, M., McCallion, P., & O'Donovan, M. A. (2017). Activities of daily living and transition to community living for adults with intellectual disabilities. *Scandinavian Journal of Occupational Therapy*, *24*(5), 357–365. doi:10.1080/11038128.2016.1227369 PMID:27707082

Kirkley, B. S. E., & Kirkley, J. R. (2004). Creating next generation blended learning environments using mixed reality, video games and simulations. *TechTrends*, *49*(3), 42–53. doi:10.1007/BF02763646

Kirkwood, A., & Price, L. (2005). Learners & learning in the twenty-first century: What do we know about students'' attitudes towards & experiences of information & communication technologies that will help us design courses? *Studies in Higher Education*, *30*(3), 257–274. doi:10.1080/03075070500095689

Klinger, M. B., & Coffman, T. (2011). Emphasizing diversity through 3D multi-user virtual worlds. In G. Kuruback & T. V. Yuzer (Eds.), *Handbook of research on transformative online education and liberation: Models for social equality*. Hershey, PA: IGI Global. doi:10.4018/978-1-60960-046-4.ch005

Klopfer, E. (2008). *Augmented learning: Research and design of mobile educational games*. Cambridge, MA: MIT Press. doi:10.7551/mitpress/9780262113151.001.0001

Klopfer, E. (2011). Forward/preface. In S. Dikkers, J. Martin, & B. Coulter (Eds.), *Mobile media learning: Amazing uses of mobile devices for learning*. Pittsburgh, PA: ETC Press.

Klopfer, E., & Sheldon, J. (2010). Augmenting your own reality: Student authoring of science-based augmented reality games. *New Directions for Youth Development*, *128*(128), 85–94. doi:10.1002/yd.378 PMID:21240956

Koehler, M., & Mishra, P. (2009). What is technological pedagogical content knowledge (TPACK)? *Contemporary Issues in Technology & Teacher Education*, *9*(1), 60–70.

Kohen-Vacs, D., Ronen, M., & Cohen, S. (2012). Mobile treasure hunt games for outdoor learning. *Bulletin of the IEEE Technical Committee on Learning Technology*, *14*(4), 24–26.

Kolb, D. A. (2014). *Experiential learning: Experience as the source of learning and development*. FT Press.

Kommers, P. A. M., Jonassen, D. H., & Mayes, J. T. (1992). *Cognitive tools for learning*. Heidelberg, Germany: Springer. doi:10.1007/978-3-642-77222-1

Koole, M., & Ally, M. 2006, Framework for the Rational Analysis of Mobile Education (FRAME) Model: Revising the ABCs of Educational Practices. In *The International Conference on Networking, International Conference on Systems and International Conference on Mobile Communications and Learning Technologies* (pp. 216-216). IEEE. 10.1109/ICNICONSMCL.2006.103

Korkmazlar, Ü. (2003). Özel Öğrenme Bozukluğu. In A. Kulaksızoğlu (Ed.), *Farklı Gelişen Çocuklar*. İstanbul: Remzi Kitabevi.

Kowalski, F. V., Kowalski, S. E., Colling, T. J., Cuba, J. G., Gardner, T. Q., Greivel, G., & Ruskell, T. (2015). Using inksurvey with pen-enabled mobile devices for real-time formative assessment: I applications in diverse educational environments. In *The Impact of Pen and Touch Technology on Education* (pp. 297–305). Cham: Springer. doi:10.1007/978-3-319-15594-4_31

Kucirkova, N. (2016). iRPD--a framework for guiding design-based research for iPad apps. *British Journal of Educational Technology*, 48598–48610.

Kukulska-Hulme, A. (2012). Mobile-Assisted language learning. In C. Chapelle (Ed.), *The encyclopedia of applied linguistics* (pp. 3701–3709). New York, NY: Wiley Online Library; doi:10.1002/9781405198431.wbeal0768

Kukulska-Hulme, A., & Shield, L. (2008). An overview of mobile assisted language learning: From content delivery to supported collaboration and interaction. *ReCALL, 20*(3), 271–289. doi:10.1017/S0958344008000335

Kurt, G. (2017). Implementing the flipped classroom in teacher education: Evidence from turkey. *Journal of Educational Technology & Society, 20*(1), 211–221.

Kushki, A., Schwellnus, H., Ilyas, F., & Chau, T. (2011). Changes in kinetics and kinematics of handwriting during a prolonged writing task in children with and without dysgraphia. *Research in Developmental Disabilities, 32*(3), 1058–1064. doi:10.1016/j.ridd.2011.01.026 PMID:21315553

Kvale, S. (1996). *InterViews: An introduction to Qualitative Research Interviewing*. Thousand Oaks, CA: Sage.

Lam, R., & Muldner, K. (2017). Manipulating cognitive engagement in preparation-to-collaborate tasks and the effects on learning. *Learning and Instruction.*

Lan, J., Sung, T., & Chang, E. (2007). *A Mobile-Divice-Supported Peer-Assited Learning System for Collaborative Early EFL Reading.* Retrieved December 15, 2012, from Language Learning & Technology: A refereed for second and foreignlanguage scholars and educators: http://llt.msu.edu/vol11num3/pdf/lansungchang.pdf

Lapointe, D., & Gunawardena, C. (2004). Developing, testing and refining of a model to understand the relationship between peer interaction and learning outcomes in computer-mediated conferencing. *Distance Education, 25*(1), 83–106. doi:10.1080/0158791042000212477

Lave, J., & Wenger, E. (1998). *Situated learning: Legitimate peripheral participation.* Cambridge, UK: Cambridge University Press.

Law, R., Chan, I. C. C., & Wang, L. (2018). A comprehensive review of mobile technology use in hospitality and tourism. *Journal of Hospitality Marketing & Management, 27*(6), 626–648. doi:10.1080/19368623.2018.1423251

Lawson, T., Çakmak, M., Gündüz, M., & Busher, H. (2015). Research on teaching practicum - a systematic review. *European Journal of Teacher Education, 38*(3), 392–40. doi:10.1080/02619768.2014.994060

Lee, I., Yamada, T., Shimizu, Y., Shinohara, M., & Hada, Y. (2005). In search of the mobile learning paradigm as we are going nomadic. In P. Kommers & G. Richards (Eds.), *Proceedings of EdMedia: World Conference on Educational Media and Technology 2005* (pp. 2888-2893). Association for the Advancement of Computing in Education (AACE). Retrieved from http://www.editlib.org/p/20521

Lee, K. (2012). The future of learning and training in augmented reality. *InSight: A Journal of Scholarly Teaching, 7*, 31-40.

Lee, I., & Shin, Y. J. (2018). Fintech: Ecosystem, business models, investment decisions, and challenges. *Business Horizons, 61*(1), 35–46. doi:10.1016/j.bushor.2017.09.003

Lee, J., & Hammer, J. (2011). Gamification in education: What, how, why bother? *Academic Exchange Quarterly, 15*(2), 146.

Leinonen, T., Keune, A., Veermans, M., & Toikkanen, T. (2016). Mobile apps for reflection in learning: A design research in K-12 education. *British Journal of Educational Technology, 47*(1), 184–202. doi:10.1111/bjet.12224

Lenhart, A., & Page, D. (2015). *Original survey report: Teens, social media & technology overview 2015.* Washington, DC: Pew Research Center. Retrieved from http://www.pewinternet.org/2015/04/09/teens-social-media-technology-2015/

Le, Q. T., Pedro, A., Pham, H. C., & Park, C. S. (2016). A virtual world based construction defect game for interactive and experiential learning. *International Journal of Engineering Education, 32*(1B), 457–467.

Lesser, E. L., & Storck, J. (2001). Communities of practice and organizational performance. *IBM Systems Journal, 40*(4), 831–841. doi:10.1147j.404.0831

Li, K. C., Lee, L. Y. K., Wong, S. L., Yau, I. S. Y., & Wong, B. T. M. (2018). Effects of mobile apps for nursing students: learning motivation, social interaction and study performance. *Open Learning: The Journal of Open, Distance and e-Learning, 33*(2), 99-114.

Liang, J. (2017). *M-learning development for enterprise Online University based on gamification.* Chinese National Centre for Philosophy and Social Science Documentation. Available at: www.ncpssd.org/literatureforregisteredusers

Lieberman, A., & Mace, D. (2010). Making practice public: Teacher learning in the 21st century. *Journal of Teacher Education, 61*(1-2), 77–88. doi:10.1177/0022487109347319

Light, B., Bagnall, G., Crawford, G., & Gosling, V. (2018). The material role of digital media in connecting with, within and beyond museums. *Convergence, 24*(4), 407–423. doi:10.1177/1354856516678587

Li, K. C., & Wang, F. L. (2012). Student engagement: Meanings, approaches and ideas for educators interested in ICT. In *International Conference on ICT in Teaching and Learning* (pp. 1-10). Springer. 10.1007/978-3-642-31398-1_1

Li-Tsang, C. W., Wong, A. S., Leung, H. W., Cheng, J. S., Chiu, B. H., Linda, F. L., & Chung, R. C. (2013). Validation of the Chinese Handwriting Analysis System (CHAS) for primary school students in Hong Kong. *Research in Developmental Disabilities, 34*(9), 2872–2883. doi:10.1016/j.ridd.2013.05.048 PMID:23816625

Little, J. W. (2012). Professional community and professional development in the learning-centered school. *Teacher learning that matters: International perspectives*, 22–46.

Little, J. W. (2002). Locating learning in teachers' communities of practice: Opening up problems of analysis in records of everyday work. *Teaching and Teacher Education, 18*(8), 917–946. doi:10.1016/S0742-051X(02)00052-5

Liu, S. (2016). The perceptions of participation in a mobile collaborative learning among pre-service teachers. *Journal of Education and Learning, 5*(1), 87. doi:10.5539/jel.v5n1p87

Liu, T. C., Wang, H. Y., Liang, J. K., Chan, T. W., Ko, H. W., & Yang, J. C. (2003). Wireless and mobile technologies to enhance teaching and learning. *Journal of Computer Assisted Learning, 19*(3), 371–382. doi:10.1046/j.0266-4909.2003.00038.x

Livingstone, S. (2007). Evaluating the online risks for children in Europe. *Telos, 73*, 52–69.

Livingstone, S. (2008). Taking risky opportunities in youthful content creation: Teenagers' use of social networking sites for intimacy, privacy and self-expression. *New Media & Society, 10*(3), 393–411. doi:10.1177/1461444808089415

Livingstone, S. (2009). *Children and the Internet: Great Expectations, Challenging Realities.* Cambridge, UK: Polity Press.

Livingstone, S., & Bober, M. (2005). *UK children go online: Final report of key project findings.* London: London School of Economics and Political Science.

Livingstone, S., Haddon, L., Vincent, J., Mascheroni, G., & Ólafsson, K. (2014). *Net Children Go Mobile: The UK Report.* London: London School of Economics and Political Science.

Locke, K. (2001). *Grounded theory in management research.* London: Sage.

Lowenthal, J. N. (2010). Using mobile learning: Determinates impacting behavioral intention. The Amer. *Journal of Distance Education, 24*(4), 195–206. doi:10.1080/08923647.2010.519947

Luckin, R. (2006). The learner centric ecology of resources: A framework for using technology to scaffold learning. *Computers & Education, 50*(2), 449–462. doi:10.1016/j.compedu.2007.09.018

Luckin, R., & Stanton, D. F. (2011). Limitless or pointless? An evaluation of augmented reality technology in the school and home. *International Journal of Technology Enhanced Learning, 3*(5), 510–524. doi:10.1504/IJTEL.2011.042102

Luna-Nevarez, C., & McGovern, E. (2018). On the Use of Mobile Apps in Education: The Impact of Digital Magazines on Student Learning. *Journal of Educational Technology Systems*, 0047239518778514.

Lunden, I. (2015). *6.1B smartphone users globally by 2020, overtaking basic fixed phone subscriptions.* TechCrunch. Retrieved from https://techcrunch.com/2015/06/02/6-1b-smartphone-users-globally-by-2020-overtaking-basic-fixed-phone-subscriptions/

Mac Mahon, B., Grádaigh, S. Ó., & Ghuidhir, S. N. (2016). iTE: Student teachers using iPad on a second level initial teacher education programme. *International Journal of Mobile and Blended Learning, 8*(2), 21–34. doi:10.4018/IJMBL.2016040102

MacArthur, C. A. (2009). Reflections on research on writing and technology for struggling writers. *Learning Disabilities Research & Practice, 24*(2), 93–103. doi:10.1111/j.1540-5826.2009.00283.x

Macià, M., & García, I. (2016). Informal online communities and networks as a source of teacher professional development: A review. *Teaching and Teacher Education, 55*, 291–307. doi:10.1016/j.tate.2016.01.021

Mackh, B. M. (2018). *Higher Education by Design - Best Practices for Curricular Planning and Instruction.* Abingdon: Taylor and Francis Press. doi:10.4324/9781351133715

Magnusson, S., Krajcik, J., & Borko, H. (1999). Nature, sources, and development of pedagogical content knowledge for science teaching. In J. Gess-Newsome & N. G. Lederman (Eds.), *Pedagogical content knowledge and science education* (pp. 95–132). Dordrecht: Kluwer Academic Publishers.

Maher, D. (2018). Supporting pre-service teachers' understanding and use of mobile devices. In J. Keengwe (Ed.), *Handbook of Research on Mobile Technology, Constructivism, and Meaningful Learning* (pp. 160–177). Hershey, PA: IGI Global. doi:10.4018/978-1-5225-3949-0.ch009

Maldonado-Molina, M., Reingle, J. M., Wesley, G., Jennings, W. G., & Prado, G. (2011). Drinking and driving among immigrant and US-born Hispanic young adults: Results from a longitudinal and nationally representative study. *Addictive Behaviors, 36*(4), 381–388. doi:10.1016/j.addbeh.2010.12.017 PMID:21216535

Malone, T. W., & Lepper, M. R. (1987). Making learning fun: A taxonomy of intrinsic motivations for learning. In R. E. Snow & M. J. Farr (Eds.), *Aptitude, learning, and instruction: III. Conative and affective process analyses* (pp. 223–253). Hillsdale, NJ: Erlbaum.

Mang, F. C., & Wardley, J. L. (2012). Effective Adoption of Tablets in Post-Secondary Education: Recommendations Based on a Trial of iPads in University Classes. *Journal of Information Technology Education: Innovations in Practice, 11*, 302–317.

Mansour, E. (2016). Use of smartphone apps among library and information science students at South Valley University, Egypt. *The Electronic Library, 34*(3), 371–404. doi:10.1108/EL-03-2015-0044

Manville, B., & Foote, N. (1996). Harvest your workers' knowledge-It's a brave new world out there. Call it post-modern reengineering, and to make your organization perform you'll have to build systems that support knowledge--not. *Datamation-Highlands Ranch, 42*(13), 78–83.

Marés, L. (2012, April). *Tablets in education: Opportunities and challenges in one-to-one programs.* Academic Press.

Marsden, E., Mitchell, R., & Myles, F. (2013). *Second language learning theories*. London: Routledge.

Martyn, M. (2007). Clickers in the classroom: An active learning approach. *EDUCAUSE Quarterly*, *2*, 71–74.

Masuhara, H. (2016). Brain studies and materials for language learning. In B. Tomlinson (Ed.), *SLA research and materials development for language learning* (pp. 23–32). New York, NY: Routledge.

Mathison, C., & Gabriel, K. (2012). Designing augmented reality experiences in authentic learning environments. In P. Resta (Ed.), *Proceedings of Society for Information Technology & Teacher Education International Conference 2012* (pp. 3470-3473). Chesapeake, VA: Association for the Advancement of Computing in Education (AACE).

Matthews, K. E. (2016). Students as partners as the future of student engagement. *Student Engagement in Higher Education Journal*, *1*(1) 1-5. Retrieved from https://journals.gre.ac.uk/index.php/raise

Mayer, R. E. (2001). *Multimedia learning*. Cambridge, UK: Cambridge University Press. doi:10.1017/CBO9781139164603

McAleese, M. (2014). *New modes of learning and teaching in higher education*. Retrieved from http://ec.europa.eu/dgs/education_culture/repository/education/library/reports/modernisation-universities_en.pdf

McAleese, M., Bladh, A., Berger, V., Bode, C., Muehlfeit, J., Petrin, T., . . . Tsoukalis, L. (2013). Improving the Quality of Teaching and Learning in Europe's HEIs. *European Commission for Education, Culture, Multilingualism, Youth & Sport*. Retrieved from http://ec.europa.eu/dgs/education_culture/repository/education/library/reports/modernisation_en.pdf

McCarty, T. L. (2018). Comparing 'new speakerhood': Context, positionality, and power in the new sociolinguistic order. *Journal of Multilingual and Multicultural Development*, *39*(5), 470–474. doi:10.1080/01434632.2018.1429457

McClanahan, B. (2012, May/June). A breakthrough for Josh: How Use of an iPad Facilitated. *Reading Improvement*.

McHale, K., & Cermak, S. A. (1992). Fine motor activities in elementary school: Preliminary findings and provisional implications for children with fine motor problems. *The American Journal of Occupational Therapy*, *46*(10), 898–903. doi:10.5014/ajot.46.10.898 PMID:1463061

McLuhan, M. (1966). *Understanding Media: The Extensions of Man*. New York: McGraw-Hill.

McMillan, D. W., & Chavis, D. M. (1986). Sense of community: A definition and theory. *Journal of Community Psychology*, *14*(1), 6–23. doi:10.1002/1520-6629(198601)14:1<6::AID-JCOP2290140103>3.0.CO;2-I

McNamara, B. E. (2007). *Learning disabilities: Bridging the gap between research and classroom practice*. Upper Saddle River, NJ: Pearson Merrill Prentice Hall.

Meishar-Tal, H., & Ronen, M. (2016). Who Enjoys More? Factors Related to Enjoyment in an Educational Mobile Game. In *European Conference on Games Based Learning* (p. 442). Academic Conferences International Limited.

Meishar-Tal., Sneh, Y., & Medzini, A. (2014). Designing mobile learning activities for outdoor learning. In *First International Conference on the use of iPads in Higher Education 2014*.

Meishar-Tal, H., & Gross, M. (2014). Teaching sustainability via smartphone-enhanced experiential learning in a botanical garden. *International Journal of Interactive Mobile Technologies*, *8*(1), 10–15. doi:10.3991/ijim.v8i1.3441

Meissel, K., & Rubie-Davies, C. M. (2016). Cultural invariance of goal orientation and self-efficacy in New Zealand: Relations with achievement. *The British Journal of Educational Psychology*, *86*(1), 92–111. doi:10.1111/bjep.12103 PMID:26856404

Mercedes, B., & Radel, B. (2016). Smartphones: From distraction to attraction. *Journal of Educational Technology Systems*, *45*(1), 93–102. doi:10.1177/0047239516639720

Mercer-Mapstone, L., Dvorakova, S. L., Groenendijk, L., & Matthews, K. E. (2017b). Idealism, conflict, leadership, and labels: Reflections on co-facilitation as partnership practice. *Teaching and Learning Together in Higher Education, 1*(21). Retrieved from http://repository.brynmawr.edu/tlthe/vol1/iss21/8/

Mercer-Mapstone, L., Dvorakova, L. S., Matthews, K. E., Abbot, S., Cheng, B., Felten, P., ... Swaim, K. (2017a). A Systematic Literature Review of Students as Partners in Higher Education. *International Journal for Students as Partners, 1*(1), 143–154. doi:10.15173/ijsap.v1i1.3119

Mercer-Mapstone, L., Marquis, E., & McConnell, C. (2018). The 'Partnership Identity' in Higher Education: Moving From 'Us' and 'Them' to 'We' in Student-Staff Partnership. *Student Engagement in Higher Education Journal, 2*(1), 12–29.

Merrick, J., Greydanus, D. E., & Patel, D. R. (2014). *Intellectual disability: Some current issues*. New York: Nova Science Publishers, Inc.

Metcalf, S. J., Kamarainen, A. M., Grotzer, T. A., & Dede, C. J. (2013). Teacher perceptions of the practicality and effectiveness of immersive ecological simulations. *International Journal of Virtual and Personal Learning Environments, 4*(3), 66–77. doi:10.4018/jvple.2013070105

Meurant, C. R. (2010, December). iPad Tablet Computing to Foster Korean EFL Digital Literacy. *International Journal of u- and e- Service Science and Technology, 3*(4), 49–62.

Miles, M. B., Huberman, A. M., & Saldana, J. (2013). Qualitative data analysis. *Sage (Atlanta, Ga.)*.

Miller, M. (2002). Resilience elements in students with learning disabilities. *Journal of Clinical Psychology, 58*(3), 291–298. doi:10.1002/jclp.10018 PMID:11836710

Millwood-Hargrave, A. (2007). *Issues facing broadcast regulation*. London: Broadcasting Standards Authority.

Millwood-Hargrave, A., & Livingstone, S. (2006). *Harm and Offence in Media Content: A review of the empirical literature*. Bristol, UK: Intellect Press.

Millwood-Hargrave, A., & Livingstone, S. (2009). *Harm and offence in media content: a review of the evidence* (Revised edition). Bristol, UK: Intellect.

Minayo, M. C. D. S., Deslandes, S. F., & Gomes, R. (2001). *Pesquisa Social: teoria, método e criatividade* (18th ed.).

Ministry of Education. (2014). *Future-focused Learning in connected communities*. Wellington, New Zealand: Author.

Mishra, P., & Koehler, M. J. (2006). Technological pedagogical content knowledge: A framework for teacher knowledge. *Teachers College Record, 108*(6), 1017–1054. doi:10.1111/j.1467-9620.2006.00684.x

Mitchell, R. (2011). Alien contact: Exploring teacher implementation of an augmented reality curricular unit. *Journal of Computers in Mathematics and Science Teaching, 30*(3), 271–302.

Mock, K. (2004). *Teaching with Tablet PC's*. Anchorage, AK: Academic Press.

Mohan, M. (2009, November). *Designing for Children- With focus on 'Play + Learn'*. Bombay: Academic Press.

MONE. (2005). *Ilköğretim Türkçe dersi öğretim programi ve kilavuzu*. Ankara: Devlet Kitapları Basımevi.

MONE. (2008). *Özel eğitim ve rehabilitasyon merkezi özel öğrenme güçlüğü destek eğitim programı*. Ankara: T.C. Milli Egitim Bakanligi Özel Öğretim Kurumları Genel Müdürlüğü.

MONE. (2015). Türkçe dersi (1-8. Sınıflar) öğretim programı. Ankara: T.C. Milli Egitim Bakanligi, Talim ve Terbiye Kurulu Başkanlığı.

Monem, R. (2018). The effect of mobile technology as an active student response system on the acquisition of U.S. history content of secondary students with specific learning disabilities. *Dissertation Abstracts International Section A: Humanities and Social Sciences, 77*(1-A)(E).

Moodle. (n.d.). Retrieved from https://moodle.org/

Moore, K., & McElroy, J. C. (2012). The influence of personality on Facebook usage, wall postings, and regret. *Computers in Human Behavior, 28*(1), 267–274. doi:10.1016/j.chb.2011.09.009

Morales, L. (2011). Can the use of clickers or continuous assessment motivate critical thinking? A case study based on corporate finance students. *Higher Learning Research Communications, 1*(1), 33–42. doi:10.18870/hlrc.v1i1.31

Morgan, R. P. (2008). Exploring the pedagogical effectiveness of clickers. *Scholarly Teaching, 3*, 31–36.

Morris, J. T., Jones, M. L., & Sweatman, W. M. (2016). Wireless technology use by people with disabilities: A national survey. In J. Santiago (Ed.), *Annual International Technology and Persons with Disabilities Conference* (Vol. 4, pp. 101–114). Academic Press.

Mor, Y., & Winters, N. (2007). Design approaches in technology enhanced learning. *Interactive Learning Environments, 15*(1), 61–75. doi:10.1080/10494820601044236

Moser, D. (1991). Why Chinese is so damn hard. In V. H. Mair (Ed.), *Schriftfestschrift: Essays on writing and language in honor of john DeFrancis on his eightieth birthday, Sino-platonic papers* (pp. 59–70). Philadelphia, PA: University of Pennsylvania.

Moyer, P. S., Bolyard, J. J., & Spikell, M. A. (2002). What are virtual manipulatives? *Teaching Children Mathematics, 8*(6), 372–377.

Nadarajah, S. G., Overgaard, B. N., Pedersen, P. W., Schnatterbeck, C. G., & Rehm, M. (2017). Gamifying Navigation in Location-Based Applications. EAI Endorsed Trans. *Creative Technologies, 4*, e2.

Nagella, U. B., & Govindarajulu, D. (2009). Adaptive approaches to context aware mobile learning application. International Journal of Computer Science and Security.

Naismith, L., Lonsdale, P., Vavoula, G., & Sharples, M. (2006, September 12). *Report 11: Literature Review in Mobile Technologies and Learning.* Retrieved from Futurelab at NFER: http://www2.futurelab.org.uk/resources/documents/lit_reviews/Mobile_Review.pdf

Naylor, A., & Gibbs, J. (2015). Using iPads as a learning tool in cross-curricular collaborative initial teacher education. *Journal of Education for Teaching, 41*(4), 442–446. doi:10.1080/02607476.2015.1081718

Naylor, A., & Gibbs, J. (2018). Deep learning: Enriching teacher training through mobile technology and international collaboration. *International Journal of Mobile and Blended Learning, 10*(1), 62–77. doi:10.4018/IJMBL.2018010105

Ndlovu, M., & Hanekom, P. W. (2014). Overcoming the limited interactivity in telematic sessions for in-service secondary mathematics and science teachers. *EDULEARN14 Proceedings.*

Neri, M.C. (2009). O Tempo de Permanência na Escola e as Motivações dos Sem-Escola. Rio de Janeiro: CPS/FVG/IBRE.

Nesbit, J., Niu, H., & Liu, Q. (2019). Cognitive tools for scaffolding argumentation. In O. O. Adesope & A. G. Rud (Eds.), *Contemporary technologies in education: Maximizing student engagement, motivation and learning.* London, U.K.: Palgrave Macmillan. doi:10.1007/978-3-319-89680-9_6

Nesterko, N., Ulitsa, N., Friedrich, M., & Glaesmer, H. (2018). Do They Feel the Same Way? Health-Related Quality of Life and Satisfaction with Life in Jewish Immigrants from the Former Soviet Union in Germany and Israel. *Journal of Cross-Cultural Psychology*, *49*(4), 618–634. doi:10.1177/0022022118759223

Neumann, M. M., & Neumann, D. L. (2015). The use of touch-screen tablets at home and pre-school to foster emergent literacy. *Journal of Early Childhood Literacy*, *17*(2), 203–220. doi:10.1177/1468798415619773

Neville, M. (2010). Meaning making using new media: Learning by design case studies. *E-Learning and Digital Media*, *7*(3), 237–247. doi:10.2304/elea.2010.7.3.237

Nevin, R. (2009). Supporting 21st century learning through Google Apps. *Teacher Librarian*, *37*(2), 35.

Ng'ambi, D. (2013). Effective and ineffective uses of emerging technologies: Towards a transformative pedagogical model. *British Journal of Educational Technology*, *44*(4), 652–661. doi:10.1111/bjet.12053

Nguyen, T. S. (2017). *Conceptual changes in Higher Education Teaching and Learning: Insights from a compulsory teacher training program for higher education teachers in Vietnam* [PhD Thesis]. Retrieved from https://ses.library.usyd.edu.au/bitstream/2123/17657/1/Nguyen_tvs-thesis.pdf

Nguyen, L., Barton, S. M., & Nguyen, L. T. (2015). iPads in higher education—Hype and hope. *British Journal of Educational Technology*, *46*(1), 190–203. doi:10.1111/bjet.12137

Ng, W. (2015). *New digital technology in education, conceptualizing professional learning for educators.*

Nicolson, R., & Fawcett, A. (2011). Dyslexia, dysgraphia, procedural learning and the cerebellum. *Cortex*, *47*(1), 117–127. doi:10.1016/j.cortex.2009.08.016 PMID:19818437

Nielsen Reports. (2016). *Millennials are top smartphone users*. Retrieved from http://www.nielsen.com/us/en/insights/news/2016/millennials-are-top-smartphone-users.html

Nikou, S. A., & Economides, A. A. (2016). The impact of paper-based, computer-based and mobile-based self-assessment on students' science motivation and achievement. *Computers in Human Behavior*, *55*, 1241–1248. doi:10.1016/j.chb.2015.09.025

Nincarean, D., Alia, M. B., Halim, N. D. A., & Rahman, M. H. A. (2013). Mobile augmented reality: The potential for education. *Procedia: Social and Behavioral Sciences*, *103*(0), 657–664. doi:10.1016/j.sbspro.2013.10.385

NJCLD- National Committee on Learning Disabilities. (1994). Learning disabilities issues on definition. In Collective perspectives on issues affecting learning disabilities (pp. 61-66). Austin, TX: PRO-ED.

Noorhidawati, A., Ghalebandi, S. G., & Siti Hajar, R. (2015). How do young children engage with mobile apps? Cognitive, psychomotor, and affective perspective. *Computers & Education*, *87*, 385–395. doi:10.1016/j.compedu.2015.07.005

Novak, J. D., & Gowin, D. B. (1984). *Learning how to learn*. Cambridge, UK: Cambridge University Press. doi:10.1017/CBO9781139173469

O'Bannon, B. W., & Thomas, K. M. (2015). Mobile phones in the classroom: Preservice teachers answer the call. *Computers & Education*, *85*, 110–122. doi:10.1016/j.compedu.2015.02.010

Obonyo, C., Davis, N., & Fickel, L. (2018). Mobile learning practices in initial teacher education: illustrations from three teacher educators. In E. Langran & J. Borup (Eds.), *Proceedings of Society for Information Technology & Teacher Education International Conference* (pp. 2424-2429). Washington, DC: Academic Press.

Ogata, H., Hui, G. L., Yin, C., Ueda, T., Oishi, Y., & Yano, Y. (2008). LOCH: Supporting mobile language learning outside classrooms. *International Journal of Mobile Learning and Organisation*, *2*(3), 271–282. doi:10.1504/IJMLO.2008.020319

Osberg, K. (1992). Virtual reality and education: A look at both sides of the sword. *HIT Lab.*

Ottenbreit-Leftwich, A. T., Glazewski, K. D., Brush, T. A., Aslan, S., & Zachmeier, A. (2017). Addressing technology integration concerns: Asynchronous video mentoring between pre-service teachers and exemplary technology-using in-service teachers. *Australasian Journal of Educational Technology*, *29*(4). doi:10.14742/ajet.3246

Ouellette-Kuntz, H., Burge, P., Brownà, H. K., & Arsenaultà, E. (2010). Public attitudes towards Individuals with intellectual disabilities as measured by the concept of social distance. *Journal of Applied Research in Intellectual Disabilities*, *23*(2), 132–142. doi:10.1111/j.1468-3148.2009.00514.x

Overbay, A., Patterson, A. S., Vasu, E. S., & Grable, L. L. (2010). Constructivism and technology use: Findings from the impacting leadership project. *Educational Media International*, *47*(2), 103–120. doi:10.1080/09523987.2010.492675

Ozdamli, F., & Uzunboylu, H. (2015). M-learning adequacy and perceptions of students and teachers in secondary schools. *British Journal of Educational Technology*, *46*(1), 159–172. doi:10.1111/bjet.12136

Ozok, A. A., Dana, B., Chakraborty, J., & Norcio, F. A. (2008). A Comparative Study Between Tablet and Laptop PCs: User Satisfaction and Preferences. *Intl. Journal of Human-Computer Interation*, *24*(3), 329–352. doi:10.1080/10447310801920524

Pace, C. (1980). Measuring the quality of student effort. *Current Issues in Higher Education*, *2*, 10–16.

Palluel-Germain, R., Bara, F., Hillairet de Boisferon, A., Hennion, B., Gouagout, P., & Gentaz, E. (2007). A Visuo-haptic device – Telemaque - increases kindergarten children's handwriting acquisition. *Proceedings of IEEE World Haptics 2007*, 72–77. doi: 10.1109/WHC.2007.13

Papadakis, S., Kalogiannakis, M., & Zaranis, N. (2017). Designing and creating an educational app rubric for preschool teachers. *Education and Information Technologies*, *22*(6), 3147–3165. doi:10.100710639-017-9579-0

Papert, S. (1990). Introduction by Seymour Papert. In I. Harel (Ed.), *Constructionist learning*. Boston: MIT Laboratory.

Papert, S., & Harel, I. (1991). *Constructionism.* Westport, CT: Ablex Publishing.

Park, Y. (2011). A pedagogical framework for mobile learning: Categorizing educational applications of mobile technologies into four types. *International Review of Research in Open and Distance Learning*, *12*(2), 78–102. doi:10.19173/irrodl.v12i2.791

Parsons, D., & Adhikar, J. (2016). Bring Your Own Device to Secondary School: The Perceptions of Teachers, Students and Parents. *Electronic Journal of E-learning*, *14*(1), 66–80.

Parsons, G. (2010). Information provision for HE distance learners using mobile devices. *The Electronic Library*, *28*(2), 231–244. doi:10.1108/02640471011033594

Partnership for 21st Century Skills. (2008). *21st century skills, education & competitiveness*. Retrieved from http://www.p21.org/storage/documents/21st_century_skills_education_and_competitiveness_guide.pdf

Pascarella, E., & Terenzini, P. (2005). *How college affects students: A third decade of research* (2nd ed.). San Francisco: Jossey-Bass.

Pask, G. (1976). Styles and strategies of learning. *The British Journal of Educational Psychology*, *46*(2), 128–148. doi:10.1111/j.2044-8279.1976.tb02305.x

Pegrum, M. (2014). *Mobile learning: Languages, literacies and cultures*. Basingstoke, UK: Palgrave Macmillan. doi:10.1057/9781137309815

Pegrum, M. (2015). Teaching and Digital Technologies Mobile learning: what is it and what are its possibilities? In M. Henderson & G. Romeo (Eds.), *Teaching and Digital Technologies* (pp. 142–154). Cambridge, UK: Cambridge University Press.

Pegrum, M., Howitt, C., & Striepe, M. (2013). Learning to take the tablet: How pre-service teachers use iPads to facilitate their learning. *Australasian Journal of Educational Technology*, *29*(4), 464–479. doi:10.14742/ajet.187

Pennington, L. (2010). Measuring communication outcomes. *Developmental Medicine & Outcomes.*, *7-8*. doi:10.1111/j.1469-8749.2009.03467.x PMID:20419868

Petersen, S. A., Divitini, M., & Chabert, G. (2008). Identity, sense of community and connectedness in a community of mobile language learners. *ReCALL*, *20*(3), 361–379. doi:10.1017/S0958344008000839

Peterson-Karlan, G. R., & Parette, P. (2005). Millennial students with mild disabilities and emerging assistive technology trends. *Journal of Special Education Technology*, *4*(20), 27–38. doi:10.1177/016264340502000404

Phelps, J., Stempel, L., & Speck, G. (1985). The children's handwriting scale: A new diagnostic tool. *The Journal of Educational Research*, *79*(1), 46–50. doi:10.1080/00220671.1985.10885646

Phillips, D. (Ed.). (2000). *Constructivism in education.* Chicago: University of Chicago Press.

Piaget, J. (1977). *To understand is to invent: The future of education.* New York: Penguin.

Piaget, J., & Inhelder, B. (1969). *The psychology of the child.* New York, NY: Basic Books.

Pinto, M., Caballero-Mariscal, D., Fernández-Pascual, R., Sales, D., & Guerrero-Quesada, D. (2019). The Attitudes of Teachers-in-Training Towards Information Literacy Skills and the Inclusion of Mobile Devices in the Process of Teacher Education. In S. Kurbanoğlu (Eds.), *Information Literacy in Everyday Life: Communications in Computer and Information Science* (pp. 597–606). Springer Cham. doi:10.1007/978-3-030-13472-3_56

Pintrich, P. R., Smith, D. A. F., Garcia, T., & McKeachie, W. J. (1991). *A manual for the use of the Motivated Strategies for Learning Questionnaire (MSLQ).* Ann Arbor, MI: National Center for Research to Improve Postsecondary Teaching and Learning, University of Michigan.

Pintrich, R. P., & De Groot, V. E. (1990). Motivational and Self-Regulated Learning Components of Classroom Academic Performance. *Journal of Educational Psychology*, *82*(1), 33–40. doi:10.1037/0022-0663.82.1.33

Ploog, B. O., Scharf, A., Nelson, D. S., & Brooks, P. J. (2013). Use of computer-assisted technologies (CAT) to enhance social, communicative, and language development in children with autism spectrum disorders. *Journal of Autism and Developmental Disorders*, *43*(2), 301–322. doi:10.100710803-012-1571-3 PMID:22706582

Ponce, H. R., Mayer, R. E., Figueroa, V. A., & López, M. J. (2018). Interactive highlighting for just-in-time formative assessment during whole-class instruction: Effects on vocabulary learning and reading comprehension. *Interactive Learning Environments*, *26*(1), 42–60. doi:10.1080/10494820.2017.1282878

Porat, E., Blau, I., & Barak, A. (2018). Measuring digital literacies: Junior high-school students' perceived competencies versus actual performance. *Computers & Education*, *126*, 23–36. doi:10.1016/j.compedu.2018.06.030

Preece, J., Nonnecke, B., & Andrews, D. (2004). The top five reasons for lurking: Improving community experiences for everyone. *Computers in Human Behavior*, *20*(2), 201–223. doi:10.1016/j.chb.2003.10.015

Prensky, M. (2005). What can you learn from a cell phone? Almost anything. *Innovate (North Miami Beach, Fla.)*, *1*(5), 1–8.

Presby, B. (2017). Barriers to reducing the digital-use divide as perceived by middle school principals. Retrieved from https://search.proquest.com/docview/1887122371?accountid=26642

Prescott, J. (2016). Teaching Style and Attitudes towards Facebook as an Educational Tool. In Learning and Teaching in Higher Education. Sage.

Prescott, J., Wilson, S., & Becket G. (2013c). Students' attitudes towards Facebook and online professionalism: subject discipline, age and gender differences *Health and Social Care Education, 2*(2), 1-8. doi:10.11120/hsce.2013.00032

Prescott, J. (2014). Faculty/educators use of and attitudes towards Facebook. *Active Learning in Higher Education.* doi:10.1177/1469787414527392

Prescott, J., Stodart, M., Wilson, S., & Becket, G. (2013 a). *The experience of using Facebook as an educational tool.* Health and Social Care Education; doi:10.11120/hsce.2013.00033

Prescott, J., Wilson, S., & Becket, G. (2013b). Facebook use in the learning environment: Do students want this? *Learning, Media and Technology, 38*(3), 345–350. doi:10.1080/17439884.2013.788027

Prodanov, C. C., & Freitas, E. C. (2013). *Metodologia do trabalho científico: métodos e técnicas da pesquisa e do trabalho acadêmico.* Feevale.

Puentedura, R. (2012). *The SAMR Model: Background and exemplars.* Retrieved from http://www.hippasus.com/rrpweblog/archives/2012/08/23/SAMR_BackgroundExemplars.pdf

Puigdevall, M., Walsh, J., Amorrortu, E., & Ortega, A. (2018). 'I'll be one of them': Linguistic modes and new speakers in three minority language contexts. *Journal of Multilingual and Multicultural Development, 39*(5), 445–457. doi:10.1080/01434632.2018.1429453

Rappolt-Schlichtmann, G., Daley, S. G., Lim, S., Lapinski, S., Robinson, K. H., & Johnson, M. (2013). Universal Design for Learning and elementary school science: Exploring the efficacy, use, and perceptions of a web-based science notebook. *Journal of Educational Psychology, 105*(4), 1210–1225. doi:10.1037/a0033217

Rap, S., & Blonder, R. (2016). Let's Face(book) it: Analyzing interactions in social network groups for chemistry learning. *Journal of Science Education and Technology, 25*(1), 62–76. doi:10.100710956-015-9577-1

Rathbone, A. L., & Prescott, J. (2017). The use of mobile applications and SMS messaging as physical and mental health interventions: A systematic review. *Journal of Medical Internet Research, 19*(5). doi:10.2196/jmir.7740

Rathbone, A., Clarry, L., & Prescott, J. (2017). Assessing the Efficacy of mHealth Applications Using the Basic Principles of CBT: A Systematic Review. *JMIR Mental Health.* doi:10.2196/jmir.8598

Regenbrecht, H., Baratoff, G., & Wilke, W. (2005). Augmented reality projects in the automotive and aerospace industries. *IEEE Computer Graphics and Applications, 25*(6), 48–56. doi:10.1109/MCG.2005.124 PMID:16315477

Reid, G., Elbeheri, G., & Everatt, J. (2015). *Assessing children with specific learning difficulties: A teacher's practical guide.* New York, NY: Routledge. doi:10.4324/9781315693873

Reinders, H., & Pegrum, M. (2016). Supporting language learning on the move. an evaluative framework for mobile language learning resources. In B. Tomlinson (Ed.), *SLA research and materials development for language learning* (pp. 116–141). New York, NY: Routledge.

Reinders, H., & White, C. (2010). The theory and practice of technology in materials development and task design. In N. Harwood (Ed.), *English language teaching materials: Theory and practice* (pp. 58–80). New York, NY: Cambridge University Press.

Reisman, J. E. (1993). Development and reliability of the research version of the Minessota Handwriting Test. *Physical & Occupational Therapy in Pediatrics, 13*(2), 41–55. doi:10.1080/J006v13n02_03

Repke, L., & Benet-Martínez, V. (2018). The (Diverse) Company You Keep: Content and Structure of Immigrants' Social Networks as a Window Into Intercultural Relations in Catalonia. *Journal of Cross-Cultural Psychology, 49*(6), 924–944. doi:10.1177/0022022117733475

Resnick, L. B. (1989). Introduction. In L. B. Resnick (Ed.), *Knowing, learning, and instruction: Essays in honor of Robert Glaser* (pp. 1–24). Hillsdale, NJ: Erlbaum.

Restivo, M. T., Alves, G. R., Dziabenko, O., & García-Zubía, J. (2013). Acquisition of higher-order experimental skills through remote and virtual laboratories. *IT Innovative Practices in Secondary Schools: Remote Experiments, 1*, 321–247.

Richardson, J., & Lenarcic, J. (2008). Text Messaging as a Catalyst for Mobile Student Administration: The" Trigger" Experience. *International Journal of Emerging Technologies & Society, 6*(2).

Richards, R. G. (1998). *The writing dilemma: Understanding dysgraphia.* Riverside, CA: Richards Educational Therapy Center.

Rieger, R., & Gay, G. (1997). Using mobile computing to enhance field study. In R. Hall, N. Miyake, & N. Enyedy (Eds.), *Computer Supported Collaborative Learning Conference (CSCL '97)* (pp. 215-223). Charlottesville, VA: AACE. 10.3115/1599773.1599800

Ritchhart, R., Church, M., & Morrison, K. (2011). *Making thinking visible; How to promote engagement, understanding and independence for all learners.* San Francisco, CA: Wiley.

Ritz, J. (2009). A new generation of goals for technology education. *Journal of Technology Education., 2*(20), 50–64.

Rivers, D. J. (2009). Utilizing the quick response (QR) code within a Japanese EFL environment. *JALT CALL Journal, 5*(2), 15–26.

Rochadel, W., do Canto, J. Z., da Silva, J. B., Delgado, M. A. G., & Rebollo, M. A. G. (2016). Ensino a distância na educação básica: A integração pedagógica de jogos digitais em ambientes virtuais. *Revista E-Tech: Tecnologias para Competitividade Industrial, 9*(1), 32–54.

Rodrigues, A. L. (2016). A integração pedagógica das tecnologias digitais na formação ativa de professores. In *IV Congresso Internacional TIC e Educação 2016* (pp. 708-720). Instituto de Educação da Universidade de Lisboa.

Rogers, C. R. (1983). *Freedom to learn for the 80's.* Merrill Publishing.

Romani, C., Tsouknida, E., di Betta, A. M., & Olson, A. (2011). Reduced attentional capacity, but normal processing speed and shifting of attention in developmental dyslexia: Evidence from a serial task. *Cortex, 47*(6), 715–733. doi:10.1016/j.cortex.2010.05.008 PMID:20655514

Romi, S. (2001). Child and youth care in Israel: Trends and dilemmas in training and in therapeutic intervention programs. *Journal of Child and Youth Care Work, 15-16*, 171–184.

Romi, S., Savicki, V., Grupper, E., & Caspi, R. (2007). Occupational burnout among child welfare workers: A work-setting comparison. *International Journal of Child and Family Welfare, 10*(3-4), 93–109.

Rosenberg, M. (1965). *Society and the adolescents Self Image.* Princeton, NJ: University Press. doi:10.1515/9781400876136

Rosenblum, S., Dvorkin, A. Y., & Weiss, P. L. (2006). Automatic segmentation as a tool for examining the handwriting process of children with dysgraphic and proficient handwriting. *Human Movement Science, 25*(4), 608–621. doi:10.1016/j.humov.2006.07.005 PMID:17011656

Rosenblum, S., Weiss, P. L., & Parush, S. (2003). Product and process evaluation of handwriting difficulties. *Educational Psychology Review, 15*(1), 41–81. doi:10.1023/A:1021371425220

Rossing, J. P., Miller, W., Cecil, A. K., & Stamper, S. E. (2012). iLearning: The future of higher education? Student's perceptions on learning with mobile tablets. *The Journal of Scholarship of Teaching and Learning*, *12*(2), 1–26. Retrieved from http://josotl.indiana.edu/ article/view/2023/1985

Rossiter, M. J., & Rossiter, K. R. (2009). *Immigrant youth and crime: Stakeholder perspectives on risk and protective factors*. Working Paper No. WP02-09.

Roundtree, R. (2011). Article. *Mobile Technology*, 919-921.

Rowe, K. J. (2006). *Effective teaching practices for students with and without learning difficulties: Constructivism as a legitimate theory of learning and of teaching*. Background paper to keynote address presented at the NSW DET Office of Schools Portfolio Forum, Wilkins Gallery, Sydney, Australia. Retrieved from: http://research.acer.edu.au/learning_processes/10

Russell-Bennett, R., Rundle-Thiele, S. R., & Kuhn, K. A. (2010). Engaging marketing students: Student operated businesses in a simulated world. *Journal of Marketing Education*, *32*(3), 253–263. doi:10.1177/0273475310377758

Ryan, R. M., & Deci, E. L. (2017). *Self-Determination Theory. Basic psychological needs in motivation, development, and wellness*. The Guilford Press.

Sabah, N. (2016). Exploring students' awareness and perceptions: Influencing factors and individual differences driving m-learning adoption. *Computers in Human Behavior*, *65*, 522–533.

Saidin, N. F., Halim, N. D. A., & Yahaya, N. (2015). A review of research on augmented reality in education: Advantages and applications. *International Education Studies*, *8*(13), 1–8. doi:10.5539/ies.v8n13p1

Saitta, E. K. H., Gittings, M. J., & Geiger, C. (2011). Learning dimensional analysis through collaboratively working with manipulatives. *Journal of Chemical Education*, *88*(7), 910–915. doi:10.1021/ed100019f

Salih, M. A. M., Abdul-Kahar, R., Zahari, W. A. M. W., Khalid, H. M., & Rahim, K. A. (2015). Strengthening Jawi writing for dyslexia students through online games-analysis of e-Jawi games online in Malaysia. *Journal of Education and Human Development*, *4*(2), 147–151. doi:10.15640/jehd.v4n2a17

Sampayo-Vargas, S., Cope, C. J., He, Z., & Byrne, G. J. (2013). The effectiveness of adaptive difficulty adjustments on students' motivation and learning in an educational computer game. *Computers & Education*, *69*, 452–462. doi:10.1016/j.compedu.2013.07.004

Sandberg, J., Maris, M., & de Geus, K. (2011). Mobile English learning: An evidence-based study with fifth graders. *Computers & Education*, *57*(1), 1334–1347. doi:10.1016/j.compedu.2011.01.015

Sang, G., Valcke, M., Braak, J., & Tondeur, J. (2010). Student teachers' thinking processes and ICT integration: Predictors of prospective teaching behaviors with educational technology. *Computers & Education*, *54*(1), 103–112. doi:10.1016/j.compedu.2009.07.010

Sarrab, M., Elbasir, M., & Alnaeli, S. (2016). Towards a quality model of technical aspects for mobile learning services: An empirical investigation. *Computers in Human Behavior*, *55*, 100–112. doi:10.1016/j.chb.2015.09.003

Sauer, A. L., Parks, A., & Heyn, P. C. (2010). Assistive technology effects on the employment outcomes for people with cognitive disabilities: A systematic review. *Disability and Rehabilitation. Assistive Technology*, *5*(6), 377–391. doi:10.3109/17483101003746360 PMID:20392190

Sayed, E., Neven, A. M., Zayed, H. H., & Sharawy, M. I. (2011). ARSC: Augmented reality student card -An augmented reality solution for the education field. *Computers & Education*, *56*(4), 1045–1061. doi:10.1016/j.compedu.2010.10.019

Scheer, A., & Plattner, H. (2012). Transforming constructivist learning into action : Design Thinking in education. *Design and Technology Education: An International Journal, 17*(3), 8–19.

Schilit, W. N. (1995). *A system architecture for context-aware mobile computing* (Doctoral dissertation). Columbia University, New York, NY.

Schmitz, B., Klemke, R., & Specht, M. (2012). Effects of mobile gaming patterns on learning outcomes: A literature review. *International Journal of Technology Enhanced Learning, 4*(5-6), 345–358. doi:10.1504/IJTEL.2012.051817

Schnorr, C. I., Freeman-Green, S., & Test, D. W. (2016). Response cards as a strategy for increasing opportunities to respond: An examination of the evidence. *Remedial and Special Education, 37*(1), 41–54. doi:10.1177/0741932515575614

Schugar, H. R., Smith, C. A., & Schugar, J. T. (2013). Teaching with interactive picture Ebooks in grades K- 6. *The Reading Teacher, 66*(8), 615–624. doi:10.1002/trtr.1168

Schunk, D. H. (2015). *Learning theories: An educational perspective* (7th ed.). Upper Saddle River, NJ: Pearson Merrill/Prentice Hall.

Schwarz, L. R. (2002). *LD and the English language learner*. Cambridge, MA: Lesley University.

Schweibenz, W. (2006). Accessibility- Why and how to include people with disabilities. *14th International BOBCATSSS Symposium*, 19–30.

Seale, J. (2009). Doing student voice work in higher education: An exploration of the value of participatory methods. *British Educational Research Journal, 36*(6), 995–1015. doi:10.1080/01411920903342038

Serafin, A. S., Adjorlu, N., Nilsson, A. N., Thomsen, L., & Nordahl, R. (2017). Considerations on the use of virtual and augmented reality technologies in music education. In *2017 IEEE Virtual Reality Workshop on K-12 Embodied Learning through Virtual Augmented Reality (KELVAR)* (pp. 1–4). 10.1109/KELVAR.2017.7961562

Seufert, M., Hoýfeld, T., Schwind, A., Burger, V., & Tran-Gia, P. (2016). *Group-based communication in WhatsApp*. Paper presented at the IFIP Networking Conference (IFIP Networking) and Workshops. 10.1109/IFIPNetworking.2016.7497256

Sevinç, Y., & Dewaele, J.-M. (2018). Heritage language anxiety and majority language anxiety among Turkish immigrants in the Netherlands. *The International Journal of Bilingualism*. Prepublished August 05, 2016. doi:10.1177/0123456789123456

Sfard, A. (1998). On two metaphors for learning and the dangers of choosing just one. *Educational Researcher, 27*(2), 4–13. doi:10.3102/0013189X027002004

Sfard, A. (2006). Participation discourse on mathematics learning. In J. Maass & W. Schlöglmann (Eds.), *New mathematics education research and practice* (pp. 153–170). Rotterdam, The Netherlands: Sense.

Sfard, A. (2008). *Thinking as communicating*. Cambridge, UK: Cambridge University Press. doi:10.1017/CBO9780511499944

Shakespeare, T., & Watson, N. (2001). The social model of disability: An outdated ideology? In Research in Social Science and Disability (Vol. 2, pp. 9–28). Academic Press.

Shamir-Inbal, T., & Blau, I. (2016). Developing Digital Wisdom by students and teachers: The impact of integrating tablet computers on learning and pedagogy in an elementary school. *Journal of Educational Computing Research, 54*(7), 967–996. doi:10.1177/0735633116649375

Shamir-Inbal, T., & Blau, I. (2017). Which pedagogical parameters predict the general quality of ICT integration from the perspective of elementary school leaders? *Computers in the Schools, 34*(3), 168–191. doi:10.1080/07380569.2017.1347427

Sharaabi-Naor, Y., Kesner, M., & Shwartz, Y. (2014). enhancing students' motivation to learn chemistry. *Sisyphus-Journal of Education*, *2*(2), 100–123.

Sharma, S. K., & Sharma, M. (2019). Examining the role of trust and quality dimensions in the actual usage of mobile banking services: An empirical investigation. *International Journal of Information Management*, *44*, 65–75. doi:10.1016/j.ijinfomgt.2018.09.013

Sharples, M., Lonsdale, P., Meek, J., Rudman, P., & Vavoula, G. N. (2007). An evaluation of MyArtSpace: A mobile learning service for school museum trips. *Paper presented at the mLearn Conference*, Melbourne, Australia.

Sharples, M., Milrad, I., Sánchez, A., & Vavoula, G. (2009). Mobile learning: Small devices, big issues. In N. Balacheff, S. Ludvigsen, & T. de Jong et al. (Eds.), *Technology Enhanced Learning: Principles and Products* (pp. 233–249). Springer. doi:10.1007/978-1-4020-9827-7_14

Sharples, M., & Spikol, D. (2017). Mobile Learning. In E. Duval, M. Sharples, & R. Sutherland (Eds.), *Technology Enhanced Learning* (pp. 89–96). New York: Springer International Publishing. doi:10.1007/978-3-319-02600-8_8

Sharples, M., Taylor, J., & Vavoula, G. (2010). A theory of learning for the mobile age. In B. Bachmair (Ed.), *Medienbildung in neuen Kulturräumen* (pp. 87–99). Wiesbaden: VS Verlag für Sozialwissenschaften. doi:10.1007/978-3-531-92133-4_6

Sheldon, K. M., Abad, N., & Hinsch, C. (2011). A two-process view of Facebook use and relatedness need-satisfaction: Disconnection drives use, and connection rewards it. *Psychology of Popular Media Culture*, *1*(S), 2–15. doi:10.1037/2160-4134.1.S.2 PMID:21280967

Shelton, B. E., & Hedley, N. R. (2004). Exploring a cognitive basis for learning spatial relationships with augmented reality. *Technology, Instruction. Cognition and Learning*, *1*(4), 323–357.

Shen, R., Wang, M., & Pan, X. (2008). Increasing interactivity in blended classrooms through a cutting-edge mobile learning system. *British Journal of Educational Technology*, *39*(6), 1073–1086. doi:10.1111/j.1467-8535.2007.00778.x

Shulman, L. S. (1986). Those who understand: Knowledge growth in teaching. *Educational Researcher*, *15*(2), 4–14. doi:10.3102/0013189X015002004

Shu, Z. M. (2008a). Chinese language multimedia and digital teaching. In Y. X. Cai, Z. M. Shu, L. F. Chen, X. Y. Zhang, S. Z. He, & M. D. Lai (Eds.), *Introduction to teaching Chinese as a foreign language* (pp. 281–299). Taipei, Taiwan: Sanmin.

Shu, Z. M. (2008b). Chinese language training and teaching material for listening and speaking. In Y. X. Cai, Z. M. Shu, L. F. Chen, X. Y. Zhang, S. Z. He, & M. D. Lai (Eds.), *Introduction to teaching Chinese as a foreign language* (pp. 231–257). Taipei, Taiwan: Sanmin.

Silva, *L.J.* (2017). *Internet das coisas. Curso de Engenharia Elétrica e Telemática, Unisul*. Palhoça.

Silva, I. N., da Silva, K. C. N., Lotthammer, K. S., Bilessimo, S. M. S., & da Silva, J. B. (2017). Capacitação docente semipresencial para docentes de escolas básicas da rede pública. *SUCEG-Seminário de Universidade Corporativa e Escolas de Governo*, *1*(1), 510–518.

Sinopec i-Academic. (2018). Available at: https://sia.sinopec.com/learn/

Siu, D. N. (2018). Google translate is using AI to power a robust offline mode. *Mashable*. Retrieved from https://mashable.com/2018/06/12/google-translate-offline-mode-ai/#abl8MTBhPsqg

Skills for Learning. (2018). The University of Salford. Available at: https://www.salford.ac.uk/skills-for-learning

Smits-Engelsman, B. C., Niemeijer, A. S., & Van Galen, G. P. (2001). Fine motor deficiencies in children diagnosed as DCD based on poor grapho-motor ability. *Human Movement Science*, *20*(1), 161–182. doi:10.1016/S0167-9457(01)00033-1 PMID:11471395

Smits-Engelsman, B. C., & Van Galen, G. P. (1997). Dysgraphia in children: Lasting psychomotor deficiency or transient developmental delay? *Journal of Experimental Child Psychology*, *67*(2), 164–184. doi:10.1006/jecp.1997.2400 PMID:9388804

Soloway, E., & Norris, C. (2018). Mobile Learning in K-12: Roadblocks to Adoption. In Second Handbook of Information Technology in Primary and Secondary Education (779-783). Springer.

Spallek, J., Zeeb, H., & Razum, O. (2010). Prevention among immigrants: The example of Germany. *BMC Public Health*, *10*(1), 92. doi:10.1186/1471-2458-10-92 PMID:20181232

Special Olympics. (2009). *Status and prospects of persons with intellectual disability*. Health, Well-being and Opportunity for People with Intellectual Disability – A Global Imperative.

Spörlein, C., & Kristen, C. (2018). Educational Selectivity and Language Acquisition among Recently Arrived Immigrants. *The International Migration Review*. doi:10.1177/0197918318798343

Squire, K., & Klopfer, E. (2007). Augmented reality simulations on handheld computers. *Journal of the Learning Sciences*, *16*(3), 371–413. doi:10.1080/10508400701413435

Stake, E. R. (1995). *The Art of Case Study Research*. Thousand Oaks, CA: Sage Publications.

Stanojevic, L., & Rakic, B. (2018). Mobile technology in Higher Education – A students' perspective on learning with mobile computing devices. In Z. Nikitovic, D. Markovic, & C. D'Arconte (Eds.), *Implementation of Information technology: IT, Marketing, education and business working together for business success*. London: Silver and Smith Publishers. Retrieved from http://www.eee-conference.com/_img/arhiva/2018/knjiga_ii_-_it_marketing_prelom.pdf#page=8

Statista. (2018). *Number of mobile phone users worldwide from 2015 to 2020*. Available at: https://www.statista.com/statistics/274774/forecast-of-mobile-phone-users-worldwide/

Statistics and facts about mobile app usage. (2015). Retrieved from https://www.statista.com/topics/1002/mobile-app-usage/

Stowell, J. R. (2015). Use of clickers vs. mobile devices for classroom polling. *Computers & Education*, *82*, 329–334. doi:10.1016/j.compedu.2014.12.008

Strauss, A., & Corbin, J. (1998). *Basics of qualitative research: Techniques and procedures for developing grounded theory* (2nd ed.). Thousand Oaks, CA: Sage Publications.

Sultana, N., & Hayhoe, S. (2013). *Assistive Technology for Students with Special Needs*. Retrieved September 17, 2013, from The Higher College of Technology - Sharjah Colleges: http://shct.hct.ac.ae/events/edtechpd2013/articles/Sultana-Hayhoe.pdf

Sung, H. Y., Hwang, G. J., & Chang, Y. C. (2016). Development of a mobile learning system based on a collaborative problem-posing strategy. *Interactive Learning Environments*, *24*(3), 456–471. doi:10.1080/10494820.2013.867889

Sung, Y.-T., Chang, K.-E., & Liu, T.-C. (2016). The effects of integrating mobile devices with teaching and learning on students' learning performance: A meta-analysis and research synthesis. *Computers & Education*, *94*, 252–275. doi:10.1016/j.compedu.2015.11.008

Suresh, S., & Venkatammal, P. (2010). Antecedents of organizational citizenship behavior. *Journal of the Indian Academy of Applied Psychology*, *36*(2), 276–286.

Sursock, A. (2015). Trends 2015: Learning and Teaching in European Universities. Brussels, Belgium: The European University Association (EUA). Retrieved from http://eua.be/Libraries/higher-education/trends-2015.pdf?sfvrsn=0

Swan, P., & Marshall, L. (2010). Revisiting mathematics manipulative materials. *Australian Primary Mathematics Classroom, 15*(2), 13–19.

Taitelbaum, D., Mamlok-Naaman, R., Carmeli, M., & Hofstein, A. (2008). *Evidence for teachers' change while participating in a continuous professional development programme and implementing the inquiry approach in the chemistry laboratory*. Academic Press.

Takacs, Z. K., Swart, E. K., & Bus, A. G. (2015). Benefits and pitfalls of multimedia and interactive features in technology-enhanced storybooks: A meta-analysis. *Review of Educational Research, 85*(4), 698–739. doi:10.3102/0034654314566989 PMID:26640299

Tanimoto, S., Thompson, R., Berninger, V. W., Nagy, W., & Abbott, R. D. (2015). Computerized writing and reading instruction for students in grades 4–9 with specific learning disabilities affecting written language. *Journal of Computer Assisted Learning, 31*(6), 671–689. doi:10.1111/jcal.12110 PMID:26858470

Tanyag, R. P., & Atienza, R. O. (2015, September). Implicit palm rejection using real-time hand model filters on tablet devices. *Proceedings of Next Generation Mobile Applications, Services and Technologies, 2015 9th International Conference on,* 347-352. 10.1109/NGMAST.2015.45

Tapscott, D. (1999a). Educating the net generation. *Educational Leadership, 56*(5), 611.

Tapscott, D. (2009b). *Grown up digital: How the net generation is changing your world.* New York: McGraw-Hill.

Tarng, W., & Ou, K. L. (2012). A study of campus butter fly ecology learning system based on augmented reality and mobile learning. In *IEEE Seventh International Conference on Wireless, Mobile and Ubiquitous Technology in Education (WMUTE),* Takamatsu, Japan (pp. 62–66). IEEE. 10.1109/WMUTE.2012.17

TeachEx. (2017). TeachEx – Teaching Excellence in Israel – brochure. Retrieved from https://www.teachex.eu/wp-content/uploads/Teachex-Brochure-Final-1.pdf

Thomas, K., & O'Bannon, B. (2013). Cell phones in the classroom: Preservice teachers' perceptions. *Journal of Digital Learning in Teacher Education, 30*(1), 11–20. doi:10.1080/21532974.2013.10784721

Thomée, S., Härenstam, A., & Hagberg, M. (2011). Mobile phone use and stress, sleep disturbances, and symptoms of depression among young adults-a prospective cohort study. *BMC Public Health, 11*(1), 66. doi:10.1186/1471-2458-11-66 PMID:21281471

Thompson, C. A., & Dekkers, A. (2009, December). *Accreditation, Video Resources and Interactivity, Tablet PC's and their Advantages*. Academic Press.

Thompson, C., & Dekkers, A. (2009). *Accreditation, Video Resources and Interactivity, Tablet PC's and their Advantages.* Retrieved October 2, 2012, from Australasian Association for Engineering Education: http://aaee.com.au/conferences/AAEE2009/PDF/AUTHOR/AE090089.PDF

Toh, C., Miller, S., & Simpson, T. (2015). The impact of virtual product dissection environments on student design learning and self-efficacy. *Journal of Engineering Design, 26*(1–3), 48–73. doi:10.1080/09544828.2015.1018149

Toledo, J. A., Collazos, C. A., Cantero, M. O., & Redondo, M. A. (2018). Collaborative strategy with augmented reality for the development of algorithmic thinking. In V. Agredo-Delgado & P. H. Ruiz (Eds.), *Human-Computer Interaction: 4th Iberoamerican Workshop* (pp.70-82). New York: Springer.

Tomlinson, B. (2016). Applying SLA principles to whole-class activities. In B. Tomlinson (Ed.), *SLA research and materials development for language learning* (pp. 33–49). New York, NY: Routledge. doi:10.4324/9781315749082

Tondeur, J., Pareja Roblin, N., van Braak, J., Voogt, J., & Prestridge, S. (2017). Preparing beginning teachers for technology integration in education: Ready for take-off? *Technology, Pedagogy and Education, 26*(2), 157–177. doi:10.1080/1475939X.2016.1193556

Tong, V. C. H., Standen, A., & Sotiriou, M. (Eds.). (2018). *Shaping Higher Education with Students: Ways to Connect Research and Teaching*. London: UCL Press. Retrieved from http://discovery.ucl.ac.uk/10044671/1/Shaping-Higher-Education-with-Students.pdf doi:10.2307/j.ctt21c4tcm

Totten, S., Sills, T., Digby, A., & Russ, P. (1991). *Cooperative learning: A guide to research*. New York: Garland.

Traxler, J. (2009). Current state of mobile learning. *Mobile Learning: Transforming the Delivery of Education and Training, 1*, 9-24.

Traxler, J., & Kukulska-Hulme, A. (2005). Mobile Learning in Developing Countries. Commonwealth of Learning.

Traxler, J. (2007). Defining, discussing and evaluating mobile learning: The moving finger writes and having writ..... *The International Review of Research in Open and Distributed Learning*, *8*(2). doi:10.19173/irrodl.v8i2.346

Tschannen-Moran, M. (2009). Fostering teacher professionalism in schools: The role of leadership orientation and trust. *Educational Administration Quarterly*, *45*(2), 217–247. doi:10.1177/0013161X08330501

Tschimmel, K. (2012). Design Thinking as an effective Toolkit for Innovation. In *Proceedings of the XXIII ISPIM Conference: Action for Innovation: Innovating from Experience* (pp. 1–20). Barcelona: Academic Press.

Tseng, C. C., Lu, C. H., & Hsu, W. L. (2007). A mobile environment for Chinese language learning. In M. Smith & G. Salvendy (Eds.), Human interface 2007 (vol. 4558, pp. 485-489). Berlin, Germany: Springer. doi:10.1007/978-3-540-73354-6_53

Twietmeyer, M. (2014). *GPS: Is it affecting our sense of place? A case study* (Doctoral dissertation). California State University, Northridge, CA.

U.S. Department of Education, Office of Educational Technology. (2017). Reimagining the Role of Technology in Education: 2017 National Education Technology Plan Update. Author.

U.S. Department of Education. (2010a). *A blueprint for reform: The reauthorization of the Elementary and Secondary Education Act*. Retrieved from http://www2.ed.gov/policy/elsec/leg/blueprint/blueprint.pdf

U.S. Department of Education. (2010b). *Transforming American education: Learning powered by technology*. National Education Technology Plan 2010. Retrieved from http://www.ed.gov/technology/netp-2010

UNESCO. (2008). *ICT competency standards for teachers: policy framework*. Paris: UNESCO.

Unicom, C. (2015). *Social Responsibility Report of 2015*. Available at: https://bit.ly/2BWFfl1

Unicom, C. (2016). *Social Responsibility Report of 2016*. Available at: https://www.unglobalcompact.org/participation/report/cop/create-and-submit/active/324231

Unicom, C. (2017). *Social Responsibility Report of 2017*. Available at: https://www.chinaunicom.com.hk/en/csr/reports/csr2017.pdf

Uriarte, F. A. (2008). Knowledge Management: Konsep, Arsitektur dan Impelementasi Introduction to Knowledge Management: A brief introduction to the basic elements of knowledge management for non-practitioners interested in understanding the subject. In Sense of community in a professional learning community of chemistry teachers: A study of an online platform for group communication. Springer.

van Galen, G. P., & de Jong, W. P. (1995). Fitts' law as the outcome of a dynamic noise filtering model of motor control. *Human Movement Science, 14*(4-5), 539–571. doi:10.1016/0167-9457(95)00027-3

Van Hartingsveldt, M. J., De Groot, I. J. M., Aarts, P. B. M., & Nijhuis-Van Der Sanden, M. W. G. (2011). Standardized tests of handwriting readiness: A systematic review of the literature. *Developmental Medicine and Child Neurology, 53*(6), 506–515. doi:10.1111/j.1469-8749.2010.03895.x PMID:21309763

Van Praag, B., & Sanchez, H. S. (2015). Mobile technology in second language classrooms: Insights into its uses, pedagogical implications, and teacher beliefs. *ReCALL, 27*(3), 288–303. doi:10.1017/S0958344015000075

Van Wyk, M., & Van Ryneveld, L. (2016). Introducing mobile videos for academic support. *Mobile Learning* Futures–sustaining quality research and practice in mobile learning. In *Proceedings of the15th World Conference on Mobile and Contextual Learning, mLearn 2016* (pp. 273–282). Sydney, Australia: The University of Technology, Sydney. Retrieved from https://www.researchgate.net/profile/Sue_Gregory2/publication/309485997_The_Use_of_Wearable_Technologies_in_Australian_Universities_Examples_from_Environmental_Science_Cognitive_and_Brain_Sciences_and_Teacher_Training/links/58784fbc08ae6eb871d19331/The-Use-of-Wearable-Technologies-in-Australian-Universities-Examples-from-Environmental-Science-Cognitive-and-Brain-Sciences-and-Teacher-Training.pdf#page=273

Vaughan, N. D. (2010). A blended community of inquiry approach: Linking student engagement and course redesign. *Internet and Higher Education, 13*(1-2), 60–65. doi:10.1016/j.iheduc.2009.10.007

Vaughan, N., & Lawrence, K. (2013). Investigating the role of mobile devices in a blended pre-service teacher education program. *Canadian Journal of Higher Education, 43*(3), 56–77.

Veiga, I. P. A. (2003). Inovações e projeto político-pedagógico: Uma relação regulatória ou emancipatória? *Cadernos CEDES.*

Vinu, P., Sherimon, P., & Reshmy, K. (2011). Towards Pervasive Mobile Learning – the vision of 21st century. *Procedia - Social and Behavioral Science.*

Viotti. E., Salerno, M. & de Negri, J.A. (2005). Inovação e Competitividade. In *Brasil: o estado de uma nação* (pp. 1–40). São Paulo: Instituto de Pesquisa Econômica Aplicada.

Vygotsky, L. S. (1978). *Mind in society.* Cambridge, MA: Harvard University Press.

Vygotsky, L. S. (1978). *Mind in society: The development of higher psychological processes.* Cambridge, MA: Harvard University Press.

Walker, R. (2013). I don't think I would be where I am at now: Pupil perspectives on using mobile devices for learning. *Research in Learning Technology, 21*(2216). doi:10.3402/rlt.v21i0.22116

Wang, X. (2012). Augmented reality: A new way of augmented learning. *eLearn, 10.*

Wang, M., Koda, K., & Perfetti, C. A. (2003). Alphabetic and nonalphabetic L1 effects in English word identification: A comparison of Korean and Chinese English L2 learners. *Cognition, 87*(2), 129–149. doi:10.10160010-0277(02)00232-9 PMID:12590041

Wang, S. S., Moon, S. I., Kwon, K. H., Evans, C. A., & Stefanone, M. A. (2010). Face-off: Implications of visual cues on initiating friendship on Facebook. *Computers in Human Behavior, 26*(2), 226–234. doi:10.1016/j.chb.2009.10.001

Ward, N. D., Finley, R. J., Keil, R. G., & Clay, T. G. (2013). Benefits and limitations of iPads in the high school science classroom and a trophic cascade lesson plan. *Journal of Geoscience Education, 61*(4), 378–384.

Warschauer, M. (2011). Tablet Computers in Education. Eventually Tablets will Facilitate more Personalized and Interactive Learning. *Educational Technology Debate – Exploring Learning in Developing Countries*. Retrieved from http://edutech-debate.org/tablet-computers-in-education/eventually-tablets-will-facilitate-more-personalized-and-interactive-learning/

Watson, S. (2018). *How mild intellectual disability is defined*. Retrieved from https://www.thoughtco.com/mild-intellectual-disability-3110889

Webster, J., & Watson, R. T. (2002). Analyzing the past to prepare for the future: Writing a literature review. *Management Information Systems Quarterly, 26*(2), xiii–xxiii.

Wedemeyer, C. A. (2009). *Learning at the back door: Reflection on non-traditional learning in the lifespan*. Charlotte, NY: Information Age Publishing.

Weiser, O., Blau, I., & Eshet-Alkalai, Y. (2018). How do medium naturalness, teaching-learning interactions and Students' personality traits affect participation in synchronous E-learning? *The Internet and Higher Education, 37*, 40–51. doi:10.1016/j.iheduc.2018.01.001

Wenger, E. (2011). *Communities of practice: A brief introduction*. Academic Press.

Wenger, E. (1999). *Communities of practice: Learning, meaning, and identity*. Cambridge University Press.

Wenger, E. (1999). *Learning in doing: Social, cognitive and computational perspectives*. New York: Cambridge University Press.

Werder, C., & Otis, M. (Eds.). (2010). *Engaging student voices in the study of teaching and learning*. Sterling, VA: Stylus.

Werder, C., Thibou, S., & Kaufer, B. (2012). Students as co-inquirers: A requisite theory in educational development. *Journal of Faculty Development, 26*(3), 34–38.

West, D. M. (2013). Mobile learning: Transforming education, engaging students, and improving outcomes. *Brookings Policy Report*, 1-7.

West, W., & Vosloo, S. (2013). *UNESCO policy guidelines for mobile learning*. Paris: UNESCO. Retrieved from http://unesdoc.unesco.org/images/0021/002196/219641e.pdf

White, E., Wepner, S., & Wetzel, D. (2003). Accessible Education Through Assistive Technology. *Technological Horizons in Education (T.H.E), 30*(7), 24-30.

Wilkinson, K., & Barter, P. (2016). Do mobile learning devices enhance learning in higher education anatomy classrooms? *Journal of Pedagogic Development*.

Williams, C. (2014). *Dysgraphia*. Retrieved from https://wiki.rit.edu/display/msse704012135/Dysgraphia

Williams, C. (2002). Learning on-line: A review of recent literature in a rapidly expanding field. *Journal of Further and Higher Education, 26*(3), 263–272. doi:10.1080/03098770220149620

Wilson, A. S., Broadbent, C., McGrath, B., & Prescott, J. (2017). Factors associated with player satisfaction and educational value of serious games. In Serious Games and Edutainment Applications (vol. 2). Springer. doi:10.1007/978-3-319-51645-5_23

Winter, S., Richter, K. F., Baldwin, T., Cavedon, L., Stirling, L., Duckham, M., ... Rajabifard, A. (2011). Location-based mobile games for spatial knowledge acquisition. *Cognitive Engineering for Mobile GIS*, 1-8.

Wise, C. J., Toto, R., & Lim, K. Y. (2006). *Introducing Tablet PCs: Initial Results From the Classroom.* Retrieved August 17, 2013, from Frontiers in Education Clearing House: http://fieconference.org/fie2006/papers/1683.pdf

Wong, L. H., Chin, C. K., Tan, C. L., & Liu, M. (2010). Students' personal and social meaning making in a chinese idiom mobile learning environment. *Journal of Educational Technology & Society, 13*(4), 15–26.

Woodcock, B., Middleton, A., & Nortcliffe, A. (2012). Considering the Smartphone Learner: Developing innovation to investigate the opportunities for students and their interest. *Student Engagement and Experience Journal, 1*(1), 1–15. doi:10.7190eej.v1i1.38

World Health Organization. (2007). *Atlas: Global Resources for Persons with Intellectual Disabilities.* Geneva: World Health Organization.

World Health Organization. (2011). World report on disability 2011. *American Journal of Physical Medicine Rehabilitation, 91.*

World Health Organization. (2018). *Assistive technology.* Retrieved from http://www.who.int/en/news-room/fact-sheets/detail/assistive-technology

Wright, S. (2015). *Deep Learning Isn't about Technology.* Retrieved hhttp://plpnetwork.com/2012/09/24/deeper-learning-technolgy

Wu, S., Chang, A., Chang, M., Yen, Y. R., & Heh, J. S. (2010, April). Learning historical and cultural contents via mobile treasure hunting in the five-harbor district of Tainan, Taiwan. In *2010 6th IEEE International Conference on Wireless, Mobile and Ubiquitous Technologies in Education (WMUTE)* (pp. 213-215). IEEE.

Wujec, T. (2010). *Build a tower, build a team. TED ideas worth spreading.* Retrieved from https://www.ted.com/talks/tom_wujec_build_a_tower?utm_source=email&source=email&utm_medium=social&utm_campaign=ios-share

Wu, T. T., Sung, T. W., Huang, Y. M., Yang, C. S., & Yang, J. T. (2011). Ubiquitous English learning system with dynamic personalized guidance of learning portfolio. *Journal of Educational Technology & Society, 14*(4), 164–180.

Yadav, S., & Chakraborty, P. (2017a). Children aged two to four are able to scribble and draw using a smartphone app. *Acta Paediatrica: Nurturing the Child, 106*(6), 991–994. doi:10.1111/apa.13818 PMID:28271551

Yadav, S., & Chakraborty, P. (2017b). Using smartphones with suitable apps can be safe and even useful if they are not misused or overused. *Acta Paediatrica (Oslo, Norway), 107*(3), 384–387. doi:10.1111/apa.14042 PMID:28833547

Yeap, J. A., Ramayah, T., & Soto-Acosta, P. (2016). Factors propelling the adoption of m-learning among students in higher education. *Electronic Markets, 26*(4), 323–338. doi:10.100712525-015-0214-x

Yearwood, E. L., Crawford, S., Kelly, M., & Moreno, N. (2007). Immigrant youth at-risk for disorders of mood: Recognizing complex dynamics. *Archives of Psychiatric Nursing, 21*(3), 62–171. doi:10.1016/j.apnu.2007.02.006 PMID:17556109

Yeunjoo, L., & Vega, L. A. (2005). Assistive Technology: Perceived Knowledge, Attitudes, and Challenges of AT Use in Special Education. *Journal of Special Education Technology, 20*(2), 60–62. doi:10.1177/016264340502000208

Yiğiter, S. (2005). *Sınıf öğretmenlerinin özel öğrenme güçlüğüne ilişkin bilgi düzeyleri ile özel öğrenme güçlüğü olan çocuklarin kaynaştirilmasina yönelik tutumlari arasindaki ilişkinin incelenmesi* (Unpublished master's thesis). Marmara University, İstanbul, Turkey.

Yılmaz, S. (2008). *Zihinsel engelli öğrencilerin kavram öğretiminde kullanılan etkileşimli eğitim CD'lerinin görsel tasarım ilkelerine ve içeriğe uyumu* (Unpublished master's thesis). Ondokuzmayıs University, Samsun, Turkey.

Yin, K. Y., & Fitzgerald, R. (2015). Pocket learning: A new mobile learning approach for distance learners. *International Journal of Mobile Learning and Organization, 9*(3), 271–283. doi:10.1504/IJMLO.2015.074215

Young, B., Ellobody, E., & Hu, T. W. (2012). 3D visualization of structures using finite-element analysis in teaching. *Journal of Professional Issues in Engineering Education and Practice, 138*(2), 131–138. doi:10.1061/(ASCE)EI.1943-5541.0000093

Young, M. R., Klemz, B. R., & Murphy, J. W. (2003). Enhancing learning outcomes: The effects of instructional technology, learning styles, instructional methods, and student behavior. *Journal of Marketing Education, 25*(2), 130–142. doi:10.1177/0273475303254004

Yu-Chang, H., & Yu-Hui, C. (2013). Mobile app design for teaching and learning: Educators' experiences in an online graduate course. *International Review of Research in Open and Distance Learning, 14*(4).

Yuen, S. C., Gallayanee, Y., & Johnson, E. (2011). Augmented Reality: An overview and five directions for AR in education. *Journal of Educational Technology Development and Exchange, 4*(1), 119–140. doi:10.18785/jetde.0401.10

Zadok, Y., & Tal, H. M. (2015). Engaging students in class through mobile technologies–Implications for the learning process and student satisfaction. *Research Highlights in Technology and Teacher Education, 2015*, 105.

Zainuddin, Z., & Attaran, M. (2016). Malaysian students' perceptions of flipped classroom: A case study. *Innovations in Education and Teaching International, 53*(6), 660–670. doi:10.1080/14703297.2015.1102079

Zhang, Y. (2000). Technology and the writing skills of students with learning disabilities. *Journal of Research on Computing in Education, 32*(4), 467–478. doi:10.1080/08886504.2000.10782292

Zilka, C. G. (2015). Social competence of children at risk: Similarities and differences among the various assessors. In E. Grupper & S. Romi (Eds.), Children and adolescents at risk in Israel: The voice of young people and issues faced by child and youth care workers (Vol. 2, pp. 113–185). Tel Aviv, Israel: MOFET Institute. (in Hebrew)

Zilka, C. G. (2014). *Empowering parents in the social media age — The Three Element Way (Hebrew).* Butan-Galim.

Zilka, C. G. (2016). Do online friendships contribute to the social development of children and teenagers? The bright side of the picture. *Journal of the Humanities and Social Sciences, 6*(8), 102–112.

Zilka, C. G. (2017a). Children in front of screens: Alone or in company? Desktop or hybrid computer? Children's viewing and browsing habits. *Journal of Educational Technology, 14*(3), 1–14. doi:10.26634/jet.14.3.13855

Zilka, C. G. (2017b). The Elements Way: Empowering parents, educators, and mentors in the age of new media. *Issues in Informing Science and Information Technology, 14*, 101–119. doi:10.28945/3702

Zilka, G. C. (2018). Always with them: Smartphone use by children, adolescents, and young adults – characteristics, habits of use, sharing, and satisfaction of needs. *Universal Access in the Information Society*. doi:10.100710209-018-0635-3

Zimmerman, B. J. (2002). Becoming a self-regulated learner: An overview. *Theory into Practice, 41*(2), 64–70. doi:10.120715430421tip4102_2

Zohar, R., Bagno, E., Eylon, B.-S., & Abrahamson, D. (2017). Motor skills, creativity, and cognition in learning physics concepts. *Journal of Functional Neurology, Rehabilitation, and Ergonomics*, *7*(3), 67–76.

Zorlu, A., & Hartog, J. (2018). *The Impact of Language on Socioeconomic Integration of Immigrants*. IZA Discussion Paper No. 11485. Available at SSRN: https://ssrn.com/abstract=3170274

Zydney, J. M., & Warner, Z. (2016). Mobile apps for science learning: Review of research. *Computers & Education*, *94*, 1–17. doi:10.1016/j.compedu.2015.11.001

About the Contributors

Alona Forkosh Baruch, PhD, is senior faculty at Levinsky College of Education and a researcher at Tel Aviv University School of Education. Her research focuses on transformations in the information era, mobile learning, social media in education, technology in K12 and higher education (including teacher education), innovative pedagogical practices using technology. She has published dozens of academic papers, books and book chapters, and papers in international conferences on these topics. Her expertise in ICT (Information and Communication Technologies) implementation in the education system is recognized nationally and internationally and is reflected in an academic record including extensive research experience in national as well as international research networks, consortia and international expert summits.

Hagit Meishar Tal, PhD, is a senior leacturer and the Head of Instructional Technologies BA Program in Holon Institute of Technology (HIT) in Israel. She is also a course coordinator in the MA Program in Learning Technologies in The Open University of Israel. Her main fields of research are collaborative learning, mobile learning, game-based learning, Social media as learning environmets and distance learning. She has published many papers in prestigious peer reviewed journals, as well as chapters in academic books and in international conferences proceedings.

* * *

Çığıl Aykut is an Associate Professor of the Department of Special Education at Gazi University, Ankara, Turkey. She earned her BS in Special Education in Turkey and MS in Special Education and received her PhD in Special Education at Gazi University. Her research focuses on, instructional methods in teaching children with special needs, learning strategies in special education.

Cory Bennett is an Associate Professor of Mathematics Education and a consultant for international schools. His previous educational experiences include teaching middle school mathematics, working as a middle school mathematics coach, new teacher mentor, and a K-12 district level instructional resource teacher in culturally, ethnically, and linguistically diverse communities. He is dual certified to teach in the elementary grades (K-8) as well as secondary mathematics (6-12) and is a National Board Certified Teacher in mathematics.

Simone Meister Sommer Bilessimo is a professor at the Federal University of Santa Catarina – Campus Araranguá and researcher at the Remote Experimentation Lab (RExLab).

Ron Blonder is a professor of science education in the Department of Science Teaching, Weizmann Institute of Science. She earned her doctoral degree in 1999 in the field of chemistry and nanotechnology at the Hebrew University of Jerusalem. The goal of Prof. Blonder's research group is to bring the chemistry of tomorrow into the high school chemistry lessons of today. The research addresses three issues: nano-scale science and technology education; science teachers' self-efficacy; and, chemistry teachers' professional development. Her efforts to enhance chemistry teachers' professional development has reached a large number of chemistry teachers via the National Chemistry Teachers' Center, the Rothschild-Weizmann program, the Israel Chemical Society, and EU-projects. Prof. Blonder is married with four children. Professional site: http://www.weizmann.ac.il/st/blonder/.

Yun Chen has obtained four degrees over her academic career including BSc, PGcert, MSc and PhD, all of which were awarded with good. She held an Overseas Research Scholarship (ORS) from the UK government (Universities UK) whilst studying for her PhD. Yun has four years teaching experience in information system development, programming languages, website/mobile application design, research methodologies at both Bachelor and Master's levels. The subjects she teaches in are all about how to apply Information Technologies into business fields. Yun is also an experienced research fellow working for EU and UK funded projects, responsible for building complex information systems. And because of her educational background, she has done some comparative research on digitalisation and technology application between the UK and China. Yun is currently the programme leader for BSc Digital Business in the University of Salford.

Teresa Coffman is a Professor of Education at the University of Mary Washington in Fredericksburg, Virginia where she teaches graduate courses in curriculum and instruction, educational theory and research, and technology integration to pre-service teachers as they work toward initial licensure and a Master's degree in Education. Her research interests are in the areas of educational technology, inquiry-based learning and collaboration, distance learning, and global education. She holds a Ph.D. and a Master of Arts in Education. Her professional background includes teaching at the high school level, as well as serving as a technology coordinator and Director of Academic Technology in K-12 education. She has also consulted on technology issues to educators and previously worked in the telecommunications field prior to academia.

Anat Cohen is a tenured-track senior academic staff member at Tel-Aviv University's School of Education; Head of the Learning and Technology program in The Department of Education in Mathematics, Science and Technology; Deputy Chair holder of technology and education affairs in the UNESCO Chair in Technology, Internationalization and Education (TIE); A research and pedagogical coordinator of Web-Supported Academic Instruction at Tel-Aviv University (Virtual TAU); one of PIs of a research project founded by the Education Ministry's chief scientist: Integrating Mathematical Applets in the Teaching Sequence. Dr. Cohen has vast experience in research and teaching in the field of learning and cyber technologies after 14 years of development of online learning materials, training of academic staff, characterization of learning-management features fit to the University's needs, as well as research activities focusing on research areas such as social networks and privacy perception on the cyber space, internet implementation in higher education, Cost-Effectiveness of Web-based Learning, innovative pedagogical practices using ICT, open educational resources (Learning object repositories and MOOCs), learning analytics and educational data mining. She has published over 90 research papers in leading peer-reviewed journals and national and international conference proceedings.

Duncan Cross is a Senior Lecturer in the School of Education and Psychology at the University of Bolton. Dr Cross teaches courses in teaching and learning in higher and professional education, and supervises doctoral students. Dr Cross is a Fellow of the Society for Education and Training with Qualified Teacher Learning and Skills status, and a Principal Fellow of the Higher Education Academy.

Osnat Dagan, Ph.D. in technology education from School of Education at Tel-Aviv University, Israel. Her Ph.D. was focus on problem solving in design and technology in Junior High School students. Osnat currently works as a lecturer and as the head of ICT academic unit at Beit Berl College. The main aim of this unit is to enhance the lecturers to integrate pedagogy with ICT in their teaching methods. She is, also coordinating ERASMUS+ IN2IT project in her college and involve in developing teacher training modules for faculty in ERASMUS+ TeachEx project. Her academic and professional interests are: enhancing design and technology, the use of innovation pedagogy with ICT technology and develop thinking skills and problem solving skills of learners using constructivist methods. She participated in technology education conferences and her researches is concentrated in these subjects. Previously, Osnat was a teacher in school for 20 years; she developed curriculum materials for science and technology education and for integrating ICT. Osnat served as the pedagogical manager in World ORT in Israel and as R&D coordinator at World ORT London international center. In this position, at Israel Osnat ran pedagogical projects such as the 1000 smart classes in the Negev and Galilee, robotics in the kindergarten, science days, etc. In London, she coordinated the use of ICT and the technology education in all ORT school across the globe (60 countries).

Daniel Edmondson is currently a PhD candidate at The University of Bolton (UoB), studying the application of knowledge through collaborative inter-disciplinary research projects, designed to support undergraduate student transition between 2nd and 3rd year of study. Daniel has also held past roles such as a non-exec Director and Trustee of the UoB Students' Union and a student member to the UoB Senate. Further research positions on a HEFCE funded Undergraduate research project and student research assistant into STEM Education.

Orit Ezra (MA) holds a Master's degree in East Asian Studies from Haifa University, and a second Master's degree in Learning and Technologies from Tel Aviv University. During her studies she embarked on the course of learning the Chinese language and as part of the process she traveled and lived in Taiwan to further improve her proficiency. Her research focuses on mobile technologies for language learning, particularly in Chinese as a second language. She is planning to expand her research in this area in the future as a PhD student. Orit also holds a Bachelor of Science in Industrial Engineering and has been working for more than 10 years as a professional training consultant for training departments of multinational companies where she has been involved in managing, developing and delivering large-scale training projects in diverse cultural settings worldwide.

Jennifer Gallup focuses on the underrepresentation of individuals with disabilities in mathematics and science. Specifically, she has focused on soft-skills needed to be successful in STEM, and the integration and effective implementation of technology to support students with disabilities. Her work revolves around understanding how technology and current virtual environments can support communication and collaboration skills for students with disabilities.

Sinan Hopcan is currently working at Computer Education and Instructional Technology - Istanbul University - Cerrahpasa in Istanbul. He received Phd from Middle East Technical University. His research interests are technology integration into special education, educational software, adaptive learning systems.

Pippa Iliff is currently a PhD candidate at the University of Bolton, researching the application of Coaching Psychology to improve university student mental health with a technology intervention. Previously, held the positions of HEFCE student researcher and Student trustee of the student union.

Rotem Israel-Fishelson is a Ph.D. candidate in the Department of Mathematics, Science, and Technology Education, at Tel Aviv University's School of Education. She has earned her MSc. in Media Technology (2016) from Linnaeus University, and B.A. in Instructional Design (2012) from Holon Institute of Technology. She was a content producer in the Math & Science Department at Time to Know, and is currently a lecturer at the Faculty of Instructional Technologies at the Holon Institute of Technology. She has published several scientific articles. Her main research interests are the acquirement of Computational Thinking and Creativity, and Technology Enhanced Learning (TEL).

Necdet Karasu is Associate Professor of the Department of Special Education at Gazi University, Ankara, Turkey. He earned his BS in Special Education in Turkey and MS in Special Education from State University of New York. He completed his PhD at Indiana University, Bloomington, the USA. His research focuses on, instructional methods in teaching children with special needs, use of instructional technologies in special education and assistive technology.

Mary Beth Klinger is a Professor of Business and Management at the College of Southern Maryland in Leonardtown, MD where she teaches undergraduate courses in business, management, leadership, organizational behavior, small business and entrepreneurship, and marketing. Her research interests are in the areas of knowledge management, leadership, innovation, and global education. She holds a Ph.D. in Organization and Management, a Master's in Business Administration, and a Master's in International Management. Her professional background includes educational consulting, employment in private industry in logistics, as well as federal service in the U.S. Office of Personnel Management.

Karen Lotthammer has a Master's degree in educational technologies at Postgraduate Program in Information and Communication Technologies - Federal University of Santa Catarina.

Carolyne Nekesa Obonyo is a PhD candidate and a Research Assistant in the College of Education at the University of Canterbury, Christchurch. While in the USA, Carolyne worked as a Teaching Assistant in the department of Educational Leadership & Technology at California State University, San Bernardino. She received her MA in Instructional Technology from California State University, San Bernardino in 2015. Carolyne's current research focuses on mobile learning practices in initial teacher education in New Zealand. Her research interests lie in the areas of classroom technology integration, the use of mobile technologies in preservice teacher preparation, teacher education, digital literacy, and e-Learning.

Julie Prescott is a Reader in Psychology and is the programme lead for the undergraduate Psychology, Psychotherapy and Counselling Programme at the University of Bolton. Julie is a Senior Fellow of the Higher Education Academy, a Chartered Psychologist, Associate Fellow of the British Psychological Society and holds APA membership. Julie's current research area focuses on technology and health/mental health, with a particular interest in young people and online counselling as well as how people gain support and use online technologies for their health and mental health support.

Noa Ragonis is the head of the Center for Teaching Enhancement (CTE) at Beit Berl College which includes the Information Communication and Technology (ICT) academic unit. She is a computer science (CS) lecturer at Beit Berl College, and at the Department of Education in Science and Technology, at the Technion. In the last two decades she is involve in the professional development of in-service and pre-service CS teachers, and in the integration of ICT tools into the academic teaching and learning processes. She has published research papers mainly focuses on cognitive aspects of teaching and learning of CS. She authored seven high school text books and teachers guides on different content of the Israeli curriculum, and co-authored the book: Guide to Teaching Computer Science, An Activity-Based Approach (2015, 2nd, Springer).

Beverly Ray is Professor of Education in the College of Education at Idaho State University. Her research interests focus on the use of current and emerging technologies as teaching and learning tools. Dr. Ray taught K-12 Social Studies and has offered professional development for K-12 teachers in the areas of social science education and technology integration for several decades in the United States and abroad.

Tami Seifert is currently a senior lecturer at Kibbutzim College of Education, Tel-Aviv. She has a PhD in Urban Studies and Instructional Design and Technology from Old Dominion University (Virginia, USA) in 2001. She earned both her B.Sc and M.Sc from Tel-Aviv University. Between 2003 and 2005 she completed a post-doctorate from Tel-Aviv University on implementation of handheld computers in education. She served as head of the Department of Educational Computing, the head of Academic ICT and vice-director of the Teacher Training for Graduates Program at Kibbutzim College of Education (2007-2014). Courses Tami has taught in the past five years include: Innovative Technologies in Education, Online Education, The Design of Online Instruction: From Planning to Assessment, Social Networking in Educational Contexts, Application of Video in Teaching. Courses Tami has taught in the past five years include: Innovative Technologies in Education, Online Education, The Design of Online Instruction: From Planning to Assessment, Social Networking in Educational Contexts, Application of Video in Teaching. She promotes and advises digital literacy among faculty.

Isabela Nardi da Silva has a degree in Information and Communication Technologies and currently holds a Master's Degree in Information and Communication Technologies from the Federal University of Santa Catarina, Brazil. She is a researcher at the Remote Experimentation Laboratory - REXLAB.

Juarez Bento da Silva has a degree in Business Administration from the Catholic University of Rio Grande do Sul (1991), specialization in Networking and Telecommunications at the UFSC(1999), Masters in Computer Science from UFSC (2002), PhD in Engineering and Knowledge Management, UFSC(2007). He is currently Associate Professor at UFSC and coordinator of the Remote Experimentation Lab (RExLab). He is currently Associate Professor at the Federal University of Santa Catarina (UFSC) and coordinator of the Remote Experimentation Lab (RExLab). He has experience in Computer Science with emphasis in Computer Systems and acting on the following topics: remote experimentation, 3D virtual worlds, embedded systems, computers in education and digital inclusion. The main interest in research and current focus is the development and use of new information and communication technologies in teaching and learning. Currently we seek to develop and implement new devices for remote testing and work on the integration of hardware (remote experiments) in 3D virtual worlds, using open-source project. This is the link to my resume for more details: http://lattes.cnpq.br/1594563006260546.

Karmel Silva holds a degree in Psychology, specialization in Health Management and a Masters in Information and Communication Technologies with emphasis on Technology and Innovation. Participates in the Research Group RExLab of the Federal University of Santa Catarina, focusing on the integration of technologies in basic education and continuing training for teachers. Acting professionally, she held positions such as: Human Resources Analyst, Organizational Psychologist and Professor at Senac, acting with active learning methodologies and supervising the Legal Apprentice program. Has experience in organizing academic events, developing research and writing research projects and extension. It works mainly with the following subjects: people management, technology integration in education, social innovation, entrepreneurship, management, m-learning.

Saniye Tugba Tokel is an associate professor of Computer Education and Instructional Technology Department at Middle East Technical University. She received B.A. and M.S. degrees in Computer Education and Instructional Technology from the Middle East Technical University and Ph.D. in Educational Technology Program at the Texas A&M University. Her research interests are virtual reality, 3D virtual worlds, simulations, serious games, e-learning and innovative learning technologies.

Ruth Waldman holds bachelor's and master's degrees in chemistry and a teaching certificate. She has been teaching high school level chemistry for 20 years, serving as coordinator of chemistry studies in her school. Ruth has also served as regional instructor for chemistry teachers in behalf of the ministry of education. She is currently a doctoral student at the Weizmann Institute of Science, specializing in the professional development of teachers in teacher communities, investigating the development of knowledge, a sense of community and identity among the teachers participating in the communities.

Yunpu Zhang completed her Bachelor Degree majored in International Economics in Nankai University, China. Before joined China Nuclear Engineering Corporation Limited in 2014, she has worked for another public company China National Chemical Engineering Corporation Limited for 18 years. Ms Zhang has engaged in different fields of management including accounting, financing, disclosure and corporate governance which leads to her deep understanding of Chinese SOEs. From 2006 to 2008, she was the panel member of Internal Control Standards Committee which provided consultation for Ministry of Finance of China. Yunpu's current position is deputy director of Board of Directors Office and she is working as a visiting scholar in the University of Salford for the year of 2018.

Gila Cohen Zilka, Ph.D., Director of the Department for Teaching Social Science and Communication at Bar-Ilan University; Head of the program for training mentors to work with children at risk, Achva Academic College, Israel. Zilka's research focuses mainly on ICT, children in the digital environment, and communication in the diversified media. Zilka published two books that offers ways to empower children, parents, and educators in our age. For parents, "Empowering Parents in the Social Media Age – The Three-Element Way," and for educators and mentors, "Empowering Educators and Mentors in the Social Media Age – The Three-Element Way."

Veronika Zimon is currently a self-employed English teacher on a leave without pay from The Ministry of Education. She has an M.Teach in education from Kibbutzim college of Education in 2014. She earned her B.A in English Linguistics from The Hebrew University, Jerusalem in 2011. Between the years 2013-2017 she taught English in an elementary school in Modi'in and in the year 2017 she was also the English coordinator at her school. She has brought her fifth grade pupils to one of the top places in the English Meitzav in 2016. Throughout her teaching she has always incorporated different means of technology in her lessons. Areas of interest include: Innovative Technologies in Education and especially in the English teaching field, Developing new and innovative teaching materials and Application of video in Teaching.

Index

I

K

L

M

N

P

R

S

www.ingramcontent.com/pod-product-compliance
Lightning Source LLC
LaVergne TN
LVHW062117300926
849161LV00010B/2477